Developments in Applied Phycology

Volume 11

Series Editor

Michael A. Borowitzka, Algae R&D Centre, School of Veterinary and Life Sciences, Murdoch University, Murdoch, WA, Australia

Aims and Scope

Applied Phycology, the practical use of algae, encompasses a diverse range of fields including algal culture and seaweed farming, the use of algae to produce commercial products such as hydrocolloids, carotenoids and pharmaceuticals, algae as biofertilizers and soil conditioners, the application of algae in wastewater treatment, renewable energy production, algae as environmental indicators, environmental bioremediation and the management of algal blooms. The commercial production of seaweeds and microalgae and products derived there from is a large and well established industry and new algal species, products and processes are being continuously developed.

The aim of this book series, Developments in Applied Phycology, is to present state-of-the-art syntheses of research and development in the field. Volumes of the series will consist of reference books, subject-specific monographs, peer reviewed contributions from conferences, comprehensive evaluations of large-scale projects, and other book-length contributions to the science and practice of applied phycology.

Prospective authors and/or editors should consult the Series Editor or Publishing Editor for more details. Series Editor: Michael A. Borowitzka - M.Borowitzka@murdoch.edu.au Publishing Editor: Éva Loerinczi - eva.loerinczi@springer.com

Alan T. Critchley • Anicia Q. Hurtado •
Iain Charles Neish
Editors

Tropical Phyconomy Coalition Development

Focus on Eucheumatoid Seaweeds

Springer

Editors
Alan T. Critchley
Verschuren Centre for Sustainability
in Energy and Environment
Cape Breton, NS, Canada

Anicia Q. Hurtado
Integrated Services for the Development of Aquaculture
and Fisheries (ISDA)
University of the Philippines Visayas
Iloilo, Philippines

Iain Charles Neish
PT Sea Six Energy Indonesia
Bali, Indonesia

ISSN 2543-0599 ISSN 2543-0602 (electronic)
Developments in Applied Phycology
ISBN 978-3-031-47805-5 ISBN 978-3-031-47806-2 (eBook)
https://doi.org/10.1007/978-3-031-47806-2

This Springer imprint is published by the registered company Springer Nature Switzerland AG
The registered company address is: Gewerbestrasse 11, 6330 Cham, Switzerland

Paper in this product is recyclable.

Preface

At the time of writing, farmed, red macroalgae, particularly of the genera *Eucheuma* and *Kappaphycus* (collectively referred to as the eucheumatoid seaweeds), provide the raw material foundation for sea vegetable, carrageenan, and agricultural, nutrient value chains that yield leading marine products' exports for Indonesia and the Philippines. The eucheumatoids are also a basis for substantial Chinese seaweed products manufacturing. In addition, several tropical jurisdictions around the world produce eucheumatoid seaweeds and aspire to build value chains based on their cultivation.

By 2020, it was evident to the founders of Phyconomy.org that the development of phyconomy for eucheumatoid seaweeds would benefit from increased collaboration among researchers of the global phyconomist community. That led to creation of the Tropical Phyconomy Collaboration Development's (Phyconomy.org) first eucheumatoid seaweeds biology webinar on July 7–8, 2021 (a.k.a. TPCD-1).

For the TPCD-1 webinar, the website Phyconomy.org was created and an accessible repository of more than 40 video presentations with textual abstracts was made available online so webinar presenters, attendees, and the public could view and read content at their convenience and interact with presenters through a linked, online forum. During the July 7–8 webinar, presenters responded to questions from attendees, and the webinar was recorded for retention as online, website content.

This work complements TPCD-1 online material with a hard-copy book. The chapters of this book were written not only to preserve TPCD-1 content in printed form but also to present further development of thinking based on lessons learned during the webinar. Some of this material indicates the direction in which collaboration has moved since TPCD-1, and at the time of writing, a TPCD-2 was in preparation with intent to move tropical phyconomy collaboration toward tangible actions.

Bali, Indonesia

Iain Charles Neish

Obituary: Farewell to Indonesian Seaweed Hero Prof. Jana Tjahjana Anggadiredja (16/6/1954–17/7/2021)

Professor Jana Tjahjana Anggadiredja, colloquially known as Prof. Jana, is one of our national heroes, especially for those of us in the Indonesian seaweed community. Therefore, I was very proud to meet Prof. Jana when we were both assigned to attend the 23rd International Seaweed Symposium (ISS), in Jeju Island, South Korea, from April 28 to May 3, 2019. There were five of us in the delegation representing the Tropical Seaweed Innovation Networks (TSIN) forum and sponsored by the United Nations Industrial Development Organization (UNIDO) SMART-Fish Indonesia Program. I was then representing the Indonesian Ministry of Education and Culture, alongside a representative from the Indonesian Ministry of Marine Affairs and Fisheries, and three representatives of the UNIDO SMART-Fish Indonesia Program, one of whom was Prof. Jana, the others being Pak Sudari and a UNIDO staff member.

Who is Prof. Jana, and why do we consider him a seaweed hero? Born in Sumedang, West Java, Indonesia, on June 16, 1954, Professor Jana Tjahjana Anggadiredja passed away on July 17, 2021, just days before he was scheduled to join us at the Tropical Phyconomy Coalition Development (TPCD-1) workshop. Prof. Jana had a multidisciplinary educational background, which served him well in his wide-ranging professional activities. After graduating with a bachelor's degree in pharmacy (1980) and pharmaceutical science (1981) from the Faculty of Mathematics and Natural Sciences, Padjadjaran University in Bandung, he continued his education in the Conservation Biology Study Program at the University of Indonesia in Jakarta. His master's thesis title was "Seaweed diversity on the Warambadi seashore of Sumba Island and its Utilization" (1998), and his doctoral dissertation title was "Diversity of antibacterial substances from selected Indonesian seaweeds" (2004). His publication record includes 11 articles in international scientific journals, 37 in Indonesian scientific journals, and 4 seaweed-related books. A further 43 papers presented in various seminars or scientific forums were not published, and Prof. Jana served as the editor for 14 books including scientific proceedings and other research-based works.

Prof. Jana was active in promoting the seaweed sector and developing seaweed science to the end of his life. One testament to his devotion to the seaweed community right to the end is that he recorded and sent his TPCD-1 contribution during his final illness. He was an active member of the Indonesian Seaweed Development Team and engaged in a wide range of research and development activities. These included seaweed cultivation methods, production processes, and product development. The latter included the development of seaweed hydrocolloids for use in various food, pharmaceutical, and other industrial products. Prof. Jana was keen to keep learning as well as to pass on his knowledge and experience. He attended fieldwork and leadership training courses around the world, including at the Institute of Marine Sciences, North Carolina University at Chapel Hill, USA (1986); Hawaii University, USA (1986); Shanghai Institute of Materia Medica, China (2004); Middle Level Leadership School LAN-RI, Indonesia (1997); R & D Management, Crown Agents United Kingdom, UK (2002); and National Resilience Institute (LEMHANAS-RI), Indonesia (2012). Prof. Jana was active in outreach, including conducting training and/or visits to various seaweed hydrocolloid industries around the world, such as Ina Agar Industry, Japan (1985); FMC-USA (1986); MCPI Corporation, Philippines (1986); B & V SA, Italy (1989); Copenhagen Pectin Factory (KPF),

Denmark (1992); Spindal SA, France (2001); and Shanghai Brilliant Gum Co. Ltd, China (2004).

As a career civil servant, Prof. Jana worked for many years at the Indonesian Agency for the Assessment and Application of Technology (BPPT). Recruited in 1982, he attained the rank of leading expert researcher in the Pharmaceutical and Food Technology division in 1999, and became a Research Professor in 2005. Structural positions held by Prof. Jana included Staff Assistant to the Minister of Research and Technology for Agriculture, and in 2010 as Deputy Head of BPPT for Natural Resource Development Technology. In 2018, while still serving as Research Professor at BPPT, he was also active as a teaching expert in the field of natural resources at the Indonesian National Resilience Institute (LEMHANAS-RI), Chairman of the Indonesian Seaweed Society (ISS), a member of the Indonesian Seaweed Industry Association (ASTRULI), a member of the Executive Council of the Asia-Pacific Phycology Association (APPA), and the UNIDO Lead Expert for Seaweed Value Chain in the SMART-Fish Indonesia Program.

Prof. Jana strove tirelessly and with a catching enthusiasm to promote and develop the potential of seaweed and seaweed-based industries worldwide, but especially in his native country, Indonesia. He cared deeply about people and the future of our world. He was also a generous man, who gave so much to those around him, sharing what he knew and helping others to develop their potential. For me, his example is an inspiration to follow in his footsteps, helping to unlock and develop the potential of seaweeds for the benefit of mankind and to address the challenges humanity faces.

Note: Asmi Citra Malina A. R. Tassakka is currently the Director for Innovation and Intellectual Property Rights at Hasanuddin University.

Asmi Citra Malina A. R. Tassakka

Center of Excellence for Development
and Utilization of Seaweeds
Hasanuddin University
Makassar, South Sulawesi, Indonesia

Faculty of Marine Science and Fisheries
Hasanuddin University
Makassar, South Sulawesi, Indonesia

Acknowledgments

The Tropical Phyconomy Coalition Development, TPCD 1, was hosted at the Univertsitas Hasanuddin, Makassar, Indonesia.

The co-editors of this book would like to thank the following:

The team at Springer Nature, Applied Phycology, including Professor Michael Borowitzka, Eva Loerinczi, and especially Bibhuti Sharma who saw promise in the book proposal and showed tremendous patience in the preparation of the final drafts.

Of course, there would be no book without the inputs of all the co-authors of the 26 chapters included between these covers, so thanks go to everyone for finally getting this project across the line.

There would not have been a TPCD 1 workshop without the generous funding provided by GENIALG an EU Horizon 2020 program and, in particular, its leader Philippe Potin, Research Director, Station Biologique de Roscoff, France.

Prior to the workshop, all first authors provided their presentations to be uploaded to the web site: www.phyconomy.org. This was established by RA Narayanan and Nelson Vadassery, Sea 6 Energy Pvt. Ltd., and it is maintained and publicly accessible. The same two IT wizards also set up the web site for registration and the message board for interaction with presenters and ensured its ongoing smooth running. Thanks gents! PT Sea Six Energy, Indonesia has been a major force in support of www.phyconomy.org. Their vision and support is acknowledged and greatly appreciated.

For the workshop at UNHAS, we would like to thank: Team leader, Jamaluddin Jompa; Steering Committee, Rohani Ambo-Rappe, Aisjah Farhum; Executive Chair, Kasmiati; Vice Chairs, Asmi Citra Malina A. R. Tassakka, Nadiarti Nurdin Kadir, Mahyudin; Secretary, Amanda Pricella Putri; Promotion and Logistics Section, Widyastuti Umar, Muchlis, Muh. Iqbal Tawakkal, Mustamin; Publications and International Relations, Abigail Mary Moore, Inayah Yasir.

Thanks again, one and all, looking forward to TPCD 2 and beyond.

Alan T. Critchley
Anicia Q. Hurtado
Iain C. Neish

Contents

1 Ten Guidelines for Phycosecurity Implemented as Biosecure Ecosystem Services Management of Tropical Seaweed Farms

Iain Charles Neish

Abstract

Lessons learned during the Phyconomy.org Topical Phyconomy Collaboration Development (TPCD) eucheumatoid seaweeds biology webinar of July 7–8, 2021, indicated an apparent need for structured ongoing collaboration toward tropical seaweed industry biosecurity in a specific way termed "phycosecurity," which was defined as *"biosecure phyconomy that results from procedures implemented against harmful biological, chemical and socioeconomic disruptions that influence ecosystem services of seaweed populations in multibiomass ecoscapes*." Phycosecurity implementation was postulated to function in a context of four elements of multibiomass ecoscape systems, namely, (1) ecosystem services from seaweed crops that are balanced with those from other ecoscape biomass per phycosecure management decisions; (2) socioecoeconomics that valorize seaweed ecosystem services per societal norms, mores, and folkways that institutionalize phycosecurity; (3) spatial management of ecoscapes done according to integrated coastal management principles that embrace phycosecurity; and (4) information technologies that provide quantified knowledge, information, tools, and solutions that enable phycosecurity. Ten policy guidelines for phycosecurity networking were proposed with the understanding that they will evolve as phycosecurity networks develop.

Keywords

Biosecurity · Phycosecurity · Ecosystem services · Seaweed · Management

1.1 Introduction

For purposes of this chapter, the term *phycosecurity* was defined as *"biosecure phyconomy that results from procedures implemented against harmful biological, chemical and socioeconomic disruptions that influence ecosystem services of seaweed populations in multibiomass ecoscapes*."

Phycosecurity concepts were developed from the body of lessons learned during the Phyconomy.org Tropical Phyconomy Collaboration Development (TPCD) eucheumatoid seaweed biology webinar of July 7–8, 2021 (see the audiovisual references section of the present document). The nucleus of the chapter was Neish (2021a, b, c).

For the TPCD webinar, the website Phyconomy.org was created, and an accessible repository of more than 40 video presentations with textual abstracts was made available online so webinar presenters, attendees, and the public could view and read content at their convenience and could interact with presenters through a linked online forum. During the July 7–8 webinar, presenters responded to questions from webinar attendees, and the webinar was recorded for retention as online website content.

Figure 1.1 depicts an overview of the conceptual framework within which phycosecurity policies tend to be developed and implemented.

1.2 Seaweed Farms as Subsets of Multibiomass Systems

Phycosecurity is a term uniquely applicable to seaweed farming. It is a concept that combines notions of seaweed farm biosecurity with notions of how seaweed farming fits in to the overarching concept of multibiomass systems in Socio-Ecological Production Ecoscapes (SEPE). Multibiomass provisioning (a.k.a. production) systems exist because of four functions, as depicted in Fig. 1.2, namely, (1) ecosystem services provision, (2) socioecoeconomic valorization,

I. C. Neish (✉)
PT Sea Six Energy Indonesia, Denpasar, Bali, Indonesia
e-mail: Ineish@sea6energy.com

A. T. Critchley et al. (eds.), *Tropical Phyconomy Coalition Development*, Developments in Applied Phycology 11,
https://doi.org/10.1007/978-3-031-47806-2_1

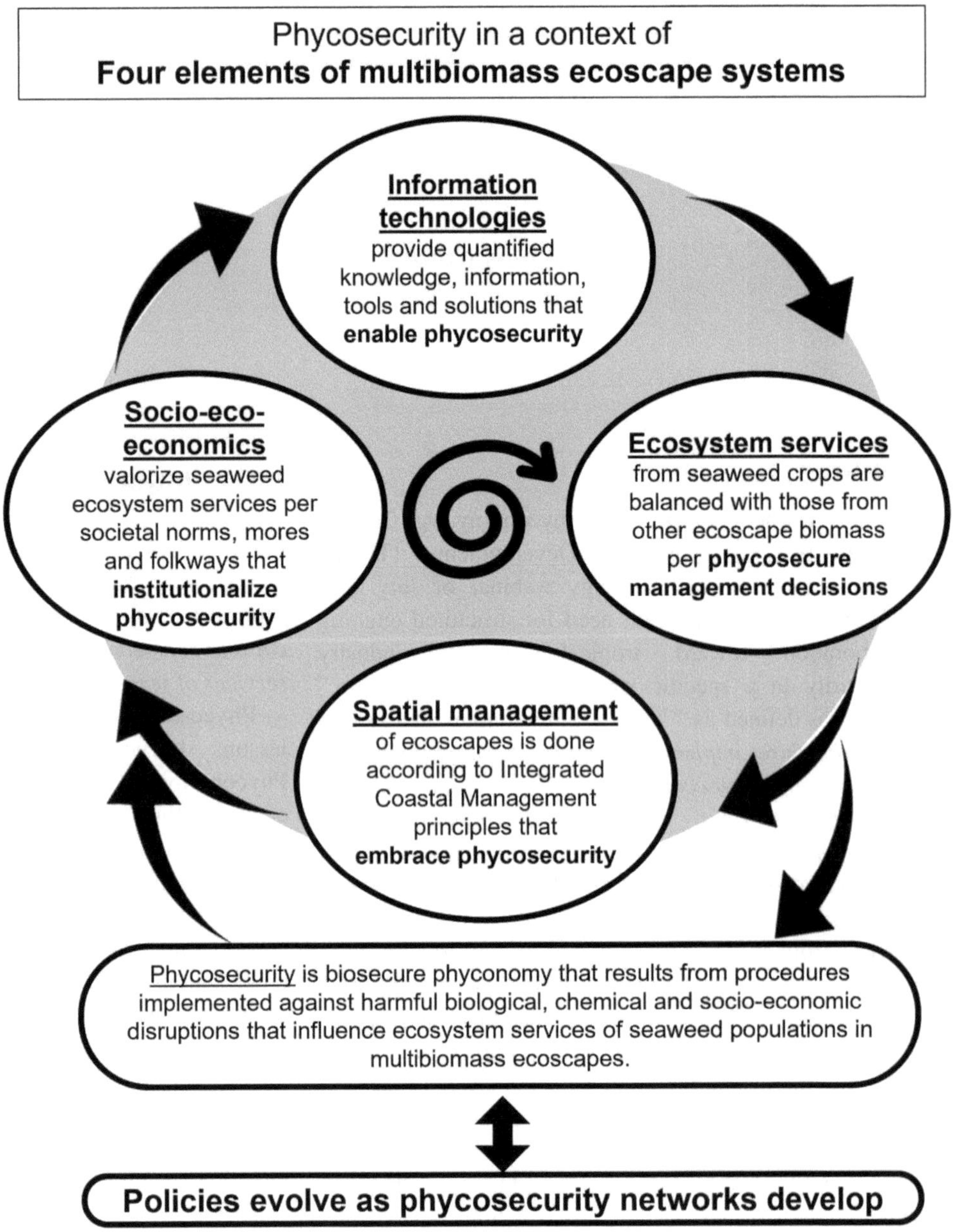

Fig. 1.1 Four elements comprise the framework of multibiomass ecoscape systems. Multibiomass systems include integrated, sustainable, multifunctional, multitrophic production and management of photoautotrophs, heterotrophs, and extractive organisms. In that context phycosecurity measures are implemented

(3) ecoscape spatial management, and (4) information technology utilization.

Berndes et al. (2007) explained how multifunctional biomass production systems are located, designed, integrated, and managed to provide specific environmental services, in addition to biomass supply. They concluded "*that the environmental benefits from a large-scale establishment of multifunctional biomass production systems could be substantial. Given that suitable mechanisms to put a premium on the provided environmental services can be identified and implemented, additional revenues can be linked to biomass production systems, and this could enhance the socioeconomic attractiveness and significantly improve the competitiveness of the produced biomass on the market.*"

Phycosecurity pertains to seaweed biomass that is produced for the variety of ecosystem services included (but not limited to) those listed in Figs. 1.3 and 1.4. Biomass production systems desirable from a phycosecurity perspective can be described using several adjectives including "integrated," "sustainable," "antifragile," "multitrophic," and "multifunctional," but the term "multibiomass" embraces all such concepts.

1.2.1 Ecoscape Spatial Management

All seaweed farms, in their context as multibiomass provisioning systems, are in SEPE that should be managed within

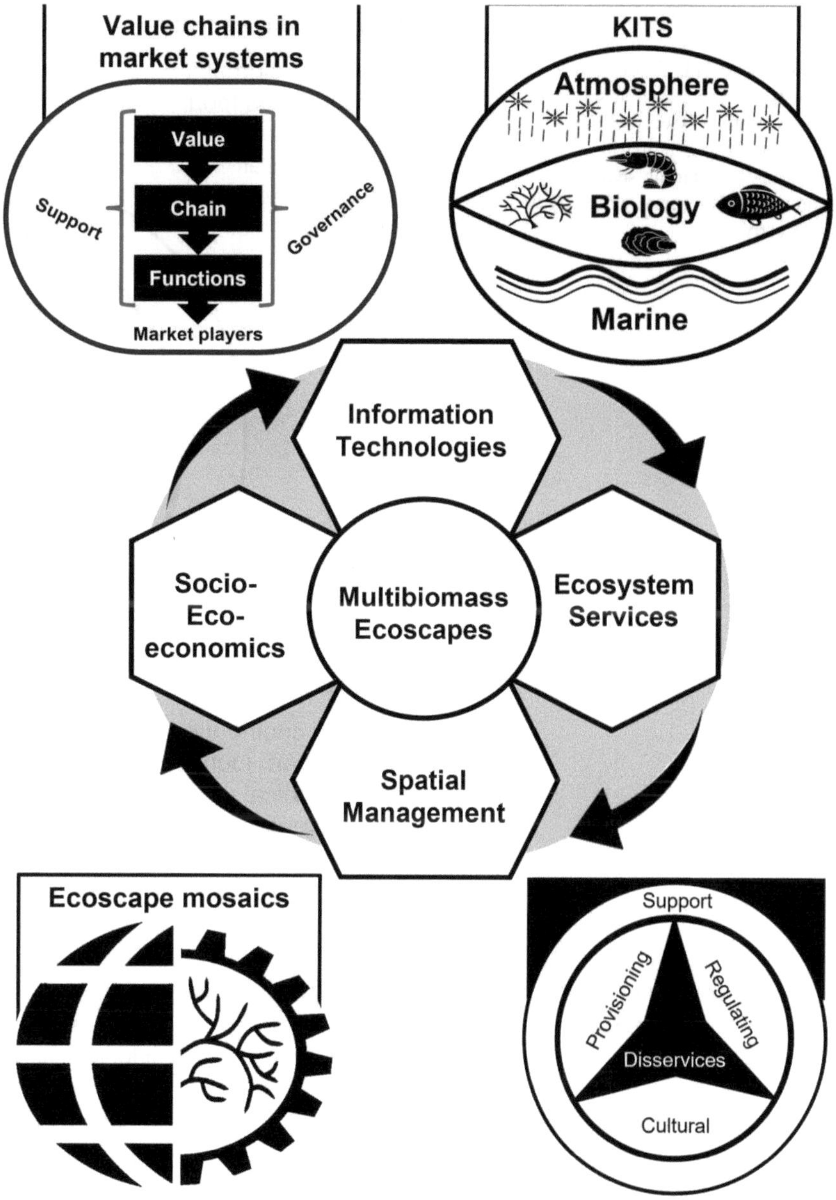

Fig. 1.2 The four multibiomass ecoscape elements of Fig. 1.1 are each associated with sets of conceptual frameworks that guide toward actions and implementations, including phycosecurity measures

Integrated Coastal Area Management (ICAM) plans. The crucial need for ICAM is defined by the diverse, often competing, and sometimes conflicting uses that humanity has for highly valued estuarine and marine coastal areas. Tenure rights for seaweed farming venues are a core requirement for all seaweed farmers, so ICAM issues are crucial to phycosecurity.

An ecoscape is "*the multidimensional landscape of a social-economic-natural complex ecosystem combining geographical patterns, hydrological process, biological vitality, anthropological dynamics and aesthetic contexts*" (Douglas et al. 2011).

A SEPE is a geographic space where people live and work in the same place. SEPE are "*biodiverse, dynamic mosaics of managed socioecological systems that produce bundles of ecosystem services for human well-being.*" (after JSSA 2010). All seaweed production systems are spatially placed within SEPE.

Ecoscape spatial management can be facilitated by satellite imagery and imagery from manned aircraft or drones, as shown in Fig. 1.10. Such imagery must be complemented by so-called "ground truthing" within physical confines of a defined SEPE. One useful aspect of ground truthing is the development of ecoscape mosaics developed in collaboration with people who live and work within the SEPE. Figure 1.5 shows an example of a mosaic drawing produced during ICAM activities of practicing seaweed farmers from eastern Indonesia.

A crucial component for ground-truthing and ecoscape mosaic development is the myriad of essential data that can

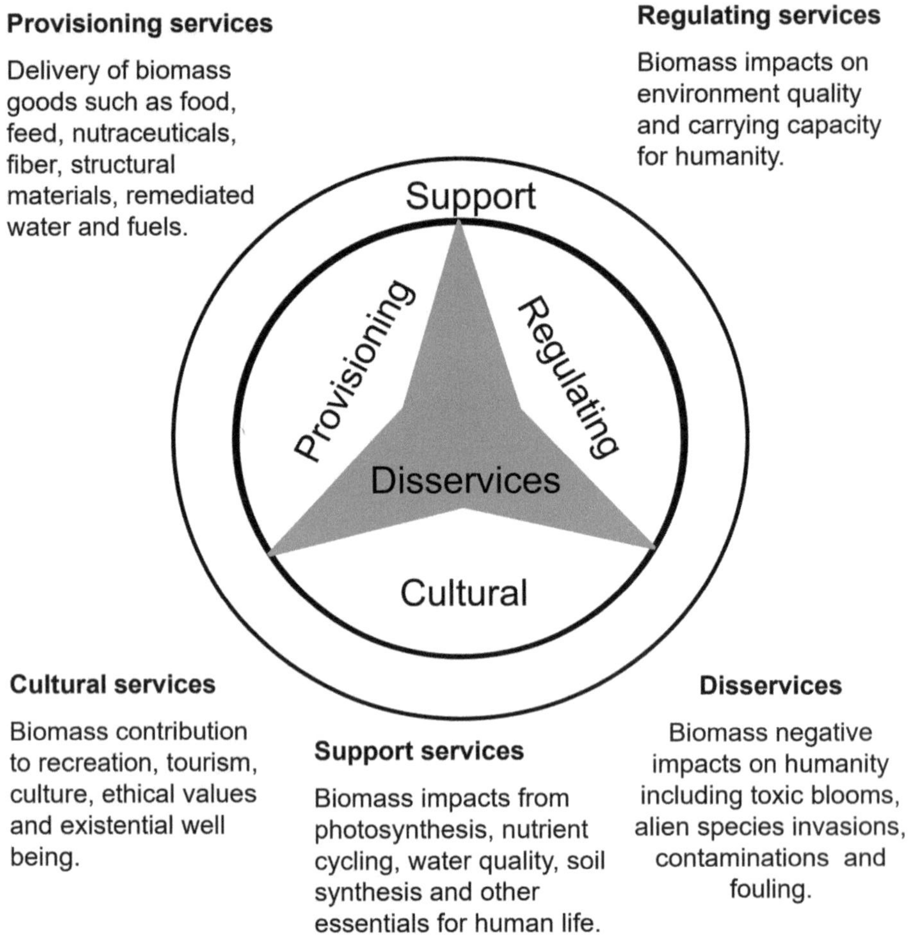

Fig. 1.3 Ecosystem services are the human gains obtained from ecosystems. This figure shows four types of ecosystem service, and potential disservices that photoautotrophs in marine ecoscapes, including seaweeds, can inflict on humanity directly or indirectly

be produced using information technology systems of the types described in Sect. 1.2.4.

1.2.2 Ecosystem Services

A goal of ICAM is to allocate spatially distributed entities in a way that enables integrated, sustainable delivery of ecosystem services while establishing balanced socioecoeconomic benefits to humanity. Note that multibiomass production systems include both cultivation systems and stewardship systems for natural ecosystems.

According to the Millennium Ecosystem Assessment (MEA), ecosystem services are defined as "*the human gains obtained from ecosystems*" (Carpenter et al. 2006; Cotas et al. 2023). They are divided into four categories designated as supporting, provisioning, regulating, and cultural services. The Common International Classification of Ecosystem Services (CICES) differs from the MEA classification as it considers supporting services to be organism functions and recognizes only provisioning, regulating, and cultural services per Cotas et al. (2023). The present document adopted the MEA classification system and added the fifth category designated as "*ecosystem disservices*" (Fig. 1.3).

Figure 1.4 shows examples of ecosystem services from photoautotrophs in marine ecoscapes. These include seaweeds and all other photosynthesizing marine and aquatic organisms.

Support services are those required for the development of all other ecosystem services, especially including photosynthesis and other primary productivity processes, so they are valorized through derivative ecological services. Among those, the primary goal of phyconomy systems is to optimize balances among provisioning services that yield financial returns from biomass production. In other words, the prime goal of phyconomy is to produce and sell seaweed crops.

Regulating services comprise ecological processes, such as climate regulation, habitat provision, oxygen production, and geological impacts, whereas cultural services are nonmaterial benefits that people derive from ecosystems, including

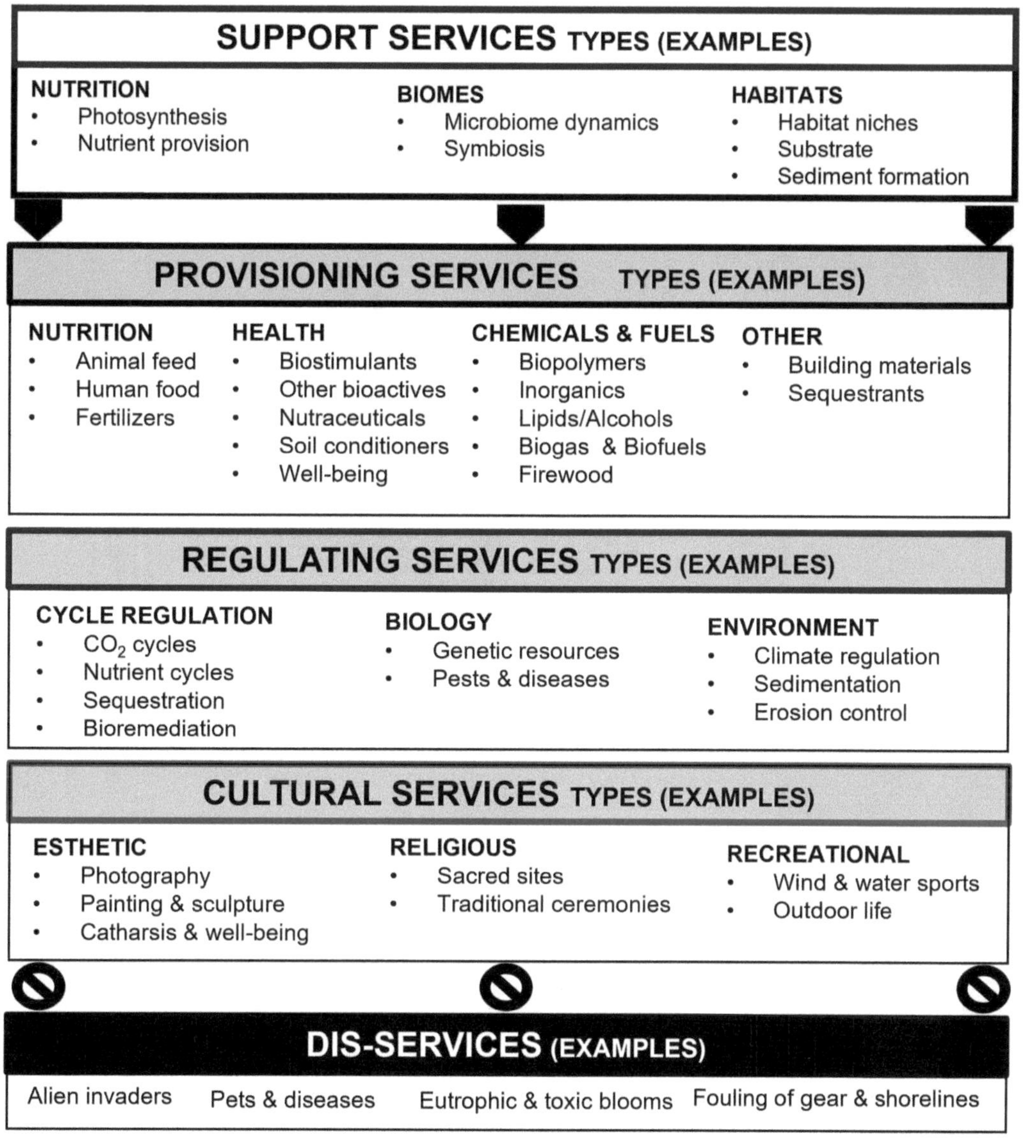

Fig. 1.4 Examples of ecosystem services from photoautotrophs in marine ecoscapes. These embrace all photosynthesizing marine organisms including seaweeds

cognitive development, recreation, and esthetic experiences. Such services are generally valorized through nonfinancial costs and benefits per ecoeconomic principles.

It is also necessary to minimize impacts from disservices that photoautotrophs may inflict. For phyconomists, control of disservices is an inevitable cost of farm operations.

From a phycosecurity perspective, production of seaweed biomass as cash crops from seaweed farms is the key provisioning service to be quantified. That must be achieved in a context of systematic balance with services from other SEPE organisms. Methods for doing that are still a work in progress. Many ecosystem regulating and cultural services have intangible value that is difficult to quantify in financial terms, as discussed below.

1.2.3 Socioecoeconomics

Ecoeconomics is an overarching concept that embraces both legacy economies and distributed ledger (a.k.a. "crypto") economies. "*Eco-economic costs and benefits are quantified and managed comprehensively with respect to both financial and ecological balances of goods and services*" (Brown 2001). Financial accounting for phycosecurity in multibiomass systems must be done with full ecoeconomic considerations in mind. Methods of value chain ecoeconomic accounting are still a work in progress, especially with ecosystem services and disservices that are difficult to valorize.

Value chains (VC) are the backbones of market systems. Early VC literature (e.g.: Porter 1985) elucidated how VC

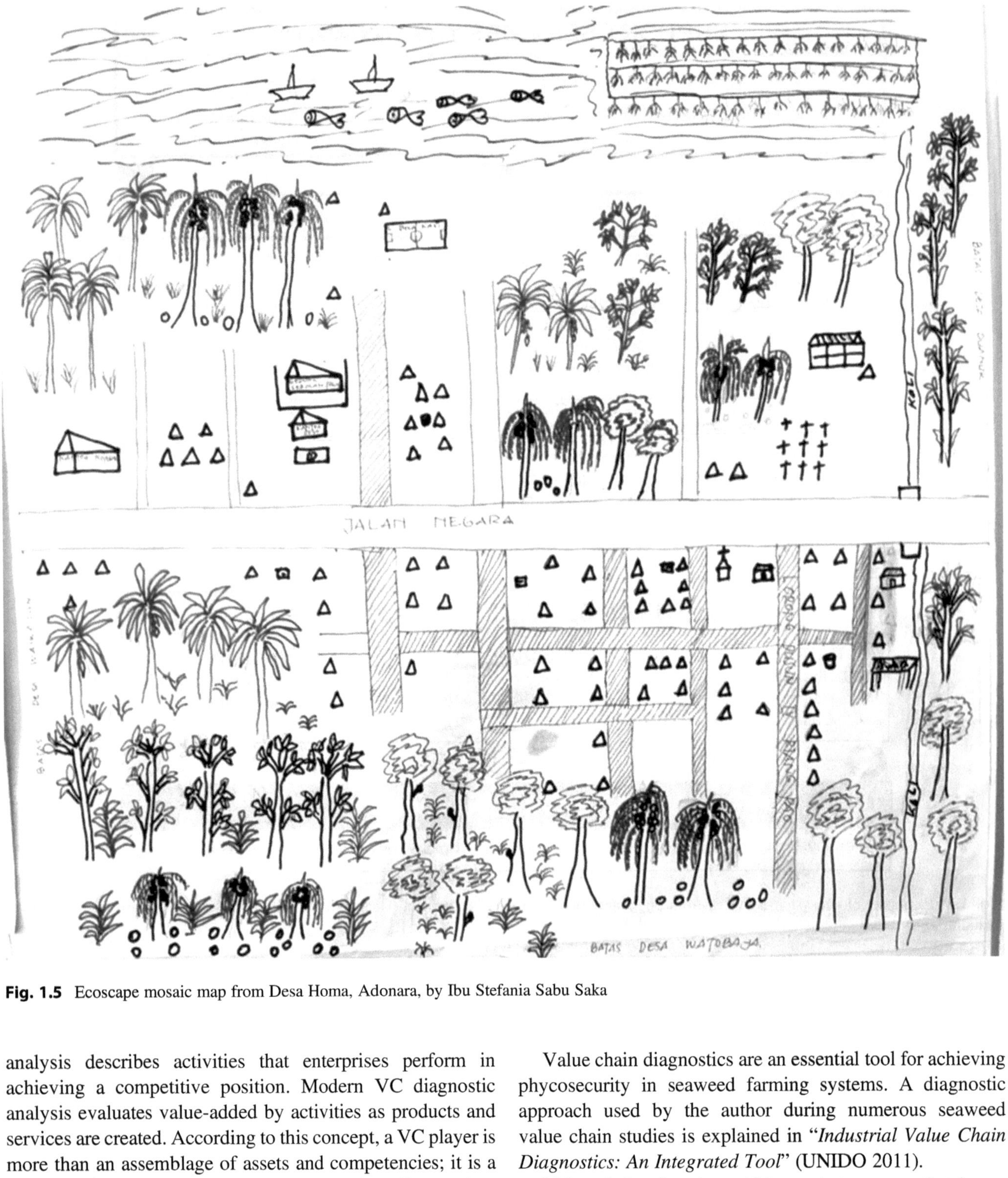

Fig. 1.5 Ecoscape mosaic map from Desa Homa, Adonara, by Ibu Stefania Sabu Saka

analysis describes activities that enterprises perform in achieving a competitive position. Modern VC diagnostic analysis evaluates value-added by activities as products and services are created. According to this concept, a VC player is more than an assemblage of assets and competencies; it is a set of functional systems that produces goods and/or services for which customers are willing to pay a price. In the case of multibiomass ecoscapes, value chains involve ecosystem services over which enterprises have influence but do not necessarily have control. That complicates accounting functions.

Value chain diagnostics are an essential tool for achieving phycosecurity in seaweed farming systems. A diagnostic approach used by the author during numerous seaweed value chain studies is explained in "*Industrial Value Chain Diagnostics: An Integrated Tool*" (UNIDO 2011).

Value chains function within market systems that incorporate elements of phycosecurity (Fig. 1.6). Market system actors execute both primary and secondary functions and activities. Useful analyses of how value chains fit in market systems include Herr and Muzira (2009) and ILO (2016).

Primary market system functions are directly concerned with the creation or delivery of products or services, many of

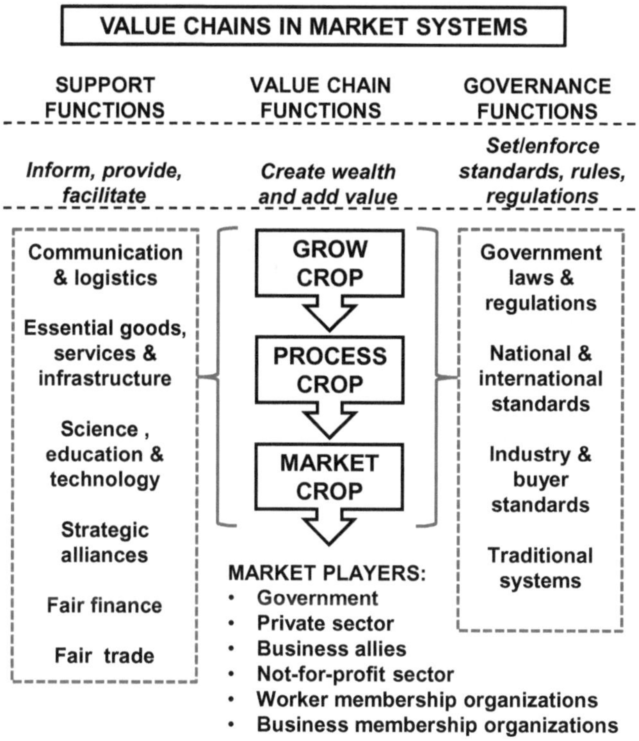

Fig. 1.6 A schematic depiction of how value chains fit within market systems (After Herr and Muzira 2009 and ILO 2016)

which can be described in terms of ecosystem services. For example, in a seaweed value chain, functions include farming of seaweed; logistics operations; manufacturing of value-added products such as carrageenan, biostimulants, and food; provision of ingredient solutions; and marketing, sales, and service. Each primary function is linked to secondary activities that enable their effectiveness and efficiency. Secondary activities include interactions among market system actors as follows:

1. Market players help to provide an enabling environment for system functions
2. Support-function providers inform, provide, and facilitate value chain functions
3. Governance-function entities set and/or enforce standards, rules, and regulations

Value chain concepts are based on a process view of VC actors. In multibiomass ecoscapes, those processes are linked with ecosystem processes. Organizations such as seaweed enterprises are systems and subsystems with inputs, transformation processes, and outputs. The advent of ecoeconomic value chain accounting has introduced problems that are still seeking solutions. From an ecosystem services perspective, efforts must be made to valorize significant regulating and cultural services when possible. The use of "credit" schemes such as carbon and nitrogen credits is an example of such valorization.

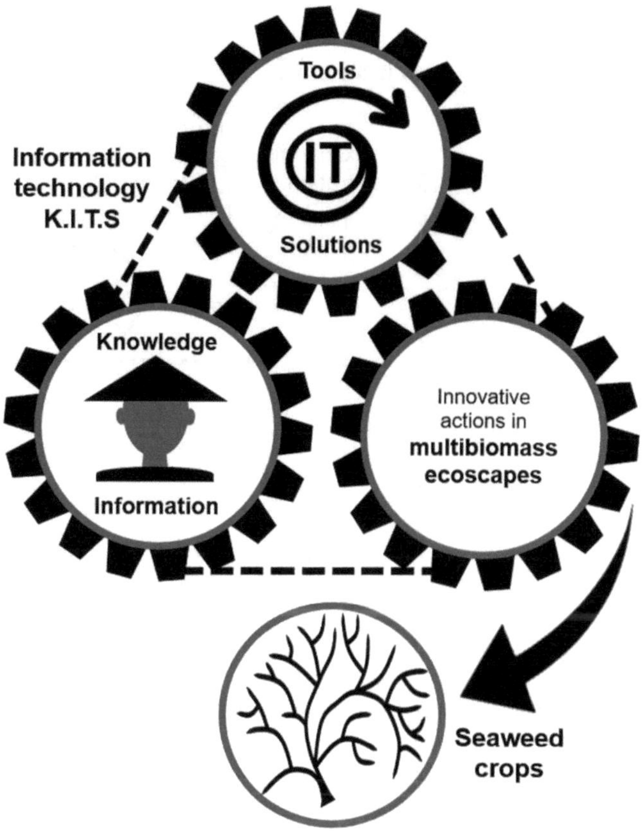

Fig. 1.7 Information technology knowledge, information, tools, and solutions (KITS) are the basis for quantification and analysis of multibiomass ecoscape systems that produce seaweed crops

From a phycosecurity perspective, it is essential to quantify seaweed passage through value chains and market systems in a way that is relatable to seaweed ecosystem services in multibiomass systems. That must be achieved in a context of systematic balance with socioecoeconomic factors that involve humans with other SEPE organisms.

1.2.4 Information Technologies (IT)

All functions that drive multibiomass systems depend on information technologies that provide knowledge, information, tools, and solutions (KITS) to system managers (Fig. 1.7). Without reliable, traceable KITS, effective, phycosecure system optimization and management is impossible. Sad to say, however, commercial tropical marine phyconomy systems struggle in the face of a severe paucity of KITS at all ecosystem service and socioecoeconomic levels.

Integration of precision technology into adaptive phyconomy systems for extensive tropical red seaweed

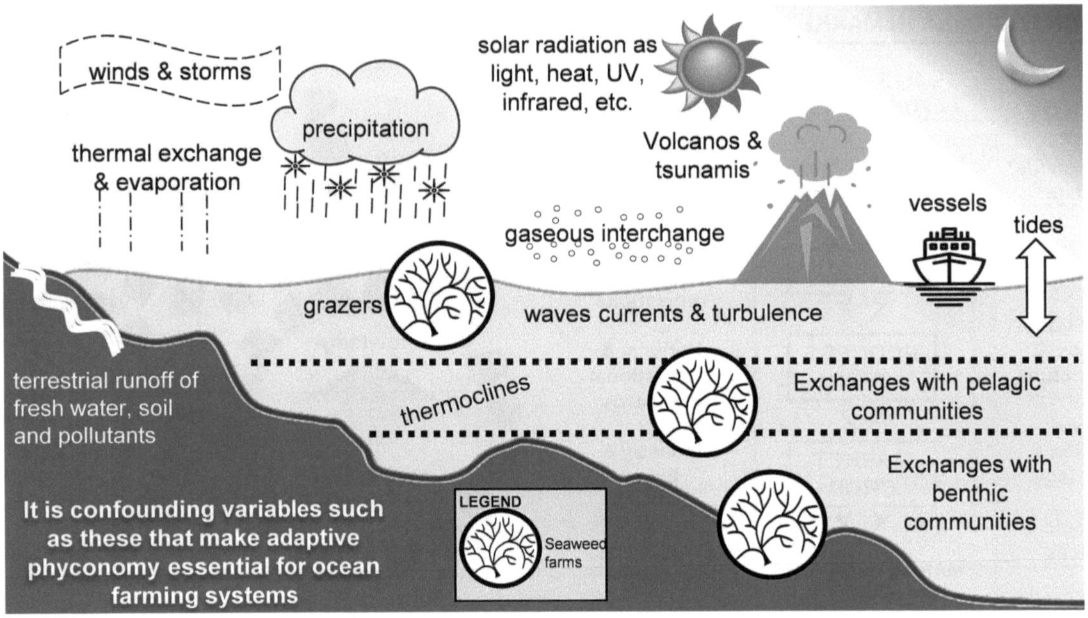

Fig. 1.8 Some key parameters that impact on multifunctional biomass production systems. Many of these are uncontrolled (i.e., "confounding") variables over which the phyconomist has low or no influence and abilities to predict parameter behavior range from high (e.g., tides) to low (e.g., tsunamis)

farming is a topic addressed in a chapter by Vadassery and Neish in the present volume. Adaptive phyconomy is a low-cost, low-control approach to seaweed biomass production that is practiced in situations, where the biology of algal crop organisms is poorly understood and/or where confounding variables render precision technologies impractical. Figure 1.8 shows some key parameters that impact on marine multibiomass systems. Many of these are uncontrolled, confounding variables over which the phyconomist has low or no influence. Abilities to predict parameter behavior range from high (e.g., tides) to low (e.g., tsunamis).

Precision phyconomy is a high-cost, high-control approach to seaweed biomass production that is practiced in situations, where the biology of crop organisms is well understood and culture conditions can be controlled.

With respect to extensive ocean farming of tropical red seaweeds, adaptive phyconomy methods prevail among coastal seaweed farmers, even as the science and technology sector aspires to apply precision techniques to commercial seaweed farming systems, especially with respect to phycosecurity, farm mechanization, information technology, tissue culture techniques, and molecular taxonomy as applied to cultivar development and propagation. As phyconomy systems evolve, precision techniques will progressively displace adaptive approaches to yield extensive farm systems that are intermediate on the adaptation-to-precision curve (see Chap. 19).

Comprehensive data acquisition, processing, and distribution through "Internet of Things" (IoT) systems of Information Technology (IT), as depicted in Fig. 1.10, is a crucial foundation for precision phyconomy because *control is a goal even though influence may be the most that can be achieved.* For adaptive phyconomy of seaweed crops in marine waters, however, control over environmental parameters is seldom feasible. Figure 1.9 shows four key considerations to be considered during design of information systems in adaptive phyconomy systems. All occur on a continuum of difficulty (hence cost) that ranges from low to high. The extent to which phycosecurity can be strengthened depends significantly on moving toward the "high" end of these scales.

Figures 1.10 and 1.11 show IT elements required for phycosecure adaptive phyconomy systems in coastal waters. These elements generate the KITS required by adaptive IT systems. From a phycosecurity perspective, salient features of adaptive phyconomy IT systems are:

1. Accuracy and cost of parameter measurement, analysis, and distribution are key to the practicality of operating commercial seaweed multibiomass provisioning systems.

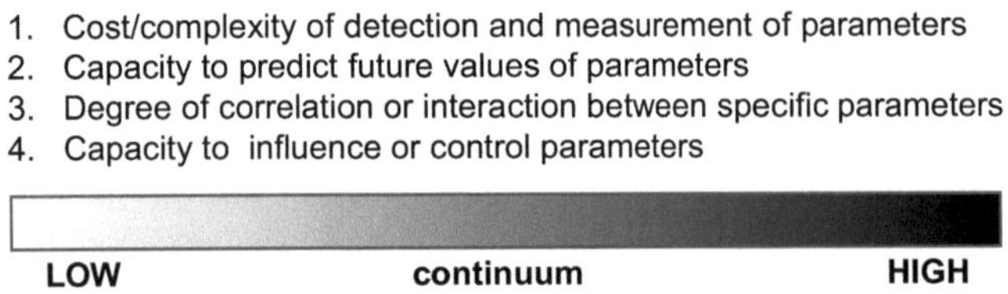

Fig. 1.9 Four key factors to be considered during design of information technology and phycosecurity systems. All occur on a continuum of difficulty (hence cost) that ranges from low to high

Fig. 1.10 The "Internet of Things" (IoT) comprises tools enabling IT KITS development and implementation in phycosecure systems

A paucity of cost-effective marine data drones, buoys, and fixed stations has been a problem for phycosecurity development that was in the process of being solved by progress with IoT technologies at the time of writing (Fig. 1.11).

2. Lack of comprehensive, cost-effective data has resulted in a situation where the field of adaptive phyconomy is figuratively drowning in a sea of untested hypotheses. Fluctuations in biomass productivity and seasonal variations in crop health have typically been attributed to influences of environmental parameters that have not been measured to a level that enables researchers to define correlations. Examples of needed, but generally unobtained data include those related to sea temperature, pH, salinity, water movement, and nutrient composition.
3. Lack of comprehensive, cost-effective data has also severely limited the capacity of phyconomists to predict parameter behavior. The main predictive tool of adaptive phyconomy has been intuition developed by farmers and phyconomists through long experience. Phycosecurity can be strengthened as IT systems develop predictive tools based on identified correlations and on experimental clarification of interactions and influences between and among multibiomass system parameters. For example, accurate tide predictions are available in much of the world, and IT KITS systems can quantify farm area tidal current speed and direction forecasts that are linked to tide predictions. With such data, it will be possible to test the myriad of hypotheses that pertain to seaweed growth and health issues that may be linked to water movement.
4. For phycosecurity, it may not be possible to control many parameters in ocean multibiomass production systems, but with adequate data, it is possible to adapt to some parameters, for example, by placing farms at favorable locations and at best water column levels. In some cases, it may even be possible to exert influence sufficient to guide some parameters. Farm habitat structures may be designed to impact on water flows and light levels reaching crops, while actions such as artificial upwelling can distribute nutrients and other water qualities in ways that enhance biomass productivity.

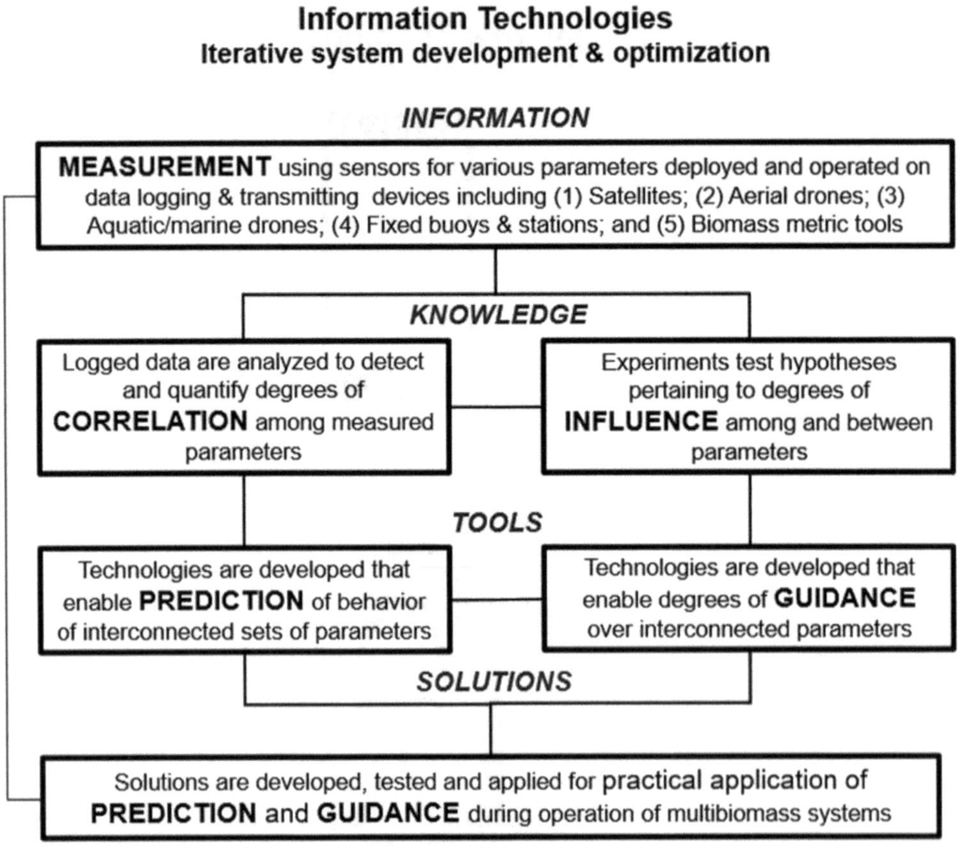

Fig. 1.11 Information technologies (IT) embed knowledge, information, tools, and solutions that are applied to phycosecurity with increasing impacts as iterative system development and optimization takes place

5. Biometric parameters in IT KITS systems are a special case because many can only be measured by hands-on contact with biomass, such as by the use of crop-logging procedures. There are no sensors for most of those, and data collection involves measurement of farm operation parameters, including biomass planted, biomass harvested, proximate analysis of biomass composition, and the like.

A key component of IT KITS relevant to phycosecurity is the set of IoT tools depicted in Fig. 1.10. Development of the IoT was rapidly evolving at the time of writing as artificial intelligence (AI) algorithms for information technologies were proliferating and costs of IoT hardware were going down. Costs of sensors and processor units potentially suitable for marine uses had reached the point where many could be purchased online for less than the price of a cup of coffee. Satellite imagery was becoming increasingly available at diminishing cost, and manned aircrafts for aerial imagery were being replaced by drones affordable even by young hobbyists. Seagoing drones were becoming available, affordable, and adapted to systems suitable for managing phycosecurity in multibiomass marine ecoscapes.

1.2.5 An Operational Definition of "Phycosecurity"

Consideration phyconomy as discussed in Hurtado et al. (2019) led to a notional definition of phyconomy as "*a branch of applied phycology that comprises systems of art, science and technology applied to biosecure production systems that yield crops of algae.*"

In a context of phycosecurity in multibiomass systems, it is more useful to define phyconomy operationally as "*a branch of applied phycology that strives to manage seaweed ecosystem services valorized per ecoeconomic principles in phycosecure, multibiomass, socioecological production ecoscapes (SEPE) that are subject to integrated coastal area management (ICAM).*"

Phycosecurity may therefore be operationally defined as "*biosecure phyconomy that results from procedures implemented against harmful biological, chemical and socio-economic disruptions that influence seaweed ecosystem services valorized per ecoeconomic principles in phycosecure, multibiomass, socioecological production ecoscapes (SEPE) that are subject to integrated coastal area management (ICAM).*"

1.3 Brief Overview of Seaweed Farming Biosecurity Concepts

Cottier-Cook et al. (2016) was a seminal paper in the field of seaweed farming biosecurity. It was, in effect, a "petition" paper authored by 31 prominent researchers from around the world. That policy brief was prepared under auspices of the United Nations University United Nations University Institute for Water, Environment and Health (UNU-INWEH) and the Scottish Association for Marine Science (SAMS), and it was titled "*Safeguarding the future of the global seaweed aquaculture industry*." It stated that "*The need for evidence-based policy decision making and sector management is paramount across . . . policy recommendations, which should be acknowledged as essential components of establishing the balance between economic growth and ocean health and incentivized by policymakers*."

The UNU-INWEH/SAMS policies were used as a starting point for drafting of guidelines for a proposed Tropical Phycosecurity Network (TroPhyNet) that can support tropical seaweed farm biosecurity developments.

1.3.1 Eight UNU-INWEH/SAMS Policy Recommendations per Cottier-Cook et al. (2016)

TroPhyNet guidelines were drafted with the intention that all eight of the following UNU-INWEH/SAMS policy recommendations should be integrated into them:

1. Establish centres of research excellence to develop and identify new indigenous cultivars, specifically chosen for their disease resistance, high yields and ability to meet consumer preferences. To undertake pathogen profiling of key farmed seaweeds to inform risk assessments for trade of seed stock and propagules and to study the interactions of specific genetic variants within a particular geographical location.
2. Establish national seed banks which are responsible for maintaining a high health status of seed stock and where disease-resistant strains can be held for use by seaweed farmers following a disease outbreak. These could be part-funded by the government, industry and potentially nongovernment organizations.
3. Maintain the genetic diversity in wild stocks by preventing the introduction of nonindigenous species and encouraging the development of local indigenous cultivars.
4. Exercise the precautionary approach when introducing new or nonindigenous cultivars to the marine environment.
5. Focus on developing and enhancing biosecurity programs through capacity building, including training in quarantine procedures and farm management practices and incentivize the development of diagnostics to rapidly detect disease and nonindigenous species, to enable adaptive risk management and better evaluation measures to be taken.
6. Incentivize long-term investment in the industry, potentially through part-government funded insurance policies to safeguard the business against natural disasters and disease outbreaks.
7. Incentivize the integration of seaweed, and other extractive species, with finfish in integrated multitrophic aquaculture (IMTA) systems to both reduce the eutrophication of the water column and benthic enrichment effects of finfish aquaculture and to minimize space used for aquaculture purposes in the coastal zone.
8. Develop assessment tools for evaluating spatial planning issues in relation to aquaculture (including seaweed) and to enable risk-based analysis of spatial management options to support the licencing process and facilitate future investments in infrastructure/insurance schemes to ensure the sustainable growth of this industry.

1.3.2 Quantification at the Core of Phycosecurity

The UNU-INWEH/SAMS policies specified that seaweed biosecurity measures should be evidence-based, and the present chapter similarly asserts that scientifically obtained data should be at the core of phycosecurity (i.e., biosecure phyconomy). Unfortunately, at the time of writing, necessary data were lacking throughout the field of tropical seaweed farming. That view was reinforced by Campbell et al. (2019), who pointed out that development of effective seaweed biosecurity frameworks is dependent on the availability of evidence-based research. They asserted that evidence applicable to seaweed farm biosecurity was limited. Consequently, it was difficult to align to well-established aquatic animal health and terrestrial crop biosecurity frameworks that were systemically interrelated in seaweed farming systems. They concluded that "*it is crucial that the low unit value (though high overall volume) of sea-weed aquaculture should not be conflated with a perceived low risk of pathogen transfer, and low impact of subsequent economic and ecological impact in receiving nations*."

Collection of evidence, in the form of environmental and biological parameter databases, involved several issues, among which the following seemed prominent:

1. Farm production data were top-secret assets of private enterprises and so were generally unavailable to multibiomass ICAM managers.
2. There were few, if any, publicly available logged biometric data for seaweeds or other organisms in multibiomass systems.
3. Affordable systems were not readily accessible for monitoring and distribution of relevant meteorological, oceanographic, and other environmental data.
4. Therefore, correlation of seaweed production with environmental parameters seemed effectively impossible; and useful prediction, guidance, and control tools were not feasible.

The paucity of data for application to phycosecurity issues was brought into stark perspective in Andrefouet (2021).

During that study, remote sensing satellite imagery and farmer interviews were used to examine relationships among tourism, volcanic activity, pandemic impacts, and the extent of seaweed farming around Nusa Lembongan, Bali, Indonesia. The paper explained that since 2003, farmable areas of Lembongan ranged from being fully planted to being partially planted to being virtually devoid of operating seaweed farms. Employment generally drifted toward tourism jobs when tourism was thriving and then back to seaweed farming, when volcanic activity or pandemic lockdowns drove downturns in tourism.

It was at Nusa Lembongan that Indonesian commercial eucheumatoid seaweed farming first succeeded in the mid-1980s; first with *Eucheuma* and then with *Kappaphycus*. The author had been periodically involved in Lembongan seaweed farming since1986 and could attest to the fact that phyconomy practices at Lembongan showed little or no evolution during the 35+ years from industry beginnings to the time of writing. Farm productivity had risen and fallen not only between years but also seasonally within years. No systems had been established for tracking crop productivity in relation to environmental or socioeconomic parameters. No ground-truthing capacity had been developed to facilitate data gathering through remote sensing by satellites, drones, or manned aircraft. The paucity of data had regulated explanations of crop productivity variation to the realm of pure speculation. This seemed to be the case for all, or most of Indonesia at the time of writing.

It is hard to imagine effective phycosecurity in the face of weak databases. Systems for comprehensive data collection and distribution must therefore be a core asset of any initiative such as the proposed TroPhyNet.

1.3.3 Further Biosecurity Policy Recommendations from Global Seaweed STAR

Since the seminal policy paper of Cottier-Cook (2016), seaweed aquaculture biosecurity issues received substantial attention, especially during studies supported by the Global Seaweed STAR initiative through 2021 (e.g., Campbell et al. 2019; Kambey et al. 2020, 2021a, b; Suyo et al. 2021a, b). Six further policy recommendations for government entities developed from that work (Kambey et al. 2020, 2021a), and they are quoted as follows:

1. Support further research to develop a strong evidence base, upon which national strategic decisions can be made on the management of the sector.
2. Establish seaweed-specific regulations and policies, providing appropriate management strategies that can be effectively enforced.
3. Establish a national database reporting what species of seaweeds are being produced and where any pest and disease outbreaks occur. This should be followed up with regular evaluation, so that the risks can be assessed by national government and each district, where possible.
4. Support for farmers to invest in the biosecurity management of seaweed cultivation systems including health monitoring equipment, training on management procedures, regional facilities for farmers to use for quarantine of seedlings or crop stock and surveillance systems.
5. Development of a risk assessment procedure for the expansion of farms into new and diseases-free areas.
6. Clarification on who the competent authority will be to regulate and support the seaweed industry.

From a market system perspective (per Fig. 1.6), the cited biosecurity recommendations seem to emphasize governance functions more than support, value chain, or market player functions. The author proposes that any phycosecurity initiative should be compatible with the cited policy recommendations and should collaborate with governance entities, but it should fundamentally be a support function initiative that is firmly grounded with value chain participants and market system players, especially including seaweed farmers in socioecological production ecoscapes.

1.3.4 Some Key Hypotheses That Require Rigorous Testing

Justification for applied phycology directed toward biosecurity issues is often tied to issues surrounding food security, crop health, environmental impacts, and climate change. These are all issues that received consideration through the cited TPCD workshops. Unfortunately, testing of many key hypotheses related to such issues lack data sufficient for evidence-based biosecurity development. Some examples include:

1. Hypotheses that postulate cause-and-effect relationships between crop productivity and climate change have been difficult to test, because historical baseline data were meagre and available IT KITS were inadequate for tracking climate changes that were taking place in seaweed farming areas. Planetary climate change is a widely acknowledged phenomenon that is commonly cited as a threat to biosecurity of global multibiomass, yet there seemed to be sparse evidence of direct climate change impacts on tropical seaweed crops at the time of writing. Baseline data seemed lacking from early years of seaweed farming in the 1970s and 1980s, and systems seemed to be lacking for tracking climate trends relevant to phycosecurity since then. Even if data were available, there seemed to be sparse knowledge of how climate parameters correlate with phycosecurity issues. It is only with development of IT KITS and better knowledge of

seaweed physiology that the existence and impacts of climate change on phycosecurity can be known.

2. Hypotheses contending that vegetative red seaweed cultivars suffer loss of vigor (senescence) over time require testing, given that evidence-based demonstration of that phenomenon seems weakly supported.
3. Likewise, hypotheses contending that seaweed farm sites suffer depleted productivity due to habitat degradation over time require further testing, given that evidence-based demonstration of that phenomenon also seems weakly supported.
4. Issues surrounding crop health, including the nature of ice-ice syndrome among eucheumatoid seaweeds, are at an early stage of understanding. Links between seaweed diseases and environmental parameters suffer from lack of environmental data.

Promotion of research collaborations that lead to testing of these and many other hypotheses should be a key function of any phycosecurity initiative that develops.

1.4 Toward a Tropical Phycosecurity Network

A perceived need to enhance phycosecurity for eucheumatoid seaweed crops was the driver for creation of Phyconomy.org and undertaking of the Tropical Phyconomy Collaboration Development (TPCD) eucheumatoid seaweeds biology webinar of July 7–8, 2021. During that event, the author presented a proposal for a *"Best Bibit Network"* (BeBiNet). During the TPCD event, it became clear that rather than standing for Best Bibit Network, the proposed initiative could better be called *"Best Biosecurity Network."* Further consideration of TPCD presentations led to the concept of Tropical Phycosecurity Network, hence "TroPhyNet."

At the time of writing, TroPhyNet did not exist as an organization beyond the Phyconomy.org website, but several collaborations were forming among participants in the TPCD eucheumatoids workshop, and discussions were underway with respect to how a collaborative effort such as TroPhyNet could be structured, funded, and brought to fruition. What follows are thoughts of the author based on what was learned from the TPCD workshop.

1.4.1 Spatial Emphasis on the Coral Triangle Region

Spatial geographic emphasis of TroPhyNet would be on the Coral Triangle (CT) region, with secondary engagement in other tropical regions around the world. Within the CT, collaborations have already commenced among phyconomists from Indonesia, Malaysia, and the Philippines, so development can commence from such initiatives. There is also Indian involvement in those collaborations. TroPhyNet headquarters would probably be based in Indonesia.

1.4.2 Disciplinary Emphasis on Data-Driven Innovations in Phyconomy Within Multibiomass Production Systems

Recommendation 7 of 8 in Cottier-Cook (2016) was as follows: *"Incentivize the integration of seaweed, and other extractive species with finfish in integrated multitrophic aquaculture (IMTA) systems."* Phyconomy comprises the systems of art, science, and technology applied to the production systems that yield crops of algae. The focus of TroPhyNet would be on biosecure phyconomy that results from procedures implemented against harmful biological, chemical, and socioeconomic disruptions that influence seaweed ecosystem services valorized per ecoeconomic principles in phycosecure, multibiomass, socioecological production ecoscapes (SEPE) that are subject to integrated coastal area management (ICAM). The overarching concept of multibiomass provisioning systems embraces IMTA but goes beyond it to include all coastal region terrestrial and aquatic biomass. Phycosecurity must be developed in the broader context of multibiomass systems, because many seaweed applications involve human food, animal feed, crop care, and well-being applications. Within seaweed production ecoscapes and in ecoscapes remote from the sea, phycosecurity links into the diverse biomass production systems of both land and sea. Aquaculture, agriculture, horticulture, silviculture, hydroponic systems, and other biomass production systems are interlinked in complex ways within the concept of multibiomass.

1.4.3 Development of Affordable Biological and Environmental Information Technology Systems

Recommendation 8 of 8 in Cottier-Cook (2016) advocated the development of assessment tools for evaluating spatial planning issues in relation to aquaculture (including seaweed) and for enabling risk-based analysis of spatial management options. A fundamental tool set required for such actions is the Internet-of-Things tools and solutions depicted in Fig. 1.10. Arguably, such tools and solutions are a crucial element of phycosecurity systems because without reliable and comprehensive data, impacts of phycosecurity systems cannot be quantified. Access to systems for comprehensive data collection and distribution must therefore be central to any initiative such as the proposed TroPhyNet.

1.4.4 Socioeconomic Emphasis on Profitable Value Chain Foundation Links

Socioeconomic emphasis of TroPhyNet would be on seaweed farming in socioecological production ecoscapes. It is there that value chain foundation links function, including seaweed mariculture (phycoculture) from planting to harvesting. VC and market system links beyond harvesting should be considered insofar as market demand requirements and crop prices influence phyconomy in multibiomass systems.

Profitable farm enterprises are the foundation for all seaweed-based value chains, so institutional interchange with farmers should take a reflexive (a.k.a. "bottom-up") approach that emphasizes relational value chain governance within local, regional, and global market systems. Institutional interventions in seaweed value chains often seem to underplay the fundamental profit-seeking motives of farmers. TroPhyNet must recognize that seaweed farmers are entrepreneurs and enterprise managers who need profits for survival and growth.

1.4.5 Ecological Emphasis on Quantified Seaweed Ecosystem Services Within Multibiomass Systems

TroPhyNet ecological focus should be on optimization of seaweed ecological services that are integrated into multibiomass production systems. Emphasis should be on provisioning services that result in commercial seaweed biomass production, through holistic production ecoscape management. Strategies should emphasize ecological approaches to management of production ecoscapes within integrated coastal area management systems. Actions should encourage and support profitable integration of seaweed farming with provisioning services from other multibiomass organisms, both marine and terrestrial, and should implement stewardship of regulating and cultural services of natural ecosystems.

1.4.6 Spatial Tenure in Legally Sanctioned Systems with Endemic Origins

TroPhyNet spatial tenure emphasis should be on legally sanctioned systems with endemic origins that are consistent with norms, folkways, and mores of societies in SEPE. Actions should result in the development of monitoring and assessment tools for evaluating spatial planning issues, in relation to seaweeds in multibiomass systems, and to enable risk-based analysis of management options to support licensing processes and facilitate investments in sustainable industry growth.

1.4.7 Organizational Emphasis on Collaboration Development

TroPhyNet organizational emphasis should be on collaboration development and alliance formation. It should engage with and strengthen existing institutions and facilities, not strive to replace them. TroPhyNet should engage with Coral Triangle and global centers of research, development, and technical excellence to develop and operate coordinated, phycosecure networks of quarantine facilities, cultivar banks, nurseries, and distribution systems for development and introduction of endemic and nonendemic seaweed cultivars and distribution of cultivar biomass in ways that are responsive both to farmers' needs and to market requirements.

1.4.8 Cultivar Emphasis Initially on Tropical Red Seaweeds

The first four of eight policy recommendations in Cottier-Cook et al. (2016) dealt with aspects of cultivar biomass management and TroPhyNet guidelines should embrace those policies.

TroPhyNet cultivar emphasis would initially be on eucheumatoid seaweeds and *Gracilaria* with ongoing expansion to include all commercially useful tropical red, green, and brown seaweed species. It should enhance phycosecurity by diversifying the multibiomass species base and discouraging monocrop strategies. Cultivars should be bred and selected from cultivar strains already extant on commercial seaweed farms and from endemic or indigenous wild stocks. Genetic modification techniques should be avoided if market forces oppose them.

TroPhyNet should promote stewardship of wild seaweed stocks and natural ecosystems to maintain genetic diversity in wild stocks. It should prevent invasive introduction of nonendemic species and should encourage the development of local endemic cultivars.

1.4.9 Phycosecurity Emphasis on Farm Enterprise Capacity Development

Per recommendation 5 of Cottier-Cook et al. (2016), TroPhyNet should focus on farm enterprise capacity development and enhancement through reflexive programs that include training in quarantine procedures and farm management practices. It should foster development of effective diagnostics and evaluation systems to rapidly detect diseases and biodiversity disturbances and should catalyze development and implementation of practical risk management measures. There should be special emphasis on phyconomy applications for seaweed-based crop-care products.

1.4.10 Financial Security Emphasis on Risk Mitigation

Recommendation 6 in Cottier-Cook (2016) was "*Incentivize long-term investment in the industry, potentially through part-government funded insurance policies to safeguard the business against natural disasters and disease outbreaks.*" In line with that, TroPhyNet's financial phycosecurity emphasis should be on risk mitigation through development and promotion of IT-based environmental prediction systems, understanding of environmental impacts on crops, development of best phyconomy practices (including crop health care), crop insurance, and other facilities that will mitigate seaweed farm risk.

1.5 Conclusions Presented as Ten Guidelines Toward a Tropical Phycosecurity Network

In conclusion, the following ten proposed TroPhyNet summary guidelines were presented:

1. *Spatial emphasis on ecoscapes of the Coral Triangle region* with secondary engagement in other tropical regions around the world, focusing on collaboration among international centers of phyconomy excellence.
2. *Disciplinary emphasis on data-driven innovations in phyconomy* within multibiomass ecoscape production systems. Precision phyconomy methodologies should be integrated into adaptive phyconomy systems.
3. *Development and implementation of affordable biological and environmental information technology systems.* Arguably, such systems are the backbone of phycosecurity systems, because without reliable and comprehensive data, impacts of phycosecurity cannot be quantified.
4. *Socioeconomic emphasis on eco-economically profitable seaweed-based value chain foundation links* in socioecological production ecoscapes. In other words, focus on seaweed farming enterprises.
5. *Ecological emphasis on quantified seaweed ecosystem services within multibiomass systems.* Emphasis should be on provisioning services that result in commercial seaweed biomass production, through holistic socioecological production ecoscape management.
6. *Spatial tenure in legally sanctioned systems with endemic origins* that are consistent with norms, folkways, and mores of societies within socioecological production ecoscape mosaics within integrated coastal area management (ICAM) plans.
7. *Organizational emphasis on collaboration development and alliance formation.* TroPhyNet should engage and strengthen existing institutions and facilities, not strive to replace them.
8. *Cultivar emphasis initially on eucheumatoids and Gracilaria,* with ongoing expansion to include all commercially useful tropical red, green, and brown seaweed species. TroPhyNet should enhance phycosecurity by diversifying the multibiomass species base and discouraging monocrop strategies.
9. *Phycosecurity emphasis on farm enterprise capacity development* and enhancement through reflexive programs that include IT-driven farm extension services, including training in quarantine procedures and farm management practices.
10. *Financial security emphasis on risk mitigation* through development and promotion of IT-based environmental prediction systems, understanding of environmental impacts on crops, development of best phyconomy practices (including crop health care), crop insurance, and other facilities that will mitigate seaweed farm risk.

Acknowledgments The author has had the privilege of working with thousands of seaweed farmers and with numerous colleagues from both the private and the public sectors. For the present chapter, special thanks are extended to contributors to the Phyconomy.org Tropical Phyconomy Collaboration Development eucheumatoid seaweeds workshop of July 2021. I hope that this chapter does justice to their work, and I apologize for any errors or omissions. I am especially indebted to PT Sea Six Energy Indonesia, which retained the author as a company director at the time of writing and has been a major force in support of Phyconomy.org.

References

Andrefouet S, Dewantamac MI, Ampoub EE (2021) Seaweed farming collapse and fast changing socioecosystems exacerbated by tourism and natural hazards in Indonesia: a view from space and from the households of Nusa Lembongan Island. Ocean Coastal Manage 207: 105586? Available online 8 March 2021 0964-5691/© 2021 Elsevier Ltd. All rights reserved

Berndes G, Börjesson P, Ostwald M, Palm M (2007) Multifunctional biomass production systems – an overview with presentation of specific applications in India and Sweden. Published online in Wiley InterScience (www.interscience.wiley.com); DOI: https://doi.org/10.1002/bbb.52; Biofuels, Bioprod Bioref 2:16–25

Brown LR (2001) Eco-economy: building an economy for the earth. WW Norton & Co., New York © 2001 Earth Policy Institute

Campbell I, Kambey CSB, Mateo JP, Rusekwa SB, Hurtado AQ, Msuya FE, Stentiford GD, Cottier-Cook EJ (2019) Biosecurity policy and

legislation for the global seaweed aquaculture industry. Appl Phycol. https://doi.org/10.1007/s10811-019-02010-5
Carpenter SR, DeFries R, Dietz T, Mooney HA, Polasky S, Reid WV, Scholes RJ (2006) Millennium ecosystem assessment: research needs. Science 2006(314):257–258
Cotas J, Gomes L, Pacheco D, Pereira L (2023) Ecosystem services provided by seaweeds. Hydrobiology 2023(2):75–96. https://doi.org/10.3390/hydrobiology2010006
Cottier-Cook, EJ, Nagabhatla, N, Badis, Y, Campbell, M, Chopin, T, Dai W, Fang, J, He P, Hewitt C, Kim GH, Huo Y, Jiang Z, Kema G, Li X, Liu F, Liu H, Liu Y, Lu Q, Luo Q, Mao Y, Msuya FE, Rebours C, Shen H, Stentiford GD, Yarish C, Wu H, Yang X, Zhang J, Zhou Y, Gachon CMM (2016) Safeguarding the future of the global seaweed aquaculture industry. United Nations University (INWEH) and Scottish Association for Marine Science Policy Brief. 12 pp
Douglas I, Goode D, Houck M, Wang R (2011) The Routledge handbook of urban ecology. Accessed 27 Jan 2020. https://doi.org/10.4324/9780203839263.ch3
Herr ML, Muzira T (2009) Value chain development for decent work: a guide for private sector initiatives, governments and development organizations. ISBN: 978-92-2-122488-4. International Labour Office, Geneva
Hurtado AQ, Neish IC, Critchley AT (2019) Phyconomy: the extensive cultivation of seaweeds, their sustainability and economic value, with particular reference to important lessons to be learned and transferred from the practice of eucheumatoid farming. Phycologia 58(5): Special issue on seaweed aquaculture
ILO (2016) Value chain development for decent work: how to create employment and improve working conditions in targeted sectors. International Labour Office, 2nd ed. ILO, Geneva, 2015 ISBN: 978-92-2-130509-5 (print); 978-92-2-130510-1 (web pdf)
JSSA (2010) Japan Satoyama Satoumi assessment (2010) Satoyama-Satoumi ecosystems and human Well-being: socioecological production landscapes of Japan – summary for decision makers. United Nations University, Tokyo, Japan. [SDM-EN_24Feb2011, PDF]
Kambey CSB, Campbell I, Sondak CFA, Nor ARM, Lim PE, Cottier-Cook EJ (2020) An analysis of the current status and future of biosecurity frameworks for the Indonesian seaweed industry. J Appl Phycol 32(4)
Kambey CSB, Campbell I, Cottier-Cook EJ, Nor ARM, Kassim A, Saded A, Lim PE (2021a) Evaluating biosecurity policy implementation in the seaweed aquaculture industry of Malaysia, using the quantitative knowledge, attitude, and practices (KAP) survey technique. Mar Pol 134:104800
Kambey CSB, Lim PE, Cottier-Cook EJ, Campbell I, Poong SW, Kassim A (2021b) Standard operating procedure of eucheumatoid cultivation using biosecurity-based approach. Copyright © Institute of Ocean and Earth Sciences (IOES), Universiti Malaya, 2021
Porter ME (1985) Competitive advantage: creating and sustaining competitive performance. The Free Press, Collier Macmillan, New York, London
Suyo JGB, Le Massonc V, Shaxson L, Maria Rovilla J, Luhan MRJ, Hurtado AQ (2021a) Navigating risks and uncertainties: risk perceptions and risk management strategies in the Philippine seaweed industry. Available online 27 January 2021 0308-597X/ © 2021 The Authors. Published by Elsevier Ltd. This is an open access article under the CC BY-NC-ND licence
UNIDO (2011) Industrial value chain diagnostics: an integrated tool. United Nations Industrial Development Organization (UNIDO)

Audio and Visual References at Phyconomy.org (Workshop TPCD1)

Neish IC (2021a) BeBiNet – a window to real problems of eucheumatoid farmers
Neish IC (2021b) Integration of precision technology into adaptive Phyconomy systems for extensive tropical red seaweed farming
Neish IC (2021c) Ten seaweed industry development lessons from Indonesia to the world
Suyo JG, Le Masson V, Shaxson L, Luhan MR, Hurtado A (2021b) Keeping afloat: stakeholders' insights into sustainability challenges and solutions in the Philippine seaweed industry

Diversity of Eucheumatoids in the Philippines

2

Bea A. Crisostomo and Michael Y. Roleda

Abstract

Eucheumatoids have been farmed in the Philippines for more than half a century. Throughout this long history, cultivars have been developed and distributed across the country. Most of the available reports on Philippine eucheumatoids focus on the farmed species *Eucheuma denticulatum*, *K. alvarezii* and *Kappaphycus striatus*. Meanwhile, little is known about the native populations of unutilized eucheumatoid species. This chapter discusses what is known regarding the diversity and distribution of farmed and wild eucheumatoid species in the Philippines. Recommendations on future research studies are also made in response to some issues identified in this review.

Keywords

Carrageenan industry · Commercially cultivated haplotype · Cultivars · *Eucheuma denticulatum* · Genetic variation · *Kappaphycus alvarezii* · Phenotypic variations · Seaweed farming · Strains · Wild populations

2.1 Introduction

Eucheumatoids are carrageenan-producing seaweeds from the family Solieriaceae (Gigartinales, Rhodophyta). The domestication of eucheumatoids began in the 1960s, when the increasing demand for carrageenan could no longer be satisfied by seaweed foraging alone. The first eucheumatoid farms were established in the southern islands of the Philippines (Doty 1973; Parker 1974; Trono 1974), and the industry soon spread throughout the country, as well as globally to over 30 countries (Bixler and Porse 2011; Valderrama et al. 2013).

Due to their morphological plasticity, the identification and classification of eucheumatoids has been very challenging. Conspecifics may exhibit very different morphologies, while different species may appear morphologically similar (Doty 1985, 1988; Doty and Norris 1985). Eucheumatoids were initially classified into the single genus *Eucheuma* J. Agardh (Agardh 1847), but through intensive morphological studies, two other genera were recognized, namely, *Kappaphycus* Doty (Doty 1988) and *Betaphycus* Doty ex P.C. Silva (Doty 1995; Silva et al. 1996). These classifications were based on several characteristics, such as branching patterns, cystocarp structures and cellular arrangement. In the following years, biochemical distinctions were used to support the classification of carrageenophytes. Notable differences in the carrageenan composition among genera were recognized (Chopin et al. 1999; Santos 1989). Members of *Eucheuma*, *Kappaphycus* and *Betaphycus* mainly produced iota, kappa and beta carrageenan, respectively. Even with these morphological and biochemical differences between genera, the identification of each species remained difficult. Fortunately, with the aid of molecular identification techniques, taxonomists were able to more confidently differentiate eucheumatoids.

The development of standard DNA markers for eucheumatoids and subsequent barcoding studies shed light on the ever-increasing discoveries regarding the genetic diversity of this group of seaweeds (Geraldino et al. 2006; Saunders 2005; Tan et al. 2012; Yang et al. 2008; Zuccarello et al. 1999). The mitochondrial markers *cox*1 (or its 5′ end region COI-5P) and *cox*2-3 spacer are the most widely used for genetic studies of eucheumatoids (Dumilag et al. 2023a,

Author Contributions
BA Crisostomo: Methodology, Investigation, Data Curation, Formal Analysis, Validation, Visualization, Writing-Original Draft.
MY Roleda: Funding acquisition, Project administration, Supervision, Conceptualization, Methodology, Investigation, Validation, Resources, Writing-Review and Editing.

B. A. Crisostomo · M. Y. Roleda (✉)
Algal Ecophysiology Laboratory (AlgaE Lab), The Marine Science Institute, University of the Philippines, Quezon City, Philippines
e-mail: bcrisostomo@msi.upd.edu.ph; myroleda@up.edu.ph

A. T. Critchley et al. (eds.), *Tropical Phyconomy Coalition Development*, Developments in Applied Phycology 11,
https://doi.org/10.1007/978-3-031-47806-2_2

2018, 2016a, b; Lim et al. 2014; Roleda et al. 2021; Tan et al. 2022). The plastid *rbc*L and the nuclear internal transcribed spacer (ITS) are also used, albeit less frequently, especially for establishing the distinction of new species and genera (Doty 1995; Dumilag and Zuccarello 2022; Fredericq et al. 1999; Núñez-Resendiz et al. 2019; Tan et al. 2014; Zuccarello et al. 2006). The availability of genetic data has led to the recognition of four additional genera in recent years. Three genera segregated from *Eucheuma sensu lato*: *Tacanoosca* J. N. Norris, P. W. Gabrielson and D. P. Cheney (Norris 2014), *Eucheumatopsis* Núñez-Resendiz, Dreckmann and Sentíes (Núñez-Resendiz et al. 2019) and *Mimica* Santiañez and M. J. Wynne (Santiañez and Wynne 2020). Meanwhile, one genus segregated from *Kappaphycus*: *Kappaphycopsis cottonii* Dumilag and Zuccarello (Dumilag and Zuccarello 2022).

Except for *Tacanoosca*, all the other eucheumatoid genera have been recorded in the Philippines. Based on the updated checklist of Philippine seaweeds, 19 eucheumatoid species have been recorded in the Philippines (Lastimoso and Santiañez 2020). These records should be treated with discretion, however, since some had been recorded at least four decades ago with no new reports in recent years. For instance, *Eucheumatopsis isiforme* was recorded from Negros Oriental and Sulu in the 1970s, but there have been no further reports on this species in the following decade until the present (Reyes 1972; Silva et al. 1987; Velasquez et al. 1975). More recent studies also show that *Eucheumatopsis isiforme* has a western Atlantic distribution as opposed to the other *Eucheuma* species (with which *Eucheumatopsis isiforme* was formerly grouped) with an Indo-Pacific distribution (Cheney 1988; Doty 1988; Núñez-Resendiz et al. 2019; Wynne 2017). Another questionable record is that of *Mimica amakusaensis*. Dumilag and Zuccarello (2022) assert that this should be regarded as an error, since there is neither a voucher specimen nor a local report to support this record, which was probably solely based on the hypothesis of Doty (1988, p. 185) that the distribution range of this species included the Philippines. On the other hand, *Kappaphycus malesianus* should be included in the Philippine species checklist, as recent DNA barcoding works confirmed the presence of *K. malesianus* in the Philippines (Dumilag et al. 2023a, 2016a). In addition, only eight other species from the Philippines have been genetically confirmed: *Betaphycus gelatinus*, *Eucheuma denticulatum*, *Eucheuma perplexum*, *Kappaphycopsis cottonii*, *Kappaphycus alvarezii*, *Kappaphycus inermis*, *Kappaphycus striatus* and *Mimica arnoldii* (Dumilag et al. 2020, 2018, 2016b, 2014; Dumilag and Lluisma 2014; Dumilag and Zuccarello 2022; Lim et al. 2014; Roleda et al. 2021; Zuccarello et al. 2006). Most of the eucheumatoid records had been identified only by morphology, which may be problematic due to the morphological plasticity of this group of seaweeds. The lack of genetic data from Philippine records also remains a bottleneck to the confirmation of their taxonomic identities.

2.2 Diversity of Eucheumatoid Cultivars

In over five decades of eucheumatoid farming, the industry has spread to 58 provinces throughout the Philippines and has become an important livelihood in coastal communities (Ask and Azanza 2002; Bixler and Porse 2011). The major farming sites are in the provinces of Sulu and Tawi-Tawi (BARMM), the Zamboanga Peninsula (Region IX), Southern Palawan (Region IV-B), and the Central Visayas regions (Trono and Largo 2019; Valderrama et al. 2013). The seedstock distribution facilitated by public and private entities, as well as the free exchange of seedstocks within and among farming communities, resulted in the spread of various cultivars across the country (Dumilag et al. 2023b; Quiaoit et al. 2016).

Dumilag et al. (2023b) reported a total of 66 cultivars that were recorded in the country. The majority of these cultivars were attributed to a single species, while some cultivar names were redundantly used for different species. Furthermore, they postulate that some cultivar names may have originally referred to a single cultivar but may have been given a different name when it was distributed to other areas. They also found that cultivar names were mostly based on the characteristics of the seaweeds, which may include their physical appearance, such as colour and branching pattern, or their history, such as their source locality and the person who discovered them. In a few cases, cultivars were given names that cannot be directly attributed to their characteristics (Dumilag et al. 2023b). Thallus colour was the most often used trait to describe different variants of the same cultivar (Dumilag et al. 2023b). Colour variations of eucheumatoid species have been well documented, and most of the documented Philippine cultivars have thalli in shades of brown, green and red, while a few were bicoloured or multicoloured (Dumilag et al. 2023b).

Currently, only *K. alvarezii*, *K. striatus*, *K. malesianus* and *E. denticulatum* are commercially cultivated in the country, although several cultivars have not yet been molecularly identified (Dumilag et al. 2023b, 2016a, b; Lim et al. 2014). Based on the abovementioned survey by Dumilag et al. (2023b), the majority of the identified cultivars belonged to the genus *Kappaphycus*: 27 as *K. alvarezii*, 15 as *K. striatus* and three as *K. malesianus*. Meanwhile, only six cultivars were identified as *E. denticulatum*. They also observed that *K. alvarezii* and *K. striatus* cultivars were the most widely farmed as a result of the higher demand for kappa-carrageenan compared to the iota-carrageenan-producing *E.*

denticulatum (Bindu and Levine 2011; Glenn and Doty 1990). Furthermore, they found that the most widely distributed *K. alvarezii* cultivars were the 'Cottonii', the 'Tambalang' and the 'Giant' cultivars; for *K. striatus*, these were the 'Sacol' and the 'Vanguard' cultivars; and for *E. denticulatum*, this was the 'Spinosum' cultivar (Dumilag et al. 2023b). On the other hand, *K. malesianus* was farmed exclusively in Tawi-Tawi (Dumilag et al. 2023b) and Zamboanga (Dumilag et al. 2023a). Dumilag et al. (2023b) also reported that 40 eucheumatoid cultivars were farmed in only three regions (BARMM, Region IX and Region XIII) in the southern Philippines. Moreover, the highest number of cultivars were farmed in Tawi-Tawi, Zamboanga City and Zamboanga del Sur, and Palawan.

Despite their diverse phenotypes, the genetic diversity of these cultivars appeared to be very low (Dumilag et al. 2023b). Based on the mitochondrial *cox*2-3 spacer sequences, most of the cultivars were of the commercially cultivated haplotypes: haplotype 3 for *K. alvarezii* (some with haplotypes KALV-1 and KALV-2), haplotype 89 for *K. striatus* (some with haplotype 117), haplotype MY216 for *K. malesianus* and haplotype 13 for *E. denticulatum*. The seemingly low diversity might have been due to the selected DNA marker. There has been very little variability in the *cox*2-3 spacer sequences in Philippine *K. striatus* and *E. denticulatum*, having only three and one haplotypes, respectively (Table 2.1; Dumilag et al. 2016a, b; Lim et al. 2014; Roleda et al. 2021).

There appeared to be no correlation between the genotypes and the phenotypes of these cultivars. Nonetheless, even with the unclear correlation between the genotypes and the phenotypes, the physical attributes of these cultivars appeared to be consistent; hence, farmers are able to distinguish between various cultivars. While the limited information from the *cox*2-3 spacer did not resolve the distinction between cultivars, other regions of the genome may be more suitable in differentiating the cultivars.

2.3 Diversity of Wild Eucheumatoid Populations

Eucheumatoid species have been recorded in at least 29 provinces in the Philippines. Based on the compiled records (Fig. 2.1; Table 2.2), Ilocos Norte and Cebu had the highest eucheumatoid species diversity, with 11 and 9 recorded species, respectively. Most of these reports, however, were made before 1987 (Silva et al. 1987) and have not yet been supported by molecular data. Some species (e.g., *Eucheuma horridum* and *E. leeuwenii*) have only been recorded in one or two locations each, with no additional reports in recent years (Silva et al. 1987). This raises doubts about the accuracy of these records. To date, there are only eight eucheumatoid species distributed in 12 provinces with supporting genetic data (Fig. 2.1; Table 2.2).

Table 2.1 Mitochondrial haplotypes identified from Philippine eucheumatoids

Species	COI-5P	cox2-3 spacer	COI-5P-cox2-3
Betaphycus gelatinus	1[b]	–	–
Eucheuma denticulatum	1[c]	1[a,c]	1[c]
Eucheuma perplexum	1[h]	–	–
Eucheuma sp.	1[b]	–	–
Kappaphycus alvarezii	6[c,e,f,i]	7[a,c,e,f,i]	13[c,e,f,i]
Kappaphycus inermis	1[d,e]	1[d,e]	1[e]
Kappaphycus malesianus	6[e, k]	2[e]	2[e]
Kappaphycus striatus	10[c,e,f,i]	3[a,c,e,f,i]	9[c,e,f,i]
Kappaphycus sp.	1[c]	1 [c]	1[c]
Kappaphycopsis cottonii	29[g]	11[g]	38[g]
Mimica arnoldii	2[l]	4[j]	2[l]
Total	59	30	67

[a] Silva et al. 1987; [b] Dumilag et al. 2014; [c] Lim et al. 2014; [d] Dumilag and Lluisma 2014; [e] Dumilag et al. 2016a; [f] Dumilag et al. 2016b; [g] Dumilag et al. 2018; [h] Dumilag et al. 2020; [i] Roleda et al. 2021; [j] Dumilag and Zuccarello 2022; [k] Dumilag et al. 2023a; [l] Crisostomo, independent analysis

Eucheuma and *Kappaphycus* were the most widely distributed genera, having been recorded in 34 and 22 provinces, respectively (Fig. 2.1; Table 2.2). Of these, *E. denticulatum* ($n = 15$) and *K. striatus* ($n = 13$) were the most common species. It is uncertain, however, whether these were native to the area or were introduced, since seedstocks from both of these species had been distributed across the country for cultivation (Dumilag et al. 2023b). Recent studies supported by genetic data confirm the presence of wild *Kappaphycus* species in nine provinces (Dumilag et al. 2016a, b; Lim et al. 2014; Roleda et al. 2021). In contrast, genetically confirmed wild specimens of *E. denticulatum* have been recorded only in Bohol (Lim et al. 2014).

There are currently 67 reported haplotypes from Philippine eucheumatoids based on the concatenated *COI*-5P and *cox*2-3 spacer sequences (*COI*-5P-*cox*2-3; Table 2.1). Meanwhile, the total number of *COI*-5P haplotypes ($n = 59$) was almost double that of cox2-3 spacer haplotypes ($n = 30$). Moreover, the *COI*-5P sequences of Philippine *K. striatus* and *Kappaphycopsis cottonii* were more divergent than their cox2-3 spacer sequences (Dumilag et al. 2023a, 2018, 2016a, b; Lim et al. 2014; Roleda et al. 2021). For the other eucheumatoids, there was little difference between the numbers of identified *COI*-5P and *cox*2-3 spacer sequences (Table 2.1).

The majority of the above haplotypes were identified from *Kappaphycopsis cottonii*. The intraspecific diversity of *Kappaphycopsis cottonii* was shown to be higher than that

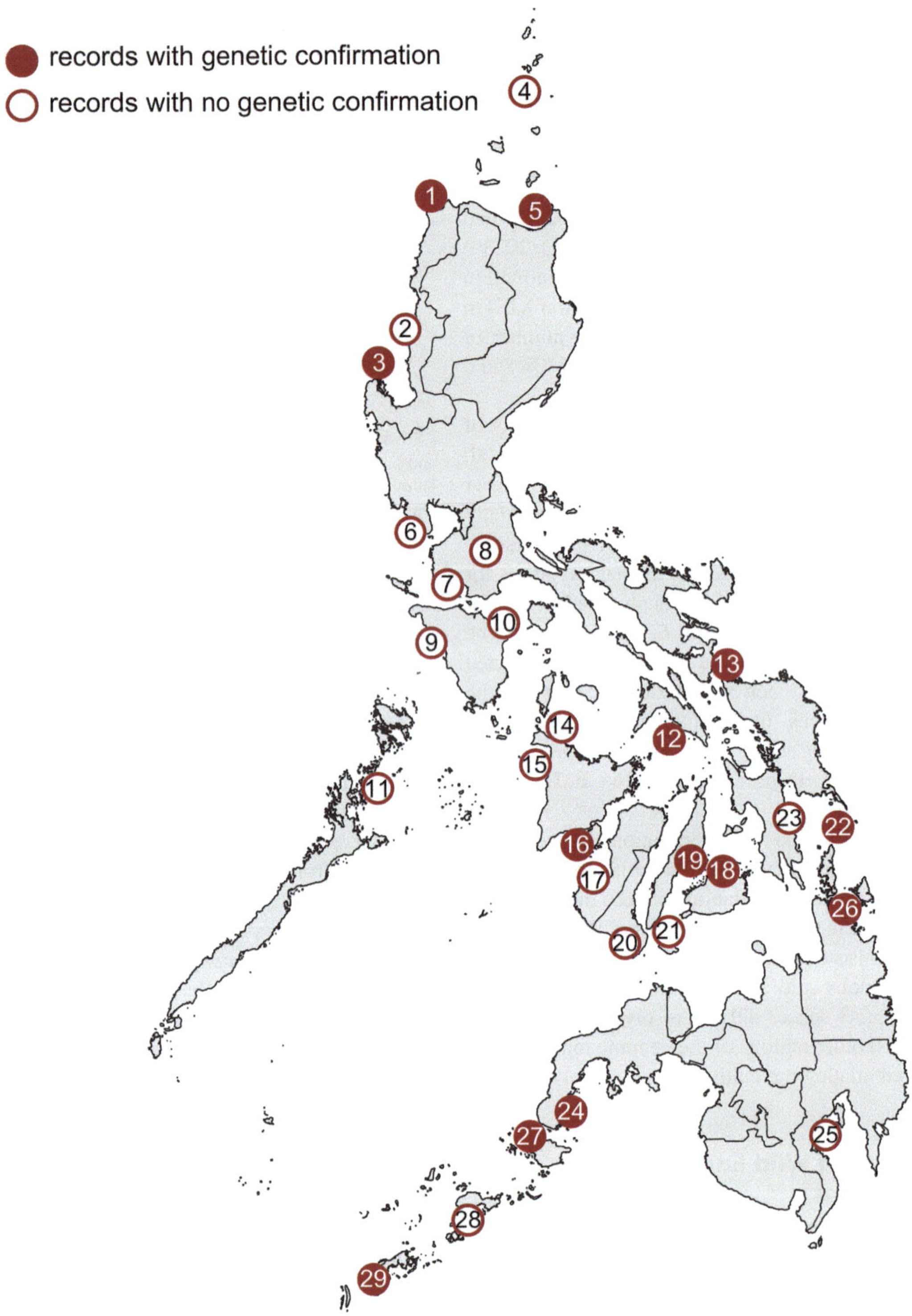

Fig. 2.1 Distribution of wild eucheumatoids in the Philippines. Numbers represent reported provinces: 1—Ilocos Norte; 2—La Union; 3—Pangasinan; 4—Batanes; 5—Cagayan; 6—Bataan; 7—Batangas; 8—Rizal; 9—Occidental Mindoro; 10—Oriental Mindoro; 11—Palawan; 12—Masbate; 13—Sorsogon; 14—Aklan; 15—Antique; 16—Guimaras; 17—Negros Occidental; 18—Bohol; 19—Cebu; 20—Negros Oriental; 21—Siquijor; 22—Eastern Samar; 23—Leyte; 24—Zamboanga; 25—Davao; 26—Surigao del Norte; 27—Basilan; 28—Sulu; 29—Tawi-Tawi. Note: Province #8 (Rizal), as reported in Silva et al. (1987), is very likely a mistake because Rizal Province is mostly landlocked and open only to a large shallow freshwater body, Laguna de Bay, between the provinces of Laguna to the south and Rizal to the north

Table 2.2 List of wild eucheumatoids recorded in the Philippines. Numbers denote the provinces presented in Fig. 2.1

Species	Distribution
Betaphycus gelatinus	1[a], 3[a], 5[a], 8[a], 13[b], 19[a], 24[a]
Eucheuma crassus	1[a], 19[a], 20[a]
Eucheuma crustiforme	1[a], 7[a], 19[a], 23[a]
Eucheuma denticulatum	1[a], 2[a], 3[a], 5[a], 6[a], 9[a], 11[a], 14[a], 18[a,c], 19[a], 21[a], 24[a], 25[a], 28[a], 29[a]
Eucheuma edule	3[a], 19[a]
Eucheuma horridum	1[a]
Eucheuma leeuwenii	19[a]
Eucheuma perplexum	1[h]
Eucheuma serra	4[a], 5[a], 11[a], 12[a], 13[a], 23[a]
Eucheuma sp.	1[b]
Eucheumatopsis isiformis	20[a], 28[a]
Kappaphycopsis cottonii	1[g,j], 3[a], 5[g, j], 10[a], 11[a], 12[g, j], 13[a, g, j], 14[a], 17[a], 18[a], 19[a,g,j], 23[a], 24[a], 27[g,j], 28[a], 29[a]
Kappaphycus alvarezii	18[c], 22[e,i], 26[f], 29[a]
Kappaphycus inermis	1[d,e]
Kappaphycus malesianus	24[k], 29[e,k]
Kappaphycus procrusteanus	15[a], 28[a]
Kappaphycus striatus	1[a], 3[a,i], 4[a], 7[a], 11[a], 12[a], 16[c], 18[a], 19[a], 24[a], 26[f], 28[a], 29[a]
Kappaphycus sp.	16[c]
Mimica arnoldii	1[a,j], 3[a], 5[j], 10[a], 11[a], 13[j], 17[a], 18[a], 19[a], 28[a], 29[a]

Note: The record of *Betaphycus gelatinus* in Rizal province (#8) as reported in Silva et al. (1987) is very likely an error. See note in Fig. 2.1
Letters in superscript denote references: [a] Silva et al. 1987; [b] Dumilag et al. 2014; [c] Lim et al. 2014; [d] Dumilag and Lluisma 2014; [e] Dumilag et al. 2016a; [f] Dumilag et al. 2016b; [g] Dumilag et al. 2018; [h] Dumilag et al. 2020; [i] Roleda et al. 2021; [j] Dumilag and Zuccarello 2022; [k] Dumilag et al. 2023a

of *Kappaphycus* species (Dumilag et al. 2018). Moreover, 33 of 38 *COI*-5P-*cox*2-3 *Kappaphycopsis cottonii* haplotypes have been identified from individuals collected from a single province, Sorsogon (Dumilag et al. 2018). In contrast, there have been only 26 *COI*-5P-*cox*2-3 haplotypes reported from the entire *Kappaphycus* genera despite the wider sampling scope (Tables 2.1 and 2.2). On the other hand, sampling efforts for *E. denticulatum* seemed lacking despite its commercial importance. The only wild *E. denticulatum* was collected from Bohol, and it had the globally cultivated haplotype EDA (Lim et al. 2014). Meanwhile, four cox2-3 spacer haplotypes have been reported from *M. arnoldii*, which were collected from Ilocos Norte, Cagayan and Sorsogon (Dumilag and Zuccarello 2022). The authors did not discuss the *COI*-5P sequences, but upon inspection, two *COI*-5P (GenBank: KX173464 and KX173467) and two *COI*-5P-*cox*2-3 haplotypes were observed (Crisostomo, independent analysis).

Aside from the commercial *COI*-5P-*cox*2-3 haplotypes KALV-A5, KSA and MY216, the rest of the *Kappaphycus* haplotypes identified from the Philippines seemed to be exclusive to the country (Dumilag et al. 2016a, b; Lim et al. 2014; Roleda et al. 2021). Half of the Philippine *Kappaphycus* haplotypes were exclusive to wild-collected specimens (Dumilag et al. 2016a, b; Lim et al. 2014; Roleda et al. 2021). Conversely, the widely cultivated haplotypes KALV-A5 and KSA have also been detected from wild-collected *K. alvarezii* and *K. striatus*, respectively (Lim et al. 2014). Furthermore, two locally cultivated *K. striatus* haplotypes, KSTR-H1 and KSTR-I1, were also present in supposedly wild populations (Dumilag et al. 2016b; Roleda et al. 2021). It is possible that cast away commercial strains from farms established themselves in the nearby reefs. However, there remains a possibility that seedstocks for cultivation were sourced from the surrounding reefs, which is commonly practiced by local seaweed farmers (Crisostomo, personal observation).

As was observed in farmed cultivars, there was no apparent correlation between the morphologies and the mitochondrial haplotypes of the wild eucheumatoids. For instance, Roleda et al. (2021) observed high phenotypic variation among natural *K. alvarezii* populations in Eastern Samar. However, the genetic variation of the *COI*-5P and *cox*2-3 spacer sequences was not able to explain the phenotypic variability of the samples. Nonetheless, an apparent geographic localization was observed for several haplotypes (including data from Dumilag et al. 2016a). Similarly, Dumilag et al. (2018) reported that the mitochondrial haplotypes of *Kappaphycopsis cottonii* did not correspond to the five morphologies observed for the species. The haplotypes seemed to be geographically segregated instead. More recently, Dumilag and Zuccarello (2022) genotyped two varieties of *M. arnoldii*—*Mimica arnoldii* var. *alcyonida* and *Mimica arnoldii* var. *arnoldii*, along with a ‘morphotype 3’, which exhibited an intermediate morphology. The variety names were designated based on the morphologies of the specimens, but the analysis of the *cox*2-3 spacer sequences was not able to support the morphologically assigned groupings. Rather, the *M. arnoldii* samples appeared to be geographically grouped. This biogeographical segregation of the *cox*2-3 spacer haplotypes was also observed by Zuccarello et al. (2006) and Halling et al. (2012). Both studies showed that native populations of *Kappaphycus* and *Eucheuma* from Hawaii and from Africa formed separate clusters that were distinct from the globally cultivated strains. Similarly, mitochondrial haplotypes in several terrestrial plants were found to be geographically structured and were consequently used in uncovering the geographical origins of some crops (Bock et al. 2014; Cheng et al. 2011; Sanjur et al. 2002).

2.4 Summary

The southern Philippine islands had the highest variety of cultivars, and most of the reported cultivars are farmed only in the south. *Kappaphycus alvarezii* is the most widely farmed species. Meanwhile, *K. malesianus* is farmed only in Zamboanga and Tawi-Tawi. Despite the wide variety of farmed cultivars, their genetic diversity seemed to be very low. In contrast, wild eucheumatoid populations had high genetic diversities. Highly diverse populations were mostly reported in the northern Philippines. Lastly, the mitochondrial markers used were not able to establish a linkage between the morphology and the haplotype of Philippine eucheumatoids. Instead, most of the reported haplotypes seemed to be geographically structured.

2.5 Recommendations

2.5.1 Expanded Survey of Philippine Eucheumatoids with Supporting Genetic Data

Updated distribution and diversity records are severely lacking for wild eucheumatoid populations. With the exception of the *Kappaphycus* and the *Kappaphycopsis* genera, the rest of the eucheumatoids were recorded in one to three locations in recent years. Thus, the records before 1847 have not yet been confirmed. In addition to broadening the sampling scope, we recommend that the previously identified sites be revisited since these locations are likely to be habited by eucheumatoid populations. Moreover, there is a need to confirm the presence of at least eight species recorded in the Philippines (listed in Lastimoso and Santiañez 2020). Aside from confirming species identities, the expansion of available genetic information would be very useful in establishing the taxonomic placements of eucheumatoid species, as well as their varieties (e.g., *Eucheuma* sp. and *Kappaphycus* sp. in Dumilag et al. 2014 and Lim et al. 2014, respectively).

2.5.2 Genome-Wide Analysis May Provide a Better Understanding of Phenotypic Variation

Based on the barcoding studies reviewed, the current genetic markers used for eucheumatoids are unable to correlate phenotypic variations with genetic variations. A broader perspective of the genome may be able to address this. The *K. alvarezii* draft genome (GenBank: GCA_002205965.3; Jia et al. 2020 unpublished) and the mitochondrial genomes of *K. malesianus, K. striatus, E. denticulatum* and B. *gelatinus* (Crisostomo et al. 2023; Li et al. 2018; Tablizo and Lluisma 2014) are valuable resources that can be used to develop new single-nucleotide polymorphism (SNP) markers. Compared to barcoding markers based on a few genes, genome-wide SNPs may prove to be more useful in the identification of cultivars, as was applied in rice (Sato et al. 2010), soy (Achard et al. 2020) and ginseng (Sun et al. 2011). This technique has also been useful in the detection of deleterious mutations in wheat (Fu et al. 2019). The identification of beneficial SNPs may also be utilized in the selection, breeding and possibly genetic engineering of superior cultivars (Hrbáčková et al. 2020; Thomson 2014).

2.5.3 Sustainable Solutions Are Needed by the Seaweed Industry to Lessen the Risks to Endemic Species

The spread of nonnative eucheumatoid strains from farms to the surrounding areas has been documented not only in the Philippines but also in countries where commercial strains have been introduced (Conklin and Smith 2005; Dumilag et al. 2016b; Halling et al. 2012). This is concerning, as commercial eucheumatoid strains were found to be predominant in some natural communities in Africa (Tano et al. 2015), Hawaii (Conklin et al. 2009) and India (Chandrasekaran et al. 2008). However, some studies report that the invasive potential of the commercially cultivated *K. alvarezii* strain is remote due to the limited occurrence of fertile tetrasporophytes and cystocarpic gametophytes and the presence of herbivores that may restrict its establishment in the wild (Castelar et al. 2009; Mandal et al. 2010). Nevertheless, as with all aquaculture activities, there is a need for sustainable solutions that reduce the negative impact of seaweed farming on native marine species while also supporting the livelihood of seaweed farmers (Brakel et al. 2021).

2.5.4 Improvement of Genetic Variation in Cultivated Eucheumatoids

Vegetative propagation has been the preferred method in the cultivation of eucheumatoids since their first domestication. Consequently, most of the cultivars to date are genetically similar. This clonality of eucheumatoid cultivars can be observed on a global scale (Halling et al. 2012; Lim et al. 2014; Zuccarello et al. 2006). Low genetic diversity in crops heightens the vulnerability of farms to disease outbreaks. In terrestrial crops, increasing genetic diversity has been shown to improve pest and disease tolerance in agroecosystems (Hajjar et al. 2008). Moreover, genetic diversity was shown to correlate with the resilience of kelp forests during environmental stress (Wernberg et al. 2018). The high genetic diversity of wild eucheumatoids may be utilized in developing new cultivars (Hinaloc and Roleda 2021).

2.5.5 Cultivation of Underutilized Eucheumatoid Species

As previously mentioned, increasing the diversity of seaweed cultivars may improve the sustainability of seaweed cultivation. In particular, *Kappaphycopsis cottonii* may be used as a substitute for *Kappaphycus* species, since it also produces kappa-carrageenan (Santos 1989). Meanwhile, *Eucheumatopsis isiforme* may be used as an alternative source of iota-carrageenan in the Americas and the Caribbean, where this species naturally occurs (Freile-Pelegrín and Robledo 2008). The cultivation of seaweeds that are native to the area may lessen the risks of species invasion brought about by introducing nonnative seedstocks. Trial cultivation of *Eucheumatopsis isiforme* had already begun at the University of Connecticut-Stamford in an effort to develop alternative sources of carrageenan (Umanzor et al. 2020).

Declarations Competing interests. The authors declare no competing interests.

Funding Primary data presented in this review were results from studies subsidized by the UPMSI inhouse research grant and research funding received from the University of the Philippines-UP System Enhanced Creative Work and Research Grant (ECWRG 2019-09-R) and Balik PhD Program (OVPAA-BPhD-2019-06), UP-Diliman, Office of the Vice Chancellor for Research and Development (OVCRD) Outright Research Grant Project No. 202039 ORG. Co-funding was also received from UKRI GCRF Global-SeaweedSTAR programme—Projects Grant no. GSS/RF/015 and GSS/RF/047 and the CHED-LAKAS-funded project "Phytochemical Characterization of Macroalgae for Food and High Value Products (PhycoPRO)". Additional funding was also received from a Safe Seaweed Coalition grant (SecureFuture; n°LS249100) and from Sea6 Energy Private Limited.

References

Achard F, Butruille M, Madjarac S, Nelson PT, Duesing J, Laffont J-L, Nelson B, Xiong J, Mikel MA, Smith JSC (2020) Single nucleotide polymorphisms facilitate distinctness-uniformity-stability testing of soybean cultivars for plant variety protection. Crop Sci 60:2280–2303. https://doi.org/10.1002/csc2.20201

Agardh JG (1847) Nya alger från Mexico. Öfversigt af Kongl. Vetenskaps-Adademiens Förhandlingar, Stockholm, pp 5–17

Ask EI, Azanza RV (2002) Advances in cultivation technology of commercial eucheumatoid species: a review with suggestions for future research. Aquaculture 206:257–277. https://doi.org/10.1016/S0044-8486(01)00724-4

Bindu MS, Levine IA (2011) The commercial red seaweed *Kappaphycus alvarezii*—an overview on farming and environment. J Appl Phycol 23:789–796. https://doi.org/10.1007/s10811-010-9570-2

Bixler HJ, Porse H (2011) A decade of change in the seaweed hydrocolloids industry. J Appl Phycol 23:321–335. https://doi.org/10.1007/s10811-010-9529-3

Bock DG, Kane NC, Ebert DP, Rieseberg LH (2014) Genome skimming reveals the origin of the Jerusalem Artichoke tuber crop species: neither from Jerusalem nor an artichoke. New Phytol 201:1021–1030. https://doi.org/10.1111/nph.12560

Brakel J, Sibonga RC, Dumilag RV, Montalescot V, Campbell I, Cottier-Cook EJ, Ward G, Le Masson V, Liu T, Msuya FE, Brodie J, Lim P, Gachon CMM (2021) Exploring, harnessing and conserving marine genetic resources towards a sustainable seaweed aquaculture. Plants People Planet 3:337–349. https://doi.org/10.1002/ppp3.10190

Castelar B, Reis RP, Moura AL, Kirk R (2009) Invasive potential of *Kappaphycus alvarezii* off the south coast of Rio de Janeiro state, Brazil: a contribution to environmentally secure cultivation in the tropics. Bot Mar 52:283–289. https://doi.org/10.1515/BOT.2009.002

Chandrasekaran S, Nagendran A, Pandiaraja D, Krishnankutty N, Balasubramanian K (2008) Bioinvasion of *Kappaphycus alvarezii* on corals in the Gulf of Mannar, India. Curr Sci 94:1167–1172

Cheney DP (1988) The genus *Eucheuma* J. Agardh in Florida and the Caribbean. In: Abbott IA (ed) Taxonomy of economic seaweeds: with reference to some Pacific and Caribbean species. California Sea Grant Program, La Jolla, pp 209–219

Cheng D, Yoshida Y, Kitazaki K, Negoro S, Takahashi H, Xu D, Mikami T, Kubo T (2011) Mitochondrial genome diversity in *Beta vulgaris* L. ssp. *vulgaris* (leaf and garden beet groups) and its implications concerning the dissemination of the crop. Genet Resour Crop Evol 58:553–560. https://doi.org/10.1007/s10722-010-9598-9

Chopin T, Kerin BF, Mazerolle R (1999) Phycocolloid chemistry as a taxonomic indicator of phylogeny in the Gigartinales, Rhodophyceae: a review and current developments using Fourier transform infrared diffuse reflectance spectroscopy. Phycol Res 47:167–188. https://doi.org/10.1046/j.1440-1835.1999.00170.x

Conklin EJ, Smith JE (2005) Abundance and spread of the invasive red algae, *Kappaphycus* spp., in Kane'ohe Bay, Hawai'i and an experimental assessment of management options. Biol Invasions 7:1029–1039. https://doi.org/10.1007/s10530-004-3125-x

Conklin KY, Kurihara A, Sherwood AR (2009) A molecular method for identification of the morphologically plastic invasive algal genera *Eucheuma* and *Kappaphycus* (Rhodophyta, Gigartinales) in Hawaii. J Appl Phycol 21:691–699. https://doi.org/10.1007/s10811-009-9404-2

Crisostomo BA, Dumilag RV, Roleda MY, Lluisma AO (2023) The complete mitochondrial genome of a wild-collected *Kappaphycus malesianus* (Solieriaceae, Rhodophyta). Mitochondrial DNA Part B 8:359–363. https://doi.org/10.1080/23802359.2023.2183728

Doty MS (1973) Farming the red seaweed, *Eucheuma*, for carrageenans. Micronesica 9:59–73

Doty MS (1985) *Eucheuma alvarezii*, sp. nov. (Gigartinales, Rhodophyta) from Malaysia. In: Abbott IA, Norris JN (eds) Taxonomy of economic seaweeds: with reference to some Pacific and Caribbean species. California Sea Grant College Program, La Jolla, pp 37–45

Doty MS (1988) Prodromus Ad Systematica Eucheumatoideorum: a tribe of commercial seaweeds related to *Eucheuma* (Solieriaceae, Gigartinales). In: Abbott IA (ed) Taxonomy of economic seaweeds: with reference to some Pacific and Caribbean species. California Sea Grant Program, La Jolla, CA, pp 159–207

Doty MS (1995) *Betaphycus philippinensis* gen. Et sp. nov. and related species (Solieriaceae, Gigartinales). In: Abbott IA (ed) Taxonomy of economic seaweeds: with reference to some Pacific and Caribbean species. California Sea Grant Program, La Jolla, pp 237–245

Doty MS, Norris JN (1985) *Eucheuma alvarezii*, sp. nov. (Gigartinales, Rhodophyta) from Malaysia. In: Abbott IA, Norris JN (eds) Taxonomy of economic seaweeds: with reference to some Pacific and Caribbean species. California Sea Grant College Program, La Jolla, pp 47–61

Dumilag RV, Lluisma A (2014) Resolving the phylogenetic affinities of *Kappaphycus inermis* within the genus *Kappaphycus* (Gigartinales, Solieriaceae) using mitochondrial and plastid markers. Phytotaxa 162:223. https://doi.org/10.11646/phytotaxa.162.4.5

Dumilag RV, Zuccarello GC (2022) Phylogeny and genetic diversity of the Philippine eucheumatoid genus *Mimica* (Solieriaceae, Rhodophyta), and the proposal for *Kappaphycopsis* gen. nov. to include the anomalous species of *Kappaphycus, K. cottonii*. Phycologia 61:496–503. https://doi.org/10.1080/00318884.2022.2079283

Dumilag RV, Liao LM, Lluisma AO (2014) Phylogeny of *Betaphycus* (Gigartinales, Rhodophyta) as inferred from COI sequences and morphological observations on *B. philippinensis*. J Appl Phycol 26:587–595. https://doi.org/10.1007/s10811-013-0063-y

Dumilag RV, Orosco FL, Lluisma AO (2016a) Genetic diversity of *Kappaphycus* species (Gigartinales, Rhodophyta) in The Philippines. Syst Biodivers 14:441–451. https://doi.org/10.1080/14772000.2016.1157643

Dumilag RV, Salvador RC, Halling C (2016b) Genotype introduction affects population composition of native Philippine *Kappaphycus* (Gigartinales, Rhodophyta). Conserv Genet Resour 8:439–441. https://doi.org/10.1007/s12686-016-0591-2

Dumilag RV, Gallardo WGM, Garcia CPC, You Y, Chaves AKG, Agahan L (2018) Phenotypic and mtDNA variation in Philippine *Kappaphycus cottonii* (Gigartinales, Rhodophyta). Mitochondrial DNA Part A 29:951–963. https://doi.org/10.1080/24701394.2017.1398745

Dumilag RV, Lin S-M, Zuccarello GC, Kraft GT (2020) The identity of *Eucheuma perplexum* (Solieriaceae, Gigartinales) and its distinction from *Eucheuma serra* as exemplified by a proposed new epitype. Phycologia 59:497–505. https://doi.org/10.1080/00318884.2020.1797375

Dumilag RV, Crisostomo BA, Aguinaldo Z-ZA, Lluisma AO, Gachon CMM, Roleda MY (2023a) Genetic diversity of *Kappaphycus malesianus* (Solieriaceae, Rhodophyta) from The Philippines. Aquat Bot 187:103649. https://doi.org/10.1016/j.aquabot.2023.103649

Dumilag RV, Crisostomo BA, Aguinaldo Z-ZA, Hinaloc LAR, Liao LM, Roa-Quiaoit HA, Dangan-Galon F, Zuccarello GC, Guillemin M-L, Brodie J, Cottier-Cook EJ, Roleda MY (2023b) The diversity of eucheumatoid seaweed cultivars in The Philippines. Rev Fish Sci Aquac 31:47–65. https://doi.org/10.1080/23308249.2022.2060038

Fredericq S, Freshwater DW, Hommersand MH (1999) Observations on the phylogenetic systematics and biogeography of the Solieriaceae (Gigartinales, Rhodophyta) inferred from rbcL sequences and morphological evidence. Hydrobiologia 398:25–38. https://doi.org/10.1023/A:1017077831840

Freile-Pelegrín Y, Robledo D (2008) Carrageenan of *Eucheuma isiforme* (Solieriaceae, Rhodophyta) from Nicaragua. J Appl Phycol 20:537–541. https://doi.org/10.1007/s10811-007-9270-8

Fu Y-B, Peterson GW, Horbach C, Konkin DJ, Beiles A, Nevo E (2019) Elevated mutation and selection in wild emmer wheat in response to 28 years of global warming. Proc Natl Acad Sci 116:20002–20008. https://doi.org/10.1073/pnas.1909564116

Geraldino PJL, Yang E-C, Bu S-M (2006) Morphology and molecular phylogeny of *Hypnea flexicaulis* (Gigartinales, Rhodophyta) from Korea. Algae 21:417–423. https://doi.org/10.4490/ALGAE.2006.21.4.417

Glenn EP, Doty MS (1990) Growth of the seaweeds *Kappaphycus alvarezii, K. striatum* and *Eucheuma denticulatum* as affected by environment in Hawaii. Aquaculture 84:245–255. https://doi.org/10.1016/0044-8486(90)90090-A

Hajjar R, Jarvis DI, Gemmill-Herren B (2008) The utility of crop genetic diversity in maintaining ecosystem services. Agric Ecosyst Environ 123:261–270. https://doi.org/10.1016/j.agee.2007.08.003

Halling C, Wikström S, Lilliesköld-Sjöö G, Mörk E, Lundsor E, Zuccarello GC (2012) Introduction of Asian strains and low genetic variation in farmed seaweeds: indications for new management practices. J Appl Phycol 25:89–95. https://doi.org/10.1007/s10811-012-9842-0

Hinaloc LAR, Roleda MY (2021) Phenotypic diversity, growth and sexual differentiation in the progeny of wild *Kappaphycus alvarezii* (Gigartinales, Florideophyceae). Phycologia 60:547–557. https://doi.org/10.1080/00318884.2021.1946307

Hrbáčková M, Dvořák P, Takáč T, Tichá M, Luptovčiak I, Šamajová O, Ovečka M, Šamaj J (2020) Biotechnological perspectives of omics and genetic engineering methods in alfalfa. Front Plant Sci 11:592. https://doi.org/10.3389/fpls.2020.00592

Jia S, Wang G, Liu G, Qu J, Zhao B, Jin X, Zhang L, Yin J, Liu C, Shan G, Wu S, Song L, Liu T, Wang X, Yu J (2020) High-quality de novo genome assembly of *Kappaphycus alvarezii* based on both PacBio and HiSeq sequencing. bioRxiv. https://doi.org/10.1101/2020.02.15.950402. Accessed 28 June 2023

Lastimoso JML, Santiañez WJE (2020) Updated checklist of the benthic marine macroalgae of The Philippines. Philipp J Sci 150:29–92

Li Y, Liu N, Wang X, Tang X, Zhang L, Meinita MDN, Wang G, Yin H, Jin Y, Wang H, Liu C, Chi S, Liu T, Zhang J (2018) Comparative genomics and systematics of *Betaphycus*, *Eucheuma*, and *Kappaphycus* (Solieriaceae: Rhodophyta) based on mitochondrial genome. J Appl Phycol 30:3435–3443. https://doi.org/10.1007/s10811-018-1450-1

Lim PE, Tan J, Phang SM, Nikmatullah A, Hong DD, Sunarpi H, Hurtado AQ (2014) Genetic diversity of *Kappaphycus* Doty and *Eucheuma* J. Agardh (Solieriaceae, Rhodophyta) in Southeast Asia. J Appl Phycol 26:1253–1272. https://doi.org/10.1007/s10811-013-0197-y

Mandal SK, Mantri VA, Haldar S, Eswaran K, Ganesan M (2010) Invasion potential of *Kappaphycus alvarezii* on corals at Kurusadai Island, Gulf of Mannar, India. Algae 25:205–216. https://doi.org/10.4490/algae.2010.25.4.205

Norris JN (2014) Marine algae of the northern gulf of California II: Rhodophyta. Smithsonian Institution Scholarly Press. https://doi.org/10.5479/si.19382812.96

Núñez-Resendiz ML, Dreckmann KM, Sentíes A, Wynne MJ, León-Tejera HP (2019) *Eucheumatopsis isiformis* gen. & comb. nov. (Solieriaceae, Rhodophyta) from the Yucatan Peninsula, to accommodate *Eucheuma isiforme*. Phycologia 58:51–62. https://doi.org/10.1080/00318884.2018.1517536

Parker HS (1974) The culture of the red algal genus *Eucheuma* in The Philippines. Aquaculture 3:425–439. https://doi.org/10.1016/0044-8486(74)90009-X

Quiaoit HAR, Uy WH, Bacaltos DGG, Chio PBR (2016) Seaweed area GIS-based mapping. Production support system for sustainable seaweed farming in The Philippines 2016 report. Cagayan de Oro (Philippines): Xavier University Press

Reyes AY (1972) A survey of the littoral benthic algae of the coastal areas of Dumaguete City. Philipp J Sci 99:131–163

Roleda MY, Aguinaldo Z-ZA, Crisostomo BA, Hinaloc LAR, Projimo VZ, Dumilag RV, Lluisma AO (2021) Discovery of novel haplotypes from wild populations of *Kappaphycus* (Gigartinales, Rhodophyta) in The Philippines. Algae 36:1–12. https://doi.org/10.4490/algae.2021.36.2.18

Sanjur OI, Piperno DR, Andres TC, Wessel-Beaver L (2002) Phylogenetic relationships among domesticated and wild species of *Cucurbita* (Cucurbitaceae) inferred from a mitochondrial gene: implications for crop plant evolution and areas of origin. Proc Natl Acad Sci 99:535–540. https://doi.org/10.1073/pnas.012577299

Santiañez WJE, Wynne MJ (2020) Establishment of *Mimica* gen. nov. to accommodate the anaxiferous species of the economically important red seaweed *Eucheuma* (Solieriaceae, Rhodophyta). Phytotaxa 439:167–170. https://doi.org/10.11646/phytotaxa.439.2.8

Santos GA (1989) Carrageenans of species of *Eucheuma* J. Agardh and *Kappaphycus* Doty (Solieriaceae, Rhodophyta). Aquat Bot 36:55–67. https://doi.org/10.1016/0304-3770(89)90091-0

Sato H, Endo T, Shiokai S, Nishio T, Yamaguchi M (2010) Identification of 205 current rice cultivars in Japan by dot-blot-SNP analysis. Breed Sci 60:447–453. https://doi.org/10.1270/jsbbs.60.447

Saunders GW (2005) Applying DNA barcoding to red macroalgae: a preliminary appraisal holds promise for future applications. Philos Trans R Soc B Biol Sci 360:1879–1888. https://doi.org/10.1098/rstb.2005.1719

Silva PC, Meñez EG, Moe RL (1987) Catalogue of the benthic marine algae of The Philippines. Smithsonian Institution Press 27:1. https://doi.org/10.5479/si.1943667X.27.1

Silva PC, Basson PW, Moe RL (1996) Catalogue of the benthic marine algae of the Indian Ocean. Univ Calif Publ Bot 79:1–1259

Sun H, Wang H-T, Kwon W-S, Kim Y-J, In J-G, Yang D-C (2011) A simple and rapid technique for the authentication of the ginseng cultivar, Yunpoong, using an SNP marker in a large sample of ginseng leaves. Gene 487:75–79. https://doi.org/10.1016/j.gene.2011.05.021

Tablizo FA, Lluisma AO (2014) The mitochondrial genome of the red alga *Kappaphycus striatus* ("Green Sacol" variety): complete nucleotide sequence, genome structure and organization, and comparative analysis. Mar Genomics 18:155–161. https://doi.org/10.1016/j.margen.2014.05.006

Tan J, Lim P-E, Phang S-M, Hong DD, Sunarpi H, Hurtado A.Q (2012) Assessment of four molecular markers as potential DNA barcodes for red algae *Kappaphycus* Doty and Eucheuma J. Agardh (Solieriaceae, Rhodophyta). PLoS One 7:e52905. https://doi.org/10.1371/journal.pone.0052905

Tan J, Lim PE, Phang SM, Rahiman A, Nikmatullah A, Sunarpi H, Hurtado AQ (2014) *Kappaphycus malesianus* sp. nov.: a new species of *Kappaphycus* (Gigartinales, Rhodophyta) from Southeast Asia. J Appl Phycol 26:1273–1285. https://doi.org/10.1007/s10811-013-0155-8

Tan P-L, Poong S-W, Tan J, Brakel J, Gachon C, Brodie J, Sade A, Lim P-E (2022) Assessment of genetic diversity within eucheumatoid cultivars in East Sabah, Malaysia. J Appl Phycol 34:709–717. https://doi.org/10.1007/s10811-021-02608-8

Tano SA, Halling C, Lind E, Buriyo A, Wikström SA (2015) Extensive spread of farmed seaweeds causes a shift from native to nonnative haplotypes in natural seaweed beds. Mar Biol 162:1983–1992. https://doi.org/10.1007/s00227-015-2724-7

Thomson MJ (2014) High-throughput SNP genotyping to accelerate crop improvement. Plant Breed Biotechnol 2:195–212. https://doi.org/10.9787/PBB.2014.2.3.195

Trono GC Jr (1974) *Eucheuma* farming in The Philippines. Univ Philipp Natur Sci Res Ctr Tech Publ, 13 p

Trono GC, Largo DB (2019) The seaweed resources of The Philippines. Bot Mar 62:483–498. https://doi.org/10.1515/bot-2018-0069

Umanzor S, Jang S, Antosca R, Critchley AT, Yarish C, Kim JK (2020) Optimizing the application of selected biostimulants to enhance the growth of *Eucheumatopsis isiformis*, a carrageenophyte with commercial value, as grown in land-based nursery systems. J Appl Phycol 32:1917–1922. https://doi.org/10.1007/s10811-020-02091-7

Valderrama D, Cai J, Hishamunda N, Ridler NB (2013) Social and economic dimensions of carrageenan seaweed farming. Fisheries and aquaculture technical paper no. 580. FAO, Rome, 204 pp

Velasquez GT, Trono GC Jr, Doty MS (1975) Algal species reported from The Philippines. Philipp J Sci 101:115–169

Wernberg T, Coleman MA, Bennett S, Thomsen MS, Tuya F, Kelaher BP (2018) Genetic diversity and kelp forest vulnerability to climatic stress. Sci Rep 8:1851. https://doi.org/10.1038/s41598-018-20009-9

Wynne MJ (2017) A checklist of benthic marine algae of the tropical and subtropical Western Atlantic: fourth revision [WWW Document]. https://www.schweizerbart.de/publications/detail/isbn/9783443510671. Accessed 7.10.22

Yang EC, Kim MS, Geraldino PJL, Sahoo D, Shin J-A, Boo SM (2008) Mitochondrial cox1 and plastid rbcL genes of *Gracilaria vermiculophylla* (Gracilariaceae, Rhodophyta). J Appl Phycol 20:161–168. https://doi.org/10.1007/s10811-007-9201-8

Zuccarello GC, Burger G, West JA, King RJ (1999) A mitochondrial marker for red algal intraspecific relationships. Mol Ecol 8:1443–1447. https://doi.org/10.1046/j.1365-294x.1999.00710.x

Zuccarello GC, Critchley AT, Smith J, Sieber V, Lhonneur GB, West JA (2006) Systematics and genetic variation in commercial shape *Kappaphycus* and shape *Eucheuma* (Solieriaceae, Rhodophyta). J Appl Phycol 18:643–651. https://doi.org/10.1007/s10811-006-9066-2

The Role of Molecular Marker Technology in Advancing Eucheumatoid Research

3

Ji Tan, Sze-Wan Poong, Claire Gachon, Juliet Brodie, and Phaik-Eem Lim

Abstract

The ability of molecular markers to detect genetic variations between organisms has led to the advancement in DNA barcoding, biodiagnostics, molecular systematics, forensics, molecular breeding, genetic modification, etc. However, the utility and prospects of molecular markers are still relatively new to seaweeds such as the commercially important eucheumatoids of *Kappaphycus* and *Eucheuma*. The present study summarizes how molecular marker technology has improved several areas of eucheumatoid research, i.e., molecular identification, systematics, genetic diversity, and bioinvasion detection. The use of molecular markers in strain selection, breeding, and conservation is also vital to improve the resilience of the eucheumatoid industry in light of pests and diseases, aging cultivars, and climate change.

Keywords

*cox*1 · *cox*2–3 spacer · Genetic · DNA marker · Systematics · Phylogeny

J. Tan
Department of Agricultural and Food Science, Faculty of Science, Universiti Tunku Abdul Rahman, Perak, Malaysia

S.-W. Poong · P.-E. Lim (✉)
Institute of Ocean and Earth Sciences, University of Malaya, Kuala Lumpur, Malaysia
e-mail: phaikeem@um.edu.my

C. Gachon
Scottish Association for Marine Science, Scottish Marine Institute, Oban, UK

Unité Molécules de Communication et Adaptation des Microorganismes, UMR 7245, Muséum National d'Histoire Naturelle, CNRS, Paris, France

J. Brodie
Natural History Museum, Research, London, UK

3.1 Introduction

A molecular marker is defined as a molecule within a biological sample that can be used for its identification. Variations in biological molecules such as DNA are often specific to species or individuals and could be utilized for a wide range of scientific research, including species identification, biodiagnostics, forensics, population genetics, molecular breeding, phylogeography, phylogenetics, genetic engineering, and so on. The use of molecular marker technology has proven to be essential in the research and development of animals and plants, although its use in seaweeds is progressing at a slower pace. This review discusses the development of molecular markers in eucheumatoid research, as well as its potential application in the future.

3.2 Overview of Eucheumatoid Farming

Red seaweeds of the genera *Kappaphycus* and *Eucheuma* are often referred to as eucheumatoids. *Kappaphycus alvarezii* and *Eucheuma denticulatum* are farmed globally for carrageenan, a hydrocolloid valued for its thickening and emulsifying properties. Carrageenan is "generally regarded as safe (GRAS)" by the US Food and Drug Administration (FDA) and is widely used as a food additive that is considered to be a vegan-friendly option when compared to gelatine (Younes et al. 2018). The market size for carrageenan is forecast to increase from USD 825 million in 2021 to USD 1248.6 million by 2028 as people look to seaweeds as a more sustainable crop of the future (Market Analysis Report 2021).

Kappaphycus alvarezii (formerly *Eucheuma cottonii*) was introduced into culture in the Philippines during the 1970s after years of field trials (Hurtado et al. 2016), and the developed cultivar displayed rapid growth and good *kappa*-carrageenan yield and gel strength. Propagules of *K. alvarezii* are produced by vegetative cuttings of mature seaweed thalli before being tied onto the culture lines, where they will be

A. T. Critchley et al. (eds.), *Tropical Phyconomy Coalition Development*, Developments in Applied Phycology 11,
https://doi.org/10.1007/978-3-031-47806-2_3

maintained and farmed for 30–45 days before reaching marketable size. The seaweeds are then harvested and sun-dried before being sent to carrageenan processing plants to be processed into alkali-treated chips, semirefined or refined carrageenan (McHugh 2003).

Driven by the market demand for carrageenan, the cultivation of *K. alvarezii* became a source of livelihood for coastal communities, which improved the lives of many in the Philippines (Hurtado et al. 2015). This success has led to the commercial introduction of eucheumatoid farming into different parts of the world, including East Africa, Fiji, India, Malaysia, Vietnam, Cambodia, Myanmar, China, and Latin America (Hurtado et al. 2014; Brakel et al. 2021). The introduction of eucheumatoid cultivars extended into the early 2000s. Throughout this period, other species of eucheumatoids, including *K. striatus* and *E. denticulatum,* were also introduced as cultivars in various parts of the world. Additionally, local species were also introduced and cultivated at a relatively smaller scale, viz., *K. malesianus* in Malaysia and *Betaphycus gelatinus* (a less common eucheumatoid) in China (Tan et al. 2013; Gao et al. 2022).

Currently, Indonesia and the Philippines are the top producers of carrageenan in the world, recording production figures of 9.8 million tons (ww) and 1.5 million tons (ww), respectively, in 2019. These two countries contribute 97.2% of global carrageenan production (Cai et al. 2021).

3.3 Challenges Affecting Eucheumatoid Farming

Despite having been farmed for the past 50 years, eucheumatoid cultivation is still plagued by numerous challenges and issues, some of which have existed for as long as its cultivation history. These include aging cultivars, diseases and pests, climate change, and challenges with taxonomy and nomenclature (Zuccarello et al. 2006; Harley et al. 2012; Datu Eraza et al. 2017; Ward et al. 2020; Cai et al. 2021).

3.3.1 Aging Cultivars

The propagules of commercial eucheumatoid cultivars (*K. alvarezii, K. striatus,* and *E. denticulatum*) have been produced vegetatively for the past five decades, leading to the possible loss of genetic vigor, which could be associated with the hypothetical Muller's rachet and Hayflick limit of cells (Hojsgaard and Hörandl 2015; Charrier et al. 2017; Lovell et al. 2017; Sheldrake 2022). The limited genetic variation of eucheumatoid cultivars has been reported since the 2000s. Efforts, however, in the development of new cultivars via selection or breeding have been slow, mostly due to the lack of resources and expertise, lack of understanding of the reproductive biology of eucheumatoids, and the difficulty in identifying eucheumatoids (Zuccarello et al. 2006; Brakel et al. 2021; Tan et al. 2022).

3.3.2 Diseases and Pests

Eucheumatoid farming is often plagued by two common diseases, namely, *ice–ice* disease (IID) and epiphytic infestation. IID refers to the phenomenon when the tissues of a eucheumatoid softens, decolorizes, and eventually disintegrates, resulting in a huge loss in gel strength and carrageenan yield. This syndrome is often associated with bacterial infections as a result of poor farm conditions and extreme fluctuations in temperature and salinity (Alibon et al. 2019; Ward et al. 2020). Common bacterial genera often associated with IID include *Vibrio*, *Pseudoalteromonas*, *Alteromonas*, *Stenotrophomonas*, *Aurantimonas*, and *Ochrobactrum* (Largo et al. 1995; Ward et al. 2020, 2022). Currently, the most practiced solution for eucheumatoids suffering from IID is to have the infected parts removed and appropriately discarded. In severe outbreaks of IID, farmers may often resort to early harvesting at all affected sites (Hughes 2017; Kambey et al. 2021; Ward et al. 2022).

Eucheumatoids may also suffer from epiphytic infestations: (1) biofouling algae loosely attached to the thallus and (2) minute red filamentous algae of the genus *Melanothamnus* (formerly *Neosiphonia* and *Polysiphonia*) that penetrate through the thallus and are strongly anchored into the tissues of the host (Franks 2010; Vairappan 2016; Ward et al. 2020). These infestations typically arise in cultivation areas with high levels of nitrate and phosphate and impact the growth performance of eucheumatoids (Vairappan et al. 2013; Nelson 2018).

To a smaller extent, farmed eucheumatoids may also be exposed to herbivory by fishes and green sea turtles, depending on the location of the farm and its surrounding environment (Tolentino-Pablico et al. 2008; Tomas et al. 2011; Kasim et al. 2017).

3.3.3 Climate Change

The rise in sea temperature, ocean acidification, and changes in climate patterns have been recorded throughout the world (Harley et al. 2012). Deviations from the optimum growth conditions of farmed eucheumatoids negatively impact their growth and productivity. For example, Kumar et al. (2020) reported a decline in the specific growth rate of the commercial *K. alvarezii* cultivar at 32 °C, as well as the inability of the cultivar to survive at 36 °C and above. *Eucheuma denticulatum* was reported to be able to tolerate up to a temperature of 33 °C (Lideman et al. 2013; Borlongan et al. 2017). On the other hand, salinity tolerance was identified as 25–45 ppt for *K. alvarezii* (Hayashi et al. 2011) and 28–33 ppt for *E. denticulatum* (Munawan et al. 2021). The development of cultivars that can better adapt to a wider range of environmental fluctuations would improve the resilience of the carrageenan industry in the face of climate change.

3.3.4 Taxonomy and Nomenclature

The taxonomy of carrageenan-producing red seaweeds, such as *Betaphycus*, *Eucheuma*, and *Kappaphycus*, has improved significantly over the years as a polyphasic approach, where genetic data are used in combination with other characteristics, such as morphology, cytology, and ecology, for identification and taxonomy. This was mainly driven by limitations of morphological identification due to the plastic nature of these seaweeds. Although the use of genetic identification has better elucidated the taxonomy of eucheumatoids, improved efforts in the genetic characterization of wild populations of eucheumatoids have revealed many potentially new species and haplotypes, some of which challenge the current delineation of genera and species (Zuccarello et al. 2006; Dumilag and Lluisma 2014; Dumilag et al. 2014, 2016a, 2018; Tan et al. 2014; Brakel et al. 2021; Roleda et al. 2021).

Attempts to review the taxonomy of eucheumatoids were also met with challenges due to the poor DNA quality of type specimens, which were insufficient for amplification and sequencing (Lim et al. 2014), while several species of eucheumatoids were never encountered again after their original description. Taxonomic resolution was also exacerbated with the commercial introduction of seaweed cultivars from one country to another, some of which may not have been documented. Although efforts have been put into the sampling of species from their type localities, progress has generally been slow and may benefit from a concerted long-term effort.

In relation to nomenclature, recommendations are available based on the *International Code of Nomenclature for algae, fungi, and plants* (ICN) (Turland et al. 2018) and the rules and recommendations for the naming of cultivated plants based on the *International Code of Nomenclature for Cultivated Plants* (ICNCP) (Spencer and Cross 2007), but this may not be specific enough, given the complexity of seaweed life histories. This is compounded by the lack of a standardized system in seaweed selection and breeding, which could be a setback in the development of new eucheumatoid cultivars and may need to be discussed and established by phycologists throughout the globe.

3.4 Molecular Marker Technology and Its Use in Eucheumatoid Research

Recently, the use of molecular markers in seaweed research has shown great promise in the selection and development of new seaweed cultivars (Hwang et al. 2019) and represents an important approach to address the aforementioned challenges in eucheumatoid farming and research. DNA is the most commonly used molecular marker because the intrinsic information of DNA is far more consistent and reliable when compared to phenotypic characteristics, which could be influenced by environmental conditions. The use of this genotypic information allows scientists to identify and distinguish individuals, as well as establish the taxonomy for all organisms on the planet, resulting in the molecular tree of life (Pace 2009). Molecular markers were initially used in seaweeds for identification and taxonomic purposes but have since been employed for a wide variety of reasons, including omics, population genetics, genetic diversity, conservation, bioinvasion detection, and modern breeding.

3.4.1 Molecular Identification of Eucheumatoids

The imperative to employ molecular markers for the identification and taxonomic delineation of eucheumatoids became apparent in the 2000s. This was due to the taxonomic confusion arising from the morphologically plastic nature of eucheumatoids, as well as the mixed use of commercial names, anecdotal local names, and incorrect scientific names, which resulted in a series of problems in the industry.

The variable appearance of both farmed and wild eucheumatoids and their general lack of distinctive morphological features hindered efforts in species identification (Zuccarello et al. 2006). Local farmers often try to name each morphotype of *Kappaphycus* and *Eucheuma* to ease coordination and communication. However, this has resulted in numerous local names across countries, and many of these have been shown to be synonymous with each other based on genetic data (Tan et al. 2013; Dumilag and Lluisma 2014; Lim et al. 2014; Datu Eraza et al. 2017). The commercial name of *Kappaphycus alvarezii* (then *Eucheuma alvarezii*), termed "cottonii," was also often mistaken for *Kappaphycus cottonii*, a different species. In addition, eucheumatoids are frequently still collectively referred to as *Eucheuma* by laymen. A simple misidentification between *Kappaphycus* and *Eucheuma* could negatively impact carrageenan processing and productivity, as *Kappaphycus* produces *kappa*-carrageenan while *Eucheuma* produces *iota*-carrageenan.

Efforts to resolve the taxonomy of eucheumatoids based on museum voucher specimens have also been challenging due to (1) morphological differences between fresh and dried specimens, (2) morphological differences between cultivars and wild samples, and (3) the lack of quality or missing herbarium specimens in some eucheumatoid species. Problems with the identification of eucheumatoids have also caused a major setback in the elucidation of the phylogenetic relationships within the eucheumatoids. For example, failed attempts to sample *K. procrusteanus* and *K. striatus* from their type localities prevent any validation of their origins or taxonomic status. This is especially true of the genus *Eucheuma*, in which less than half of the 31 taxonomically accepted species have been molecularly characterized (Guiry and Guiry 2023).

To address these issues, molecular markers have been employed over the last few decades and have proven effective in the identification and phylogenetic inference of eucheumatoids. The first extensive attempt to genetically characterize eucheumatoids was performed by Zuccarello et al. (2006) using the mitochondrial *cox*2–3 spacer and plastid RUBISCO spacer. Despite a focus on the phylogenetic reconstruction of *Kappaphycus* and *Eucheuma*, the study reported that the two molecular markers were not able to distinguish all eucheumatoid morphotypes known in cultivation. The same phenomenon was reported in Malaysia, which indicated that many farmed morphotypes were not supported by genetic data and were merely a result of morphological plasticity (Tan et al. 2013). The suitability of other genetic markers as DNA barcodes, i.e., short, universal DNA sequences used for species identification, was also assessed by Tan et al. (2012), who recommended the use of *cox*1, *cox*2–3 spacer, and *rbc*L for the DNA barcoding of eucheumatoids.

Molecular identification of *K. alvarezii* cultivars from across the globe using the *cox*2–3 spacer has shown that they were all genetically identical (i.e., *cox*2–3 spacer haplotype 3), despite showing a great degree of variation in morphology and color. Genetic results also revealed that the same cultivar has been used over 50 years of farming (de Barros-Barreto et al. 2013; Liu et al. 2013; Tan et al. 2013, 2022; Lim et al. 2014; Ratnawati et al. 2020). The same was also observed for the commercial cultivars of *K. striatus* and *E. denticulatum*. However, these genetic studies also identified other farmed eucheumatoid species, such as *K. malesianus*, as well as haplogroups of *K. alvarezii*, *K. striatus* and *E. denticulatum*, that were probably unofficially introduced by farmers.

Currently, the *cox*2–3 spacer is the standard genetic marker used for the molecular identification of eucheumatoids. This marker is able to resolve the identity of several commonly confused species or varieties, including *K. malesianus* and *K. alvarezii* (Tan et al. 2014), *K. cottonii* and *K. alvarezii* "cottonii" (Zuccarello et al. 2006), and *K. alvarezii* "tambalang" and *E. denticulatum* "endong/cacing" (Montes et al. 2008; Lim et al. 2014; Tan et al. 2014; Dumilag et al. 2016a, 2018). In recent years, molecular data were also shown to be invaluable at identifying wild specimens (Lim et al. 2014; Tano et al. 2015; Dumilag et al. 2016b, 2018; Roleda et al. 2021), which are often morphologically unidentifiable due to their small size and damaged thalli, as well as their tendency to be morphologically different from conspecific cultivars. This is especially important in facilitating cultivar selection and conservation.

3.4.2 Phylogenetic Relationships of Eucheumatoids

The phylogeny of *Kappaphycus* and *Eucheuma* was first elucidated in detail by Zuccarello et al. (2006) using the *cox*2–3 spacer and RUBISCO spacer markers. Their results showed that (1) *Kappaphycus* and *Eucheuma* were both monophyletic genera; (2) the universally used commercial cultivar of *K. alvarezii* consisted of one haplotype; (3) *K. alvarezii* and *K. striatus* were genetically distinct; (4) the local name-based morphotypes were not supported by genetic data; and (5) multiple haplogroups were recorded within *K. alvarezii*, *K. striatus*, and *E. denticulatum*. The inferred phylogeny of *Kappaphycus* and *Eucheuma* has since become the basis of considerable eucheumatoid research.

Today, genetic characterization plays a major role in the phylogenetic reconstruction of eucheumatoids, with additional molecular markers including mitochondrial *cox*1, COI-5P, *cox*2, nuclear ITS, plastid UPA and *rbc*L being introduced and used globally. However, the *cox*2–3 spacer, *cox*1 (~1407 bp), or COI-5P (~630 bp) markers are the most commonly used. The *cox*1 or COI-5P markers, which harbor more phylogenetically informative characters than the *cox*2–3 spacer, are often used in conjunction with each other to provide a better reflection of the phylogeny of eucheumatoids. A list of commonly used molecular markers for the study of eucheumatoids is summarized in Table 3.1.

The *cox*2–3 spacer phylogeny of the eucheumatoid genera *Kappaphycus*, *Eucheuma*, *Betaphycus*, and *Mimica* is shown in Fig. 3.1 (Zuccarello et al. 2006; Dumilag and Lluisma 2014; Lim et al. 2014; Tan et al. 2014; Dumilag et al. 2018). The *cox*2–3 spacer typically does not resolve intergeneric relationships well but is still commonly used in phylogenetic reconstruction due to its ease of amplification and sequencing, as well as the large database of sequenced eucheumatoid species. Representative DNA sequences of all taxonomically accepted species of *Kappaphycus* except *K. procrusteanus* have been characterized over the past two decades, and the genus was inferred to be monophyletic and

Table 3.1 Details of commonly used molecular markers in eucheumatoid studies

Genetic marker (origin)	Primer name	Forward (F)/ Reverse (R)	Primer sequence (5′→3′)	References	Tm (°C)	Common use
*cox*1 (Mitochondrial)	COXI43F	F	TCAACAAATCATAAAGATATTGGWACT	Geraldino et al. (2006)	50	Identification, DNA barcoding, phylogeny, genetic/haplotype diversity, bioinvasion detection
	COXI1549R	R	AGGCATTTCTTCAAANGTATGATA			
	C622F	F	CCTGTNTTAGCAGGWGCTATTACAATG	Yang et al. (2008)		
	C880R	R	ACAGTATACATATGATGNGCTCAAAC			
COI-5P (Mitochondrial)	GazF1	F	TCAACAAATCATAAAGATATTGG	Saunders (2005), Robba et al. (2006)	50	Identification, DNA barcoding, phylogeny, genetic/haplotype diversity, bioinvasion detection
	GazR1	R	ACTTCTGGATGTCCAAAAAAYCA			
*cox*2–3 spacer (Mitochondrial)	*Cox*2_for	F	GTACCWTCTTTDRGRRKDAAATGTGATGC	Zuccarello et al. (1999)	50	Identification, DNA barcoding, phylogeny, genetic/haplotype diversity, bioinvasion detection
	*Cox*3_rev	R	GGATCTACWAGATGRAAWGGATGTC			
*cox*2 (Mitochondrion)	*Kcox*2_F71	F	TTCAAGATCCTGCAACTCC	Tan et al. (2012)	50	Identification, DNA barcoding
	*Kcox*2_R671	R	ATTTCACTGCATTGGCCAT			
ITS (Nuclear)	ITSP1	F	GGAAGGAGAAGTCGTAACAAGG	Tai et al. (2001)	50	Identification, phylogeny
	G4	R	CTTTTCCTCCGCTTATTGATATG			
UPA (Plastid)	p23SrV-f1	F	GGACAGAAAGACCCTATGAA	Sherwood and Presting (2007)	55	Identification, phylogeny, bioinvasion detection
	p23SrV-r1	R	TCAGCCTGTTATCCCTAGAG			
*rbc*L (Plastid)	F-7	F	AACTCTGTAGAACGNACAAG	Gavio and Fredericq (2002)	50	Identification, DNA barcoding, phylogeny
	F-577	F	GTATATGAAGGTCTAAAAGGTGG	Freshwater and Rueness (1994)		
	R-753	R	GCTCTTTCATACATATCTTCC			
	R-rbcS start	R	TGTGTTGCGGCCGCCCTTGTGTTAGTCTCAC			
Rubisco spacer (Plastid)	rbcF1	F	TATACTTCTACAGACACAGCTGA	Zuccarello and West (2002)	55	Identification, phylogeny
	rbcR2-M2	R	ATTTCACACAGGAAACAGCTATGACATGTCAAATAATGGTAGTCCCCA			

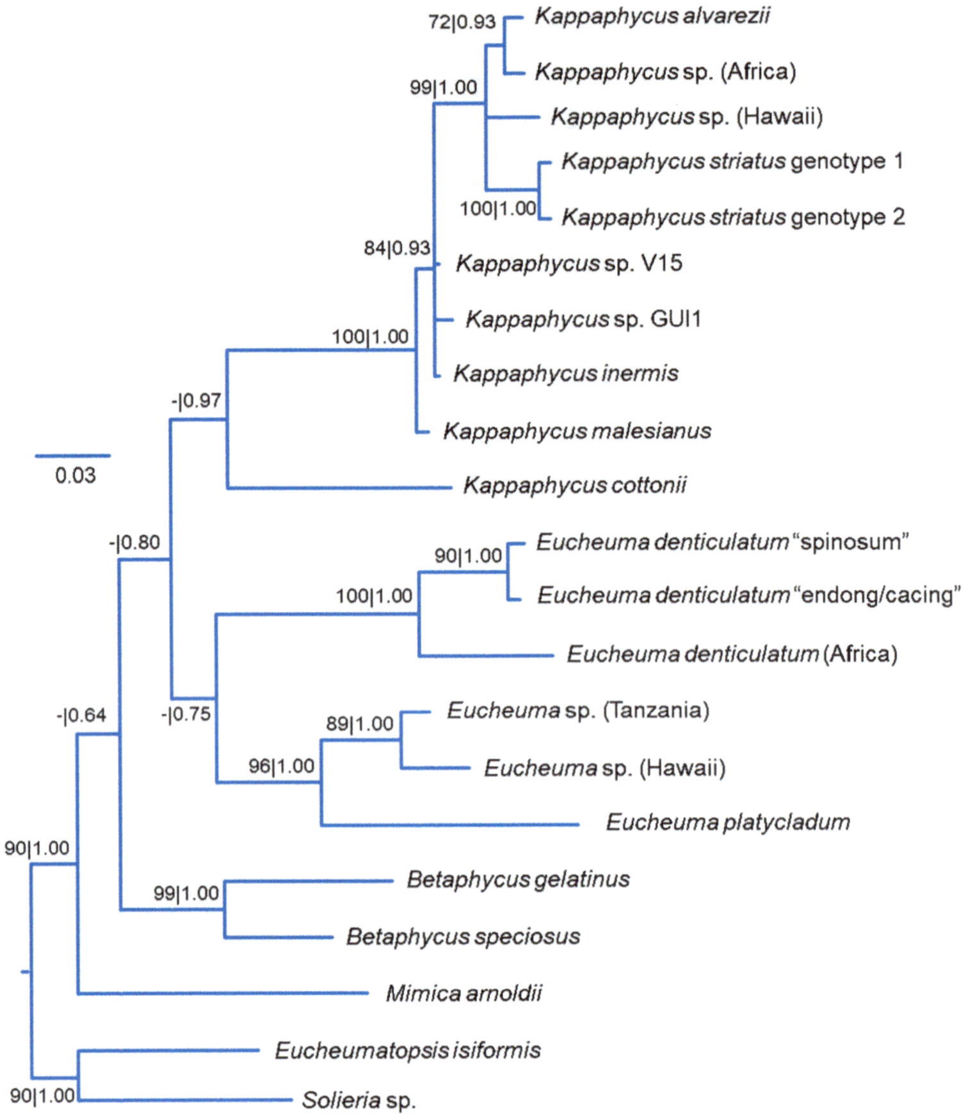

Fig. 3.1 A simplified Bayesian tree showing the phylogeny of *Kappaphycus*, *Eucheuma*, *Betaphycus*, and *Mimica* based on the *cox*2–3 spacer. Number at nodes indicates ML bootstrap support | Bayesian posterior probabilities

sister to *Eucheuma*. Multigene phylogenetic studies have shown that *Kappaphycus* spp. from Africa and Hawaii were genetically different from *K. alvarezii,* and two major subclades exist for *K. striatus*.

The taxonomy of *Eucheuma* was less well studied compared to *Kappaphycus* due to morphological plasticity, poor quality, or missing type specimens, in addition to some species that were not encountered again after their original description. *Eucheuma denticulatum* is the most commonly characterized species, because it is a commercial cultivar. Genetic analyses have shown that there are two different genotypes of *E. denticulatum* in Southeast Asia, namely, the main "spinosum" cultivar and the less common "endong/cacing" variety, which appear to be morphologically distinct (Zuccarello et al. 2006; Lim et al. 2014). The "*E. denticulatum*" from Africa forms a distinct lineage sister to its Southeast Asian counterparts and is most likely to be a native genotype. Other DNA-sequenced *Eucheuma* include *E. platycladum* from Africa, *E. perplexum* from Taiwan, and *E. serra* from Australia (based on *rbc*L and COI-5P data). The taxonomy of *Eucheuma* is undoubtedly in need of attention and revision, based on a large number of unidentified DNA sequences, as well as recent taxonomic revisions of *Eucheumatopsis isiformis* (previously *Eucheuma isiforme*) and *Mimica arnoldii* (previously *Eucheuma arnoldii*) (Núñez-Resendiz et al. 2019; Santiañez and Wynne 2020).

Our understanding of the phylogeny of *Kappaphycus*, *Eucheuma*, *Betaphycus*, and *Mimica* has also improved greatly in recent years as wild eucheumatoids were genetically characterized (Montes et al. 2008; Dumilag et al. 2016a, b; Dumilag 2018; Roleda et al. 2021). Continued efforts in the sampling and multigene characterization of wild specimens would undoubtedly shed light on the taxonomic status of several problematic taxa, such as *Kappaphycus* sp. (Africa), *Kappaphycus* sp. (Hawaii), *K. inermis-K. malesianus* complex, the multiple lineages of *K. striatus*, *E. denticulatum* (Africa), and *E. denticulatum* "endong/cacing."

3.4.3 Genetic Diversity of Eucheumatoids

The limited genetic variation of commercial eucheumatoid cultivars has prompted studies to characterize wild populations of eucheumatoids from around the world for cultivar development and conservation programs. This is typically performed by identifying haplotypes of each eucheumatoid species using either the *cox*2–3 spacer, *cox*1, or COI-5P. A list of known haplotypes is summarized in Table 3.2. Detailed haplotype networks are available in Zuccarello et al. (2006), Dumilag and Lluisma (2014), Lim et al. (2014), Dumilag et al. (2016b, 2018), and Roleda et al. (2021). Lim et al. (2014) reported a high genetic diversity of wild *Kappaphycus* and *E. denticulatum* in Southeast Asia and proposed broader and more concerted sampling efforts. This was followed by several studies on the genetic diversity of *K. alvarezii*, *K. striatus*, and *K. cottonii* in the Philippines (Dumilag et al. 2016a, 2018; Roleda et al. 2021), as well as new projects in Africa. Currently, the Philippines recorded the richest *Kappaphycus* biodiversity, with all six species being reported from there (Dumilag et al. 2016a). It is quite likely that other species and haplotypes will be discovered with more extensive sampling expeditions, especially around the less studied islands of Indonesia.

The DNA information generated for each species of eucheumatoids can also be used to estimate their inter- and intraspecific genetic distances. A summary of the genetic distance of *Kappaphycus*, *Eucheuma*, *Betaphycus*, and *Mimica arnoldii* based on the *cox*2–3 spacer, *cox*1, and COI-5P molecular markers is provided in Table 3.3. In addition to comparing the genetic difference between two samples, this information is also useful in determining suitable threshold values in delineating a species or subspecies. According to Table 3.3, the interspecific genetic distances for taxonomically accepted *Kappaphycus* spp. are 0.91–19.29% for the *cox*2–3 spacer, 4.19–10.52% for *cox*1, and 0.79–12.54% for COI-5P. Theoretically, genotypes recording a genetic difference smaller than the lower limit of the range could be considered conspecific. However, the current dataset does not represent a reliable inter- or intraspecific threshold due to missing genetic data of eucheumatoid species.

Table 3.2 Number of haplotypes of common eucheumatoids identified based on *cox*2–3 spacer, *cox*1, and COI-5P

	Number of haplotypes[a]		
Species	*cox*2–3 spacer	*cox*1	COI-5P
K. alvarezii	11	5	11
K. sp. (Africa)	3	NA	NA
K. sp. (Hawaii)	2	NA	NA
K. striatus lineage 1	2	2	6
K. striatus lineage 2	6	5	3
K. malesianus	2	4	2
K. inermis	1	1	1
K. cottonii	11	1	29
E. denticulatum (cultivar)	4	3	2
E. denticulatum "endong/cacing"	2	4	3
E. denticulatum (Africa)	4	NA	NA

[a] Haplotype information is based on published data from Zuccarello et al. (2006), Dumilag and Lluisma (2014), Lim et al. (2014), Dumilag et al. (2016b, 2018), and Roleda et al. (2021)

3.4.4 Detection of Bioinvasion

Propagules of farmed eucheumatoids could occasionally detach from a culture line and be carried away by waves, eventually settling down in areas far away from seaweed farms. These escaped seaweeds can quickly grow and outcompete other benthic organisms there, thereby disrupting the natural ecosystem (Conklin and Smith 2005). The bioinvasive nature of escaped *K. alvarezii* cultivars has been reported in Hawaii, India, the Philippines, and Tanzania (Conklin and Smith 2005; Chandrasekaran et al. 2008; Dumilag et al. 2016b; Tano et al. 2015). However, to date, there is limited evidence to indicate that escaped seaweeds can become cystocarpic in the wild.

Recent efforts to genetically characterize wild populations of eucheumatoids have also revealed insights into the extent of invasion caused by escaped cultivars. These cultivars can be identified based on their haplotypes, which are often different from wild eucheumatoid populations in an area. In Malaysia, the commercial cultivar of *K. alvarezii* (*cox*2–3 spacer haplotype 3) was collected in Sabangkat, far from its known native distribution in Karindingan. The commercial cultivar of *K. striatus* (*cox*2–3 spacer haplotype 89) was also found among wild populations of *K. malesianus* and *K. striatus* (*cox*2–3 spacer haplogroup 117) in the same area (Lim et al. 2014). Similar findings were also reported from the Philippines, where commercial *K. striatus* were taking over the indigenous species on Hoyanjog Island, Surigao del Norte, Philippines (Dumilag et al. 2016b). Despite this, the impact of these escaped farmed cultivars on natural wild populations of eucheumatoids or other marine organisms is unknown. As such, long-term monitoring and ecological investigations of affected areas are recommended.

3.5 Future Applications of Molecular Marker Technology

3.5.1 Strain Selection and Breeding

The limited genetic variation of eucheumatoid cultivars exposes them to risks of being wiped-out by environmental changes or pests and diseases. It is therefore important to develop new and better cultivars to improve the resilience of

Table 3.3 Genetic distance between different genotypes of *Kappaphycus*, *Eucheuma*, *Betaphycus*, and *Mimica arnoldii* based on the *cox*2–3 spacer, *cox*1, and COI-5P molecular markers

		1	2	3	4	5	6	7	8	9	10	11	12	13	14	15	16	17	18	19
1	*K. alvarezii*	0.30-1.52 0.07-6.11 0.16-0.79																		
2	*K.* sp. (Africa)	0.92-2.45 NA NA	0.61 NA NA																	
3	*K.* sp. (Hawaii)	2.44-3.35 NA NA	2.14-2.76 NA NA	0 NA NA																
4	*K. striatus* lineage 1	3.96-4.88 4.19-4.48 4.44-5.24	3.35-4.29 NA NA	3.66-3.96 NA NA	0.30 1.28 0.16-0.63															
5	*K. striatus* GUI2	3.96-4.88 4.41-6.61 4.76-5.08	3.35-4.29 NA NA	3.66-3.96 NA NA	0.3 0.07-1.21 0.16-0.48	0 0 0														
6	*K. striatus* lineage 2	3.35-4.88 4.34-4.48 4.44-4.92	3.05-4.29 NA NA	3.35-3.96 NA NA	0.91-1.22 0.14-1.42 0.79-1.43	0.91-1.22 1.00-1.35 1.11	0.3-0.61 0.14-0.28 0.32													
7	*K.* sp. V15	2.74-3.66 6.04-6.11 6.83-7.14	3.05-3.68 NA NA	3.35 NA NA	4.57-4.88 6.04-6.33 6.19-6.67	4.57-4.88 6.11 6.19	4.27-4.88 6.33-6.61 6.67-6.98	0 0 0												
8	*K. inermis*	3.35-4.27 NA 6.51-7.14	3.66-4.29 NA NA	3.96 NA NA	5.18-5.49 NA 6.19-6.67	5.18-5.49 NA 6.19	4.88-5.49 NA 6.67-6.98	0.61 NA 1.59	0 NA 0											
9	*K.* sp. GUI1	3.35-4.27 6.04-6.11 6.35-6.98	3.66-4.29 NA NA	3.96 NA NA	5.18-5.49 6.04-6.33 6.03-6.51	5.18-5.49 6.11 6.03	4.88-5.49 6.33-6.61 6.51-6.83	0.61 0.71 1.43	0 NA 1.11	0 0 0										
10	*K. malesianus*	3.66-4.57 5.83-6.04 6.35-6.98	3.96-4.60 NA NA	4.27 NA NA	5.49-5.79 6.11-6.40 6.03-6.51	5.49-5.79 5.97-6.11 6.03	5.18-5.79 6.18-6.61 6.51-6.83	0.91 0.07-0.21 1.11	1.52 NA 0.79-1.11	1.52 0.07-0.21 0.32-0.63	0.61 0.07-1.35 0.32									
11	*K. cottonii*	16.15-18.29 10.38-10.52 11.27-12.38	16.14-17.66 NA NA	16.76-17.98 NA NA	15.84-17.06 11.23 11.43-12.22	15.84-17.06 11.16 11.59-12.06	15.54-17.37 11.09-11.23 11.11-12.06	15.54-17.06 11.30 11.43-12.38	15.54-17.06 NA 11.43-12.38	15.54-17.06 11.44 11.59-12.38	15.23-17.36 11.51-11.80 11.27-12.54	0.3-1.82 0 0.48-2.06								
12	*E. denticulatum* "spinosum"	18.92-19.83 15.21-15.49 16.83-17.14	18.38-18.60 NA NA	17.99 NA NA	19.53-19.83 14.57-14.93 15.24-15.87	19.53-19.83 14.50-14.71 15.56	18.91-19.53 14.50-15.07 15.40-15.71	17.39 14.78-15.00 16.51	17.39 NA 16.83	17.39 14.57-14.78 16.51	17.69-17.99 14.57-14.78 16.35	19.74-20.35 14.57-14.64 16.03-16.51	0 0.14-0.21 0							
13	*E. denticulatum* "endong/cacing "	18.61-20.44 15.14-15.28 16.19-16.98	18.08-19.20 NA NA	17.69-18.60 NA NA	19.22-20.43 14.50-14.71 14.92-15.97	19.22-20.43 14.43-14.50 15.24-15.56	18.61-20.13 14.43-14.85 15.08-15.71	17.08-17.99 14.85-14.93 15.87-16.35	17.08-17.99 NA 16.19-16.67	17.08-17.99 14.64-14.71 15.87-16.35	17.39-18.60 14.64-14.71 15.71-16.19	19.44-20.96 14.29-14.36 15.71-16.67	0.30-1.82 0.36-0.78 0.16-0.63	0.61-1.82 0.07 0.32-0.48						
14	*E. denticulatum* (Africa)	22.51-23.74 NA NA	21.37-22.50 NA NA	22.21-22.52 NA NA	21.91-22.52 NA NA	21.91-22.52 NA NA	21.91-22.83 NA NA	21.59-21.90 NA NA	20.98-21.29 NA NA	20.98-21.29 NA NA	21.29-21.90 NA NA	20.04-20.95 NA NA	8.79-9.09 NA NA	8.48-9.39 NA NA	0.30-1.52 NA NA					
15	*E.* cf. *perplexum*	18.03-18.96 NA 14.41-14.73	18.14-18.95 NA NA	18.63-18.95 NA NA	18.02-18.63 NA 14.74-15.06	18.02-18.63 NA 15.06	18.32-18.94 NA 15.21-15.53	16.20-16.51 NA 13.77	16.20-16.51 NA 14.26	16.20-16.51 NA 14.42	15.59-16.21 NA 14.09-14.10	15.59-16.82 NA 13.92-14.71	14.62-14.93 NA 12.97	14.01-14.93 NA 12.65-12.97	17.08-18.00 NA NA	1.52 NA 0				
16	*E. platycladum*	19.93-21.61 NA NA	19.61-21.39 NA NA	21.16-21.91 NA NA	20.54-21.91 NA NA	20.54-21.91 NA NA	18.32-18.94 NA NA	18.70-19.47 NA NA	18.86-18.70 NA NA	18.86-18.70 NA NA	19.00-19.78 NA NA	18.96-20.66 NA NA	20.13-21.52 NA NA	19.82-22.42 NA NA	21.34-22.73 NA NA	14.98-17.68 NA NA	4.28 NA NA			
17	*E.* cf. *serra*	NA NA 15.85-16.02	NA NA NA	NA NA NA	NA NA 14.90-15.38	NA NA 15.06	NA NA 15.53-15.85	NA NA 15.36	NA NA 16.18	NA NA 15.70	NA NA 14.69-14.70	NA NA 15.68-16.64	NA NA 8.33	NA NA 8.17-8.65	NA NA NA	NA NA 12.18	NA NA NA	NA NA 0		
18	*Betaphycus* spp.	18.28-22.53 NA 12.70-16.35	18.58-21.60 NA NA	18.59-21.01 NA NA	18.28-20.39 NA 11.59-15.08	18.28-20.39 NA 11.90-14.92	18.59-21.30 NA 12.06-15.71	17.06-19.78 NA 6.67-6.98	17.06-19.78 NA 13.81-15.56	17.06-19.78 NA 13.33-14.76	16.75-20.08 NA 13.02-15.08	16.11-19.74 NA 13.49-15.87	19.76-20.34 NA 14.44-16.51	19.45-20.95 NA 14.44-16.98	20.66-21.56 NA NA	15.54-17.03 NA 13.29-15.04	18.96-22.15 NA NA	NA NA 13.94-14.88	10.62 NA 8.73	
19	*Mimica arnoldii*	20.48-22.62 13.23-13.34 13.17-13.81	20.01-21.38 NA NA	21.39-21.70 NA NA	20.48-21.39 12.78-13.17 13.33-13.81	20.48-21.39 12.68 13.65-13.81	21.39-22.01 13.08-13.28 13.33-13.81	19.56-20.17 13.44 13.02-13.17	19.56-20.17 NA 13.49-13.65	19.56-20.17 13.64 13.33-13.49	19.26-20.48 13.44-13.64 13.02-13.49	18.01-19.54 13.85 14.44-15.08	19.28-20.19 14.63-14.73 14.29-14.44	18.97-20.81 14.84-14.93 14.13-14.76	19.24-20.76 NA NA	17.74-18.36 NA 12.31-12.47	21.21-22.26 NA NA	NA NA 12.64-12.96	15.58-19.83 NA 11.90-14.60	1.53-1.83 NA 0.16-0.63

*Blue fonts indicate genetic distance based on the *cox*2–3 spacer (332 bp); green fonts indicate genetic distance based on *cox*1 (1407 bp); red fonts indicate genetic distance based on COI-5P (630 bp)

*Genetic distance was computed based on published data from Zuccarello et al. (2006), Dumilag and Lluisma (2014), Lim et al. (2014), Dumilag et al. (2016b, 2018), Montes et al. (2008), and Roleda et al. (2021) and selected DNA sequences of *E.* cf. *perplexum*, *E.* cf. *serra*, and *M. arnoldii* from GenBank

the carrageenan industry (Hwang et al. 2019). This can be done by developing new cultivars based on the rich genetic diversity of indigenous eucheumatoid populations.

The conventional selection process of potential eucheumatoid cultivars is relatively straightforward and involves the harvesting of wild specimens, followed by genetic characterization and acclimatization, mass propagation, and growth studies performed either in the sea or using indoor aquaculture facilities. The effectiveness of molecular identification for seaweed breeding is well documented in *Porphyra* and *Pyropia* (Hwang et al. 2005, 2018, 2019; Wade et al. 2020). With the majority of *Kappaphycus* species and haplotypes already documented using the *cox*2–3 spacer and *cox*1 markers, wild specimens could be genetically identified first to determine their potential as new cultivars. For instance, all wild *K. alvarezii* specimens genetically characterized to have a *cox*2–3 spacer haplotype other than haplotype 3 (global cultivar) are suitable candidates for selection. A potential cultivar is either propagated in vitro using micropropagation techniques or vegetatively propagated in seedling nurseries before being outplanted (Reddy et al. 2017; Ali et al. 2020). The selected specimens will then be assessed for traits of commercial value, including growth rates, carrageenan type and composition, carrageenan strength and yield, resistance to environmental stress, etc. Established new cultivars should typically be registered with the relevant fisheries or aquaculture departments prior to local commercial introduction.

Alternatively, the ability to breed eucheumatoids would contribute to genetic gain and offer breeders more control in selecting economically important traits. Theoretically, the breeding of a eucheumatoid species would require the culture of quality male and female gametophytes (n) with desirable traits, which are then crossed to produce a tetrasporophyte ($2n$) typically used as a cultivar (Bast 2014). In plants and animals, the selective breeding of a desirable economic trait often involves numerous generations of genetic crossing and backcrossing. Although a recent study has managed to elucidate the life history of *Kappaphycus alvarezii* in vitro for the first time (Hinaloc and Roleda 2021), our overall understanding and capability for the selective breeding of eucheumatoids is still limited by several hurdles, including (1) the lack of a robust and stable system for the long-term cultivation and maintenance of specimens; (2) poor understanding of environmental cues triggering the release of male gametes, carpospores and tetraspores; (3) the lack of an identification system to differentiate tetrasporophytes, immature male and female gametophytes; and (4) the lack of documentation of the hybridization process.

Indeed, the selection and breeding of cultivars are time-consuming processes that could take decades to establish, as shown by developments in other commercial seaweeds, such as *Saccharina*, *Pyropia*, and *Undaria* (Hwang et al. 2012, 2019; Zhang et al. 2016). To facilitate this process, phycologists are also looking into the production of hybrids, mutagenesis, and marker-assisted selection and breeding (Niwa et al. 2009; Hwang et al. 2019; Kim 2021). Considering the many challenges to eucheumatoid breeding, the current option for cultivar development would most likely have to be based on the collection and assessment of wild specimens.

3.5.2 Genetic and Genomic Characterization

The selection and breeding of eucheumatoids would benefit from a robust genetic identification system, especially in the identification of wild specimens, tetrasporophytes and gametophytes. The use of DNA markers in the selection or breeding process of a commercial crop is often referred to as marker-assisted selection or breeding. These markers are typically associated with a desirable trait in the crop, e.g., fast growth rate, carrageenan type and yield, resistance to pests and diseases, etc.

Although the use of genetic markers such as *cox*2–3 spacer and *cox*1 markers has greatly improved our understanding of the genetic diversity of eucheumatoids throughout the globe, these markers only offer limited information about qualitative commercial traits, such as carrageenan type and morphology. For example, these markers are not able to differentiate the isomorphic gametophyte and sporophyte stages of eukaryotes, which are usually identified when reproductive structures are present or by means of ploidy determination (Zitta et al. 2012). A rapid method to distinguish the different life history stages of eucheumatoids is valuable, as they display different growth rates, carrageenan strength, and yield (Doty and Santos 1978; Hayashi et al. 2007; Zitta et al. 2012). This was shown to be possible via the identification of preferential genes (e.g., expression of EF-1a only in the sporophyte stage of *Porphyra purpurea*) or candidate genes (e.g., higher GiODC expression in fertile tissues of *Grateloupia imbricata*) involved in reproduction (Liu et al. 1996; García-Jiménez and Robaina 2015).

To further improve the selection and breeding process of eucheumatoids, it is important to identify and elucidate genes that affect quantitative traits such as growth rate, thallus size, and carrageenan yield. Regions of DNA that govern quantitative traits are termed quantitative trait loci (QTL). QTL can be identified by determining whether an association exists between a molecular marker genotype and a desirable trait (Miles and Wayne 2008).

To identify QTLs for eucheumatoids, multiple strains of a species that differ genetically in the trait of interest are subjected to genotyping using molecular markers. In addition to genetic markers based on single nucleotide polymorphisms (SNPs), other markers such as sequence repeats (or microsatellites) or restriction fragment length polymorphisms (RFLPs) could also be used for QTL analysis (Complex Trait Consortium 2003; Miles and Wayne 2008). Genetic crossings are generally performed multiple times, and the phenotypic and genotypic characteristics of the offspring are characterized and scored. Markers that are genetically linked to a QTL of the trait of interest will show a significant association with the phenotype. It has been reported that a trait of interest is often expressed based on a few loci with fairly large effects, with the remainder due to loci of small effect (Mackay 2001). QTL studies require large sample sizes to ensure accuracy, which is particularly important in the study of eucheumatoids, as the link between QTLs and traits of interest are affected by sex, environment, and epistasis (Leips and Mackay 2000; Leamy et al. 2011).

Despite the promise of modern breeding methods, the fact remains that such a development is premature in the study of eucheumatoids when compared to other commercial commodities of the world (García-Jiménez and Robaina 2015). Currently, there are ongoing efforts being put into the development of new genetic markers that contribute to studies on the taxonomy, genetic diversity, selection, and breeding as well as QTL analysis of eucheumatoids. This can be achieved by high-throughput sequencing of either the mitochondrial, plastid, or nuclear genome of eucheumatoids. Annotated genomes can then be used as a source of information to identify molecular markers for primer design and QTL assessment.

The technological complexity and capital requirements associated with marker-assisted breeding and QTL analysis are likely to be a challenge for developing countries, where most eucheumatoids are farmed. This is especially true considering the long-term nature of cultivar development projects, which are only possible through international cooperation and collaboration (Charrier et al. 2017; Wade et al. 2020). However, an increased global interest in seaweeds as a sustainable food source of the future is seeing an influx of foreign support in funding, expertise, facilities, and human capital development, which would undoubtedly lead to improvements in the seaweed and carrageenan industry. For example, the multidisciplinary GlobalSeaweedSTAR project (https://www.globalseaweed.org/) has brought together eucheumatoid experts from the United Kingdom, Philippines, Malaysia, and Tanzania, which facilitated technology transfer, capacity development, and research toward sustainable seaweed aquaculture. Among the major outputs of this initiative included (1) the identification of the social, political, and institutional landscape of the local seaweed industry in developing countries, as well as policy development to promote sustainable economic growth of this industry; (2) in-depth research into the pests and diseases of eucheumatoids, with emphasis on developing new detection methods and effective biosecurity practices; and (3) the genetic characterization of eucheumatoids across the globe toward the development of new commercial strains with resistance toward climate change as well as pests and diseases. These developments have brought about an unprecedented improvement in the scope and depth of global eucheumatoid research and development.

3.5.3 Identifying Germplasm for Conservation

In recent years, coordinated efforts to document the genetic diversity of eucheumatoids have extended from Southeast Asia to other regions, including Hawaii and Africa, which also recorded rich genetic diversity and potentially new species of eucheumatoids. Notably, these studies indicated the presence of indigenous populations of eucheumatoids in Indonesia, Malaysia, the Philippines, Hawaii, and Africa, despite the use of introduced cultivars from the Philippines (Zuccarello et al. 2006; Lim et al. 2014; Dumilag et al. 2016a, 2018; Ratnawati et al. 2020; Brakel et al. 2021; Roleda et al. 2021). The documentation of the rich genetic diversity of wild *K. alvarezii*, *K. cottonii*, *K. striatus*, *K. inermis*, and *K. malesianus* in Southeast Asia called for more conservation efforts, including the establishment of marine protection areas and long-term environmental monitoring, to protect these indigenous populations from environmental changes, anthropogenic development, and bioinvasion (UP-MSI et al. 2002; Dumilag et al. 2016b). Likewise, the incidence of escaped nonnative cultivars established in the wild should also be avoided by the development and use of indigenous eucheumatoid cultivars.

It is also important for the germplasm of wild eucheumatoids to be maintained in germplasm banks (or biobanks), nurseries, or dedicated research farms, which are far less likely to be affected by environmental change and stress (Wade et al. 2020). Here, genotyping, selection, breeding, and storage of specimens can be performed to improve the quality and resilience of eucheumatoid cultivars. Alternatively, it may also be possible to adopt the "conservation through cultivation" approach used in plants, where eucheumatoids are introduced into the marine aquarium hobby as a means to preserve germplasm and promote propagation and breeding. This is especially true considering the rising popularity of using macroalgal refugia to maintain nitrate and phosphate levels in marine tanks.

3.6 Conclusion

The carrageenan industry is poised for massive demand in the future, as the world looks to seaweeds as a sustainable source of food and phycocolloids. However, despite developments in taxonomy, genetic studies and micropropagation techniques, the farming of eucheumatoids still relies heavily on the same few aging cultivars introduced since the 1970s. This calls for the development of new cultivars of better quality and resilience in the face of challenges that include ocean acidification, pests, and disease as well as global warming. This development can be achieved based on a good understanding of the triphasic life history of eucheumatoids, reliable culture techniques, marker-assisted identification, selection, and breeding. Despite the massive undertakings of eucheumatoid cultivar development, the collaborative ties that will be established between governments, global carrageenan industry players, phycologists, and farmers will undoubtedly bring eucheumatoid research and production to greater heights.

References

Ali MKM, Critchley AT, Hurtado AQ (2020) Micropropagation and sea-based nursery growth of selected commercial *Kappaphycus* species in Penang, Malaysia. J Appl Phycol 32:1301–1309

Alibon RD, Gonzales JMP, Ordoyo AET, Echem RT (2019) Incidence of ice-ice disease associated with *Kappaphycus alvarezii* in the seaweed farms in Zamboanga peninsula, Mindanao, Philippines. SSR Inst Int J Life Sci 5:2148–2155. https://doi.org/10.21276/ssr-iijls.2019.5.1.6

Bast F (2014) An illustrated review on cultivation and life history of agronomically important seaplants. In: Vitor Hugo Pomin (ed) Seaweed: mineral composition, nutritional and antioxidant benefits and agricultural uses Nova, New York, pp 39–70. ISBN: 978-1-63117-571-8

Borlongan IAG, Gerung GS, Nishihara G, Terada R (2017) Light and temperature effects on photosynthetic activity of *Eucheuma denticulatum* and *Kappaphycus alvarezii* (brown and green color morphotypes) from Sulawesi Utara, Indonesia. Phycol Res 65:69–79

Brakel J, Sibonga RC, Dumilag RV et al (2021) Exploring, harnessing and conserving marine genetic resources towards a sustainable seaweed aquaculture. Plants People Planet 3:337–349. https://doi.org/10.1002/ppp3.10190

Cai J, Lovatelli A, Aguilar-Manjarrez J, et al (2021) Seaweeds and microalgae: an overview for unlocking their potential in global aquaculture development. FAO Fisheries and Aquaculture Circular. No. 1229, Rome, 36 p

Chandrasekaran S, Arun Nagendran N, Pandiaraja D et al (2008) Bioinvasion of *Kappaphycus alvarezii* on corals in the Gulf of Mannar, India. Curr Sci 94:1167–1172

Charrier B, Abreu MH, Araujo R et al (2017) Furthering knowledge of seaweed growth and development to facilitate sustainable aquaculture. New Phytol 216:967–975. https://doi.org/10.1111/nph.14728

Complex Trait Consortium (2003) The nature and identification of quantitative trait loci. Nat Rev Genet 4:911–916

Conklin EJ, Smith JE (2005) Abundance and spread of the invasive red algae, *Kappaphycus* spp., in Kane'ohe Bay, Hawai'i and an experimental assessment of management options. Biol Invasions 7:1029–1039. https://doi.org/10.1007/s10530-004-3125-x

Datu ED, Bahron A, Alin J et al (2017) On going assessment of issues in the seaweed farming industry in Sabah, Malaysia. J Asian Acad Appl Bus 4:49–60

de Barros-Barreto MBB, Marinho LC, Reis RP et al (2013) *Kappaphycus alvarezii* (Gigartinales, Rhodophyta) cultivated in Brazil: is it only one species? J Appl Phycol 25:1143–1149. https://doi.org/10.1007/s10811-012-9952-8

Doty MS, Santos GA (1978) Carrageenans from tetrasporic and cystocarpic *Eucheuma* species. Aquat Bot 4:143–149

Dumilag RV (2018) *Betaphycus gelatinus* and *B. philippinensis* (Gigartinales, Rhodophyta) are conspecific. Phytotaxa 372:22–34

Dumilag RV, Lluisma AO (2014) Resolving the phylogenetic affinities of *Kappaphycus inermis* within the genus *Kappaphycus* (Gigartinales, Solieriaceae) using mitochondrial and plastid markers. Phytotaxa 162:223–231. https://doi.org/10.11646/phytotaxa.162.4.5

Dumilag RV, Liao LM, Lluisma AO (2014) Phylogeny of *Betaphycus* (Gigartinales, Rhodophyta) as inferred from COI sequences and morphological observations on *B. philippinensis*. J Appl Phycol 26:587–595. https://doi.org/10.1007/s10811-013-0063-y

Dumilag RV, Orosco FL, Lluisma AO (2016a) Genetic diversity of *Kappaphycus* species (Gigartinales, Rhodophyta) in The Philippines. Syst Biodivers 14:441–451. https://doi.org/10.1080/14772000.2016.1157643

Dumilag RV, Salvador RC, Halling C (2016b) Genotype introduction affects population composition of native Philippine *Kappaphycus* (Gigartinales, Rhodophyta). Conserv Genet Resour 8:439–441. https://doi.org/10.1007/s12686-016-0591-2

Dumilag RV, Gallardo WGM, Garcia CPC et al (2018) Phenotypic and mtDNA variation in Philippine *Kappaphycus cottonii* (Gigartinales, Rhodophyta). Mitochondrial DNA Part A DNA Mapping, Seq Anal 29:951–963. https://doi.org/10.1080/24701394.2017.1398745

Franks SJ (2010) Genetics, evolution, and conservation of Island plants. J Plant Biol 53:1–9. https://doi.org/10.1007/s12374-009-9086-y

Freshwater DW, Rueness J (1994) Phylogenetic relationships of some European *Gelidium* (Gelidiales Rhodophyta) species based on rbcL nucleotide sequence analysis. Phycologia 33(3):187–194. https://doi.org/10.2216/i0031-8884-33-3-187.1

Gao G, Gao L, Jian M, Jian A, He L (2022) The potential of seaweed cultivation to achieve carbon neutrality and mitigate deoxygenation and eutrophication. Environ Res Lett 17(1):014018. https://doi.org/10.1088/1748-9326/ac3fd9

García-Jiménez P, Robaina RR (2015) On reproduction in red algae: further research needed at the molecular level. Front Plant Sci 6:1–6. https://doi.org/10.3389/fpls.2015.00093

Geraldino PJL, Yang EC, Boo SM (2006) Morphology and molecular phylogeny of *Hypnea flexicaulis* (Gigartinales Rhodophyta) from Korea. Algae 21(4):417–423. https://doi.org/10.4490/ALGAE.2006.21.4.417

Guiry MD, Guiry GM (2023) AlgaeBase, World-wide electronic publication, National University of Ireland, Galway, AlgaeBase

Harley CDG, Anderson KM, Demes KW et al (2012) Effects of climate change on global seaweed communities. J Phycol 48:1064–1078. https://doi.org/10.1111/j.1529-8817.2012.01224.x

Hayashi L, De Paula EJ, Chow F (2007) Growth rate and carrageenan analyses in four strains of *Kappaphycus alvarezii* (Rhodophyta, Gigartinales) farmed in the subtropical waters of São Paulo State, Brazil. J Appl Phycol 19:393–399. https://doi.org/10.1007/s10811-006-9135-6

Hayashi L, Faria GSM, Nunes BG et al (2011) Effects of salinity on the growth rate, carrageenan yield, and cellular structure of *Kappaphycus alvarezii* (Rhodophyta, Gigartinales) cultured *in vitro*. J Appl Phycol 23:439–447. https://doi.org/10.1007/s10811-010-9595-6

Hinaloc LAR, Roleda MY (2021) Phenotypic diversity, growth and sexual differentiation in the progeny of wild *Kappaphycus alvarezii* (Gigartinales, Florideophyceae). Phycologia 00:1–11. https://doi.org/10.1080/00318884.2021.1946307

Hojsgaard D, Hörandl E (2015) A little bit of sex matters for genome evolution in asexual plants. Front Plant Sci 6. https://doi.org/10.3389/fpls.2015.00082

Hughes AC (2017) Understanding the drivers of southeast Asian biodiversity loss. Ecosphere 8. https://doi.org/10.1002/ecs2.1624

Hurtado AQ, Reis RP, Loureiro RR, Critchley AT (2014) *Kappaphycus* (Rhodophyta) cultivation: problems and the impacts of Acadian marine plant extract powder. In: Pereira L, Neto JM (eds) Marine algae. Taylor & Francis Group LLC, pp 251–299

Hurtado AQ, Neish I, Critchley T (2015) Developments in production technology of *Kappaphycus* in The Philippines: more than four decades of farming. J Appl Phycol 27:1945–1961. https://doi.org/10.1007/s10811-014-0510-4

Hurtado AQ, Lim PE, Tan J et al (2016) Biodiversity and biogeography of commercial tropical carrageenophytes in the southeast Asian region. In: Pereira L (ed) Carrageenans: sources and extraction methods, molecular structure, bioactive properties and health effects. Nova Science Publishers, Inc, New York, pp 51–74

Hwang MS, Kim S-M, Ha D-S et al (2005) DNA sequences and identification of *Porphyra* cultivated by natural seeding on the southwest coast of Korea. Algae 20:183–196

Hwang EK, Gong YG, Park CS (2012) Cultivation of a hybrid of free-living gametophytes between *Undariopsis peterseniana* and *Undaria pinnatifida*: morphological aspects and cultivation period. J Appl Phycol 24:401–408

Hwang IK, Kim SO, Hwang MS et al (2018) Intraspecific variation of gene structure in the mitochondrial large subunit ribosomal RNA and cytochrome c oxidase subunit 1 of *Pyropia yezoensis* (Bangiales, Rhodophyta). Algae 33:49–54. https://doi.org/10.4490/algae.2018.33.2.20

Hwang EK, Yotsukura N, Pang SJ et al (2019) Seaweed breeding programs and progress in eastern Asian countries. Phycologia 58: 484–495. https://doi.org/10.1080/00318884.2019.1639436

Kambey CSB, Campbell I, Cottier-Cook EJ et al (2021) Seaweed aquaculture: a preliminary assessment of biosecurity measures for controlling the ice-ice syndrome and pest outbreaks of a *Kappaphycus* farm. J Appl Phycol 33:3179–3197. https://doi.org/10.1007/s10811-021-02530-z

Kasim M, Mustafa A, Male I et al (2017) New methods on cultivation of *Eucheuma denticulatum* and *Kappaphycus alvarezii* in Indonesia. J Fish Aquat Sci 12:207–217. https://doi.org/10.3923/jfas.2017.207.217

Kim N (2021) Culture study on the hybrid by interspecific crossing between *Porphyra pseudolinearis* and *P. dentata* (Bangiales, Rhodophyta), two dioecious species in culture. Algae 26. https://doi.org/10.4490/algae.2011.26.1.079

Kumar YN, Poong SW, Gachon C et al (2020) Impact of elevated temperature on the physiological and biochemical responses of *Kappaphycus alvarezii* (Rhodophyta). PLoS One 15:1–16. https://doi.org/10.1371/journal.pone.0239097

Largo DB, Fukami K, Nishijima T (1995) Occasional pathogenic bacteria promoting *ice-ice* disease in the carrageenan-producing red algae *Kappaphycus alvarezii* and *Eucheuma denticulatum* (Solieriaceae, Gigartinales, Rhodophyta). J Appl Phycol 7:545–554

Leamy LJ, Gordon RR, Pomp D (2011) Sex-, diet-, and cancer-dependent epistatic effects on complex traits in mice. Front Genet 2:1–11. https://doi.org/10.3389/fgene.2011.00071

Leips J, Mackay TFC (2000) Quantitative trait loci for life span in *Drosophila melanogaster*: interactions with genetic background and larval density. Genetics 155:1773–1788. https://doi.org/10.1093/genetics/155.4.1773

Lideman L, Nishihara G, Noro T, Terada R (2013) Effect of temperature and light on the photosynthesis as measured by chlorophyll fluorescence of cultured *Eucheuma denticulatum* and *Kappaphycus* sp. (Sumba strain) from Indonesia. J Appl Phycol 25:399–406. https://doi.org/10.1007/s10811-012-9874-5

Lim PE, Tan J, Phang SM et al (2014) Genetic diversity of *Kappaphycus* Doty and *Eucheuma* J. Agardh (Solieriaceae, Rhodophyta) in Southeast Asia. J Appl Phycol 26(2):1253–1272. https://doi.org/10.1007/s10811-013-0197-y

Liu QY, Baldauf SL, Reith ME (1996) Elongation factor 1 alpha genes of the red alga *Porphyra purpurea* include a novel, developmentally specialized variant. Plant Mol Biol 31:77–85

Liu C, Huang X, Liu J (2013) Assessment of introduced *Kappaphycus* (Solieriaceae, Rhodophyta) species relationships in China with molecular markers. Acta Oceanol Sin 32:59–65. https://doi.org/10.1007/s13131-013-0277-5

Lovell JT, Williamson RJ, Wright SI et al (2017) Mutation accumulation in an asexual relative of *Arabidopsis*. PLoS Genet 13:1–14. https://doi.org/10.1371/journal.pgen.1006550

Mackay TFC (2001) Quantitative trait loci in *Drosophila*. Nat Rev Genet 2:11–20. https://doi.org/10.1038/35047544

McHugh DJ (2003) A guide to the seaweed industry. FAO Fisheries Technical Paper 441, Rome, 105 p

Miles CM, Wayne M (2008) Quantitative trait locus (QTL) analysis. Nat Educ 1:208

Montes HR, Pobre KFR, Lluisma AO (2008) Phylogenetic affiliation of the "Endong"/"Spaghetti" variety of *Eucheuma* as revealed by molecular data. Philipp Agric Sci 91:86–93

Munawan, Kasim M, Ruslaini (2021) Growth rate of *Eucheuma denticulatum* cultivated in horizontal net and vertical net. IOP Conf Ser Earth. Environ Sci 925(1):012018. https://doi.org/10.1088/1755-1315/925/1/012018

Nelson WG (2018) Development of an epiphyte indicator of nutrient enrichment: a critical evaluation of observational and experimental studies. Ecol Indic 79:207–227. https://doi.org/10.1016/j.ecolind.2017.04.034

Niwa K, Hayashi Y, Abe T, Aruga Y (2009) Induction and isolation of pigmentation mutants of *Porphyra yezoensis* (Bangiales, Rhodophyta) by heavy-ion beam irradiation. Phycol Res 57:194–202

Núñez-Resendiz ML, Dreckmann KM, Sentíes A et al (2019) *Eucheumatopsis isiformis* gen. & comb. nov. (Solieriaceae, Rhodophyta) from the Yucatan peninsula to accommodate *Eucheuma isiforme*. Phycologia 58:51–62. https://doi.org/10.1080/00318884.2018.1517536

Pace NR (2009) Mapping the tree of life: progress and prospects. Microbiol Mol Biol Rev 73:565–576. https://doi.org/10.1128/mmbr.00033-09

Ratnawati P, Simatupang NF, Pong-Masak PR et al (2020) Genetic diversity analysis of cultivated *Kappaphycus* in Indonesian seaweed farms using COI gene. Squalen Bull Mar Fish Postharvest Biotechnol 15:65–72. https://doi.org/10.15578/squalen.v15i2.466

Reddy CRK, Yokoya NS, Wilson YTL, Luhan MRJ, Hurtado AQ (2017) Micropropagation of *Kappaphycus* and *Eucheuma*: trends and prospects. In: Hurtado AQ, Critchley AT, Neish IC (eds) Tropical seaweed farming trends problems and opportunities, pp 91–110

Robba L, Russell SJ, Barker GL, Brodie J (2006) Assessing the use of the mitochondrial *cox*1 marker for use in DNA barcoding of red algae (Rhodophyta). Am J Bot 93:1101–1108

Roleda MY, Aguinaldo Z-ZA, Crisostomo BA et al (2021) Discovery of novel haplotypes from wild populations of *Kappaphycus* (Gigartinales, Rhodophyta) in The Philippines. Algae 36:1–12. https://doi.org/10.4490/algae.2021.36.2.18

Santiañez WJE, Wynne MJ (2020) Establishment of *Mimica* gen. nov. to accommodate the anaxiferous species of the economically important red seaweed *Eucheuma* (Solieriaceae, Rhodophyta). Phytotaxa 439:167–170. https://doi.org/10.11646/PHYTOTAXA.439.2.8

Saunders GW (2005) Applying DNA barcoding to red macroalgae: a preliminary appraisal holds promise for future applications. Philos Trans R Soc Lond Ser B Biol Sci 360(1462):1879–1888. https://doi.org/10.1098/rstb.2005.1719

Sheldrake AR (2022) Cellular senescence, rejuvenation and potential immortality. Proc R Soc B 289:20212434

Sherwood AR, Presting GG (2007) Universal primers amplify a 23S rDNA plastid marker in eukaryotic algae and cyanobaceria. J Phycol 43(3):605–608. 10.1111/jpy.2007.43.issue-3. https://doi.org/10.1111/j.1529-8817.2007.00341.x

Spencer RD, Cross RG (2007) The International Code of Botanical Nomenclature (ICBN) the International Code of Nomenclature for Cultivated Plants (ICNCP) and the cultigen. Taxon 56(3):938–940. https://doi.org/10.2307/25065875

Tai V, Lindstrom SC, Saunders GW (2001) Phylogeny of the Dumontiaceae (Gigartinales Rhodophyta) and associated families based on SSU rDNA and internal transcribed spacer sequence data. J Phycol 37(1):184–196. https://doi.org/10.1046/j.1529-8817.2001.037001184.x

Tan J, Lim P, Phang S et al (2012) Assessment of four molecular markers as potential DNA barcodes for red algae *Kappaphycus* Doty and *Eucheuma*. J Agardh PLoS One 7(12):e52905. https://doi.org/10.1371/journal.pone.0052905

Tan J, Lim P-E, Phang S-M (2013) Phylogenetic relationship of *Kappaphycus* Doty and *Eucheuma* J. Agardh (Solieriaceae, Rhodophyta) in Malaysia. J Appl Phycol 25(1):13–29. https://doi.org/10.1007/s10811-012-9833-1

Tan J, Lim PE, Phang SM et al (2014) *Kappaphycus malesianus* sp. nov.: a new species of *Kappaphycus* (Gigartinales, Rhodophyta) from Southeast Asia. J Appl Phycol 26:1273–1285. https://doi.org/10.1007/s10811-013-0155-8

Tan PL, Poong SW, Tan J et al (2022) Assessment of genetic diversity within eucheumatoid cultivars in East Sabah, Malaysia. J Appl Phycol 34:709–717. https://doi.org/10.1007/s10811-021-02608-8

Tano SA, Halling C, Lind E et al (2015) Extensive spread of farmed seaweeds causes a shift from native to nonnative haplotypes in natural seaweed beds. Mar Biol 162:1983–1992. https://doi.org/10.1007/s00227-015-2724-7

Tolentino-Pablico G, Bailly N, Froese R, Elloran C (2008) Seaweeds preferred by herbivorous fishes. J Appl Phycol 20:933–938. https://doi.org/10.1007/s10811-007-9290-4

Tomas F, Cebrian E, Ballesteros E (2011) Differential herbivory of invasive algae by native fish in the Mediterranean Sea. Estuar Coast Shelf Sci 92:34

Turland NJ, Wiersema JH, Barrie FR, Greuter W, Hawksworth DL, Herendeen PS, Knapp S, Kusber WH, Li DZ, Marhold K, May TW, McNeill J, Monro AM, Prado J, Price MJ, Smith GF (eds) (2018) International Code of Nomenclature for algae, fungi, and plants (Shenzhen Code) adopted by the Nineteenth International Botanical Congress Shenzhen, China, July 2017. Regnum Vegetabile 159. Glashütten: Koeltz Botanical Books. https://doi.org/10.12705/Code.2018

Vairappan CS (2016) Seasonal occurrences of epiphytic algae on the commercially cultivated red alga *Kappaphycus alvarezii* (Solieriaceae, Gigartinales, Rhodophyta). J Appl Phycol 18:611–617. https://doi.org/10.1007/s10811-006-9062-6

Vairappan CS, Chung CS, Matsunaga S (2013) Effect of epiphyte infection on physical and chemical properties of carrageenan produced by *Kappaphycus alvarezii* Doty (Solericeae, Gigartinales, Rhodophyta). J Appl Phycol 26:923–931. https://doi.org/10.1007/s10811-013-0126-0

Wade R, Augyte S, Harden M et al (2020) Macroalgal germplasm banking for conservation, food security, and industry. PLoS Biol 18:e3000641. https://doi.org/10.1371/journal.pbio.3000641

Ward GM, Faisan JP, Cottier-Cook EJ et al (2020) A review of reported seaweed diseases and pests in aquaculture in Asia. J World Aquac Soc 51:815–828. https://doi.org/10.1111/jwas.12649

Ward GM, Kambey CSB, Faisan JP et al (2022) Ice-ice disease: an environmentally and microbiologically driven syndrome in tropical seaweed aquaculture. Rev Aquac 14:414–439. https://doi.org/10.1111/raq.12606

Yang EC, Kim MS, Geraldino PJL, Sahoo D, Shin J-A, Boo SM (2008) Mitochondrial *cox*1 and plastid *rbc*L genes of *Gracilaria vermiculophylla* (Gracilariaceae Rhodophyta). J Appl Phycol 20(2):161–168. https://doi.org/10.1007/s10811-007-9201-8

Younes M, Aggett P, Aguilar F et al (2018) Re-evaluation of carrageenan (E407) and processed *Eucheuma* seaweed (E407a) as food additives. EFSA J 16:5238. https://doi.org/10.2903/j.efsa.2018.5238

Zhang J, Liu T, Bian D et al (2016) Breeding and genetic stability evaluation of the new *Saccharina* variety "Ailunwan" with high yield. J Appl Phycol 28:3413–3421

Zitta CS, Oliveira EM, Bouzon ZL, Hayashi L (2012) Ploidy determination of three *Kappaphycus alvarezii* strains (Rhodophyta, Gigartinales) by confocal fluorescence microscopy. J Appl Phycol 24:495–499. https://doi.org/10.1007/s10811-011-9704-1

Zuccarello GC, West JA (2002) Phylogeography of the *Bostrychia calliptera – B. pinnata* complex (Rhodomelaceae Rhodophyta) and divergence rates based on nuclear mitochondrial and plastid DNA markers. Phycologia 41(1):49–60. https://doi.org/10.2216/i0031-8884-41-1-49.1

Zuccarello GC, Burger G, West JA, King RJ (1999) A mitochondrial marker for red algal intraspecific relationships: variable within populations and maternally inherited. Mol Ecol 8:1443–1448

Zuccarello GC, Critchley AT, Smith J et al (2006) Systematics and genetic variation in commercial *Kappaphycus* and *Eucheuma* (Solieriaceae, Rhodophyta). J Appl Phycol 18:643–651. https://doi.org/10.1007/s10811-006-9066-2

UP-MSI, ABC, ARCBC, et al (2002) Marine protected areas in Southeast Asia. ASEAN Regional Center for Bidoiversity Conservation, Department of Environment and Natural Resources, Los Banos, Philippines

Reproductive Biology and Novel Cultivar Development of the Eucheumatoid *Kappaphycus alvarezii*

4

Michael Y. Roleda, Lourie Ann R. Hinaloc, Ida T. Capacio, Ma. Cecilia B. Jao, and Bea A. Crisostomo

Abstract

Kappaphycus species are globally cultivated in tropical waters as a major source of *k*-carrageenan. Since the early 1970s, *Kappaphycus* farming, through clonal propagation, has been confined to a few good-quality commercial strains of unknown ploidy and/or life history phase. After five decades of successful cultivation, the productivity of *Kappaphycus* spp. has continually declined. This has been attributed to low genetic variability, making the >50-year-old cultivars more susceptible to environmental stressors, pests, and diseases. Hence, the establishment of new cultivars with unique genetic makeup and corresponding desirable traits can provide alternative seedstocks from the currently cultivated strains. Cultivar development includes the collection of wild individuals, both vegetative and reproductive materials, for culture collection and biobanking, spore release and cultivation, selection, and breeding. Moreover, the distinction of specific life history phases, sex, and ploidy among the established cultivars can be tested for specific traits, such as vigor and fitness, productivity, and biochemistry. This initiative can avoid repeated crop failure associated with old and fatigued strains and conduct targeted farming of specific cultivars with corresponding known traits. Understanding their ecological tolerance to different abiotic factors can also mitigate risks in seaweed farming in relation to climate change. Continuous cultivar development is necessary for future crop domestication and improvement programs to sustain the livelihoods of thousands of small-scale farmers and the multi-million-dollar seaweed industry.

Keywords

Carposporophyte · Clonal propagation · Cystocarp · Gametophyte · Life history · Phenotype · Ploidy · Sexual reproduction · Tetrasporophyte

4.1 Introduction

Commercially cultivated seaweed taxa as industrial sources of carrageenan are collectively called eucheumatoids. This group, in the family Solieriaceae, includes seven genera: *Eucheuma* J. Agardh (Agardh 1847), *Kappaphycus* Doty (Doty 1988), *Betaphycus* Doty (Silva et al. 1996), *Tacanoosca* J.N. Norris, P.W. Gabrielson & D.P. Cheney (Norris 2014), *Eucheumatopsis* Núñez-Resendiz, Dreckmann & Sentíes, 2019 (Núñez-Resendiz et al. 2019), *Mimica* Santiañez & M.J. Wynne (Santiañez and Wynne 2020), and *Kappaphycopsis* Dumilag & Zuccarello (Dumilag and Zuccarello 2022). The iota-carrageenan-producing *Eucheuma denticulatum* and the kappa-carrageenan-producing *Kappaphycus alvarezii*, *K. striatus*, and *K. malesianus* are the primary cultivated species sustaining the global carrageenan industry (Dumilag et al. 2023).

Author Contributions
MY Roleda: Funding acquisition, Project administration, Supervision, Conceptualization, Methodology, Investigation, Validation, Resources, Writing-Original draft.
LAR Hinaloc: Methodology, Investigation, Data Curation, Validation, Visualization, Writing-Review and Editing.
IT Capacio: Methodology, Investigation, Data Curation, Validation, Visualization, Writing-Review and Editing.
MCB Jao: Methodology, Investigation, Data Curation, Validation, Visualization.
BA Crisostomo: Methodology, Investigation, Data Curation, Validation, Visualization, Writing-Review and Editing.

M. Y. Roleda (✉) · L. A. R. Hinaloc · B. A. Crisostomo
Algal Ecophysiology (AlgaE) Laboratory, The Marine Science Institute, University of the Philippines, Quezon City, Philippines
e-mail: myroleda@up.edu.ph; lrhinaloc@msi.upd.edu.ph; bcrisostomo@msi.upd.edu.ph

I. T. Capacio ·
M. C. B. Jao
Department of Agriculture- Bureau of Fisheries and Aquatic Resources, National Seaweed Technology Development Center, Cabid-an, Sorsogon, Philippines
e-mail: icapacio@bfar.da.gov.ph

A. T. Critchley et al. (eds.), *Tropical Phyconomy Coalition Development*, Developments in Applied Phycology 11,
https://doi.org/10.1007/978-3-031-47806-2_4

Worldwide, eucheumatoids are clonally propagated. Repetitive vegetative propagation from branch cuttings provides benefits by ensuring genetic consistency and homogenous quality and the avoidance of energetic costs associated with sexual reproduction that can compromise the growth rate. However, clonal propagation can also be detrimental to species fitness, making the entire cultivar more susceptible to disease due to a lack of genetic variation. Consequently, after decades of successful farming of different eucheumatoid species with high growth rates and good carrageenan yield and quality, they are now observed to have thinner branches and are more susceptible to epiphytism and ice-ice disease, thereby compromising carrageenan yield and quality (e.g., Largo et al. 1995a, b; Hurtado et al. 2006; Vairappan et al. 2008; and elsewhere in this book). Consequently, seaweed prices tend to be very volatile depending on the quality of the harvest. The quality of harvest and buying price influences the potential earnings of seaweed farmers, exposing them to significant economic problems. In this regard, there is an urgent need to establish new cultivars collected from wild populations to generate seedstocks with superior quality, e.g., high growth rates (Narvarte et al. 2022), resistance to disease, tolerance to a wide range of environmental factors, and high carrageenan yield and quality, compared to old, farmed cultivars.

Recently, new cultivars have been successfully established from the spores of fertile materials collected from the wild (Hinaloc and Roleda 2021), producing progeny corresponding to the next generation in the life cycle (Fig. 4.1). Current initiatives will allow us to provide a variety of new high-quality seedstocks developed from the wild to sustain the livelihoods of thousands of small-scale farmers and the global carrageenan seaweed industry. This chapter will focus on cultivar development in *Kappaphycus alvarezii*.

4.2 Life History

Kappaphycus species, similar to most taxa in the class Florideophyceae, follow a triphasic, *Polysiphonia*-like life cycle (Fig. 4.1). In this chapter, for the distinction of the three life history phases, the diploid tetrasporophyte will be designated Phase I, the haploid gametophyte Phase II, and the diploid carposporophyte Phase III. The diploid tetrasporophyte and the haploid male and female gametophytes are most often isomorphic, while the diploid carposporophyte is reduced and remains attached to the female gametophyte.

The aflagellated and cell-wall-deficient male gametes are called spermatia and are formed singly in each spermatangium (Dixon 1973). The oogonium (also carpogonium) has an extended apex with a hair-like projection called the trichogyne, which receives the spermatium during fertilization. The carposporophyte subsequently develops from the fertilized carpogonium, producing a mass of carpospores. The term cystocarp is often used synonymously with carposporophyte, but the term is used to include vegetative female gametophyte tissue such as the membranous envelope around the cystocarp called the pericarp. The carpospores released germinate to produce the diploid tetrasporophyte phase. Mature plants produce tetraspores by meiotic division within the tetrasporangium, as evidenced by the occurrence of synaptonemal complexes (Kugrens and West 1972).

4.3 Natural Populations

Many algal populations exhibit a predominance of tetrasporophyte generation (De Wreede and Klinger 1988). This predominance in *Kappaphycus alvarezii* was observed in the study by Ganzon-Fortes et al. (2013) in an offshore reef in Guiuan, Eastern Samar, Philippines. The monthly field collection of 50 *Kappaphycus* specimens conducted from June 2012 to May 2013 recorded a higher tetrasporic population (42–70%) and low gametophytic female (7.41–27.08%) and male (1.82–10.20%) population, while 8.33–36% were identified as vegetative. All reproductive phases were found throughout the year, except for the months of November, January, and March when no male gametophytes were recorded. The presence of fertile plants at the same site was also reported in other eucheumatoid reproductive biology studies (Ganzon-Fortes 2016; Hinaloc 2017; Rodrigora 2017; Roleda et al. 2017). Considering that the spermatia in red algae lack flagella, it may present a challenge for the fertilization of an egg by nonmotile sperm. However, the high number of tetrasporophytic individuals in the natural population suggests successful fertilization despite the perceived limitations of non-flagellated sperm and the lower number of male gametophytes in the natural population.

The diploid tetrasporophytes produced by genetic combination of the parental genes are considered to be more phenotypically robust with enhanced fitness compared to the haploid gametophytes that are suggested to be less equipped (genetically fit) for survival (Kain and Destombe 1995). The same has been reported for the stress physiological responses of diploid carpospores compared to the haploid tetraspores of *Gigartina skottsbergii* (Roleda et al. 2008).

In accordance with the genetic buffering hypothesis, diploids are known to have better cellular control, which makes them more robust and resilient to environmental stressors (Raper and Flexer 1970). Furthermore, because they have two copies of each gene, diploids are regarded to be more fit than haploids, because they can withstand the impact of harmful recessive mutations. Masking in diploids,

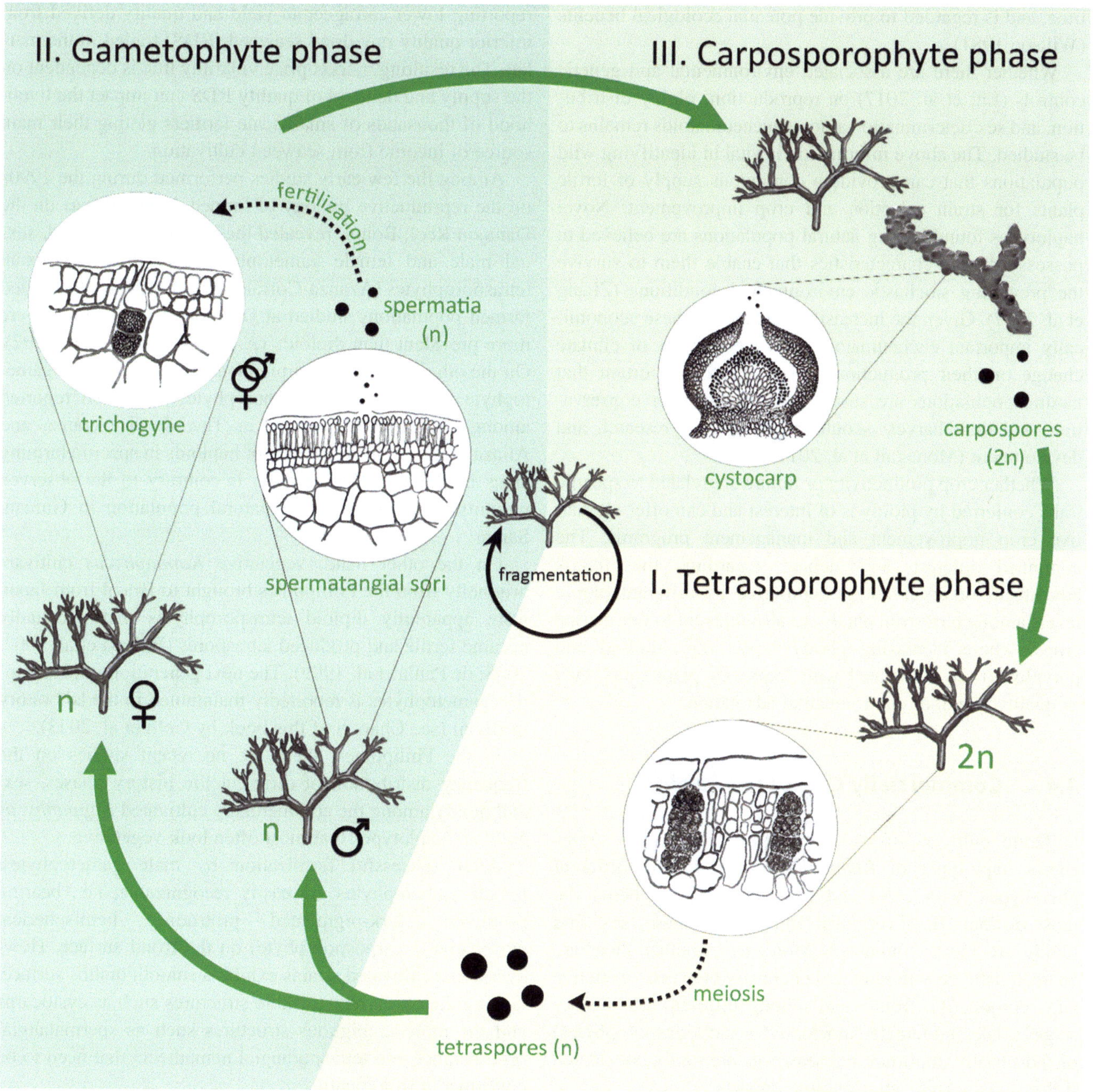

Fig. 4.1 Life cycle of *Kappaphycus*. The typical triphasic life history exhibits an isomorphic alternation of generations from the (I) diploid tetrasporophyte phase to the (II) haploid dioecious gametophyte phase. Tetrasporogenesis initiates the formation of zonate tetrasporangia through meiosis. Released tetraspores develop into male and female gametophytes. After sexual differentiation, spermatogenesis and oogenesis and subsequent fertilization produce the (III) diploid carposporophyte phase. This diminutive carposporophyte produced within the vegetative tissue (pericarp) develops attached to the female gametophyte. The prominent dark-pigmented protruding globose or hemispherical structures are called cystocarps that may be found scattered throughout the thallus of the female gametophyte. Only during this developmental phase are the cystocarpic female gametophytes easily distinguishable from the isomorphic male gametophytes and tetrasporophytes. Carpospores released develop into tetrasporophytes, completing the life cycle. Fragmentation and self-propagation of the macro thalli may occur at any stage

however, may be harmful since it allows mutations to persist through time (Otto and Marks 1996). Conversely, haploid populations can eliminate harmful mutations more quickly than diploid populations, which tend to conceal harmful alleles (Orr and Otto 1994; Hughes and Otto 1999). The alternation of generation between haploid-diploid, therefore, provides the utilization of a broader range of ecological niches, especially in environments that vary over space and time, and is regarded to provide potential ecological benefits (Willson 1981).

Whether there are associated environmental and genetic controls (Liu et al. 2017) on reproduction, ploidy distribution, and sex determination among eucheumatoids remains to be studied. The above information is vital in identifying wild populations that can provide a continuous supply of fertile plants for strain selection and crop improvement. Novel haplotypes found among natural populations are believed to possess inherent characteristics that enable them to survive the prevailing stochastic environmental conditions (Zhang et al. 2017). Given the increasing demand for these economically important eucheumatoids and the impacts of climate change on their production, it is critically important that natural populations are sustainably managed for conservation, and their harvest should be limited for research and development (Monagail et al. 2017).

Whether crop productivity or failure are related to specific traits conferred by ploidy is of interest and can offer alternative crop improvement and management programs. The continued interest and debate regarding the fitness advantages or disadvantages conferred by different ploidy levels among terrestrial plants are also relevant to our marine crops, where increasing ploidy levels (i.e., diploid and polyploids) are associated with improved plant traits such as quality, yield, or environmental adaptation.

4.4 Commercially Cultivated Strains

Different cultivars of the commercially cultivated strains and/or haplotypes of *Kappaphycus* spp. exhibit different phenotypes, with color and gross morphology being the most distinct. However, their life history phase, sex, and ploidy are mostly unknown. Moreover, whether there are characteristic growth rates and chemistry between vegetative and reproductive fronds and among different life history stages (male and female gametophytes and tetrasporophytes) of industrially important red seaweeds are rarely considered if they have significant economic impacts.

Most of the cultivars used by seaweed farmers have been vegetatively propagated since the 1970s. By avoiding the energetic cost associated with sexual reproduction, clonal propagation from vegetative cuttings can ensure genetic consistency and homogeneity in biomass quality. However, because of the lack of genetic variation, continuous vegetative propagation can also be harmful to fitness by increasing the susceptibility of the entire crop to disease outbreaks. Seaweed farmers have been reporting lower growth rates and seasonal crop failure due to greater susceptibility to epiphytism and ice-ice disease, which are mostly attributed to decreasing cultivar quality exacerbated by the impacts of climate change. Consequently, seaweed processors are also reporting lower carrageenan yield and quality derived from inferior quality raw dried seaweed (RDS) traded in the market. The resulting market price volatility that is dependent on the supply and demand of quality RDS can impact the livelihood of thousands of small-scale farmers getting their main source of income from seaweed cultivation.

Among the few early studies performed during the 1990s on the reproductive biology of farmed *Kappaphycus* on the Danajon Reef, Bohol, revealed the presence of haploid, sexual male and female gametophytes and diploid, asexual tetrasporophytes (Azanza-Corrales 1990). Among the limited farmed populations studied at Danajon Reef, haploids were more prevalent than diploids (Azanza-Corrales et al. 1992). On the other hand, cystocarpic plants (haploid female gametophyte with diploid carposporophyte) were also reported among the farmed cultivars in Tawi-Tawi (Azanza and Aliaza 1999). This prevalence of haploids in specific farming areas in Danajon Reef, Bohol, is contrary to the observed ubiquity of diploids in the natural population in Guiuan, Samar.

On the other hand, vegetative *Kappaphycus* cultivars originally from the Philippines brought to Brazil from Japan were apparently diploid tetrasporophytes that eventually became fertile and produced tetraspores (Bulboa et al. 2007, 2008; de Paula et al. 1999). The next generation of presumptive gametophytes is reportedly maintained in the laboratory in Brazil (see Chap. 8 of this book by Gelli et al. 2023).

In the Philippines, there are no recent studies on the frequency distribution of different life history phases, sex, and ploidy among the commercially cultivated *Kappaphycus* cultivars/haplotypes that most often look vegetative.

After successful fertilization by male gametophytes, female gametophytes are easily recognizable, i.e., bearing prominent dark-pigmented protruding hemispherical cystocarps (=carposporophyte) on the frond surface. However, most cultivated strains exhibit a smooth thallus surface, lacking the easily recognizable structures such as cystocarps and the more ambiguous structures such as spermatangial protuberance and tetrasporangial nemathecia that need to be confirmed histologically.

Considering that the farmed cultivars are clonally propagated and allowed to grow only, e.g., 45–50 days, it may be possible that repeated clonal propagation rendered the algal thalli to lose collective reproductive capacity or the thalli are maintained in a perpetually young (immature stage) and not able to mature enough for the reproductive structures to develop and become evident.

The growth rates of vegetative and reproductive fronds and those of the different life history stages (i.e., male and female gametophytes and tetrasporophytes) of economically important red seaweeds are rarely studied. Among the few studies published, slightly higher, but not significantly different, growth rates were reported in tetrasporophytes compared to gametophytes of *Gracilaria dura* (Sambhwani et al. 2020) and *Gracilaria chilensis* (Guillemin et al. 2014). However, the vegetative fronds of the male and female gametophytes, as well as the tetrasporophytes, showed significantly higher growth rates when compared to their respective reproductive fronds. This demonstrates the associated cost of reproduction in the triphasic life history of red seaweeds (Guillemin et al. 2014). Conversely, the growth rates of young gametophytes of *Gracilaria vermiculophylla* were significantly higher than those of tetrasporophytes; however, the same outcome was not observed in mature fronds, where the growth rates of gametophytes and tetrasporophytes did not significantly vary (Abreu et al. 2011). A recent study on the juvenile progenies of *Kappaphycus alvarezii* showed a higher growth rate in diploid tetrasporophytes (11.52 ± 1.0%) than in haploid gametophytes (9.45 ± 1.2%), both grown under hatchery conditions (Hinaloc and Roleda 2021). However, further studies are warranted to validate this observation and to test whether the same response will be observed under field conditions. Moreover, carrageenan yield and quality, among other biochemical indicators (e.g., protein, carbohydrate, and pigments), also need to be investigated.

4.5 In Vitro Spore Release and Cultivation of Different Phases of the Triphasic Life History

Recent investigations of biodiversity in Guiuan, Samar, Philippines, discovered seven novel haplotypes in the natural population of *Kappaphycus alvarezii*, which offers a reservoir of thus far unutilized wild genotypes that could be taken advantage of in the development of new cultivars with desirable traits (Roleda et al. 2021). In the wild, fertile tetrasporophytes and cystocarpic female gametophytes have been frequently encountered and used as sources of parental plants (Fig. 4.2; Hinaloc and Roleda 2021) for in vitro spore release, germination, and laboratory cultivation of sporelings (Fig. 4.3), corresponding to the next generation in the life cycle (Fig. 4.1).

For example, a green tetrasporophytic parental plant (Fig. 4.4a) produced a cohort of siblings, i.e., male and female gametophytes expressing a variety of colors, e.g., multicolored individuals (different shades of green, brown, and red; Fig. 4.4b) and individuals with different shades of yellow, green, brown, grey, black, orange, and red (Fig. 4.4c, d). Sexual differentiation became apparent after 8 months (Fig. 4.5), when fertilization already occurred among siblings and protruding cystocarps were observed among female gametophytes (Fig. 4.5a, c, e). Otherwise, distinct spermatangial nemathecia were observed among the male gametophytes (Fig. 4.5b, d, f).

Thereafter, the sexually differentiated sibling gametophytes were individually segregated and clonally cultivated in discrete aquaria for further observation. From another tetrasporophytic parental plant with a tricolor phenotype (green, brown, red; Fig. 4.6a), the resulting progeny and sibling gametophytes expressed variable phenotypes in terms of color and branching (Fig. 4.6b–f) that were very prominent in two sibling male gametophytes (Fig. 4.6e, f).

From previous cocultivation with their sibling male gametophytes, a fertilized female gametophyte with mature cystocarps was separated and clonally propagated. Interestingly, outgrowth was observed directly originating from the cystocarps (Fig. 4.6b). Further clonal propagation of the original female gametophyte, which eventually liberated all carpospores and wasted the cystocarps, together with the vegetative outgrowth, exhibited morphological changes in the branching pattern over time (Fig. 4.6c—June 16, 2021; Fig. 4.6d—February 24, 2022), suggesting phenotypic plasticity.

In the last 3 years (2020–2023), after the wasting of its cystocarps, continuous clonal propagation of the segregated female gametophyte, in the absence of a male counterpart, expressed its vegetative form as expected. Whether further long-term cultivation of an isolated female gametophyte can undergo apomictic reproduction, a direct-type life history, such as those reported in cystocarpic Helgolandian *Mastocarpus stellatus* (Roleda et al. 2004), remains to be seen. Alternatively, whether unreleased carpospores are capable of germinating in the carposporangium and generating an outgrowth directly projecting atop the cystocarp requires further study. The new outgrowth from the main axis that is female gametophyte, if emanating from the carpospore, will effectively be a tetrasporophytic branch. The occurrence of mixed life history phases in red algae has been reported in *Caloglossa leprieurii* (West et al. 2001). In the meantime, the cultivar, as shown in Fig. 4.6b–d, is provisionally classified as a presumptive female gametophyte until we are able to produce molecular tools to determine ploidy and sex.

On the other hand, the brown, cystocarpic, female gametophyte (= carposporophyte; Fig. 4.7a) produced tetrasporophyte progeny, where siblings exhibited either brownish, greenish, or greenish brown color phenotypes (Fig. 4.7b–f).

Each individual sibling, with a unique phenotype, i.e., branching and color, was separated and clonally cultivated to sizable biomass in the land-based hatchery to produce seedstock for further sea-based cultivation. Growth rates between land-based hatchery and sea-based nursery cultivation of new cultivars (i.e., the novel haplotypes) can then be compared with those of the commercially cultivated haplotypes to determine whether the new cultivars are suitable for commercial cultivation to replace the old cultivars.

Fig. 4.2 Habit of wild reproductive *Kappaphycus alvarezii* collected from a shallow reef flat in Guiuan, Eastern Samar, Philippines. Color expression likely has a genetic basis, as plants exhibiting different color phenotypes are found growing side by side exposed to the same growing conditions. (**a**, **c**, **d**) are haploid female gametophytes with prominent protruding dark-pigment cystocarps (i.e., diploid carposporophyte enclosed in the pericarp) randomly scattered throughout the thallus; (**b**) is a diploid tetrasporophyte, where tetrasporangial nemathecia—translucent whitish protuberances—can be observed. Wild vegetative thalli are typically smooth without distinctive protuberances. Thalli of commercially cultivated strains also exhibit smooth morphology (see Fig. 4.9). Scale bars = 5 cm

From the original life phases of the wild individuals (Phase I and Phase III) collected in Guiuan, Eastern Samar, the new generations (Phase II and Phase I, respectively) established that subsequently produced spores (Phase III and Phase I, respectively) in the hatchery were again induced to release spores. They germinated and were grown to maturity for the completion of the triphasic life cycle (back to Phase I and Phase II/III, respectively) under laboratory and hatchery conditions. This work is currently ongoing at the University of the Philippines, Marine Science Institute (UP-MSI) seaweed culture laboratory and gene bank and in the hatchery facility in Bolinao Marine Laboratory (BML), Pangasinan.

4.6 Vegetative Propagation

The use of micro-propagation technology (Fig. 4.8) as a tool for maintaining clonal propagules and producing seedstocks in the mariculture of economically important eucheumatoids, primarily *Kappaphycus* and *Eucheuma,* has been established (Reddy et al. 2017) and performed routinely in different

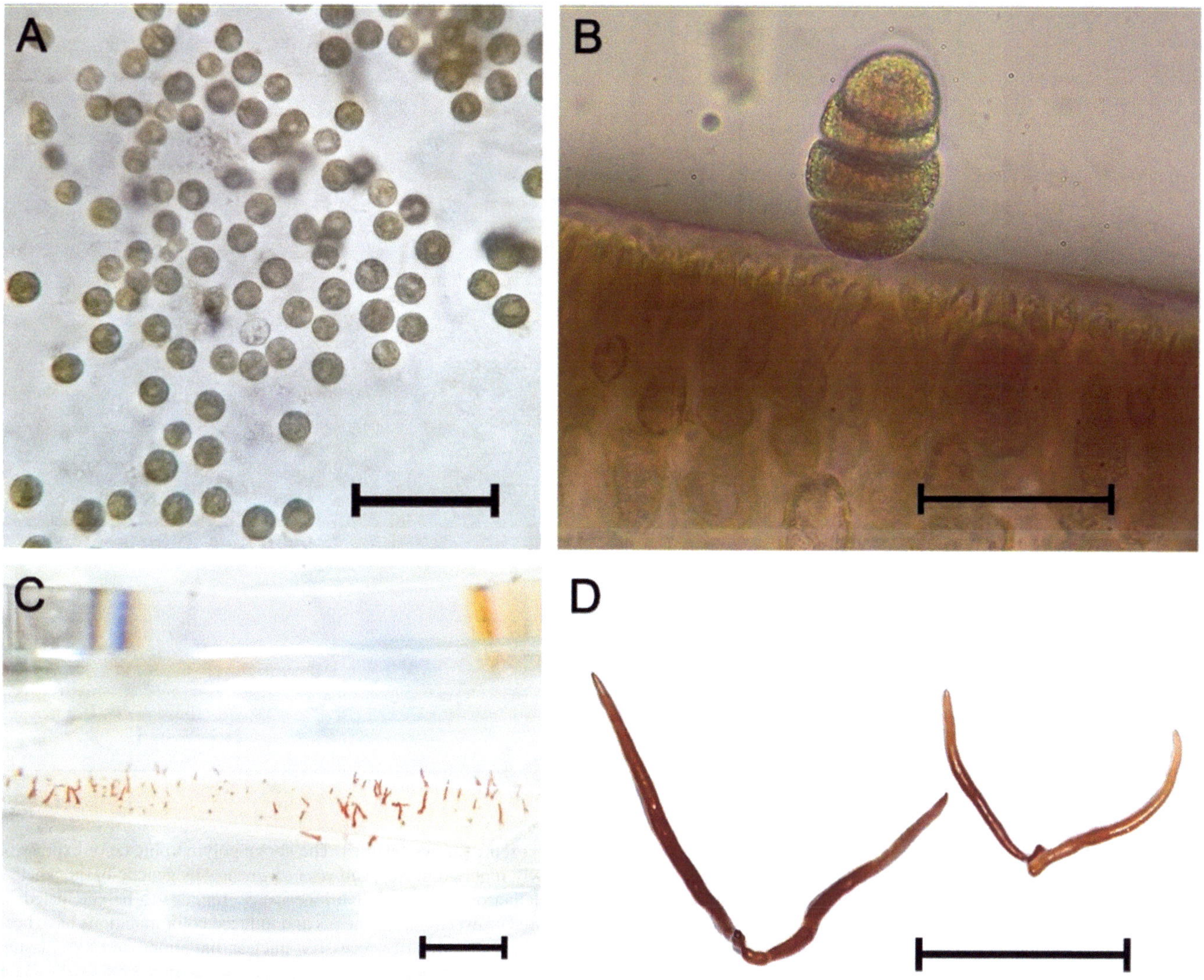

Fig. 4.3 In vitro spore release and early developmental stages of *Kappaphycus alvarezii*. (**a**) carpospore released en masse; (**b**) thin section of fertile tetrasporophytic thallus showing intact zonate tetrasporangia with four tetraspores; (**c**) 1-month-old sporelings attached on a glass slide; (**d**) detached 5-month-old tetrasporelings. Scale bars: **a** = 100 μm, **b** = 50 μm, **c** = 1 cm, **d** = 2 cm

research laboratories, e.g., the National Seaweed Technology Development Center (NSTDC) of the Department of Agriculture–Bureau of Fisheries and Aquatic Resources (DA–BFAR).

Notably, the resulting micropropagated clones exhibited changes in morphology (Fig. 4.9). Molecular analysis, however, showed that the original and micropropagated strains still belonged to the same *cox*1-*cox*2-3 haplotype (Crisostomo BA, Aguinaldo Z-Z, Roleda MY, unpublished data), suggesting morphological plasticity. Whether corresponding significant improvement in phenotypic traits, e.g., growth rates, carrageenan yield and quality, and increased resistance to pathogens, can be attributed to the micropropagated strains compared to the original plants requires comprehensive and systematic studies.

One of the main advantages of vegetative propagation is that it avoids the energy costs associated with sexual reproduction. Instead of allocating energy to developing reproductive organs, vegetative plants may focus more on other functions, such as growth and survival. In the red alga *Gracilaria chilensis*, for instance, vegetative thalli have higher photosynthetic activities as well as growth and survival rates than reproductive thalli (Guillemin et al. 2014). Asexual reproduction also provides a mechanism for self-incompatibility and for dioecious plants to bypass the barrier of finding a mate, especially in environments, where gamete mobility is limited. This mode of reproduction is favored by aquatic plants, which utilize both specialized asexual propagules and vegetative fragments for reproduction and dispersal (Li 2014). The variety of functional attributes of

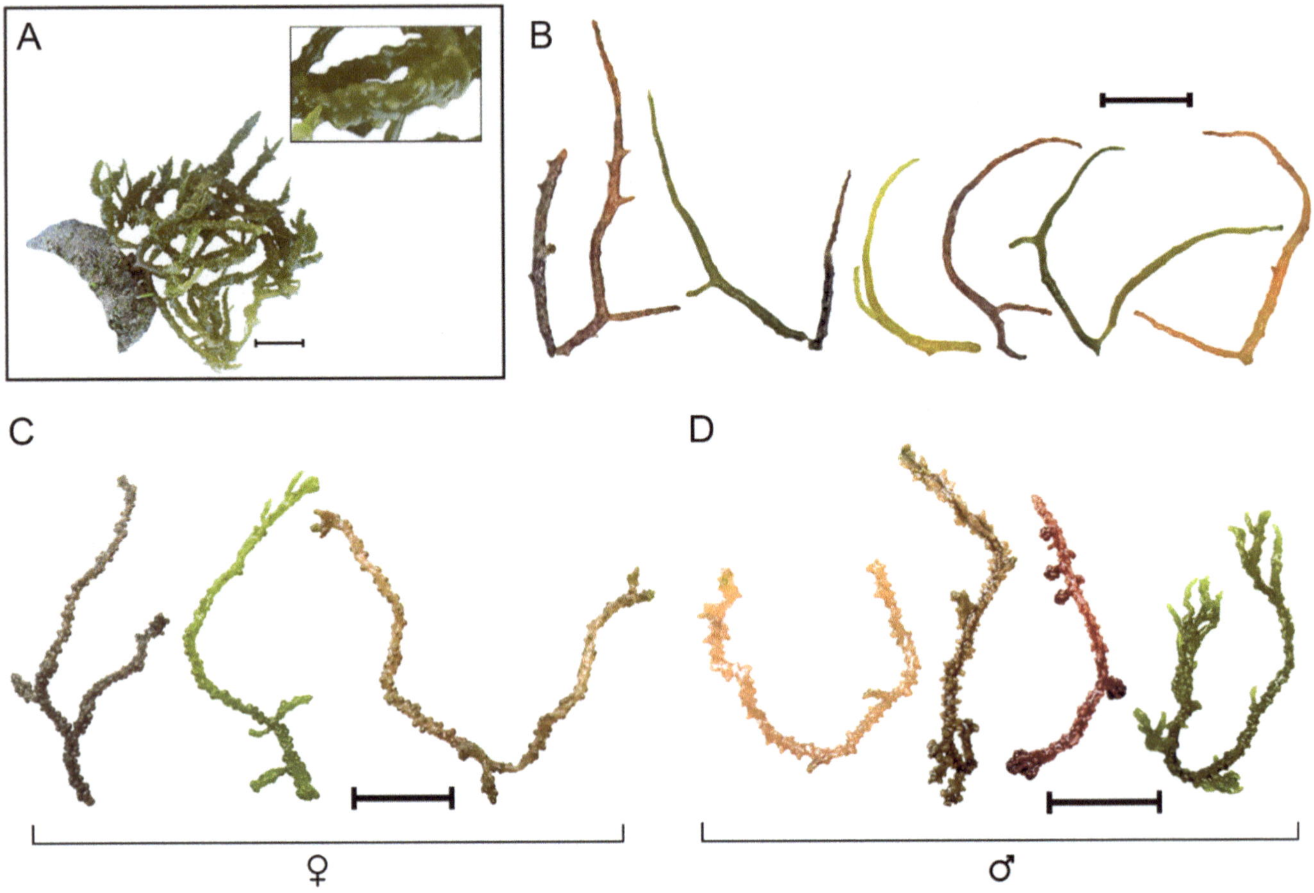

Fig. 4.4 Life cycle progression from a parental tetrasporophyte to gametophyte progeny. (**a**) green diploid tetrasporophytic parental plant, KaTR-O; (**b**, **c**, **d**) Haploid gametophytic progeny exhibiting different color phenotypes—(**b**) representative 7-month-old progeny expressing a variety of colors with some individuals showing bicolored pigmentation (i.e., reddish-brown, brownish green); (**c**) 8-month-old sexually differentiated female and (**d**) male gametophytes exhibiting distinct reproductive features. The above polymorphic color expression is newly reported in *Kappaphycus alvarezii*—its genetic basis, whether Mendelian or non-Mendelian inheritance, remains to be elucidated. In other red seaweeds, spontaneous and induced color mutations have been reported, i.e., primarily recessive nuclear mutations and cytoplasmic mutations of maternal inheritance. Scale bars: **a**–**b** = 3 cm, **c**–**d** = 5 cm

vegetative propagules may also increase the survivability and colonization success of the algae (Grace 1993; Boedeltje et al. 2007). Several successful invaders are species that favor asexual modes of reproduction (Hollingsworth and Bailey 2000; Liu et al. 2006; Ahmad et al. 2008; Castro et al. 2016; Datta et al. 2019). Clonal propagation also allows the preservation of the characteristics of the original plant. This is especially important in agronomy, where crops are selected for certain traits, such as chemical composition, which may otherwise be lost in sexually produced offspring (e.g., grapes in Bessis 2007). Moreover, some cultivars have lost the ability to sexually reproduce, thus relying entirely on vegetative propagation (e.g., saffron, banana, citrus, seedless cultivars) (Chichiriccò 1984; Heslop-Harrison and Schwarzacher 2007; Ollitrault and Navarro 2012).

Despite the abovementioned advantages of asexual reproduction, exclusive reliance on this mode of reproduction poses survival risks related to low genetic variability. Vegetative propagation may also enhance pathogen transmission, as propagules from susceptible plants may harbor pathogens (Rice 1983). In contrast, sexually reproducing susceptible parents may produce progeny that are resistant to parental pathogens (Rice 1983). Studies have shown that populations with high genetic diversities are more likely to survive disease outbreaks compared to those with low diversities (Maron et al. 2011; Schnitzer et al. 2011). The same principle is recognized and applied in agroecosystems (He et al. 2019). Furthermore, populations with high genetic diversities are less vulnerable to extreme climatic changes, as was observed in the responses of kelp forests to a marine heatwave (Wernberg et al. 2018). Thus, sexual reproduction may increase the ability of an algal population to adapt to selective pressures through the maintenance of genetic diversity (Lei 2010; Johnson et al. 2020). Aside from its role in ecosystem

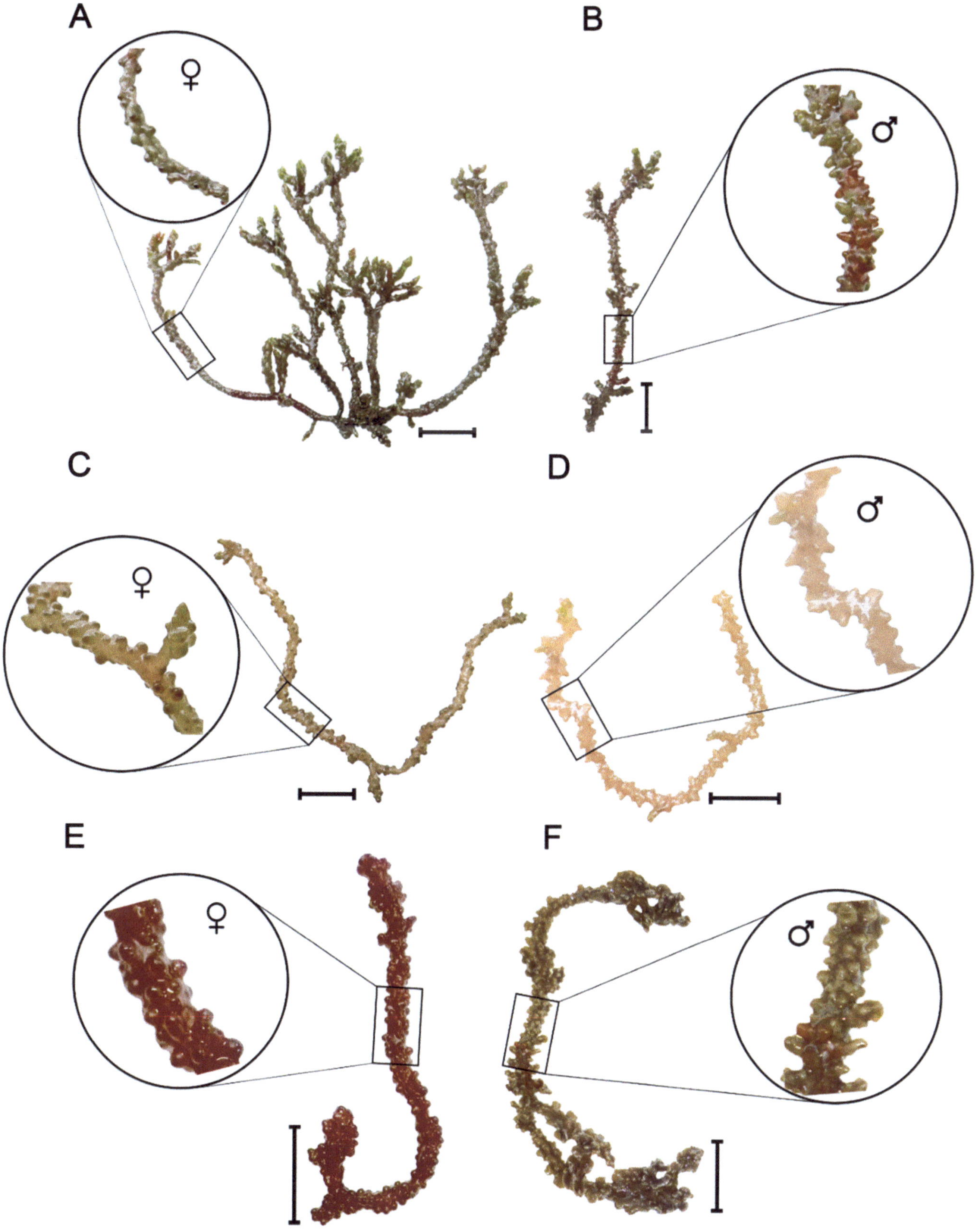

Fig. 4.5 External morphology of sexually differentiated 8-month-old gametophyte progeny from different parental tetrasporophyte plants: (**a**–**b**) KaTR-N, (**c**–**d**) KaTR-O, (**e**–**f**) KaTR-Q. The next-generation haploid female gametophytes (**a**, **c**, **e**) are characterized by the presence of numerous cystocarps (= diploid carposporophyte) scattered throughout the thalli, while the haploid male gametophytes (**b**, **d**, **f**) have distinct spermatangial protuberances. Scale bars: **a**–**b**, **e**–**f** = 3 cm; **c**–**d** = 5 cm

Fig. 4.6 Clonal propagation of individually segregated sibling gametophytes, i.e., sexually differentiated and morphologically distinct. (**a**) parental multicolored diploid tetrasporophytic plant, KaTR-N; (**b**) resultant haploid female gametophyte that was fertilized by a sibling male gametophyte, which developed mature cystocarps (= diploid carposporophyte). Subsequent clonal propagation of the individual cystocarpic plant showed development of outgrowth originating directly from the cystocarps (yellow arrows). Further cultivation of (**b**) after cystocarps were spent, the vegetative clones exhibited morphologically plastic morphotypes, as observed in (**c**) documented on June 16, 2021, and (**d**) documented on February 24, 2022 (**e, f**) haploid male gametophyte siblings expressing different morphologies and shades of colors. Scale bar = 3 cm

Fig. 4.7 Life cycle progression from a cystocarpic female gametophyte (= carposporophyte) parent to tetrasporophyte progeny. (**a**) brown carposporophyte parental plant and (**b**–**f**) resultant tetrasporophyte progeny, where siblings exhibit different habits and shades of green and brown after >1 year of cultivation in a land-based hatchery. Scale bar = 3 cm

Fig. 4.8 Micropropagation of *Kappaphycus alvarezii*. (**a**) vegetative mother plant with a smooth thalli surface acclimatized under laboratory conditions for 1 week; (**b**) preparation of explants measuring 5 mm thallus segments; (**c**) laboratory culture setup of micropropagules in flasks with moderate aeration; (**d**) plantlets with developing primordial shoots at 30 (left), 60 (center), and 90 (right) days

maintenance, sexual reproduction is also an important tool for crop domestication, as well as cultivar and hybrid development (Scarcelli et al. 2006; Meyer et al. 2012). Ultimately, the favorability of different reproduction strategies *in natura* depends on a range of factors, including but not limited to genetics, age, lifespan, competition, and environmental conditions (Zhang and Zhang 2007; Yang and Kim 2016).

Fig. 4.9 Morphological transformation observed in micropropagated *Kappaphycus alvarezii*. (**a**, **c**) original vegetative mother plants collected in two farm sites in Bulusan and Castilla, Sorsogon, respectively; (**b**, **d**) corresponding mature field-grown micropropagated clones as processed in Fig. 4.8. Scale bars = 5 cm

Declarations Competing interests. The authors declare no competing interests.

Funding The primary data presented in this review chapter were results from studies subsidized by the UPMSI inhouse research grant and research funding received from the University of the Philippines-UP System Enhanced Creative Work and Research Grant (ECWRG 2019-09-R) and Balik PhD Program (OVPAA-BPhD-2019-06), UP-Diliman, Office of the Vice Chancellor for Research and Development (OVCRD) Outright Research Grant Project No. 202039 ORG. Co-funding was also received from UKRI GCRF Global-SeaweedSTAR program—Projects Grant no. GSS/RF/015 and GSS/RF/047 and the CHED-LAKAS-funded project "Phytochemical Characterization of Macroalgae for Food and High Value Products (PhycoPRO)." Additional funding was also received from a Safe Seaweed Coalition grant (SecureFuture; n° LS249100) and from Sea6 Energy Private Limited. The eucheumatoid micropropagation initiatives at the National Seaweed Technology Development Center (NSTDC) are funded by the Bureau of Fisheries and Aquatic Resources (BFAR).

References

Abreu MH, Pereira R, Sousa-Pinto I, Yarish C (2011) Ecophysiological studies of the non-indigenous species *Gracilaria vermiculophylla* (Rhodophyta) and its abundance patterns in Ria de Aveiro lagoon, Portugal. Eur J Phycol 46:453–464. https://doi.org/10.1080/09670262.2011.633174

Agardh JG (1847) Nya alger från Mexico. Öfversigt af Kongl. Vetenskaps-Adademiens Förhandlingar, Stockholm 4:5–17

Ahmad R, Liow P-S, Spencer DF, Jasieniuk M (2008) Molecular evidence for a single genetic clone of invasive *Arundo donax* in the

United States. Aquat Bot 88:113–120. https://doi.org/10.1016/j.aquabot.2007.08.015

Azanza RV, Aliaza TT (1999) *In vitro* carpospore release and germination in *Kappaphycus alvarezii* (Doty) Doty from Tawi-Tawi, Philippines. Bot Mar 42:281–284. https://doi.org/10.1515/BOT.1999.031

Azanza-Corrales R (1990) The farmed *Eucheuma* species in Danajon reef, Philippines: vegetative and reproductive structures. J Appl Phycol 2:57–62. https://doi.org/10.1007/BF02179769

Azanza-Corrales R, Mamauag SS, Alfiler E, Orolfo MJ (1992) Reproduction in *Eucheuma denticulatum* (Burman) Collins and Hervey and *Kappaphycus alvarezii* (Doty) Doty farmed in Danajon reef, Philippines. Aquaculture 103:29–34. https://doi.org/10.1016/0044-8486(92)90275-P

Bessis R (2007) Evolution of the grapevine (*Vitis vinifera* L.) imprinted by natural and human factors. Canadian J Bot 85:679–690. https://doi.org/10.1139/b07-060

Boedeltje G, Ozinga WA, Prinzing A (2007) The trade-off between vegetative and generative reproduction among angiosperms influences regional hydrochorous propagule pressure. Glob Ecol Biogeogr 17:50–58. https://doi.org/10.1111/j.1466-8238.2007.00365.x

Bulboa C, de Paula EJ, Chow FY (2007) Laboratory germination and sea out-planting of tetraspore progeny from *Kappaphycus striatum* (Rhodophyta) in subtropical waters of Brazil. J Appl Phycol 19:357–363. https://doi.org/10.1007/s10811-006-9142-7

Bulboa C, de Paula EJ, Chow FY (2008) Germination and survival of tetraspores of *Kappaphycus alvarezii* var. *alvarezii* (Solieriaceae, Rhodophyta) introduced in subtropical waters of Brazil. Phycol Res 56:39–45. https://doi.org/10.1111/j.1440-1835.2008.00483.x

Castro S, Castro M, Ferrero V, Costa J, Tavares D, Navarro L, Loureiro J (2016) Invasion fosters change: independent evolutionary shifts in reproductive traits after *Oxalis pes-caprae* L. introduction. Front Plant Sci 7:874. https://doi.org/10.3389/fpls.2016.00874

Chichiriccò G (1984) Karyotype and meiotic behaviour of the triploid *Crocus sativus* L. Caryologia 37:233–239. https://doi.org/10.1080/00087114.1984.10797702

Datta A, Schweiger O, Kühn I (2019) Niche expansion of the invasive plant species *Ageratina adenophora* despite evolutionary constraints. J Biogeogr 46:1306–1315. https://doi.org/10.1111/jbi.13579

de Paula EJ, Pereira RTL, Ohno M (1999) Strain selection in *Kappaphycus alvarezii* var. *alvarezii* (Solieriaceae, Rhodophyta) using tetraspore progeny. J Appl Phycol 11:111–121. https://doi.org/10.1023/A:1008085614360

De Wreede RE, Klinger T (1988) Reproductive strategies in algae. In: Plant reproductive biology. Oxford University Press, pp 267–284

Dixon PS (1973) The biology of the Rhodophyta. Oliver & Boyd, Edinburgh

Doty MS (1988) Podromus ad systematica Eucheumatoideorum: a tribe of commercial seaweeds related to *Eucheuma* (Solieriaceae, Gigartinales). In: Abbott IA (ed) Taxonomy of economic seaweeds with reference to some Pacific and Caribbean species, vol 2. California Sea Grant College Program, La Jolla, pp 159–207 [Report T-CSGCP-018]

Dumilag RV, Zuccarello GC (2022) Phylogeny and genetic diversity of the Philippine euchematoid genus *Mimica* (Solieriaceae, Rhodophyta), and the proposal for *Kappaphycopsis* gen. nov. to include the anomalous species of *Kappaphycus*, *K. cottonii.* Phycologia 61:496–503

Dumilag RV, Aguinaldo Z-ZA, Crisostomo BA, Hinaloc LAR, Liao LM, Roa-Quiaoit HA, Galon-Dangan F, Zuccarello GC, Guillemin M-L, Brodie J, Cottier-Cook EJ, Roleda MY (2023) Eucheumatoid seaweed cultivar diversity in The Philippines. Rev Fish Sci Aquacult 31:47–65. https://doi.org/10.1080/23308249.2022.2060038

Ganzon-Fortes ET (2016) Sea out-planting and seedstock production of laboratory generated carrageenophyte cultivars. Technical report submitted to the Philippine Council for Agricultural and Aquatic Resources Research and Development, PCAARRD. The Marine Science Institute, College of Science, University of the Philippines Diliman, 7 pp

Ganzon-Fortes E, Saco J, Albaladejo M, Projimo V (2013) Monthly variation in the occurrences of the reproductive populations of *Kappaphycus alvarezii* (Doty) Doty ex P.C. Silva (Rhodophyta: Gigartinales) in Nabalikad Reef, Guiuan, Samar. Poster presented at the Philippines Association of Marine Science (PAMS) 12th National Symposium, University of the Philippines Visayas, Tacloban College, Tacloban City, p 106

Gelli VC, Plastino EM, Yokoya NS (2023) Cultivation and domestication of *Kappaphycus alvarezii* strains at Ubatuba Bay, São Paulo state, southeastern Brazil. In: Hurtado AQ, Critchley AT, Neish I (eds) Tropical seaweed cultivation - phyconomy - proceedings of tropical phyconomy coalition development, TPCD #1, UNHAS Univ., Makassar, Indonesia, July 7–8, 2021

Grace JB (1993) The adaptive significance of clonal reproduction in angiosperms: an aquatic perspective. Aquat Bot 44:159–180. https://doi.org/10.1016/0304-3770(93)90070-d

Guillemin ML, Valenzuela P, Gaitán-Espitia JD, Destombe C (2014) Evidence of reproductive cost in the triphasic life history of the red alga *Gracilaria chilensis* (Gracilariales, Rhodophyta). J Appl Phycol 26:569–575. https://doi.org/10.1007/s10811-013-0072-x

He H-M, Liu L-N, Munir S, Bashir NH, Wang Y, Yang J, Li C-Y (2019) Crop diversity and pest management in sustainable agriculture. J Integr Agr 18:1945–1952. https://doi.org/10.1016/s2095-3119(19)62689-4

Heslop-Harrison JS, Schwarzacher T (2007) Domestication, genomics and the future for banana. Ann Bot-London 100:1073–1084. https://doi.org/10.1093/aob/mcm191

Hinaloc LAR (2017) In vitro carpospore recruitment and germling growth in *Kappaphycus alvarezii* (Doty) Doty ex P.C. Silva (Rhodophyta: Gigartinales). M.Sc. thesis. University of the Philippines, Diliman, 73 pp

Hinaloc LAR, Roleda MY (2021) Phenotypic diversity, growth and sexual differentiation in the progeny of wild *Kappaphycus alvarezii* (Gigartinales, Florideophyceae). Phycologia 60:547–557. https://doi.org/10.1080/00318884.2021.1946307

Hollingsworth ML, Bailey JP (2000) Evidence for massive clonal growth in the invasive weed Fallopia japonica (Japanese Knotweed). Bot J Linn Soc 133:463–472. https://doi.org/10.1111/j.1095-8339.2000.tb01589.x

Hughes JS, Otto SP (1999) Ecology and the evolution of biphasic life cycles. Am Nat 154:306–320

Hurtado AQ, Critchley AT, Trespoey A, Lhonneur GB (2006) Occurrence of *Polysiphonia* epiphytes in *Kappaphycus* farms in Calaguas Is., Camarines Norte, Philippines. J Appl Phycol 18:301–306

Johnson AJ, Orth RJ, Moore KA (2020) The role of sexual reproduction in the maintenance of established *Zostera marina* meadows. J Ecol 108:945–957. https://doi.org/10.1111/1365-2745.13362

Kain JM, Destombe C (1995) A review of the life history, reproduction, and phenology of *Gracilaria.* J Appl Phycol 7:269–281

Kugrens P, West JA (1972) Ultrastructure of tetrasporogenesis in the parasitic red alga *Levringiella gardneri* (Setchell) Kylin. J Phycol 8:370–383

Largo DB, Fukami K, Nishijima T, Ohno M (1995a) Laboratory-induced development of the ice-ice disease of the farmed red algae *Kappaphycus alvarezii* and *Eucheuma denticulatum* (Soliriaceae, Gigartinales, Rhodophyta). J Appl Phycol 7:539–543

Largo DB, Fukami K, Nishijima T (1995b) Occasional pathogenic bacteria promoting ice-ice disease in the carrageenan-producing red algae *Kappaphycus alvarezii* and *Eucheuma denticulatum* (Soliriaceae, Gigartinales, Rhodophyta). J Appl Phycol 7:545–554

Lei SA (2010) Benefits and costs of vegetative and sexual reproduction in perennial plants: a review of literature. J Ariz Nev Acad Sci 42:9–14. https://doi.org/10.2181/036.042.0103

Li W (2014) Environmental opportunities and constraints in the reproduction and dispersal of aquatic plants. Aquat Bot 118:62–70. https://doi.org/10.1016/j.aquabot.2014.07.008

Liu J, Dong M, Miao SL, Li ZY, Song MH, Wang RQ (2006) Invasive alien plants in China: role of clonality and geographical origin. Biol Invasions 8:1461–1470. https://doi.org/10.1007/s10530-005-5838-x

Liu X, Bogaert K, Engelen A, Leliaert F, Roleda MY, de Clerck O (2017) Seaweed reproductive biology: environmental and genetic controls. Bot Mar 60:89–108. https://doi.org/10.1515/bot-2016-0091

Maron JL, Marler M, Klironomos JN, Cleveland CC (2011) Soil fungal pathogens and the relationship between plant diversity and productivity: soil pathogens, productivity and invasibility. Ecol Lett 14:36–41. https://doi.org/10.1111/j.1461-0248.2010.01547.x

Meyer RS, DuVal AE, Jensen HR (2012) Patterns and processes in crop domestication: an historical review and quantitative analysis of 203 global food crops. New Phytol 196:29–48. https://doi.org/10.1111/j.1469-8137.2012.04253.x

Monagail MM, Cornish L, Morrison L, Araújo R, Critchley AT (2017) Sustainable harvesting of wild seaweed resources. Eur J Phycol 52:371–390. https://doi.org/10.1080/09670262.2017.1365273

Narvarte BCV, Hinaloc LAR, Genovia TGT, Gonzaga SMC, Tabonda-Nabor AM, Roleda MY (2022) Physiological and biochemical characterization of new wild strains of *Kappaphycus alvarezii* (Gigartinales, Rhodophyta) cultivated under land-based hatchery conditions. Aquat Bot 183:103567. https://doi.org/10.1016/j.aquabot.2022.103567

Norris JN (2014) Marine algae of the northern Gulf of California, II: Rhodophyta. Smithsonian contributions to botany, no 96, pp [i]-xvi, [1]-555. Smithsonian Institution Scholarly Press, Washington, DC

Núnez-Resendiz M, Dreckmann KM, Senties A, Wynne MJ, León-Tejera HP (2019) *Eucheumatopsis isiformis* gen & comb. nov. (Solieriaceae, Rhodophyta) from the Yukatan Peninsula, to accommodate *Eucheuma isiforme*. Phycologia 58:51–62

Ollitrault P, Navarro L (2012) Citrus. In: Badenes M, Byrne D (eds) Fruit breeding. Handbook of plant breeding, vol 8. Springer, Boston, MA, pp 622–662. https://doi.org/10.1007/978-1-4419-0763-9_16

Orr HA, Otto SP (1994) Does diploidy increase the rate of adaptation? Genetics 136:1475–1480

Otto SP, Marks JC (1996) Mating systems and the evolutionary transition between haploidy and diploidy. Biol J Linn Soc 57:197–218

Raper JR, Flexer AS (1970) The road to diploidy with emphasis on a detour. Symp Soc General Microbiol 20:401–432

Reddy CRK, Yokoya NS, Yong WTL, Luhan MRJ, Hurtado AQ (2017) Micropropagation of *Kappaphycus* and *Eucheuma*: trends and prospects. In: Hurtado A, Critchley A, Neish I (eds) Tropical seaweed farming trends, problems and opportunities. Developments in applied phycology, vol 9. Springer, Cham, pp 91–110. https://doi.org/10.1007/978-3-319-63498-2_5

Rice WR (1983) Parent-offspring pathogen transmission: a selective agent promoting sexual reproduction. Am Nat 121:187–203. https://doi.org/10.1086/284050

Rodrigora LD (2017) Laboratory tetraspore recruitment and early development in wild *Kappaphycus alvarezii* (Rhodophyta: Gigartinales) for farming. M.Sc. thesis. University of the Philippines, Diliman, 72 pp

Roleda MY, van de Poll WH, Hanelt D, Wiencke C (2004) PAR and UVBR effects on photosynthesis, viability, growth and DNA in different life stages of two coexisting Gigartinales: implications on recruitment and zonation pattern. Mar Ecol Prog Ser 281:37–50

Roleda MY, Zacher K, Wulff A, Hanelt D, Wiencke C (2008) Susceptibility of spores of different ploidy levels from Antarctic *Gigartina skottsbergii* (Gigartina, Rhodophyta) to ultraviolet radiation. Phycologia 47:361–370

Roleda MY, Rodrigora-Calala LD, Hinaloc LAR, Ganzon-Fortes ET (2017) Red, blue, and green, and shades in between: diversity and plasticity in color expression in the life history of *Kappaphycus*. Phycologia 56:161

Roleda MY, Aguinaldo ZZA, Crisostomo BA, Hinaloc LAR, Projimo VZ, Dumilag RV, Lluisma AO (2021) Discovery of novel haplotypes from wild populations of *Kappaphycus* (Gigartinales, Rhodophyta) in The Philippines. Algae 36:1–12. https://doi.org/10.4490/algae.2021.36.2.18

Sambhwani K, Modi J, Singhala A, Bramhabatt H, Mishra A, Mantri VA (2020) Analysis of functional traits in female gametophytic and tetrasporophytic life phases of industrially important red alga *Gracilaria dura* (Rhodophyta: gracilariacae). J Appl Phycol 32:1961–1969. https://doi.org/10.1007/s10811-020-02116-1

Santiañez WJE, Wynne MJ (2020) Establishment of *Mimica* gen. nov. to accommodate the anaxiferous species of the economically important red seaweed *Eucheuma* (Solieriaceae, Rhodophyta). Phytotaxa 439:167–170. https://doi.org/10.11646/phytotaxa.439.2.8

Scarcelli N, Tostain S, Vigouroux Y, Agbangla C, Daïnou O, Pham J-L (2006) Farmers' use of wild relative and sexual reproduction in a vegetatively propagated crop. The case of yam in Benin. Mol Ecol 15:2421–2431. https://doi.org/10.1111/j.1365-294X.2006.02958.x

Schnitzer SA, Klironomos JN, Hillerislambers J, Kinkel LL, Reich PB, Xiao K, Rillig MC, Sikes BA, Callaway RM, Mangan SA, van Nes EH, Scheffer M (2011) Soil microbes drive the classic plant diversity-productivity pattern. Ecology 92:296–303. https://doi.org/10.1890/10-0773.1

Silva PC, Basson PW, Moe RL (1996) Catalogue of the benthic marine algae of the Indian Ocean. Univ Calif Publ Bot 79: [i]-xiv, 1-1259

Vairappan CS, Chung CS, Hurtado AQ, Soya FE, Lhonneur GB, Critchley A (2008) Distribution and symptoms of epiphyte infection in major carrageenophyte-producing farms. J Appl Phycol 20:477–483. https://doi.org/10.1007/s10811-007-9299-8

Wernberg T, Coleman MA, Bennett S, Thomsen MS, Tuya F, Kelaher BP (2018) Genetic diversity and kelp forest vulnerability to climatic stress. Sci Rep-UK 8:1–8. https://doi.org/10.1038/s41598-018-20009-9

West JA, Zuccarello GC, Kamiya M (2001) Reproductive patterns of *Caloglossa* species (Delesseriaceae, Rhodophyta) from Australia and New Zealand: multiple origins of asexuality in *C. leprieurii*. Literature review on apomixis, mixed-phase, bisexuality and sexual compatibility. Phycol Res 49:183–200. https://doi.org/10.1046/j.1440-1835.2001.00239.x. https://onlinelibrary.wiley.com/doi/10.1046/j.1440-1835.2001.00239.x

Willson MF (1981) On the evolution of complex life cycles: a review and an ecological perspective. Ann Mo Bot Gard 68:275–300. https://doi.org/10.2307/2398799

Yang YY, Kim JG (2016) The optimal balance between sexual and asexual reproduction in variable environments: a systematic review. J Ecol Environ 40:12. https://doi.org/10.1186/s41610-016-0013-0

Zhang Y, Zhang D (2007) Asexual and sexual reproductive strategies in clonal plants. Front Biol China 2:256–262. https://doi.org/10.1007/s11515-007-0036-0

Zhang H, Mittal N, Leamy LJ, Barazani O, Song BH (2017) Back into the wild: apply untapped genetic diversity of wild relatives for crop improvement. Evol Appl 10:5–24. https://doi.org/10.1111/eva.12434

A Review of the Use of Spores for the Supply of High-Quality *Kappaphycus alvarezii* Seedlings

5

Rajuddin Syamsuddin

Abstract

Seaweeds are an important source of income for coastal communities and are a leading export commodity that is a source of foreign exchange for Indonesia, where the species that occupies the top position in terms of quantity, marketed globally, is the red seaweed *Kappaphycus alvarezii*. Optimization of seaweed "seed" (seedling) production is very important for succesful cultivation of this species. In addition to thallus fragmentation, *Kappaphycus* can reproduce by spores. Several researchers have suggested various advantages of using spores during cultivar propagation in the seaweed cultivation industry. Based on their results, the author investigated the possibilities of using spore technology as an effort to enhance propagule biomass in order to meet the needs of Indonesian seaweed farmers. The results are presented in this chapter.

Keywords

Kappaphycus alvarezii · Spores · Seedlings · Cultivation

5.1 Introduction

Selected seaweeds are an important source of income for coastal communities and are a leading export commodity that is a source of foreign exchange for Indonesia. The species of red alga occupying the top position in terms of quantity marketed globally, at a good price, is the carrageenophyte *Kappaphycus alvarezii*. Optimization of "seed" (seedling/propagule) production is very important for the successful cultivation of this commodity. In addition to (asexual) thallus fragmentation, red seaweeds generally can reproduce by spores. Spores are part of the life cycle. They are single-celled, similar with the ability to develop into new thalli. For a more detailed account of the *Kappahycus* typical, triphasic life cycle see Roleda et al., Chap. 4 in this book. Several researchers have suggested the various advantages of using spores propagules for multiplication of seaweeds especially those belonging to the red algae (Rhodophyta) (Glenn et al. 1996; Kostamo 2008; Mantri et al. 2009; Forbord et al. 2020; Suda and Mikami 2020). However, it is not an easy task as can be seen in this book (see also Gelli et al., Chap. 9). This chapter presents a preliminary assessment of the possibility of using spores from *K. alvarezii* to produce seedlings in Indonesia, at times required by farmers. Salinity, temperature, and wavelength (color) of light were factors affecting the release and development of spores, while the texture of the substrate affects the number of spores that could stick to the surface which affected the growth and development of the new thalli.

5.2 Materials and Methods

5.2.1 Time and Place of Studies

A series of studies on the effect of salinity, temperature, light color, type of artificial substrata on the release, and development of *Kappaphycus alvarezii* spores was carried out in 2013 and 2014. Research was undertaken in the wet laboratory of the Faculty of Marine Sciences and Fisheries, Hasanuddin University. An experiment to grow reproductive fragments and spores into mature plants in the sea was conducted in 2015 in the coastal waters of Jonggoa, North Galesong District, Takalar Regency, South Sulawesi, Indonesia.

Experiments on salinity, temperature, light color, and type of substrata were conducted individually. The effect of salinity and temperature on spore release and development used 5 L glass jars with aeration (Fig. 5.1). The light color and

R. Syamsuddin (✉)
Department of Fisheries, Faculty of Marine Science and Fisheries, Hasanuddin University, Makassar, Indonesia

A. T. Critchley et al. (eds.), *Tropical Phyconomy Coalition Development*, Developments in Applied Phycology 11,
https://doi.org/10.1007/978-3-031-47806-2_5

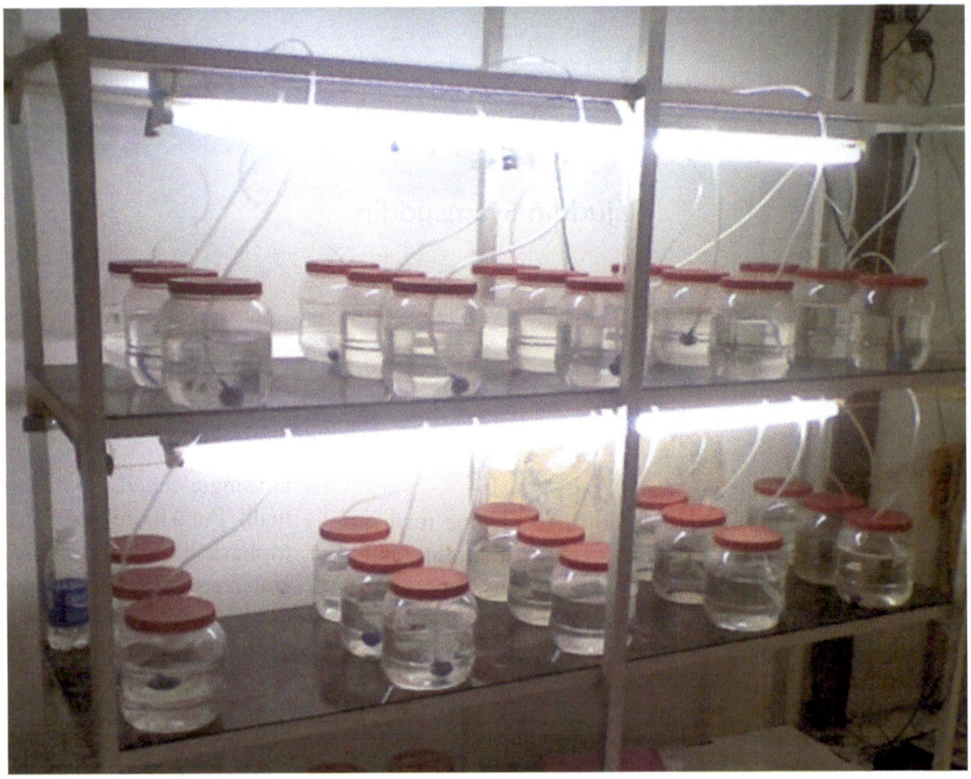

Fig. 5.1 Experimental setup with aerated 5 L jars. Reproductive thalli were maintained at selected light and temperature conditions

artificial substrata for spore settlement were conducted in aquaria 10 × 10 × 10 cm (Fig. 5.2).

Seawater was filtered to avoid fine particles that can interfere with the growth and development of the thalli reproductive sporangia and any released spores within the experimental set ups Water was changed every 3 days by as much as 30% of the volume. Each container was aerated with weak air bubbles to ensure the distribution of dissolved nutrients. Which are needed for the opening of the sporangia (cystocarps) and the release and subsequent development of any spores. In addition, the aeration also functioned to minimize attachment of any sediment to the thallus, cystocarp, and spores. Light was provided from 20-Watt fluorescent lamps. The photoperiod was 12 h light, 12 h dark.

A petri dish was placed at the bottom of each container. Two glass slides were placed as a substratum for the attachment and initial development of released spores. The test reproductive thalli were tied with a fine thread and positioned hanging 3 cm above the glass slides.

The salinity levels tested were: 20, 25, 30, 35, and 40 ppt with three replications of each. To obtain the higher salinities (i.e., up to 40 ppt), seawater was evaporated under direct sunlight. For the lower salinities (i.e., 20–30 ppt) seawater was diluted with fresh water using the formula shown in Fig. 5.3:

Salinity was measured using a hand refractometer.

A thermostat/heater was placed in a vertical position in each jar and used to obtain the range 27, 29, 31, 33, and 35°C.

In the experiment with different colors of light, e.g., white, yellow, green, blue, and red colored through plastic sheets were pasted on the walls of the aquaria (Fig. 5.2). The spore settlement substrata were: mosquito net, Dacron, and gauze (Fig. 5.4). These were attached side by side on one

Fig. 5.2 Calculation of salinity levels

$$Sn = \frac{S1V1 + S2V2}{V1 + V2};$$

Sn : desired water salinity (ppt)
S1 : diluted water Salinity (ppt)
S2 : diluting water Salinity (ppt)
V1 : volume of diluted water (L)
V2 : Volume of diluting water ((L).

Fig. 5.3 Aquaria (10x10x10cm) used to study the effects of light color. Thalli with reproductive structures were suspended in the water

glass sheet. Each of these evaluations were conducted with three triplicates.

The *Kappaphycus alvarezii* obtained from cultivators in the coastal waters of Takalar Regency, South Sulawesi Province, Indonesia. The pre-selected thallus pieces were cut to a minimum length of 5 cm, and each had 2–3 cystocarps (sporangia) (Fig. 5.5). The thalli were cleaned of any sediment by rinsing with fresh sea water and aeration for approximately 3 h before transfer to the experimental containers.

5.2.2 Growth of Reproductive Segments in the Sea

Selected fragments of thalli with cystocarps were inserted in to net material tied to 250 ml mineral water bottles (Fig. 5.6). A bottle filled with sand was placed at each corner of the raft to keep it below the surface of the water. The square raft (60 xn 60 cm) was made of standard 3 inch (7.6 cm) diameter PVC pipe Each raft was connected by nylon ropes to form a row from which the test bottles were suspended (Fig. 5.7).

Every week the development of released spores was observed and photographed. Before taking pictures, we cleaned the surface of the bottle and the net attached to the bottle from any sediment by shaking the bottle. Adhering organisms, such as small fish, shellfish, and small crabs were manually returned to the sea. Thalli with cystocarps were allowed to release spores and grow into seedlings for 16 weeks. Biomass was accumulated in sufficient quantities for 1 clump of seedlings. Morphology was compared with the seedlings used by one of the cultivators and grown together in the fisherman's area.

Fig. 5.4 Three artificial substrata used for assessing spore attachment

5.2.3 Cultivation (Planting) Seeds From Spores

Although in minimal quantities, the seeds produced from spore development in the sea were included with the cultivation of the same species carried out by one of the cultivators at the experimental location. This was done with the aim of examining the growth rate and performance of these seedlings morphologically.

5.2.4 Observations

Spores were released which adhered to the glass slides. These were observed under a microscope at 100–400X and photographed using a digital camera. Observations of the spore release started on day one after placement of the cystocarpic segments. In the following days, observations were made of spore development into juvenile thalli. Some young thalli remained attached to the glass slides, and some were attached to the petri dish due to movement by aeration.

After microscopic observation each glass slide with adherent spores was returned to the petri dish.

The success of the germinating spores was expressed as fresh weight. The young thalli were washed off the slides spraying water on the surface. The wash water was run through filter paper in a glass funnel. The thalli were retained on the filter paper. Fresh weight of the thalli was measured with a Sartorius analytical balance.

Data on the release and development of spores were analyzed descriptively.

5.3 Results and Discussion

The thalli used turned out to be carposporophytes and tetrasporophytes. The release of spores occurred both from sporangia (spore sacs) and from cystocarps (the carposporophyte), which matured and then peeled and burst. Spores were released in the form of tetraspores and carpospores. Tetraspores, present in (produced by) tetrasporangia, were haploid (1*n*) spores produced by the tetrasporophyte (diploid) phase as a result of meiosis. Carpospores are diploid (2*n*) spores (the result of mating between spermatia and egg cells that come out of the carpogonia), which under favorable environmental conditions, grow into diploid plants (Lee 1974). Only at

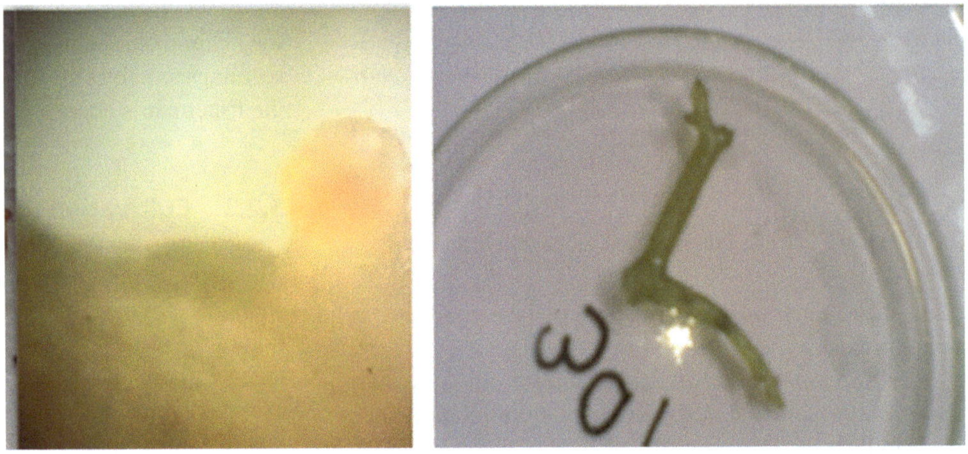

Fig. 5.5 Thalli of *Kapphycus alvarezii* with cystocarps

optimal salinity and temperature do spores develop into plantlets and then into young thalli.

5.3.1 Release and Development of Spores at Different Salinities

Tetraspores were released at salinities of 20, 25, 30, 35, and 40 ppt. Susanto and Pramesti (2001) stated that red algae release the most spores at 30 ppt salinity, released on day 20. At 20 and 25 ppt salinity, the release of spores was slower on day 19 to 20, respectively. At a salinity of 35 ppt, the spores also developed into plantlets on day 13, faster than other salinity treatments, and with thicker branch diameters. De Miranda et al. (2012) noted that the maximum growth of spores from the red alga *Hydropuntia caudata* occurred at a salinity of 35 ppt, and the development was incomplete and very slow at low salinity. Low salinity can cause morphological abnormalities of the spore cell wall, including thinness. Under normal conditions of seawater above normal (40 ppt), the released spores did not germinate into plantlets until the end of the study. Young thalli began to form on day 24 at a salinity of 30–35 ppt; these thalli developed with more branches which were thicker and longer.

Fig. 5.6 Reproductive thalli held between the water bottle and the net

5.3.2 Spores Release and Development at Different Temperatures

At all water temperatures tested, i.e., 27 to 35 °C, the spores were released. Temperatures of 33–35 °C caused the spores to release relatively quickly, which occurred on the first day of observation. The release and development of spores could be delayed due to sub-optimal temperatures (Orduna-Lonas and Robledo 1999) namely, 27–29 °C. De Miranda (2012) noted the death of spores at high temperatures (35 °C) in the subtropics and the slow development of the red alga *H. caudata* at low temperatures. The optimal temperature for the release and development of spores of *K. alvarezii* in this study was quite high, namely, 31–35 °C. The temperature of 35 °C was the best temperature for the development of spores into plantlets. Ramlov et al. (2012) noted that *Gracilaria domingensis* (Kütz.) spores did not develop at suboptimal temperatures (15–30 °C).

5.3.3 Spore Release and Development at Different Light Colors and Substrates

Photosynthetically active radiation (PAR) has high energy and is effectively absorbed by cells (Salisbury and Rose 1969). Tetraspores and carpospores were released on day 5 from the cystocarps exposed to all the tested colors (white, yellow, green, blue, and red). Five days later, the spores developed into plantlets with white, yellow, blue, and red light, none under green light. More plantlets were

Fig. 5.7 Bottles with cystocarp-bearing thalli hanging at each the corner of each (60 x 60 cm) raft. Rafts connected in lines

formed under red light. Despite their low energy, yellow and red light can also be absorbed by the photosynthetic pigments of this seaweed species, causing the release of spores and their subsequent development into plantlets.

In terms of the influence of some artificial substrata materials, Lobban and Harrison (1994) stated that the substratum is one of the determining factors in the attachment of spores to be able to successfully grow into thalli. More spores attached and developed into young thallii on the Dacron, which ranged from 12.2 to 13 g wet, followed by mosquito net as much as 4.7–5.2 g wet, and the lowest was on the Kasin gauze at 2.9–3.4 g wet. This indicated that the spores of *K. alvarezii* adhered well to the fibrous or thick substrata with pores. Yudiati et al. (2004) also succeeded in growing spores to young thalli by using a cotton substratum, which was close to the texture of Dacron. The least spores stuck to the gauze because it is made of a material that is easily destroyed, so that the attached thalli were easily detached along with the threads that made up the gauze.

5.3.4 Experiment on Development of Spore to Thallus Seeds in the Sea

The growth of spores into thalli in the sea occured eight weeks after the initial attachment of the thalli containing cystocarps. Herbivorous fish, crabs, and shellfish found on the netting may have preyed on some seaweed thalli that grew from spores that were released and attached to the net fibers. These animals are suspected to be the cause of the lack of seeds (young thalli) of *Kappaphycus* obtained during this time period. The thalli grew from the spores released from the cystocarps and were found between the mesh and attached the net and bottles. Here, the thalli have grown clearly with visible branches (Fig. 5.8).

Four weeks after manual removal of the attached pests, the thalli began to grow out between the bottles and nets, growing thicker, with more branches, averaging only 0.5 kg per raft (Fig. 5.9).

After a further 4 weeks, the thalli grew thicker (with more and longer branches) and with a larger diameter, (Fig. 5.10), characteristic of good quality *Kappaphycus* seedlings.

The thalli with thick branches were considered to be suitable seedlings, which had reached approximately 50 g. They were removed from the artificial substratum and then grown in the commercial seaweed farm of the fishermen at the research location.

5.3.5 Cultivation (Planting) Seeds from Spores

After 6 weeks of cocultivation carried out on the farm by fishermen, the thalli grown from spores were clearly superior to those the fishermen had grown from vegetative thallus cuttings. *Kappaphycus* seedlings from spores showed a good quality with a bright color with lush branches (Trono 1974). The thalli grown experimentally looked improved morphologically (Fig. 5.11) as compared to thalli which were grown from vegetative thallus fragmentation, i.e., the normal practice used by the fishermen. Such seed characteristics will have an impact on increasing the production and quality of *K. alvarezii*. The good quality seedlings were then used by the fishermen.

Figure 5.12 shows images recorded from several parts of reproduction of *Kappaphycus alvarezii*.

Asexual reproduction in the form of cuttings from the seaweed thallus will result in a wound healing response

Fig. 5.8 Thalli growing sticking through the meshes covered in fine sediment

Fig. 5.9 Thalli with lush branching after 16 weeks of growth

Fig. 5.10 The thalli were supple, flexible, with shiny and bright with pointed tips. These are the characteristics of good quality *Kappaphycus alvarezii* seedlings

Fig. 5.11 *Kappaphycus alvarezii* harvested from spore-produced seedlings (**a**) thick and darker color, (**b**) seedling produced from vegetative thallus cuttings

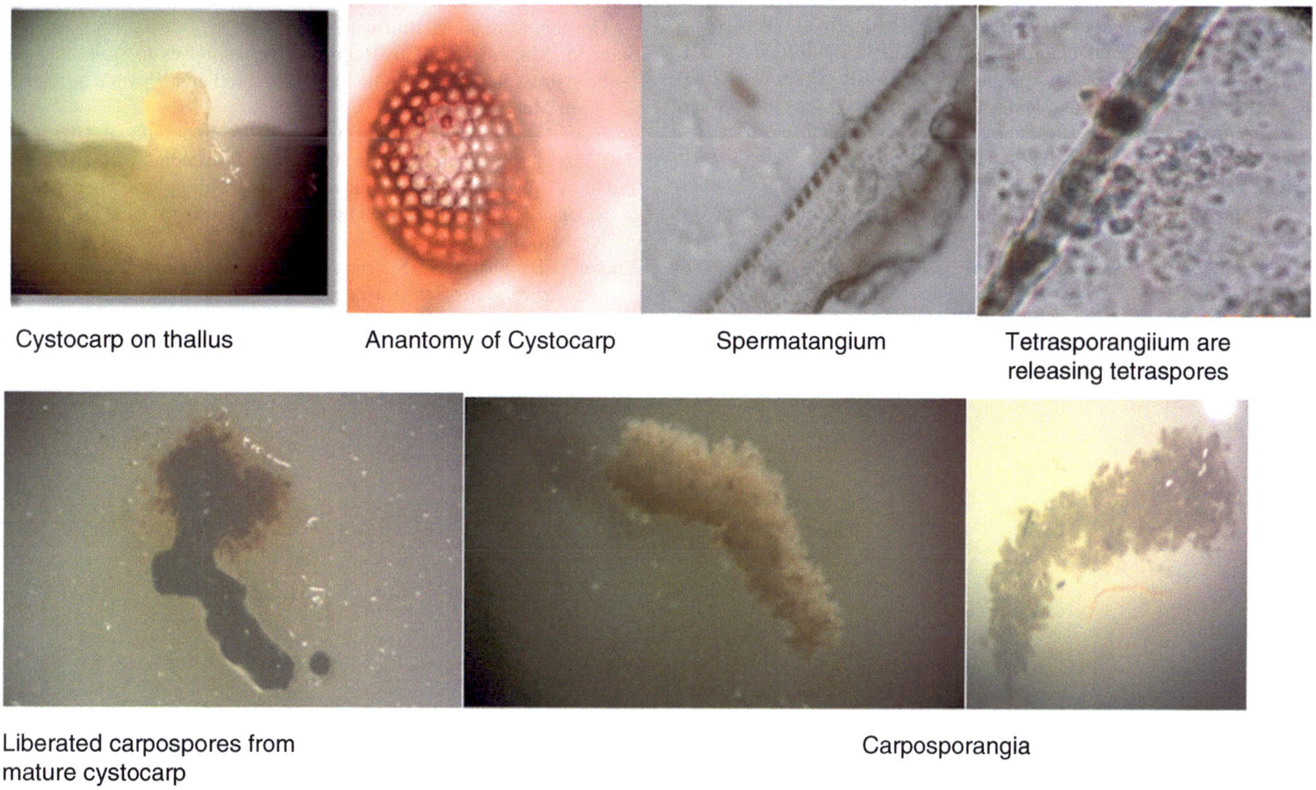

Fig. 5.12 Some pictures recorded from several parts of the sexual reproduction phase of *Kappaphycus alvarezii*, including the shape of spores

where the cuttings are taken (Suda and Mikami 2020). The yield of *K. alvarezii* from the use of reproductive fragments can be used to produce many young thalli for use in the following growing seasons. Seedling production from spores can be used effectively when there is an indication that the quality of the stock has begun to decline. The use of spores (tetraspores and/or carpospores) can provide real cultivation benefits for another red alga such as *Gracilaria dura* (Mantri et al. 2009). Cultivation techniques derived from spores can produce seaweed stocks with greater adaptability to environmental variations and ensure a sustainable seedling supply (Halling et al. 2005). In another example, tetrasporophytes cultured in the laboratory showed rapid adaptation to the marine environment with a satisfactory average growth rate of 4.7% d^{-1} (Mantri et al. 2009).

Gametophyte thalli and the stimulation of spore release is an important reproductive strategy for the cultivation of the red alga *Pyropia yezoensis*. Stimulating changes in the life cycle from vegetative growth to spore release is one strategy discovered for the successful, commercial farming of this alga (for nori) at scale (Suda and Mikami 2020). Cultivation of carpospores from the agarophyte, red alga *Gracilaria* has proven to be a more efficient method when compared to the use of vegetative thalli (Glenn et al.

1996). Further studies are required on Indonesian *Kappaphycus alvarezii* cultivation. As evidenced with the materials used for the observations in this chapter, the presence of both carpo- and tetrasporophyte generations in the Indonesian farms needs to be elucidated in order to perfect the use of isolated generations to generate tetraspores and carpospores for cultivation of the respective phases of new and improved strains and cultivars of *Kappaphycus alvarezii.*

References

De Miranda GEC, Yokoya NS, Fujii MT (2012) Effects of temperature, salinity and irradiance on carposporeling development of *Hydropuntia caudata* (Gracilariales, Rhodophyta). Rev. bras. farmacogn. vol 22 no.4 Curitiba July/Aug. 2012 Epub June 14, 2012

Forbord SK, Steinhovden B, Solvang T, Handå AR, Skjermo J (2020) Effect of seeding methods and hatchery periods on sea cultivation of *Saccharina latissima* (Phaeophyceae): a Norwegian case study. J Appl Phycol 32:2201–2212

Glenn EP, Moore D, Fitzsimmons K, Azevedo C (1996) Spore culture of the edible red seaweed, *Gracilaria parvispora* (Rhodophyta). Aquaculture 42:59–74

Halling C, Aroca G, Cifuentes M (2005) Comparison of spore inoculated and vegetative propagated cultivation methods of *Gracilaria chilensis* in an integrated seaweed and fish cage culture. Aquac Int 13:409–422

Kostamo K (2008) The life cycle and genetic structure of the red alga *Furcellaria lumbricalis* on a salinity gradient. Sci. Rep 33:1–34

Lee E (1974) Phycology. Cambridge University Press, 478 p

Lobban CS, Harrison PJ (1994) Seaweed ecology and physiology. Cambridge University Press, 366 pp

Mantri AV, Thakur MC, Kumar M, BhavanathJha CRKR (2009) The carpospore culture of industrially important red alga *Gracilaria dura* (Gracilariales, Rhodophyta). Aquaculture 297(1–4)

Orduna-Lonas, Robledo D (1999) Effects of irradiance and temperature on the release and growth of carpospores from *Gracilaria cornea.* J Agardh (Gracilariales, Rhodophyta). Bot Mar. 42:315-319

Ramlov F, de Souza JMC, Farias A, Maraschin M, Horta PA, YokoyaNS. (2012) Effects of temperature, salinity, irradiance, and nutrients on the development of carposporelings and tetrasporophytes in *Gracilaria domingensis* (Kütz.) Sonder ex Dickie (Rhodophyta, Gracilariales). Bot Mar 55:253–259

Salisbury FK, Rose C (1969) Plant physiology. Wadsworth Publ. Co. Inc. Belman. California, 764 p

Suda M, Mikami K (2020) Reproductive responses to wounding and heat stress in Gametophytic thalli of the red alga *Pyropia yezoensis.* Front Mar Sci 7:394

Susanto AB, Pramesti R (2001) Carpospore liberation of *Gracilaria gigas* Harvey from Lombok, Indonesia. Proceeding ISSM-ISTEC

Trono GC (1974) *Eucheuma* farming in The Philippines. U. P. Natural Science Research Centre, Quezon City, Philippines. Uan, J. 1990, Kiribati, pp. 10–15

Yudiati EE, Susilo S, Suryono CA (2004) Teknik Setting Spora *Gracilaria gigas* Sebagai Penyedia Benih Unggul dalam Budidaya Rumput Laut. UNDIP Semarang Ilmu Kelautan 9(1):37–40

Evaluation of a Low-Cost Prototype for Micropropagation of *Kappaphycus alvarezii* and Its Application

6

Thilaga Sethuraman, Mahalingam Selvakumar, Shanmugam Munisamy, and Doss Ganesh

Abstract

Segments of *Kappaphycus* thalli are generally cultured in liquid medium using conical flasks of different sizes under continuous agitation. This method facilitates the initial growth and produces new thalli of less than 0.5 cm within 15–20 days. However, the growth rate of thalli is very limited despite the use of several plant growth regulators, including AMPEP (*Ascophyllum* Marine Plant Extract Powder). To enhance the thallus growth rate, a simple and low-cost prototype culture tank was developed to simulate the condition of the sea for culture and maintenance of *Kappaphycus* thalli for extended times under laboratory conditions. This prototype consists of a low-cost rectangular glass tank connected to an aerator for facilitating water movement and aeration. The tank has the provision to drain the used liquid media from the bottom of the tank. This prototype has an adjustable elastic beading around the inner side of the tank wall that can be set to provide the desired water level, with a provision to tie any number of tiny nylon ropes to stably hang the microcuttings of the thalli in the tank. The water movement in the liquid culture is similar to sea waves, and since the microcuttings are hung stably at a desirable depth, the growth rate is much faster compared to conventional methods. The application of this prototype for various studies related to micropropagation and genetic improvement of *Kappaphycus* is discussed in this chapter.

T. Sethuraman · M. Selvakumar · D. Ganesh (✉)
Department of Plant Biotechnology, School of Biotechnology, Madurai Kamaraj University, Madurai, Tamil Nadu, India
e-mail: dossganesh@mkuniversity.ac.in

S. Munisamy
Research and Development Division (DSIR-Lab), AquAgri Processing Private Limited, SIPCOT Industrial Complex, Manamadurai, Tamil Nadu, India

Department of Biomarine Resource Valorisation, Division of Food Production and Society Norwegian Institute of Bioeconomy Research (NIBIO) Torggarden, Bodo, Norway

Keywords

Kappaphycus alvarezii · Eucheumatoid · Micropropagation · Low-cost prototype · Farming

6.1 Introduction

Kappaphycus, a genus of red seaweed, has tremendous potential for commercial applications, and it is cultivated extensively in the Southeast Asian region, particularly in the Philippines, Indonesia, and Malaysia and to some extent in Vietnam and Cambodia (Hayashi et al. 2017) and recently in Sri Lanka (Shanmugam et al. 2017). Additionally, India is one of the countries thriving to produce an adequate quantity of *Kappaphycus* to meet ever-increasing domestic needs, as well as contributing to global seaweed markets. It is estimated that the commercial production of *Kappaphycus* is expected to increase due to its ever-growing demands in the global market, as this seaweed is recognized as one of the key sources of carrageenan, which is used as an additive in processed food, dairy products, water gels, cosmetics, and pharmaceuticals (Campbell and Hotchkiss 2017). To enhance production, the two approaches taken are the production of high-quality seed materials for farming and the improvement of existing genetic stocks for enhanced phyconomic traits related to high biomass and tolerance to several biotic and abiotic stresses. The present crisis in *Kappaphycus* production in several seaweed-producing countries is attributed to continuous exploitation of the vegetative propagation of *Kappaphycus* and *Eucheuma* from limited genetic resources since the introduction of commercial farming in 1970 (Doty 1973; Doty and Alvarez 1981). The loss of genetic and vegetative vigor in addition to susceptibility to "ice-ice" and epi-endophytes has become a major issue in commercial farming of *Kappaphycus*.

To develop viable strategies for the propagation of elite genotypes of *Kappaphycus*, several attempts were made to regenerate plantlets from the diploid phase of spores (Azanza

A. T. Critchley et al. (eds.), *Tropical Phyconomy Coalition Development*, Developments in Applied Phycology 11,
https://doi.org/10.1007/978-3-031-47806-2_6

and Aliaza 1999; Azanza and Ask 2003; Luhan and Sollesta 2010). However, this approach was not viable primarily because of the seasonality and scarcity of fertile plants bearing spores from wild genotypes. Moreover, it was realized that the development of sporelings from *Kappaphycus* was tedious, owing to their slow germination rate under controlled conditions. Therefore, micropropagation has become a viable method to facilitate the availability of elite genotypes as seed materials for distribution to farms. A number of reports on the micropropagation of *Kappaphycus* and *Eucheuma* revealed the potential application of tissue culture techniques to generate a large number of clones of elite genotypes of *Kappaphycus* (Reddy et al. 2003; Hurtado and Biter 2007; Hayashi et al. 2008; Sulistiani et al. 2012; Yong et al. 2014; Neves et al. 2015; Luhan and Mateo 2017; Budiyanto et al. 2019; Hurtado and Critchley 2019).

A major development in the micropropagation of *Kappaphycus* is the use of a commercial seaweed supplement from *Ascophyllum nodosum*, popularly named AMPEP (*Ascophyllum* Marine Plant Extract Powder). AMPEP enhances the micropropagation of *Kappaphycus* and the subsequent development of new thalli (Hurtado et al. 2009; Yunque et al. 2011; Tibubos et al. 2017; Ali et al. 2018; Febriyanti et al. 2019; Umanzor et al. 2019; Souza et al. 2019). The benefits of AMPEP have been well studied using an *Arabidopsis* model system (Wally et al. 2013), where it has been shown to stimulate growth by augmenting the biosynthesis and accumulation of plant hormones, including cytokinin, which regulates growth. While the use of AMPEP is a breakthrough, the micropropagation of *Kappaphycus* still needs to be improved to allow for commercially viable seed production and to extend the technique to several other applications in the context of genetic improvement. Toward this objective, a simple prototype culture system was developed to improve culture conditions to further enhance thallus growth. Here, we describe the system and discuss its possible utility and implications in *Kappaphycus* research. This was previously done at the eucheumatoid seaweeds biology webinar of July 7–8, 2021 (see the audio-visual references section of the present document).

6.2 Materials and Methods

6.2.1 Sample Collection and Processing

Young *Kappaphycus* thalli at the vigorous growth phase under field conditions were collected from a commercial farm of AquAgri Processing Private Ltd. Rameswaram, Tamil Nadu. Actively growing segments measuring approximately 6 cm were cut and washed with povidone iodine (1%). The sand particles attached on the surface of the thallus were gently removed under the povidone iodide solution with a hair-brush and repeatedly washed with sterile water for the experiment.

6.2.2 Pre-acclimatize

Segments of thalli were transferred to filtered seawater in a 5 L conical flask (Borosil, India), and they were gently agitated by connecting the culture flask to an aerator. The acclimatization process was continued for 3 days under a 12 h photoperiod at 23 ± 2 °C. Only healthy thalli were recovered after 3 days of preacclimatization and used further for experiments. Approximately 500 microcuttings measuring 8–10 mm were cut and cultured in filtered seawater in a 2 L volume flask containing AMPEP (3 mg L^{-1}), IAA (3-indole acetic acid), and kinetin (0.01 mg L^{-1} each). This served as a control for comparison with the prototype culture system.

6.2.3 Preparation of the Prototype

The prototype consisted of a rectangular glass tank with an inlet at one end to connect with an aerator to facilitate water movement within the tank. The tank had the provision to drain the used liquid media from the bottom of the tank. The prototype had adjustable elastic beading around the inner wall of the glass tank to fix a glass rod for hanging thallus segments at the desired depth of seawater (Fig. 6.1).

6.2.4 Culture of the Microcuttings in the Prototype Tank

Healthy thallus segments, initially acclimatized to seawater in flask culture, were recovered and made into smaller microcuttings measuring 8–10 mm, and these segments were tied with a tiny nylon thread and hung from the glass rod into seawater containing AMPEP (3 mg L^{-1}) and IAA and kinetin (0.01 mg L^{-1} each). Seawater was aerated continuously to provide gentle agitation. Fresh seawater was added once a week for maintenance of cultures during the experiment.

6.2.5 Evaluation of the Prototype

Maintenance of thalli was continued for 30 days to observe the response of the tissues. To compare thallus growth with regular flask culture, thalli of similar nature and size were cultured in an aerated flask as a control (Fig. 6.2), and at the

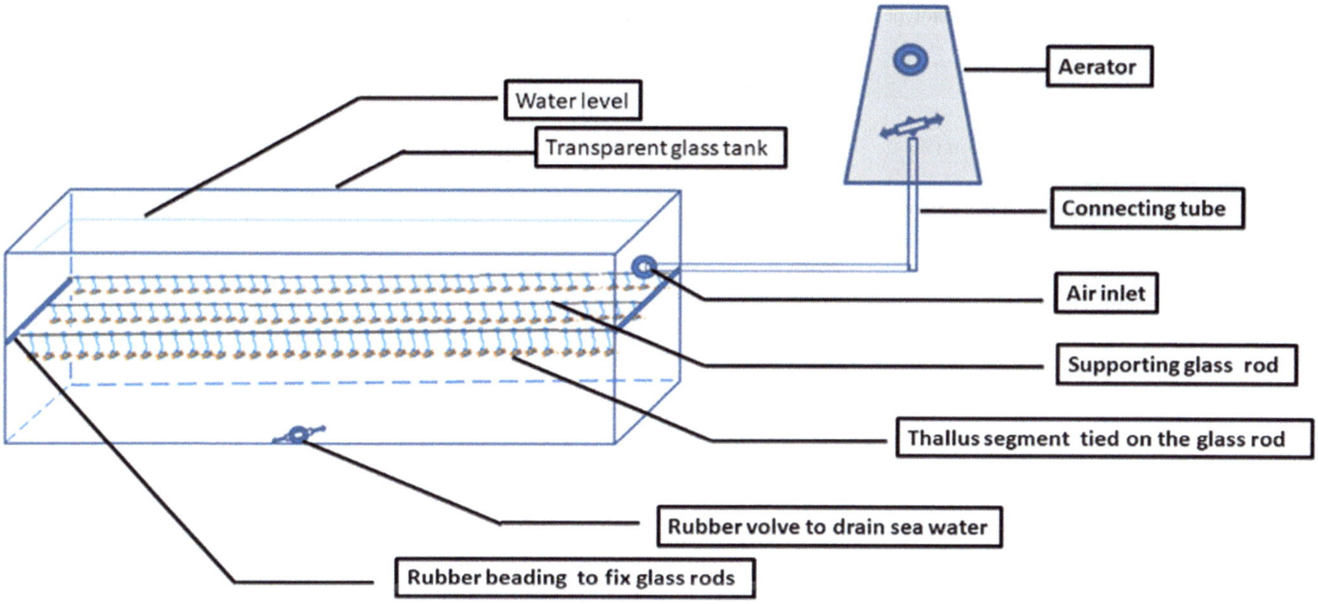

Fig. 6.1 Design of prototype tank with a provision to culture the microcuttings of *Kappaphycus* thalli under seawater in the laboratory, connected with aerator for continuous and long-term culture

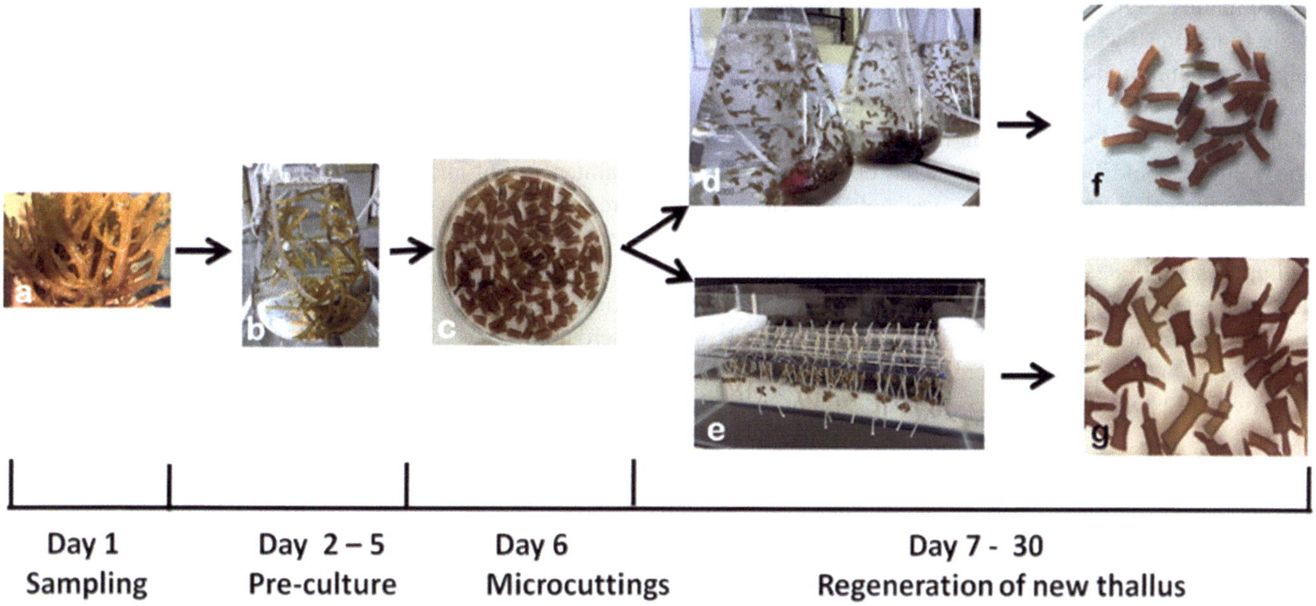

Fig. 6.2 Regeneration of new thalli of *Kappaphycus alvarezii* from flask and prototype culture systems. Comparison of growth: (**a**) sampling (day 1); (**b**) preculture of larger segments of *Kappaphycus* for acclimatization; (**c**), preparation of microcuttings; (**d** and **e**) maintenance of cultures in flask and prototype culture systems; (**f**) microcuttings of *Kappaphycus* showing the emerging new thalli; (**g**) microcuttings cultured in prototype tank showing improved thallus growth with branching

end of 30 days, the survival rate, response of culture, and development of new thalli were observed and documented. In another experiment, the prototype was evaluated for effectiveness in carrying out long-term culture by retaining the culture for 120 days. At the end of the experiment, the growth response was observed and documented.

6.2.6 Growth of Regenerated Thalli Under Field Conditions

Thalli regenerated using the prototype were transferred after 30 days to a small 2 × 2 m raft. The entire raft was covered with nylon mesh to prevent the escape of sprouting thalli. The

Table 6.1 Comparison of flask and prototype culture systems for the growth and development of microcuttings of *Kappaphycus* cultured for 30 days

Parameters	Flask culture	Miniature tank culture
Survival (%)	40	82
Pigment retention without discoloration (%)	32	82
Days taken for discoloration	03	20
Days taken for thickening of medullary tissues	14	06
Days taken for initiation of meristem growth	14	08
Percentage of responding microcutting	30	72
Length of newly formed thallus	3–4 mm	1.3–20.4 mm
No. of new thallus per/microcutting	01	2–4
Length of longest new thallus	4 mm	20.5 mm

initial weight at the end of the prototype culture was compared with the final biomass at the end of 45 days of raft culture under field conditions.

6.3 Results

6.3.1 Growth Performance

Large thallus segments, preacclimatized for an initial 3 days, were found fresh without any discoloration, as evidenced by pigment retention in more than 95% of the thallus segments. Loss of tissues due to discoloration followed by death of the thalli was commonly observed in flask culture. Only 40% of the cultures could be recovered at the end of 15 days of culture; thereafter, no mortality was noticed (Table 6.1). Interestingly, survival increased significantly in the prototype culture system (=82%). It was observed that discoloration or loss of pigmentation rapidly occurred in flask culture compared to tank culture during the initial culture period. The percentage of responding microcuttings was significantly improved (72%) in the prototype culture system compared to flask culture (only 30%). The length of newly formed thalli in flask culture was observed to be 3–4 mm at the end of 30 days of culture. However, microcuttings cultured on the prototype culture system grew significantly more, and the new thalli reached a mean length of 20.5 mm. The number of new thalli per microcutting ranged 2–4 in the prototype culture compared to only one new thallus per microcutting in the flask culture. These observations clearly showed that the prototype culture system increases thallus growth compared to flask culture (Fig. 6.2).

6.3.2 Field Acclimation

Maintenance of microcuttings under flask and prototype culture systems allowed the differential growth and development of thalli to proceed slowly with an increase in biomass to 1.8 and 3 g, respectively (Fig. 6.3). A rapid increase in biomass from 3 g to 23 g could be observed only when the microcuttings were transferred from the prototype culture system to field conditions. Microcuttings developed further and produced new branches of varying lengths under field

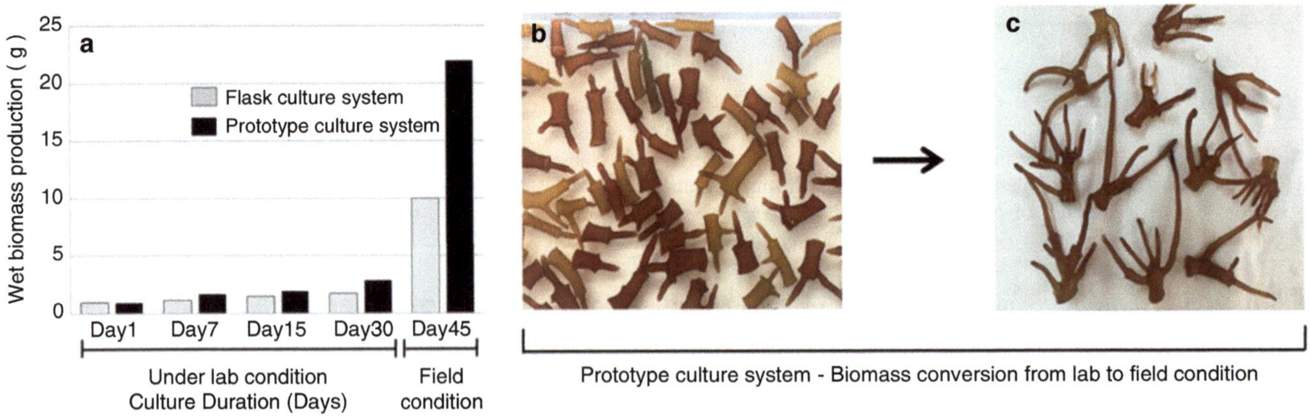

Fig. 6.3 Biomass production and field acclimatization. (**a**) Graphical representation showing improved biomass production in the prototype culture system; (**b**) healthy microcuttings with newly proliferating thalli from the prototype culture system; (**c**) conversion of sprouting thalli into healthy growing seeds ready for transfer into rafts for biomass production/seed multiplication

Fig. 6.4 Field maintenance and conversion of proliferated thallus obtained from prototype culture for biomass production: (**a**) Initial culture of thalli in small 2 × 2 m rafts; (**b**) healthy growth of thallus clumps; (**c**) closer view of thalli showing healthy development; (**d**) further multiplication in larger rafts

conditions. We found that the prototype culture system is superior to flask culture, inducing faster growth and more biomass during lab culture and after transfer to field conditions.

6.3.3 Conversion of Acclimatized Thallus Into Higher Biomass

Acclimatized thalli in a small raft (2 × 2 m size) were grown in the sea for approximately 45 days. Initially, the thalli were inserted into the nylon rope (Fig. 6.4). Acclimatized thalli grew well, and each thallus segment developed into a larger clump with a sign of vigorous and continuous vegetative growth. These clumps were recovered from the 2 × 2 m raft, broken up into smaller clumps, and transferred to a larger raft for further multiplication. We observed that the rate of multiplication was similar to that of the normal multiplication method (Fig. 6.4).

6.4 Discussion

Cultivation of *Kappaphycus* on a commercial scale is one of the major activities pursued in several seaweed-growing countries. Maintenance of elite genotypes of *Kappaphycus* under in vitro conditions in addition to continuous multiplication through micropropagation offers the scope for availability of authenticated seedlings throughout the year rather than depending upon an unauthenticated source of seedlings from unreliable sources. In addition, seedlings developed from sexual spores under natural conditions often encounter a high degree of genetic variation and interfere with genetic stability and yield (Dawes and Koch 1991; Dawes et al. 1993; Hurtado and Cheney 2003; Hurtado and Biter 2007; Hayashi et al. 2008; Sulistiani et al. 2012; Yong et al. 2014; Luhan and Mateo 2017). Micropropagation is the most applicable strategy in *Kappaphycus* cultivation, not only for the production of high-quality seed materials but also for maintaining the

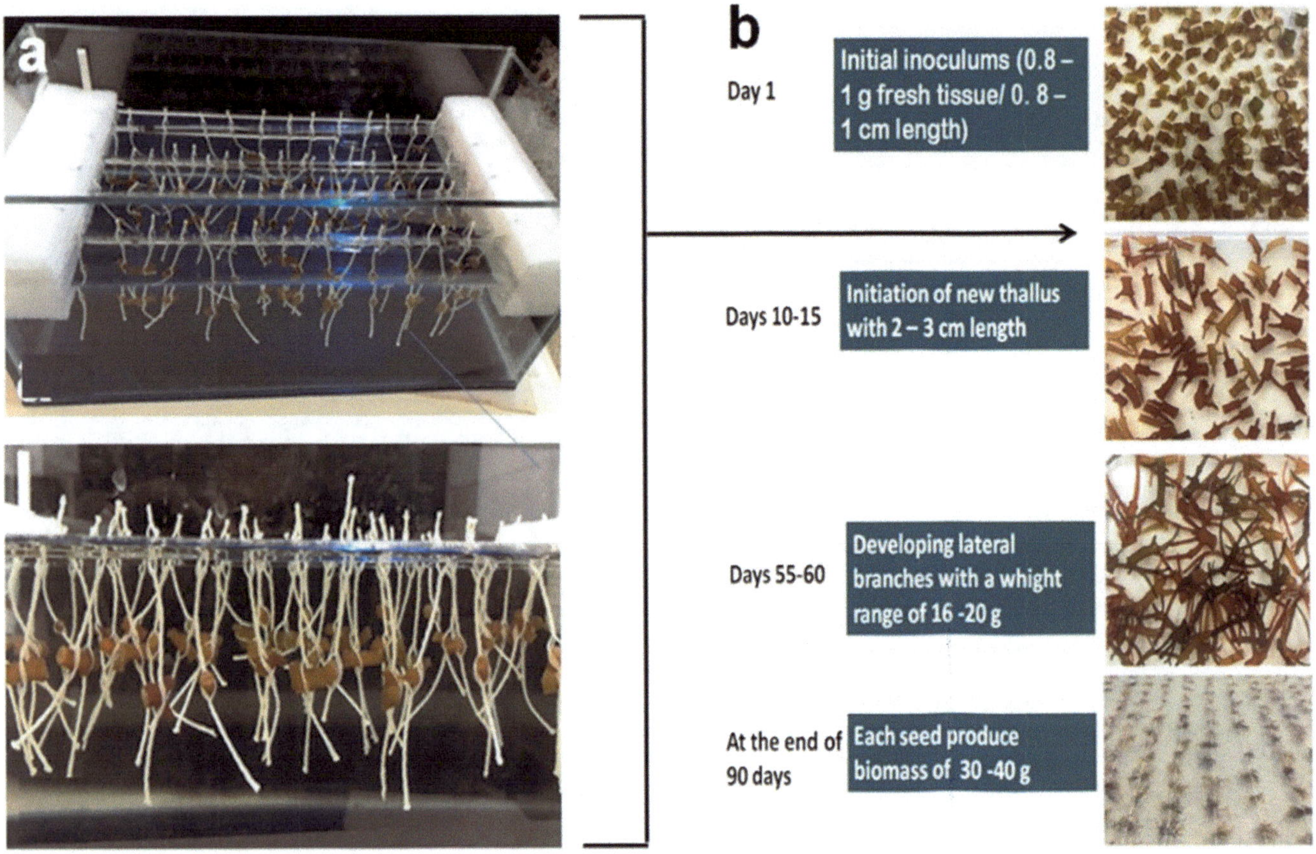

Fig. 6.5 Sequential phases of micropropagation with time for the production of biomass using the prototype culture system

genetic fidelity of propagules for commercial cultivation (Sulistiani et al. 2012; Yong et al. 2014). Most of the research work on the micropropagation of *Kappaphycus* used flask culture in seawater medium containing various plant growth hormones (Hurtado et al. 2009; Yunque et al. 2011). To improve the growth and development of thalli during micropropagation, the *A. nodosum* extract AMPEP has been used in a number of instances and proven to be effective for enhancing the growth of *Kappaphycus* (Hurtado et al. 2009; Yunque et al. 2011). This unique natural biostimulant is effective when used singly or in combination with IAA or kinetin.

Despite the use of various plant growth regulators, including AMPEP, the rate of thallus growth, particularly during the initial phase of micropropagation, is still limited, with insignificant conversion of biomass for commercial application. For example, Hurtado et al. (2009) performed an extensive study with the aim of enhancing the thallus growth of four strains of *Kappaphycus*, namely, Kapilaran, Tambalang, Adik-Adik, and sacol green, using various growth regulators in combination with AMPEP. Each variety responded differently, and the number of days taken for shoot formation in the presence of AMPEP and plant growth regulator (PGR) varied from 17–78 days, depending upon the strain. However, the biomass conversion from the day of initial culture to the end of post culture needs improvement for commercial application. In another example, Yunque et al. (2011) used eight different concentrations of AMPEP (0.001–5 mg L^{-1}), along with PGR, to enhance thallus growth in four different varieties of *K. alvarezii* and one variety of *K. striatum*. It was reported that the development of shoot primordia and the average length of newly formed thalli developed in various concentrations of AMPEP with and without PGRs was 3–8 mm despite 3 weeks of culture.

This prototype is highly useful for enhancing the growth of thalli from 1 cm to 2–3 cm within 15 days of culture. This finding is significant compared to previous reports, where Yunque et al. (2011) reported an increase in thallus length of only 8 mm at the end of 19 days using flask culture. In this study, recovery of sprouting thalli at the end of 15 days and subsequent transfer of thalli to small rafts led to profuse growth of thalli in the sea with many lateral branches of varying lengths. The profuse growth of thalli could be attributed to the influences of PGR and AMPEP during culture in the prototype tank, in addition to the natural availability of nutrients in sea water. The result is highly

Table 6.2 Comparison of certain parameters between flask culture and the prototype culture system

Parameter/ advantages	Flask culture	Prototype culture system
Explant movement	Circulatory: because all the explants are freely movable due to the placement of air ball at the bottom of the culture flask	Stagnant: since each microcutting is tied with tiny nylon thread on the glass support, movement is strictly restricted
Light intensity	Difficulty to provide uniform light intensity as the explants are circulated	Uniform light intensity can be provided, since the explants are stably hanged at a constant depth
Cell/explant damage	Tissue/explant damage often occurs due to vigorous movement/over crowed, and surface of the culture flask can damage the meristem	Normally cells are not damaged as they are individually hanged and are separated each other to facilitate gentle waves similar to that of sea
Tagging of treatments	Not possible in a single flask as they are mixed	Since it has provision to tie individual plant on a glass support, many pretreatments can be imposed in a single prototype culture tank
Replacement with fresh media	Need to take out the entire explants for replacing with fresh media	Used media can be drained without disturbing the experimental setup
Observation	Very difficult to track the observation of individual segments in response to the imposed treatment	Each segment can be tagged/labeled for observation

encouraging in the context of commercial-scale production of elite *Kappaphycus* seeds, as this method could yield 30–40 g of fresh biomass at the end of 90 days (Fig. 6.5).

The prototype culture system can be efficiently used for culture of microcuttings of *Kappaphycus* by stable immersion in sea water. By hanging the segments from a glass rod, this system provided constant light intensity to each of the segments at a fixed depth. In addition, this prototype has several other advantages (Table 6.2).

6.5 Conclusion

Micropropagation is highly reliable for seedling production of *Kappaphycus*, considering the fact that the thallus has natural regeneration potential. Natural propagation of *Kappaphycus* through simple fragmentation could be one of the adaptations to maintain its existence in nature, since its sexual reproduction through spore formation is very limited. In vitro methods of propagation have been applied to a wide range of applications in the context of genetic improvement to overcome issues related to biotic and abiotic stresses. This approach requires systematic experiments with long-term culture maintenance under laboratory conditions. In our study, microcuttings of *Kappaphycus* thalli could be successfully maintained in the prototype tank for 120 days without affecting growth during the induction of polyploidy using colchicine (unpublished data). Therefore, the prototype developed for the micropropagation of *Kappaphycus* is expected to be useful for several applications, and further improvement of this prototype culture setup is underway.

Acknowledgments We thank Mr. Abhiram Seth, Managing Director, AquAgri Processing Pvt Ltd, SIPCOT Industrial Complex, Manamadurai—630606, Tamil Nadu, India, for his constant encouragement during this study.

Declarations All authors declare that there are no conflicts of interest.

Competing Interests The authors have no competing interests.
Ethics approval not applicable.

Availability of Data The datasets generated and/or analyzed during the current study are available from the corresponding author on reasonable request.

Author Contributions Conceptualization: [Doss Ganesh]; Methodology: [Doss Ganesh, Thilaga Sethuraman]; Formal analysis and investigation: [Thilaga Sethuraman, Mahalingam Selvakumar]; Writing—original draft preparation: [Thilaga Sethuraman, Doss Ganesh]; Writing—review and editing: [Doss Ganesh]; Funding acquisition: [Shanmugam Munisamy]; Resources: [Shanmugam Munisamy]; Supervision: [Doss Ganesh].

Funding This work was supported by research funding from AquAgri Processing Private Ltd., Manamadurai, Tamil Nadu, India. Part of this research was funded by RUSA (Rashtriya Uchchatar Shiksha Abhiyan), Madurai Kamaraj University, and DST-PURSE (Department of Science and Technology-Promotion of University Research and Scientific Excellence) under a Phase II program of Madurai Kamaraj University.

References

Ali MKM, Yasir SM, Critchley AT, Hurtado AQ (2018) Impacts of *Ascophyllum* marine plant extract powder (AMPEP) on the growth, incidence of the endophyte *Neosiphonia apiculata* and associated carrageenan quality of three commercial cultivars of *Kappaphycus*. J Appl Phycol 30:1185–1195

Azanza RV, Aliaza T (1999) In vitro carpospore release and germination in *Kappaphycus alvarezii* (Doty) Doty from Tawi-Tawi, Philippines. Bot Mar 42:281–284

Azanza RV, Ask E (2003) *Kappaphycus alvarezii* (Doty) Doty carposporeling growth and development in the laboratory. Intl Seaweed Symp 17:95–99

Budiyanto B, Kasim M, Abadi SY (2019) Growth and carrageenan content of local and tissue culture seed of *Kappaphycus alvarezii* cultivated in floating cage. AACL Bioflux 12:167–178

Campbell R, Hotchkiss S (2017) Carrageenan industry market overview. In: Hurtado AQ, Critchley AT, Neish IC (eds) Tropical seaweed farming trends, problems and opportunities: focus on Spinosum and Cottonii of commerce. Springer, Dordrecht, pp 193–206

Dawes C, Koch E (1991) Branch, micropropagule and tissue culture of the red algae *Eucheuma denticulatum* and *Kappaphycus alvarezii* farmed in The Philippines. J Appl Phycol 3:247–257

Dawes CJ, Trono GC Jr, Lluisma AO (1993) Clonal propagation of *Eucheuma denticulatum* and *Kappaphycus alvarezii* for Philippine seaweed farms. Hydrobiologia 260:379–383

Doty MS (1973) Farming the red seaweed, *Eucheuma*, for carrageenans. Micronesica 9:59–73

Doty MS, Alvarez VB (1981) *Eucheuma* farm productivity. Proc Int Seaweed Symp 8:688–691

Febriyanti F, Aslan LOM, Iba W, Patadjai AB, Nurdin AR (2019) Effect of various planting distances on growth and carrageenan yield of *Kappaphycus alvarezii* (Doty) using seedlings produced from mass selection combined with tissue- cultured method. IOP Conf Series: Earth Environ Sci 278:1–8

Hayashi L, Yokoya NS, Kikuchi DM, Oliveira EC (2008) Callus induction and micropropagation improved by colchicines and phyto regulators in *Kappaphycus alvarezii* (Rhodophyta, Solieriaceae). J Appl Phycol 20:653–659

Hayashi L, Reis RP, Alves dos Santos AA, Castelar B, Robledo D, de Vega GB, Msuya FE, Eswaran K, Yasir S, Ali MJ, Hurtado AQ (2017) The cultivation of *Kappaphycus* and *Eucheuma* in tropical and subtropical waters. In: Hurtado AQ, Critchley AT, Neish IC (eds) Tropical seaweed farming trends, problems and opportunities: focus on *Kappaphycus* and *Eucheuma* of commerce. Springer, Dordrecht, pp 55–90

Hurtado AQ, Biter A (2007) Plantlet regeneration of *Kappaphycus alvarezii* var. adik by tissue culture. J Appl Phycol 19:783–786

Hurtado AQ, Cheney DP (2003) Propagule production of *Eucheuma denticulatum* (Burman) Collins et Hervey by tissue culture. Bot Mar 46:338–341

Hurtado AQ, Critchley AT (2019) Recent advances in the use of on-land nurseries for commercial production and out-planting of *Kappaphycus* seedlings, a carrageen-bearing seaweed. Institute of Ocean and Earth Sciences Monograph Series 17: taxonomy of southeast Asian seaweeds III

Hurtado AQ, Yunque DA, Tibubos K, Critchley AT (2009) Use of Acadian marine plant extract powder from *Ascophyllum nodosum* in tissue culture of *Kappaphycus* varieties. J Appl Phycol 21:633–639

Luhan MRJ, Mateo JP (2017) Clonal production of *Kappaphycus alvarezii* (Doty) Doty in vitro. J Appl Phycol 29:2339–2344

Luhan MRJ, Sollesta H (2010) Growing the reproductive cells (carpospores) of the seaweed, *Kappaphycus striatum*, in the laboratory until out-planting in the field and maturation to tetrasporophyte. J Appl Phycol 22:579–585

Neves FAS, Simioni C, Bouzon ZL, Hayashi L (2015) Effects of spindle inhibitors and phytoregulators on the micropropagation of *Kappaphycus alvarezii* (Rhodophyta, Gigartinales). J Appl Phycol 27:437–445

Reddy CRK, Kumar GRK, Siddhanta AK, Tewari A (2003) In vitro somatic embryogenesis and regeneration of somatic embryos from pigmented callus of *Kappaphycus alvarezii* (Doty) Doty (Rhodophyta, Gigartinales). J Phycol 39:610–616

Shanmugam M, Sivaram K, Rajeev E, Pahalawattarachchi V, Chandraratne PN, Asoka JM, Abhiram S (2017) Successful establishment of commercial farming of carragenophyte *Kappaphycus alvarezii* (Doty) in Sri Lank: economics of farming and quality of dry weed. J Appl Phycol 29:3015–3027

Souza JMC, Castro JZ, Critchley AT, Yokoya NS (2019) Physiological responses of the red algae *Gracilaria caudata* (Gracilariales) and *Laurencia catarinensis* (Ceramiales) following treatment with a commercial extract of the brown alga *Ascophyllum nodosum* (AMPEP). J Appl Phycol 31:1883–1888

Sulistiani E, Soelistyowati DT, Alimuddin YSA (2012) Callus induction and filaments regeneration from callus of cottonii seaweed *Kappaphycus alvarezii* (Doty) collected from Natuna Islands, Riau Islands province. Biotropia 19:103–114

Tibubos K, Hurtado AQ, Critchley AT (2017) Direct formation of axes in new plantlets of *Kappaphycus alvarezii* (Doty) Doty, as influenced by the use of AMPEPK+, spindle inhibitors and plant growth hormones. J Appl Phycol 29:2345–2349

Umanzor S, Shin S, Marty-Rivera M, Augyte S, Yarish C, Kim JK (2019) Preliminary assessment on the effects of the commercial seaweed extract, AMPEP, on growth and thermal tolerance of the kelp *Saccharina spp.* from the Northwest Atlantic. J Appl Phycol 31: 3823–3829

Wally OSD, Critchley AT, Hiltz D, Craigie JS, Han X, Zaharia LI, Abrams SR, Prithiviraj B (2013) Regulation of phytohormone biosynthesis and accumulation in *Arabidopsis* following treatment with commercial extract from the marine macroalga *Ascophyllum nodosum*. J Plant Growth Regul 32:324–339

Yong WTL, Ting SH, Yong YS, Thien VY, Wong SH, Chin WL, Rodrigues KF, Anton A (2014) Optimization of culture conditions for the direct regeneration of *Kappaphycus alvarezii* (Rhodophyta, Solieriaceae). J Appl Phycol 26:1597–1606

Yunque DAT, Tibubos KR, Hurtado AQ, Critchley AT (2011) Optimization of culture conditions for tissue culture production of young plantlets of carrageenophyte *Kappaphycus*. J Appl Phycol 23:433–438

The Importance of the Biosecurity Concept for a Resilient Eucheumatoid Aquaculture Industry

7

Cicilia S. B. Kambey, Jonalyn P. Mateo, Sadock B. Rusekwa, Adibi R. M. Nor, Calvyn F. A. Sondak, Iona Campbell, Anicia Q. Hurtado, Flower E. Msuya, Phaik Eem Lim, and Elizabeth J. Cottier-Cook

Abstract

Biosecurity is a concept supported by sustainable health management-based policies that is widely applied to the aquaculture sector for controlling and managing disease, pests, and other relevant risks in the industry. For many years, biosecurity has been applied to aquatic aquaculture management to counter the economic and ecological impact and consequences, and recently, the concept has been adapted to seaweed aquaculture sectors. In this chapter, we propose the concept of a biosecurity strategy for eucheumatoid aquaculture on-farm and at the national level. We also highlight the current use of biosecurity strategies in seaweed regulatory frameworks in the main producing countries. The implementation of current biosecurity strategies on-farm was also evaluated through farmers' knowledge, attitudes, and practices (KAP). National biosecurity strategy showed similar challenges identified from the regulatory frameworks in the main seaweed producing countries, such as unavailability of specific seaweed aquaculture policy/regulations, unclear national seaweed health management systems, lack of biosecurity applications used in the national seaweed frameworks, and poor support systems to accommodate seaweed aquaculture health management. The results of on-farm biosecurity implementation identified insufficient biosecurity knowledge and resources of seaweed farmers among the producing countries. Various strategies are recommended to overcome biosecurity challenges based on national capability.

Keywords

Eucheumatoids · Biosecurity · Biosecurity policy and regulation · Farmers' practices · Seaweed aquaculture health management

7.1 Introduction

Health management in the aquaculture sector must be applied and addressed for sustainability purposes. The definition of the biosecurity concept by the Food and Agriculture Organization (FAO) is a strategic and integrated approach that encompasses the policy and regulatory frameworks for analyzing and managing relevant risks to human, animal, and plant life and health and the environment (FAO 2007, 2020). The concept, therefore, has been applied to production sectors, including the agriculture, poultry, animal husbandry, and aquaculture industries, which need proper health management of their production system. The rapid growth of the

C. S. B. Kambey (✉)
PT. Sea Six Energy Indonesia, Bali, Nusa Dua, Indonesia

Institute of Ocean and Earth Sciences, Universiti Malaya, Kuala Lumpur, Malaysia
e-mail: ciciliasbk@sea6energy.com

J. P. Mateo
Institute of Marine Fisheries and Oceanology, College of Fisheries and Ocean Sciences, University of the Philippines Visayas, Miagao, Iloilo, Philippines

Institute of Aquaculture, College of Fisheries and Ocean Sciences, University of the Philippines Visayas, Miagao, Iloilo, Philippines

S. B. Rusekwa · F. E. Msuya
Zanzibar Seaweed Cluster Initiative, Zanzibar, United Republic of Tanzania

A. R. M. Nor
Institute for Advanced Studies, Universiti Malaya, Kuala Lumpur, Malaysia

C. F. A. Sondak
Faculty of Fisheries and Marine Science, Sam Ratulangi University, Manado, Indonesia

I. Campbell · E. J. Cottier-Cook
Scottish Association for Marine Science (SAMS), Scottish Marine Institute, Oban, Argyll, UK

A. Q. Hurtado
Institute of Aquaculture, College of Fisheries and Ocean Sciences, University of the Philippines Visayas, Miagao, Iloilo, Philippines

P. E. Lim
Institute of Ocean and Earth Sciences, Universiti Malaya, Kuala Lumpur, Malaysia

A. T. Critchley et al. (eds.), *Tropical Phyconomy Coalition Development*, Developments in Applied Phycology 11,
https://doi.org/10.1007/978-3-031-47806-2_7

global aquaculture industry has led to a dramatic increase in the spread of pathogens (e.g., pests, bacteria, viruses, and other parasites) by translocation of commercial aquatic organisms, such as fishes, shrimp, and mollusks (Soliman and Inglis 2018; Rodgers et al. 2019; Bouwmeester et al. 2021). Increasing seawater temperatures caused by global climate change have also triggered the wider distribution, expansion, and emergence of new pathogens, including diseases, parasites, pests (Bayliss et al. 2017), and nonnative species (Campbell et al. 2020; Foster et al. 2021). Focusing on pathogens, the quantification of the impacts and consequences is particularly challenging in the marine environment. Damage to economic and ecological and environmental services, however, has been highlighted in a number of studies, including a reduction in profit and revenues (Soliman and Inglis 2018; Naylor et al. 2021; Mantri et al. 2022), poor food security systems (Henriksson et al. 2021), low market accessibility (Freitas et al. 2020; Ward et al. 2021), and a lack of social welfare and ecosystem services (Paseka et al. 2020; Welsh et al. 2021).

Preventing and controlling pathogen spread is a major concern of aquaculture biosecurity starting from on-farm (Oidtmann et al. 2011) to national and international levels (FAO 2020; Campbell et al. 2020; Mantri et al. 2022), particularly when unrecognized species enter transport pathways (Sikes et al. 2018). As "prevention is better than cure," it is advisable to focus on preventing the occurrence of pathogens in the aquaculture system (Asefa and Abunna 2018). To limit the economic and ecological impact and consequences (e.g., low production, loss of capital investments, loss of native biodiversity, and change in coastal environment), a comprehensive aquaculture health management system is suggested (FAO 2007; Oidtmann et al. 2011). A well-established national health management system should incorporate a proper biosecurity strategy, whose concept can be adapted to suit farm level and be incorporated into the national regulatory frameworks, as a priority component to be considered, particularly in seaweed aquaculture. A proper concept must include a clear disease control program, risk assessment, and an active national aquaculture surveillance system, such as the national progressive management pathway program for aquaculture biosecurity (PMP/AB) (Bondad-Reantaso et al. 2018) and the disease surveillance checklist (Bondad-Reantaso et al. 2021). Cottier-Cook et al. (2022) suggested implementing these methods through seaweed aquaculture PMP/AB to limit seaweed aquaculture risks. However, higher investment costs for biosecurity intervention, estimated at 15–20% of total operating costs (Watson et al. 2009; Adams et al. 2011; Soliman and Inglis 2018), have become a major barrier, particularly in the eucheumatoid aquaculture industry, where most farmers are small-individual holders in low- to middle-income countries (Neish et al. 2017; Kambey et al. 2021a; Mateo et al. 2021; Ndawala et al. 2022). Implementing biosecurity measures, however, was shown to improve farm productivity eight times (Fasina et al. 2012) and reduce disease occurrence by up to 70% on a seaweed farm in Malaysia (Kambey et al. 2021a) and up to 50% in Tanzania (Ndawala et al. 2022). Proper biosecurity regulations and policies, therefore, can play a significant role in changing the behavior of stakeholders toward improving their practices. Appropriate biosecurity practices and interventions, such as training, can update stakeholders' knowledge of the latest policies and regulations and improve their attitude toward and awareness of biosecurity (Jia et al. 2017; Shannon et al. 2020; Campbell et al. 2022; Kambey et al. 2022)

Global eucheumatoid seaweed production is dominated by Southeast Asia, ranging from 29% to 42% total production in 2010–2018, where the red seaweeds *Kappaphycus* and *Eucheuma* are the main contributors (FAO 2020). Indonesia, the Philippines, Tanzania, and Malaysia are among the major producing countries of eucheumatoid seaweed. Despite increasing demand, several challenges remain unresolved, and currently, there remain a number of challenges for the industry (Hurtado et al. 2019). These challenges include the occurrence of ice-ice disease (IID) syndrome and the spread of the epi-endophytic filamentous algae (EFA) *Neosiphonia* or *Polysiphonia*, which lack preventative strategies in the main seaweed-producing countries (Vairappan et al. 2010; Tsiresy et al. 2016; Achmad et al. 2016; Largo et al. 2020; Ward et al. 2021; Faisan et al. 2021).

An international joint research program, named GlobalSeaweedSTAR (2017–2021), was initiated to safeguard eucheumatoid aquaculture in some of the main producing countries: Indonesia, the Philippines, Malaysia, and Tanzania. One of the key themes of the initiative was to understand, propose, and implement the best biosecurity applications by analyzing the unresolved biosecurity challenges ranging from on-farm management and national seaweed regulatory frameworks to the understanding of seaweed stakeholders' knowledge, attitudes, and practices (KAP). In this chapter, we have summarized the findings of the biosecurity investigation of the GlobalSeaweedSTAR project in the four top producing eucheumatoid countries to include published and unpublished data.

7.2 Current Seaweed Policies and Regulations in Eucheumatoid-Producing Countries

7.2.1 Indonesia

As the largest producer of eucheumatoids, Indonesia's production reached a maximum of 9.92 M fresh weight (FW), with an export value of US$324 M in 2019 (Annual report

MMAF 2019, FAO-FIGIS 2021). To achieve this high production level, the Ministry of Marine Affairs and Fisheries (MMAF), as the national competent authority, and other Indonesian policymakers (e.g., parliament, other government institutions that are relevant to aquaculture) have produced a number of policies, regulations, and legislations in a national framework to strengthen and include seaweed in the aquaculture industry of Indonesia. The seaweed regulatory framework specifically prioritizes increasing national production capacity and stimulating investment in the sector (Ministerial Regulation No. 217/PER-DJPB/2017; PPRI No 81/2015; Ministerial Regulation No. 107/2016 from the Ministry of Labour; and Law No 21/2019). No seaweed aquaculture policies or regulations have addressed health management, particularly to overcome the main challenges such as the detection, prevention, and control of EFA and IID at the farm level. The updated National Quarantine Law No. 21/2019 still does not have any fundamental changes for aquaculture health management and how to mitigate risks. The latest ministerial regulation No 1/KEPMEN-KP/2019 about seaweed aquaculture guidelines has not been designed for seaweed aquaculture health management with insufficient disease information and mitigation. To date, there is a low awareness of mitigating the occurrence of EFA and IID syndrome, which affect productivity on many farms, as well as invasive nonnative species. This inaction has led to a low recognition of the importance of formulating biosecurity guidelines in Indonesia (Kambey et al. 2020).

Focusing on health management, there are no specific regulations concerning seaweed aquaculture health management; thus, infrastructures and procedures are also unavailable (Ward et al. 2020, 2021; Cottier-Cook et al. 2022). For example, there are no quarantine protocols for seaweed and their examination when transferring seaweed between farms and when introducing new seed stocks. Furthermore, no surveillance or tracking of these movements has been observed. Consequently, EFA has spread extensively throughout the productive areas in Indonesia (Mulyaningrum et al. 2019; Simatupang et al. 2021; Mariño et al. 2019), which is often accompanied by the presence of IID, which continually infects entire seaweed farms and nearby areas (Ward et al. 2020, 2021). Currently, disease-free seaweed is rare on seaweed farms in Indonesia (Zamroni 2021).

Indonesia has eleven (11) national regulating instruments incorporating the biosecurity concept and its application (Table 7.1). Most of these instruments have a binding statute borne by the national law, with a strong legal baseline that has to be implemented. However, there are several challenges identified in the seaweed framework that make the policies and regulations ineffective as proposed, such as (1) unspecific allocation of seaweed aquaculture in the national framework, (2) limited variety of biosecurity approaches used in the national seaweed aquaculture framework, (3) limited scientific information and lack of knowledge regarding seaweed aquaculture health management, (4) unclear precautionary guidance and instructions in each instrument, (5) insufficient evidence of biosecurity hazards in seaweed aquaculture, and (6) limited health management components in the current seaweed framework (Kambey et al. 2020). With an insufficient context of regulations, the national seaweed aquaculture system has therefore become inconsistent and ineffective in controlling seaweed aquaculture risks. The existence of regulations has resulted in insufficient biosecurity knowledge, and the resources of seaweed farmers have been greatly influenced by current knowledge, practices, and research innovation. The limitation of biosecurity applications as related to aquaculture health management concepts may become the deciding factor in the national production and sustainability of this industry (Cai et al. 2021; Cottier-Cook et al. 2021, 2022).

7.2.2 Philippines

Similar to Indonesia, farming eucheumatoid seaweed is still the best option for the daily activities of coastal communities in the Philippines. This was proven by the contribution of seaweed in the national aquaculture production figures as the main contributor from the early 1970s to the present (PSA 2018). In recent years, the production reached 1.49 Mt. FW (FAO FishStatJ 2021) and became the top commodity exported by the country (Mateo et al. 2020). However, there has been a sharp decline in seaweed production in the period of 2011–2012, which has not yet recovered to the present (PSA 2013, 2016; FAO 2020). Continuous disturbance by pest and disease outbreaks, limited availability of healthy seedlings on farms, and a significant increase in natural disaster events have worsened national seaweed productivity and decreased the contribution of the industry to national domestic income (Hurtado et al. 2019; Faisan et al. 2021).

Various government institutions, such as the Philippine Bureau of Fisheries and Aquatic Resources (BFAR) and the Bureau of Agriculture and Fisheries Standards (BAFS), are policymakers for seaweed aquaculture, while the Department of Agriculture is the national competent authority of the industry. By having multiple regulatory institutions responsible for the seaweed aquaculture industry, their responsibility in developing and managing the seaweed industry as a whole has been unfocused to the needs. Even after more than fifty (50) years of commercialization of eucheumatoid seaweed farming in the Philippines (Mateo et al. 2020), there are no binding regulations that have been produced specifically dedicated to seaweed aquaculture; thus, the biosecurity concept and general health management of this sector have been

Table 7.1 Indonesia's seaweed aquaculture regulations, biosecurity description, strengths, and weaknesses regarding biosecurity concerns (data are modified from Kambey et al. 2020)

Regulation	Description	Strengths	Weaknesses
National Fisheries Law No. 31/2004 and Amendment of Fisheries Law No. 45/2009	Regulates most aspects of fisheries and aquaculture activities, including prevention of pests/pathogens/invasive nonnative species, environmental impacts, food safety and food quality, licensing, zoning, and other enforcement aspects	Binding regulation, use penalties	Seaweed not explicitly mentioned, unspecific for regulating seaweed aquaculture, prevention measures not related to seaweed aquaculture risk, not using precautionary principles, information based on general knowledge
National Quarantine Law (Law No. 16/1992)	Regulates all matters concerning quarantine of aquatic organisms, terrestrial animal, and plant	Binding regulation, use penalties and precautionary principle	Seaweed not explicitly mentioned, not specific mitigating seaweed aquaculture challenges such as no quarantine procedure related to seaweed aquaculture
Fish Quarantine Regulations (PPRI No. 15/2002)	Fish quarantine protocols regarding disease and pathogens	Binding regulation, use diseases and pathogen detection procedure, precautionary principle. Information used based on experiential evidence	Seaweed not explicitly mentioned, no procedure for detecting seaweed disease and pathogens
Fish Quarantine Requirements (MMAF-PER No. 10/MEN/2012)	Additional guidance for fish quarantine protocols and requirements regarding pest, pathogen, and invasive species management	Binding regulation, use scientific and experiential-evidence, and use precautionary principle	Seaweed not explicitly mentioned, no specific requirement for investigating and preventing disease and pathogen
Aquaculture Farm Management Regulations (PPRI No. 28/2017)	Regulation of fish cultivation, utilization and preservation of stocks, management of infrastructure and facilities, product quality control, environmental impact management, enforcement, and monitoring requirements	Binding regulation, using precaution, include prevention and detection methods for fisheries aquaculture	No explicit mention of seaweed, methods and procedures, cannot apply in seaweed farm, information used based on general knowledge
Regulation of GMO Organisms (PPRI No. 21/2005)	GMO management including research and development cultivation, assessment, utilization, and control	Binding regulation, explicit use of precaution	Seaweed not explicitly mentioned, information used base on general knowledge
Food Safety Regulations (MMAF-KEP No. 02/MEN/ 2007)	Provides guidance for fish cultivation with a focus on utilization of fish for feeds, fertilizer, probiotics, and disinfection, relative to food safety	Binding regulation	Seaweed not explicitly mentioned includes precaution but unclear. Information used base on general knowledge
SNI No. 7672/2011	Procedure standard for seedling requirement and seed health checked	Explicit mention of seaweed, clear procedure in providing seaweed seed requirement, information provided by expert and stakeholders opinion	Voluntary guideline and not a binding regulation, poor implementation to farmer
SNI No. 7579/2010	Procedure standard for seaweed outplanting methods	Explicit mention of seaweed, clear procedure provides information of seaweed outplanting methods, and information provides by scientific evidence, and expert opinion	Voluntary guideline, not a binding regulation, poor implementation to farmer
SNI No. 2690/2015	Procedure standard at postharvest process and recommended methods for drying seaweed	Explicit mention of seaweed, clear information regarding postharvest procedures, and information provided by scientific evidence and expert opinion	Voluntary guideline, not a binding regulation, poor implementation to farmer

poorly formulated. Quarantine Administrative Order No. 1/1981 and Executive Order 292 Agriculture 1987 provide disease and pest detection mechanisms, but unfortunately, these regulations are not specific to seaweed aquaculture. The seaweed regulations produced by the government and other institutions are focused on increasing the national seaweed quality, quantity, and, ultimately, adequate food security to Filipinos (e.g., the Code of Good Aquaculture Practices (GAqP) PNS/BAFS 135:2014, the PNS/BAFS 208:2017, and the PNS/BAFS 208:2021). Consequently, these regulations are limited in providing a proper seaweed aquaculture health management system and biosecurity guidelines to mitigate current seaweed aquaculture risk hazards. Since no specific regulations are in place governing seaweed aquaculture, quarantine processes are also poorly delivered to farmers, as in Indonesia. Therefore, certain biosecurity regulations are very important for further development of the industry considering the economic (e.g., trade restriction) and environmental risks (e.g., pest/pathogen, climate changes) often faced by the industry (Hurtado et al. 2019; Mateo et al. 2020; Suyo et al. 2021; Faisan et al. 2021).

Of the seven regulation instruments in the seaweed aquaculture framework of the Philippines, most are related to general fisheries aquaculture with a focus on fish and shrimp (Table 7.2). There are several biosecurity challenges that can be highlighted as barriers in seaweed aquaculture, such as (1) a lack of seaweed policies and regulations, (2) limited binding policies for seaweed aquaculture, (3) limited biosecurity approaches used in the regulatory framework, (4) a lack of support from competent authorities, (5) insufficient guidance for the use of the precautionary principle in national regulation, and (6) limited involvement of experts in seaweed framework development (Mateo et al. 2020). These results indicate a clear gap in the current seaweed biosecurity legislation and policy in the Philippines that may have negative consequences for farm productivity, farmer practices, and the wider marine environment.

7.2.3 Malaysia

Malaysia has been developing the seaweed aquaculture industry based on policies and regulations that include a biosecurity concept in the framework (Sekaran 2014; Manap and Pauzi 2020). The concept of biosecurity in national policies and regulations at the farm level has supported the industry in managing production and aquaculture risks under the Department of Fisheries Malaysia (DOFM). Currently, the production of seaweed reaches 180,000 T FW and produces approximately 18,204 T dry weight of carrageenan (Lim et al. 2021). A code of practice, Malaysian Standards 2467: 2012, has been developed as a national standard for eucheumatoid seaweed farming practice, which includes disease prevention measures, such as quarantine, disease and pest detection, farm risk management, quality control of the seaweed crop, and associated risks for seaweed aquaculture based on environmental factors (Department of Standard Malaysia 2012). The biosecurity policy for managing good seaweed aquaculture practices at the on-farm level was further strengthened with the introduction of the Malaysian Good Agricultural Practices Certification program (MyGAP/2014) and the manual guideline of eucheumatoid farming practices (DOFM 2012). The objective of MyGAP was to standardize crop quality to gain better recognition of Malaysia's seaweed products in domestic and international markets. However, the objectives of the certification program were not achieved and implemented due to a poor understanding of the crop management standards by the seaweed stakeholders, particularly the farmers, despite being written in the local language (Nor et al. 2017).

The seaweed industry has continued to decline in production from 2013 to the present day (Lim et al. 2021), which indicates that the policy framework was not able to reverse the main challenges faced by the industry (Nor et al. 2017; Kambey et al. 2021b). No proper strategy has been made for maintaining high productivity on seaweed farms by reducing grazing and IID/EFA outbreaks. These factors have remained the main cause of biomass loss since 2005 (Vairappan et al. 2010; Eraza et al. 2017; Ward et al. 2021). Recently, IID has affected the quality of seaweed, and it remains a persistent threat in the productive regions in Semporna, Kunak, Kudat, and Lahad Datu (Vairappan et al. 2010; Eraza et al. 2017; Kambey et al. 2021a, b, Lim et al. 2021). Similar to Indonesia and the Philippines, the insufficiency of the seaweed aquaculture framework in Malaysia has been highlighted, particularly the lack of a health management strategy. Unclear information on practical guidelines to detect pests and diseases, including the EFA, and practical guidance for farmers to handle the infected crop, no clear procedure for seaweed quarantine and movement between farms have also contributed to this loss of production. In addition, no incentives have been given to farmers or seaweed companies to encourage good farm management measures, the subsidy program (e.g., free seedling, free farm equipment, and facilities) is available regardless of their practices. There have also been no further prevention strategies for reducing farm-level biosecurity risks; for example, only limited research has been conducted to date to develop new cultivar resistance to environmental changes, there is no innovation methods to minimize the effects of IID and grazers, and no improvement on farm surveillance systems by government officers (Table 7.3). The majority of existing seaweed regulations in Malaysia are also based on nonbinding regulations and thus serve as only voluntary guidelines; therefore, the contribution of these regulations to farmers' practices is limited.

7.2.4 Tanzania

The eucheumatoid species *Eucheuma denticulatum*, *Kappaphycus alvarezii*, and *Kappaphycus striatus* have been cultivated in Tanzania since the first establishment of farming in 1989 (Msuya 2006; Hedberg et al. 2018). The three eucheumatoid strains were introduced from the Philippines for improving coastal community livelihood, albeit for commercial purposes (Tano et al. 2015; Hedberg et al. 2018). Currently, the cultivation of seaweeds has expanded to other areas of Tanzania, including Zanzibar, and supports approximately 31,000 farmers, particularly

Table 7.2 Philippine's seaweed aquaculture regulations, biosecurity description, strengths, and weaknesses regarding biosecurity concerns (data are modified from Mateo et al. 2020)

Regulation	Description	Strengths	Weaknesses
Presidential Decree 143 3 (Plant Quarantine Decree)	Plant Quarantine Decree provided practical measures to be used to limit the import/export of diseases and pests	Binding regulation	Not specific for seaweed aquaculture, no explicit mention of precautionary principle
Quarantine Administrative Order No.1(otherwise known as "BPI Quarantine Administrative Order No. 1, Series of 1981")	Regulations for quarantine restrictions, import/export of live plant products, nursery stocks, vegetative parts used in propagation, and other plant products declared as prohibited/ restricted under "Special Quarantine Orders"	Binding regulation	Mention of algae but unclear mention of seaweed aquaculture, which are not covered under the special quarantine orders. No explicit mention of precautionary
Executive Order 292 Agriculture (Administrative Code) Title IV-Agriculture p.67	General provisions on accelerating agricultural development and the production of agricultural crops, fisheries, and livestock by optimizing the use of resources and by applying modern farming systems and technology to attain food security for domestic use and export	Binding regulation, including prevention methods	No specific procedure and process for seaweed aquaculture, no explicit of precautionary
Republic Act 10654 (Amendment of RA 8550—The Philippines Fisheries Code of 1998)	Specifies regulations for monitoring import/export of fishery/aquatic resources (including seaweed), aims to prevent, deter, and eliminate illegal, unreported, and unregulated fishing (as amended in 2015). Provide requirement to all fish and fishery products to be examined for fish pests or diseases detection and that the quality should meet international standards. Includes measures for surveillance, implemented of Hazard Analysis and Critical Control Point (HACCP)	Binding regulation, including prevention and detection methods for diseases and for food security	No explicit mention of seaweed makes the procedures too general and not mitigating seaweed aquaculture challenges on diseases. This regulation is mostly to cover seafood products
Code of Good Aquaculture Practices (GAqP) (PNS/BAFS 135:2014)	Adopted from the "Good Aquaculture Practice Farmers Guidance Workbook" of BFAR and developed under the European Union (EU) Trade Related Technical Assistance (TRTA) Project. Aims to protect aquaculture farms/projects, which include seaweed farms from environmental risks associated with production	Explicit mention of seaweed, clear procedure in providing seaweed seed requirement, information provide by expert and stakeholder opinion	Not a binding regulation
Code of Good Aquaculture Practices (GAqP) Organic Aquaculture (PNS/BAFS 112: 2016)	Equates to existing Codex Standard relative to food safety. Identifies the requirements to be documented as organic aquaculture including parallel productions, site selection, ecosystem interactions, fertilization, breeding and hatchery management, and aquatic plants	Explicit mention of seaweed, clear procedure in providing seaweed seed requirement, information provided by expert and stakeholder opinion	Not a binding regulation

(continued)

Table 7.2 (continued)

Regulation	Description	Strengths	Weaknesses
Code of Good Aquaculture Practices (GAqP) for Seaweed (PNS/BAFS 208:2017)	Provides guidance on practices to prevent or reduce the risk of hazards during farming, harvesting, and postharvest handling of seaweed, including guidance on site selection, sanitation, management, diseases, harvesting and transport, traceability and record keeping, and aspects on farmers' welfare such as personnel health, hygiene, and child labor	Explicit mention of seaweed, clear procedure in providing seaweed seed requirement, information provided by expert and stakeholder opinion	Not a binding regulation

Table 7.3 Malaysia's seaweed aquaculture regulations, biosecurity description, strengths, and weaknesses regarding biosecurity concerns (data are modified from Kambey et al. 2021b)

Regulation	Description	Strengths	Weaknesses
Fisheries Act 1985	National Fisheries Legislation	Binding regulation, use only penalty as biosecurity approach	Not specific for seaweed aquaculture, use general prevention and detection approach, poor implementation to farmer
Malaysia Good Agricultural Practices (MyGAP) Guidelines (2014)	Certification of aquaculture product	Procedures for farm certification are clear and information used from scientific literature, expert, and stakeholder opinion	No explicit mention of seaweed, not a legal binding regulation which cannot implement to all stakeholders, use general prevention approach, and cannot mitigate the current and the crucial seaweed aquaculture risks
Malaysian Standards; Seaweed Code and Practices No. 2467 (2012)	National standard of seaweed cultivation practices	This regulation specifies for seaweed aquaculture, information used from scientific literature, expert, and stakeholder opinion	No legal binding regulation and information regarding the prevention and detection procedures are unclear and unspecific, e.g., mitigating EFA and IID, grazer, and infected seaweed stock
Seaweed Farming Manual (2012)	Seaweed farm guidelines for *Kappaphycus/Eucheuma*	This regulation specifies for seaweed aquaculture with prevention; detection is clearly informed, information used from scientific literature, expert, and stakeholder opinion	Not a legally binding regulation, this manual practice guideline has poor implementation for farmers

women who typically run farms (Rusekwa et al. 2020; Campbell et al. 2022). Significant farmer engagement in eucheumatoid aquaculture has had a considerable impact on education and/or socioeconomic improvement (Msuya 2020).

The production of eucheumatoids in Tanzania increased to 170,000 T FW in 2016 and then declined in 2017 by 30% over recent years, mainly due to multiple factors, including increasing seawater temperatures and lack of a good farming practice (Largo et al. 2020). These factors have led to the occurrence of IID and massive infestations of the EFA throughout the eucheumatoid aquaculture industry in Tanzania and elsewhere (Cottier-Cook et al. 2016; Msuya et al. 2022).

No policies and regulations are mandated to minimize seaweed aquaculture challenges and non-explicated mention seaweeds in Tanzania (Rusekwa et al. 2020; Ndawala et al. 2021). Biosecurity regulations and practices for aquaculture are governed through two competent authorities, one governing mainland Tanzania (Government of Tanzania or GoT) and the other governing the semiautonomous region of the Zanzibar Islands (Revolutionary Government of Zanzibar or RGoZ) (Rusekwa et al. 2020). The two ministries of agriculture, each with respective Departments of Marine Aquaculture (with approval by their parliaments), are responsible for developing draft policies before their implementation across the seaweed industry. In both departments, plant quarantine and phytosanitary measures already exist. However, neither of them is specific to seaweed aquaculture, especially in the movement of live seaweed, so there is a lack of control over the seaweed/plants or plant products that can be introduced into or exported from the country (Ndawala et al. 2021; GoT 1998). In contrast, a majority of biosecurity policies have been developed for the agriculture and finfish aquaculture sectors (Ndawala et al. 2021). The two governments, however, have shown recent interest in assisting the development of the seaweed industry in their respective regions through a few updated informal policies that have been produced and that act as voluntary guidelines. These guidelines aim to provide support for seaweed farmers to access appropriate farming equipment and seedling/propagules and to promote best farming practices (Rusekwa et al. 2020; Campbell et al. 2022) (Table 7.4). Currently, no specific regulations concerning seaweed aquaculture exist for

Table 7.4 Tanzania's seaweed aquaculture regulations, biosecurity description, strengths, and weaknesses regarding biosecurity concerns (data are modified from Rusekwa et al. 2020)

Regulation	Description	Strengths	Weaknesses
Tanzania Mariculture Guidelines Source Book (GoT 2001)	Guidance book for promoting sustainability development of mariculture on the Tanzania mainland and promoting coastal Aquaculture management	Explicit mention of seaweed	Not binding regulation. There are no biosecurity approach and precaution methods that have been designed. Source of information mainly based on general knowledge
Fisheries Act No. 22 of 2003, Tanzania (GoT 2003b)	An Act to regulate fishing and the fishing industry and aquaculture development in mainland Tanzania. The Act includes the section of aquacultural development, quality and standard management, enforcement, offences, and penalties	Binding regulation, use penalty and prevention as biosecurity approach, use information based on expert opinion and precaution	This regulation does not explicitly mention seaweed and seaweed aquaculture, but as aquatic flora. The preventative approach is unclear for the fisheries aquaculture
Environmental Management Act (GoT 2004)	This regulation provides a legal institutional framework for sustainable management of environment including impact and risk assessments, prevention and control of pollution, waste management, environmental quality standards	Binding regulation, use information based on expert opinion, have precaution and penalty for biosecurity approach to manage the environmental system	This regulation does not explicitly mention seaweed to seaweed aquaculture environment management
Seaweed Strategic Development Plan (GoT 2005)	A strategic plan for seaweed and seaweed aquaculture to provide a framework for development of the seaweed industry in Tanzania, including expanding the seaweed aquaculture, market acceptability, financial stabilities, social awareness, and environmentally sustainable	Explicit mention of seaweed in the regulation and information used based on experiential-based study	Not binding regulation, none of biosecurity approaches used, not having precaution principles
Seaweed Cluster Initiative (2006)	This initiative aims to address seaweed problems and tap the scientific information for the benefit of the farmers and the country to increase seaweed production through modifying the farming technique and adding value to the produced seaweed	Explicit mention of seaweed, information used based on experiential-based study	Not binding initiative, none of biosecurity approaches has been specified, and not having any precautions
Tanzania National Research and Development Policy (GoT 2010b)	This policy aims to provide guidance to research in the public and private sectors, policy- and decision-makers, as well as development partners in addressing present and future national research challenges for socioeconomic development	Explicit mention of seaweed	Not binding regulation, none of biosecurity approaches and precaution has been used, and information used are general knowledge
Fisheries Act No. 7 of 2010 (GoT 2010a)	This regulation provides information for promoting and regulating general fisheries (includes aquatic flora, e.g., seaweed aquaculture), conserving, licensing, and general fisheries management	Binding regulation, information used based on expert opinion, use penalty and prevention as biosecurity approached, and precaution	No explicit mention of seaweed but aquatic plants. This regulation focuses on general fisheries sector only. The preventative measures used are presented in general terms and unclear for seaweed aquaculture
Zanzibar Agricultural Transformation Initiative (RGoZ 2010)	This initiative becomes a strategic intervention toward agricultural transformation and commercialization in Zanzibar to create a good environment for production, processing, and marketing of agricultural products in the next 10 years. The interventions include public and private sector investment upon capitalizing on the opportunities provided by domestic and export markets and consequently addressing challenges facing agricultural development	Explicit mention of seaweed	Not binding regulation, none of biosecurity approach has been used, not having precaution, and information used are general knowledge

(continued)

Table 7.4 (continued)

Regulation	Description	Strengths	Weaknesses
Aquaculture Development Strategy and Plan of Zanzibar (RGoZ 2014)	This strategy aims to develop a strategy and good plan for aquaculture industry, which includes seaweed aquaculture	Explicit mention of seaweed	Not binding regulation, none of biosecurity approach has been used, not having precaution, and information used are general knowledge
National Fisheries Policy of Tanzania (GoT 2015)	This policy aims to guide the fisheries industry toward achieving a clear vision and mission of the sector. The long-term objective is to achieve sustainable fisheries that will lead to food security, poverty reduction, and increase in the national income	Binding regulation, information based on expert opinion, using precaution	Not explicit on seaweed, and using preventative approach, but the process is unclear for seaweed aquaculture
Tanzania Research Agenda (Priorities) (2016–2020) (GoT 2016)	A document that specifies research needs of the country. It strives to build a society that utilizes scientific evidence to inform its policy and decision-making processes. This document also includes various agenda for seaweed aquaculture research	Explicit mention of seaweed	Not binding regulation, none of biosecurity approach has been used, not having precaution, and information used are general knowledge
Zanzibar Research Agenda (2015–2020) (RGoZ 2015)	This document strives to build a society that utilizes scientific evidence to inform its policy and decision-making processes. This document also includes various research agenda for development of seaweed aquaculture	Explicit mention of seaweed	Not binding regulation, none of biosecurity approach has been used, not having precaution, and information used are general knowledge
Tanzania Aquaculture Development Strategy (GoT 2018)	The strategy aims to develop and sustain the aquaculture industry including seaweed aquaculture	Explicit mention of seaweed	Not binding regulation, none of biosecurity approach has been used, not having precaution, and information used are general knowledge

biosecurity and health management in Tanzania. Consequently, there is no national biosecurity strategy established to control the introduction, movement, and export of seaweed. Therefore, seaweed farmers have continued to transfer crops from one region to another without considering the impacts. Without the involvement of farmers, processors, buyers, and exporters in the value chain, the development of an adaptive approach to the biosecurity concept provides no guarantee for the successful implementation of these regulations (Reed and Curzon 2015; Rusekwa et al. 2020).

7.3 Biosecurity Capacity of Seaweed Farmers and Policy Implementation

As a basic concept in aquaculture management, biosecurity is included as one of the standards for a health system management strategy, which has a long-term investment (Stentiford et al. 2020). As mentioned in the previous sections, the major producing countries of eucheumatoids, such as Indonesia, the Philippines, Malaysia, and Tanzania, do not have specific seaweed aquaculture biosecurity regulations pertaining to aquaculture management. To date, the implementation of the current regulations and policies that concern on-farm biosecurity practices has received poor attention by the government and competent authorities in these producing countries, while aquaculture risks have significantly increased and occurred in an unpredicted manner (Rusekwa et al. 2020; Zamroni 2021; Suyo et al. 2021). As a consequence, understanding the biosecurity risks faced by farmers and practicing proper health management at eucheumatoid farms are unknown and understudied (Campbell et al. 2022). To assess the current capacity of seaweed farmers to understand and practice biosecurity applications at the farm level, it is important to identify the effectiveness of seaweed regulatory framework implementation, especially those that remain unknown and undocumented. Identification of the biosecurity knowledge, attitude, and practices (KAP) of seaweed farmers was conducted by a parallel survey in Indonesia, the Philippines, Malaysia, and Tanzania within the period of 2018 to 2021. The quantitative KAP survey has been widely used in various policy assessments (Ritter et al. 2017; Jia et al. 2017), and the method was fitted to assess the capacity of seaweed stakeholders towards biosecurity. The seaweed aquaculture biosecurity components at the farm level used in the survey are shown in Table 7.5. In addition, evaluating biosecurity capacity implementation through the KAP also aimed to identify

Table 7.5 The biosecurity components used in the KAP survey

Biosecurity component	Subcategory	Category
Identify the causative agent of disease/pathogenic pest at farm (e.g., infectious diseases, epiphytes, grazers, parasites, suboptimal environment, invasive species)	Understanding	Knowledge
Identify the activities that may spread disease/pathogenic pests at farm (diseases introduction) (e.g., share farm equipment with infected farm, multiple use of equipment in all seasons and conditions, use of infected seaweed, unknown health condition of the seaweed, no monitoring of the seawater parameters)	Awareness	Knowledge
Considered to be practiced on the prevention, detection and control measures at farm (e.g., detection of disease and pest on seedling, isolation/farm quarantine process, monitoring condition of seaweed including infected crop, using healthy seedling, managing stocking density, and crop separation by source and age)	Internal biosecurity	Attitude
Considered to be practiced on the prevention, detection, and control measures within farms (e.g., water monitoring, avoid sharing equipment with diseased farm, and disinfection process on farm equipment)	External biosecurity	Attitude
Practicing the prevention, detection, and control measures at farm (e.g., detect unhealthy seedling, isolate/use farm quarantine process, monitor seaweed condition including the infected crop, do crop management including the disposal of farm waste on land and unused chemicals)	Internal biosecurity	Practice
Practicing the prevention, detection, and control measures within farms (e.g., monitor seawater parameters, keep natural disinfection process on farm equipment, do surveillance/monitoring the farm system and the risks)	External biosecurity	Practice

gaps in biosecurity knowledge and practices of farmers, measuring similarity or diversity in biosecurity challenges, subsequently highlighting the best strategy for system improvement. Therefore, the identification of specific factors can be captured for the global biosecurity seaweed aquaculture industry (Campbell et al. 2022; Mantri et al. 2022). Poor, fair, and good KAP indices defined by scores of <50%, 50–75%, and >75%, respectively (Jia et al. 2017; Mateo et al. 2021), were used. Below are the results of the KAP assessment in each eucheumatoid-producing country assessed:

7.3.1 Biosecurity KAP of Indonesian farmers

Farmers in the South Sulawesi region were selected for the biosecurity KAP survey, mainly, because this region is the top producer of eucheumatoid seaweed in Indonesia. Among high-production areas in South Sulawesi, the Takalar region was chosen for this interview because it has consistently produced the highest amount of seaweed since 2015 (Indonesian Fisheries Statistics Year Book 2017). In 2022, the Takalar area booked seaweed production of approximately 1.5 MT FW (dkp.sulselprov.co.id). Thirty individual farmers from five villages in Takalar (Punaga, Puntondo, Laikang, Maleleya, and Boddia) participated in this survey (Fig. 7.1). The results highlighted that farmers' knowledge was identified as moderate or fair (index score 63.81%) in understanding and awareness of biosecurity risks at the farm. Most of the farmers lacked an understanding of the biosecurity risks, with very few recognizing epiphytic/endophytic pests (e.g., the EFA, parasites) and invasive species. The awareness of the introduction pathways of pests and diseases to the farm was identified to be poor among farmers, as well as the proper ways of sanitizing processes (e.g., using undisinfected and unclean farm ropes in multiple growing cycles) and the recognition of healthy seedlings. Those same findings were also highlighted by Zamroni and Yamao (2011) and Mariño et al. (2019) in different farm areas in Indonesia.

In contrast, most farmers had a good attitude toward biosecurity practices, with an average score of 84%. Most of them also agreed with the benefits of the biosecurity concept either external or internal. For example, 91% of farmers agreed with the biosecurity management practice concept, 87% agreed with the process of disinfection of farm equipment, 76% agreed with the prevention of disease and pest occurrence in the farms through quarantine processes of new seedlings, 85% agreed on removing the infected crop from the farm, and 95% agreed on the importance of regular monitoring of seawater parameters. However, the majority of the farmers were not able to apply their biosecurity attitudes to regular practices, as indicated by their fair score (60.78%). Farmers mainly contributed to the lack of biosecurity practices at the internal biosecurity level. For example, only 60% of farmers isolated and monitored new seeds, of which 90% of seedlings had come from a nearby farm (NF)/own seed, while a trader considered a "healthy seed" by visual examination. Approximately 43% of farmers removed the infected crop in the farm system but disposed it within the same farm area. Most of the farmers (50%) removed the entire diseased crop but rarely monitored the environmental parameters (30%), because they lacked devices and relied only on their own experience per season.

Based on the sociodemographic level (Table 7.6), primary school is the highest education level of most seaweed farmers in the Takalar region. Ninety percent of farmers in this area did not have any proper seaweed aquaculture training from the competent local authority and government officer. Therefore, their knowledge came mainly from their elders, who passed on their traditional practices and knowledge to the

Fig. 7.1 Farmer's KAP interview in Takalar Region, South Sulawesi, Indonesia (Photo courtesy CSB Kambey)

Table 7.6 Demographic status of eucheumatoid farmers

Demographic factors	Indonesia	Philippines	Malaysia	Tanzania
Education	60% up to elementary school and 40% middle to high school	49.5% up to elementary school, 31.7% middle and high school	80% up to elementary school, 20% middle and high school	59.6% up to elementary school, 33.7% middle school
Gender	47% female, 53% male	29% female, 71% male	13% female, 87% male	87% female, 13% male
Farm training status	87% untrained	72% untrained	70% untrained	Mostly get trained
Occupation	37% as full-time seaweed farmer	Mostly as full-time seaweed farmer	26.8% as full-time seaweed farmer	Mostly as full-time seaweed farmer
Experience	77% ≥ 10 years	69% ≥ 10 years	40% ≥ 10 years	Not informed

(Data source: Mateo et al. 2021; Kambey et al. 2021b; Campbell et al. 2022; unpublished data of GSSTAR)

next generations. Farmers, therefore, worked based on their own understanding, and as a consequence, seaweed farms were poorly managed (Zamroni and Yamao 2011; Mulyaningrum et al. 2019; Mariño et al. 2019; Zamroni 2021). Examples include no data on loss of harvest due to diseases, lack of traceability of seed, no proper detection of pests on new seedlings, and reuse of the infected crop at multiple growing cycles. Thc improvement of seaweed aquaculture health management by farmers, therefore, needs the assistance of competent local authorities or government officers through regular monitoring of cultivated farms/areas (Simatupang et al. 2021; Rimmer et al. 2021) and training (Zamroni 2021). The establishment of group discussions to provide access and updated information with regard to risk hazards, environmental factors, and proper farm management is highly recommended (Kambey et al. 2020, 2021a, b).

7.3.2 Biosecurity KAP of Philippine farmers

In the Philippines, a KAP survey was conducted among farmers in four top seaweed aquaculture areas, namely, Tawi-Tawi, Palawan, Zamboanga, and Bohol. A total of 120 farmers were interviewed from July 2018 to October 2019 (Fig. 7.2). According to Quiaoit et al. (2018), the majority of the provinces in the Philippines (90%) are seaweed farming areas due to their demographic position as an archipelagic country. Hence, this country pioneered commercial eucheumatoid production in the 1970s and was the lead producer until Indonesia overtook the Philippines in terms of total production in 2007 (Porse and Rudolph 2017). The number of farmers in each of the productive areas is estimated at 60,000–65,000 families in Tawi-Tawi, 5567 in Palawan, and 25,000 in Zamboanga (AQ Hurtado, pers. Comm), while seaweed farmers in Bohol have not been estimated.

The results showed that farmers from Tawi-Tawi had the highest KAP scores, with Zamboanga having the lowest scores. In general, the Philippines farmers across the regions had fair knowledge in either understanding biosecurity risks or awareness of the introduction of pests and diseases at the farm level, with an average score of 61% (Mateo et al. 2021). The lowest score of biosecurity knowledge was detected in understanding parasites and invasive species, while most farmers lacked awareness in managing the introduction of pathogens on the farm (Campbell et al. 2022). For example, the sharing of farm equipment among the farmers was common practice, and there was poor disinfection of the equipment, even though there was a disease outbreak event. Uncontrolled incidences of diseases in the nursery areas also weakened the awareness scores. Similar to Indonesia, the biosecurity attitude among famers in the Philippines scored good (82%) in both the external and internal biosecurity parts. For example, farmers agreed to the prevention of diseases and pests through disinfection and cleaning processes on farm equipment, as well as removal of epiphytes and pests and regular monitoring of the farm system and environmental parameters (Mateo et al. 2021). However, the biosecurity practices score remained with a fair index (62.3%). Insufficient practices by farmers were recorded for both external and internal biosecurity practices, such as a lack of preventing disease and pest introduction by isolating the new seed. Most farmers disposed of their farm wastes within the farms by just shaking the ropes and allowing the infected crop, epiphytes, and other entangled waste to fall into the water. Poor disinfection and cleaning of the equipment (e.g., main ropes) and poor practices of monitoring water quality/parameters worsened the biosecurity practice scores. Farmers did not change their farming practices even when there was a disease outbreak. Government surveillance systems to monitor seaweed farms were wanting. The KAP assessment highlighted how biosecurity challenges were addressed by farmers in the Philippines. Likewise, it suggested how the KAP survey could be used as a tool by policymakers and scientists to address gaps in biosecurity management practices. It also recommended the formulation of an independent seaweed aquaculture regulation and the use of science-based decisions in formulating policies (Hurtado et al. 2021).

7.3.3 Biosecurity KAP of Malaysian farmers

In Malaysia, 67 farmers from the three most productive areas contributed to the KAP survey. Thirty (30) individual farmers in the Semporna area, 23 in Kunak, and 14 farmers in Tawau, who were either individual or worker farmers, participated in the interviews (Fig. 7.3). The total number of farmers in Malaysia reached approximately 1000 (DOFM 2018; Lim et al. 2021). However, the number of Malaysian farmers has decreased significantly in the last decade due to many inactive farms in the Semporna area caused by various factors, including disease and pest outbreaks, grazing, and unavailability of healthy seedlings (Asri et al. 2021; Kambey et al. 2021a; Lim et al. 2021).

The KAP results confirmed that farmers have a fair knowledge index with a score of 54.5%, which consisted of a fair understanding of biosecurity risks (64%) and a poor awareness of the introduction of pests and diseases on the farm (46%). Insufficient knowledge was found on the lack of recognizing the parasites and the presence of invasive species on the farm, while poor awareness was detected by the reuse of infected seedlings, grazing by animals on the infected seed, and the failure to recognize healthy seed. The attitude of farmers was identified as fair (index score of 65.3%) for internal and/or external biosecurity. Farmers, however, in the producing areas showed poor biosecurity practices, mostly also in the internal part (index score of 46%), while in the external part, it was fair (index score of 58%). The prevention practices were indicated with poor attention by the farmers in Semporna, Kunak, and Tawau. For example, the majority of the farmers in the Semporna area reused the disease/infected crop because healthy seed was rare and hardly found within the area. Similarly, farmers in Kunak and Tawau chose to grow new seedlings together with the infected seaweed crop. Farmers also lacked the practices of detecting measures, as well as the isolation and quarantine of new propagules. For external biosecurity practices, most farmers in Semporna and Kunak did not disinfect their farm equipment, in contrast with the farmers in Tawau, who usually sundried ropes as a natural way of disinfecting their equipment. This KAP result identified that the implementation of current policies and regulations is imperative and suggested the need to improve the system by active surveillance of farm management practices, which must be conducted by competent authorities, who will conduct regular training programs for all farmers, both domestic and migrant seaweed workers (Kambey et al. 2021b; Asri et al. 2021).

7.3.4 Biosecurity KAP of Tanzanian farmers

Most seaweed farmers are located in the Zanzibar islands (Unguja and Pemba) in Tanzania. The number of farmers in this area is estimated to be 15,000–25,000 (Msuya 2006;

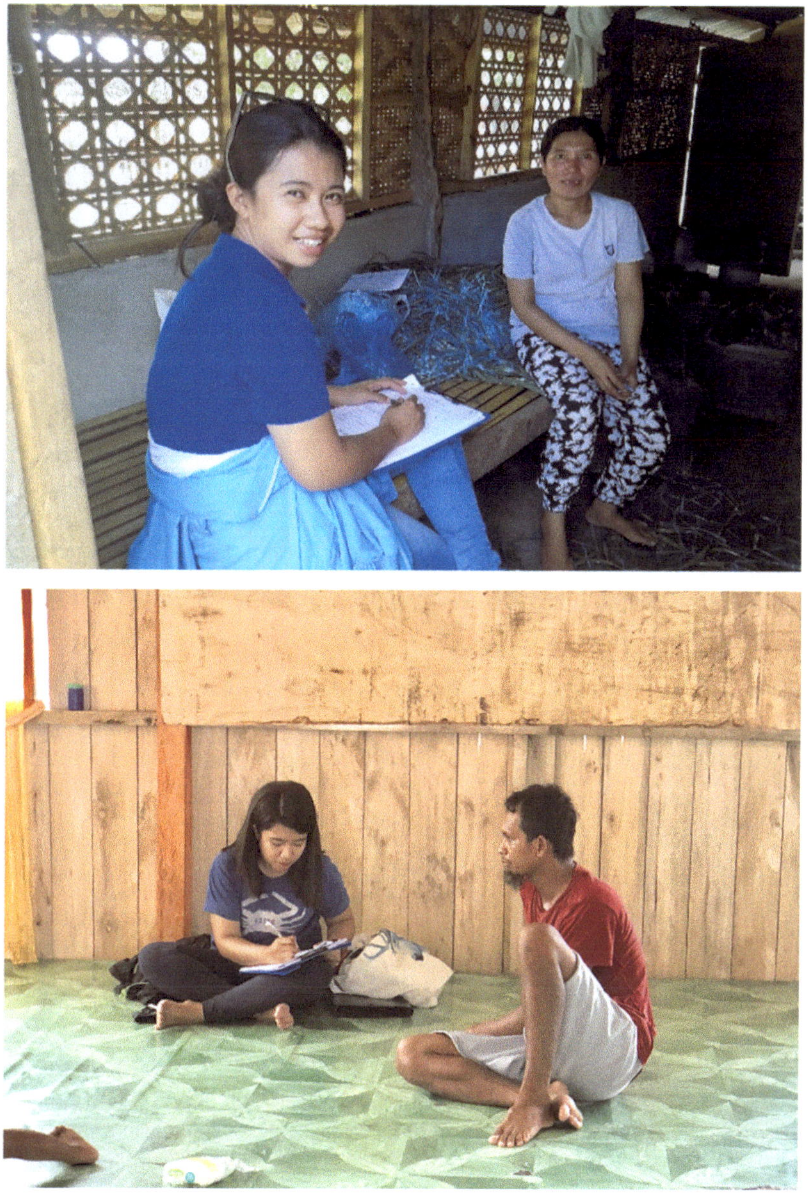

Fig. 7.2 Farmer's KAP interview in Bohol and Tawi-Tawi Region, Philippines (Photo courtesy JP Mateo)

Rusekwa et al. 2020). In mainland Tanzania, two coastal areas of Lindi and Mtwara were also recorded as top productive areas of farming eucheumatoids. The number of farmers in these two areas was estimated to be approximately 6000, mainly consisting of individual farmers who work for a company (FE Msuya 2021 pers. Comm). The survey on KAP in 2018–2021 was conducted with 89 farmers in four areas, including Unguja, Pemba, Lindi, and Mtwara.

Campbell et al. (2022) studied the KAP of farmers in Tanzania and found that most of the farmers have a fair index for biosecurity knowledge (index score 50%). The majority of the farmers (44%) had a good understanding of biosecurity risks, while awareness of the introduction of pests and diseases at the farm level was distributed equally among farmers (33–38%). Farmers' awareness of how to properly disinfect their equipment and how to keep a healthy nursery system by culturing healthy seedlings is becoming a critical point of managing biosecurity in Tanzania (Msuya 2020). Unlike farmers in Indonesia and the Philippines, the biosecurity attitude of Tanzanian farmers was of fair index with a score of 50%, which is consistently reflected in their knowledge index. In the biosecurity practices, most of the farmers (85%) scored poorly in either the external or the internal biosecurity practices. From the results, it was confirmed that disease and pest management in Tanzania was poorly managed by farmers, which correlated with the absence of biosecurity regulations in the country. With most seaweed farmers being women, it is advisable to include

Fig. 7.3 Farmer's KAP interview in Samporna Region, Sabah, Malaysia (Photo courtesy A. Asri)

Table 7.7 The key factors of insufficient on-farm biosecurity practices in the eucheumatoid aquaculture of the four study countries

Factors	Insufficient implementation by farmer	Country
Biosecurity risks at farm	Lack knowledge and awareness of seaweed health status and risks, knowledge mainly driven by local thought such as diseases caused by seasonal only	All production countries assessed
Sanitation	Mainly disinfect tying ropes only, while other ropes are left undisinfected	All production countries assessed
Monitored the environmental parameters	The environmental parameters such as temperature and salinity are not monitored	All production countries assessed
Isolation and quarantine of seaweed	Lack of practice	All production countries assessed
Used healthy crop	The infected seed (EFA infestation, bleached thallus) in multiple growing cycle is often reused	Indonesia, Malaysia, Tanzania
Seaweed crop management at farm	Lack managing crop density, particularly while in suboptimal condition. Uncontrolled disposal of the infected crop within farm area. Never clean the crop from epiphytes/pests/biofilm while growing out. No detection of health status of seedling at initial preparation stage	All production countries assessed
Nursery system	Mostly, there is no nursery system for the eucheumatoid seaweed	All production countries assessed
Chemical treatment at farm	Using fertilizer to grow seaweed	Indonesia
Farm monitored	Regular farm monitor not in daily basis	All production countries assessed
Government monitor and consistency of national surveillance system	No regular monitoring by government and unavailable national surveillance system from the government and the competent authority	All production countries assessed

gender resilience in the component of the national biosecurity strategy of seaweed aquaculture regulations of Tanzania, which would strengthen biosecurity awareness among farmers. Through the KAP tool, evidence of insufficient management of the health system in eucheumatoid aquaculture in the four countries was found, as shown in Table 7.7. In Fig. 7.4, the concept of biosecurity for seaweed aquaculture can be summarized as follows.

7.4 Conclusion and Recommendations

Insufficient seaweed regulations and biosecurity contexts in the national seaweed aquaculture system make the industry lack resilience to changes either environmentally, economically, or socially. In general, the current farm practices of farmers in the majority of eucheumatoid producing countries

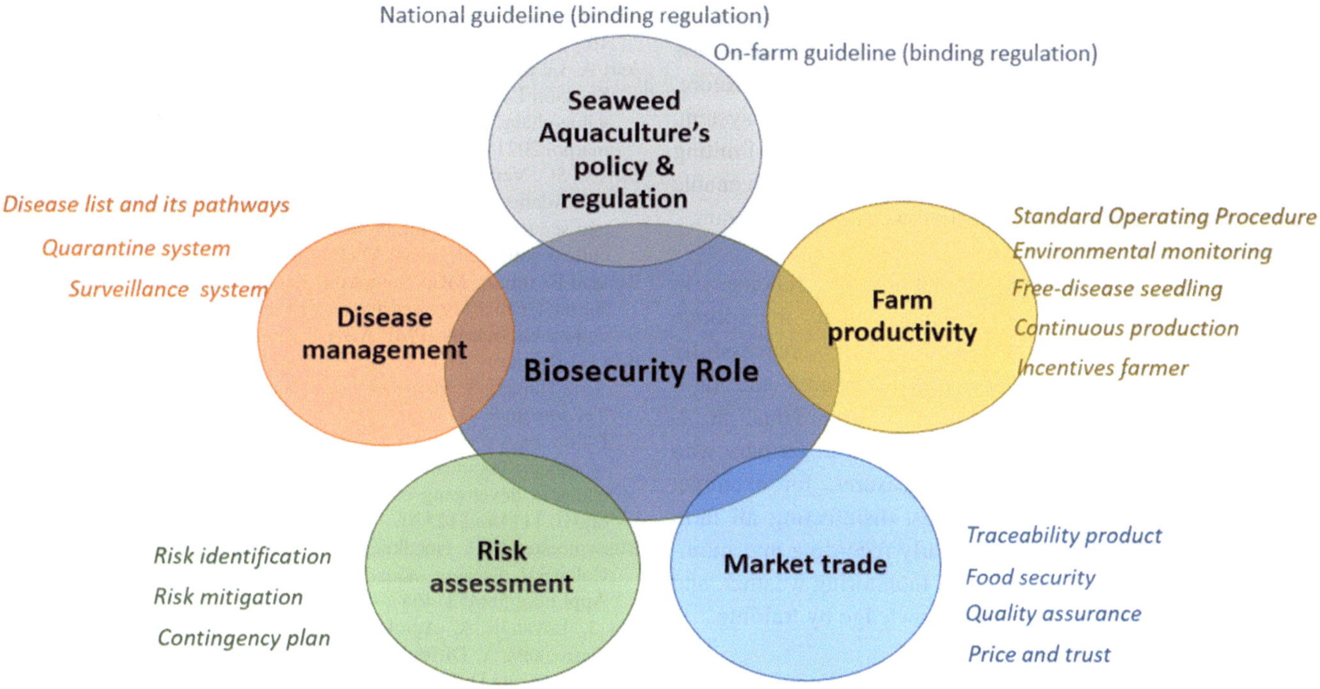

Fig. 7.4 Concept of biosecurity for seaweed aquaculture at the national and on-farm levels

have been affected by the low-income nature of seaweed farming, to which the farmer knowledge was limited (mostly having a low level of education) and the capacity to receive the updated information was minimal. Therefore, the implementation of such new techniques, timely procedures in activities, and additional operational costs, including proper pest and disease management measures and biosecurity applications, were minimally undertaken. This could facilitate a reconciliation and a compromise between the science, policy, and practice knowledge gap, which could benefit from improving and safeguarding the seaweed industry. Various biosecurity risks occur continuously at eucheumatoid farms; therefore, strategies for improving the health management practices of farmers in the most productive countries are important. The recommended biosecurity strategies are as follows:

1. *Produce Specific Seaweed Aquaculture Regulations*
   The establishment of seaweed-specific regulations and policies is beneficial to the industry and will provide appropriate management strategies that can effectively be enforced on a national scale. The introduction of legally binding seaweed aquaculture regulations, aquaculture risk mitigation processes, and a health management strategy, plus the development of new technology and innovative concepts of implementation in the eucheumatoid aquaculture industry, will also secure the sustainable future of the sector.
2. *Implementation of Seaweed Aquaculture Risk Assessment*
   Risk assessment is an important biosecurity application at the national level for evaluating and identifying risks at the farm level based on the occurrence of hazards. At present, there are no assessments focusing on seedling management, minimal protection of wild seaweed stocks, and, subsequently, a scarcity of wild eucheumatoid seaweeds, which partially contributes to the low genetic diversity in farming cultivars. All of these factors combined with the lack of biosecurity measures are leading to declining production in many of the leading seaweed-producing countries. Identifying farm risks enables the mitigation process to be more accurately planned and allows the inclusion of the precautionary approach in the health management strategy. It is important that this risk assessment can be adapted to all eucheumatoid-producing countries, since they are all facing similar risks.
3. *Implementation of Seaweed Quarantine System*
   An appropriate quarantine system is imperative to detect the introduction and transfer of pathogens within and between farms. The introduction of a robust quarantine system will minimize the spread of any pathogens that have been unintentionally introduced.
4. *Provide a Eucheumatoid Pathogen List*
   Currently, eucheumatoid pathogens (e.g., bacteria, viruses, parasites, grazers, biofouling organisms) are poorly studied in the main producing countries, particularly Indonesia and the Philippines. As a consequence, the

list of endemic and exotic eucheumatoid pathogens is lacking, which may have led to a lack of awareness amongst the seaweed farmers and stakeholders. Therefore, there is poor monitoring of the seaweed health system, which specifically focuses on primary yield-limiting pathogens. It is crucial that this list is developed to enable the implementation of an effective monitoring program.

5. *Incentivization for Farmers*
   There are no existing regulations or schemes in eucheumatoid producing countries that incorporate incentives to reward farmers for successfully incorporating effective biosecurity measures and farm management practices on their farms. Thus, it is recommended that incentives be offered to farmers who adapt and practice biosecurity measures, for example, cultivating only healthy seedlings, disinfecting all farm equipment by sun drying, regularly recording and monitoring disease infection at farms, monitoring weather, and updating disease management knowledge by training.
6. *Standardization of Farm Practices*
   Standardizing farming practices from the preparation to harvesting stage will secure the industry from inappropriate practices by the farmers (e.g., use of crops infected with IID or EFA in multiple growing cycles, poor disinfection of farm equipment, mixed farming of healthy and unhealthy seaweed crops, improper disposal of infected crop, poor skills in the monitoring of water quality). Since farm certification may be costly to individual seaweed farmers, providing evidence to demonstrate the benefits of introducing biosecurity measures and standard protocols and training for farm operations to guide farmers is highly recommended to effectively implement the health management strategy and minimize the eucheumatoid cultivation risks.

Acknowledgments This work was part of the GlobalSeaweedSTAR project supported by the Global Challenge Research Fund (GCRF) Biotechnology and Biological Sciences Research Council Grant 2007 No. BB/P027806/1 and University Grant No. IF015-2019. The authors would like to acknowledge to all seaweed farmers who contributed to the survey and Mr. Azam Asri for providing a photo to be used in this publication.

References

Achmad M, Alimuddin A, Widyastuti U, Sukenda S, Suryanti E, Harris E (2016) Molecular identification of new bacterial causative agent of ice-ice disease on seaweed *Kappaphycus alvarezii*. PeerJ Prep 4. https://doi.org/10.7287/peerj.preprints.2016v1

Adams C, Shumway S, Whitlatch R, Getchis T (2011) Biofouling in marine molluscan shellfish aquaculture: a survey assessing the business and economic implications of mitigation. J World Aquacult Soc 42:242–252

Asefa A, Abunna F (2018) Maintenance of fish health in aquaculture: review of epidemiological approaches for prevention and control of infectious disease of fish. Vet Med Intern 5432497. https://doi.org/10.1155/2018/5432497

Asri A, Le Masson V, Montalescot V, Lim PE, Nor AMR, Hussin H, Shaxson L (2021) The role of migrants in the Malaysian seaweed value-chain. Mar Policy 134:104812. https://doi.org/10.1016/j.marpol.2021.104812

Bayliss SC, Verner-Jeffreys DW, Bartie KL, Aanensen DM, Sheppard SK, Adams A, Feil EJ (2017) The promise of whole genome pathogen sequencing for the molecular epidemiology of emerging aquaculture pathogens. Front Microbiol 8:121

Bondad-Reantaso MG, Sumption K, Subasinghe R, Lawrence M, Berthe F (2018) Progressive management pathway to improve aquaculture biosecurity (PMP/AB)1. FAO Aquac Newslett 58:9–1

Bondad-Reantaso MG, Fejzic N, MacKinnon B, Huchzermeyer D, Seric-Haracic S, Mardones FO, Mohan CV, Taylor N, Jansen MD, Tavornpanich S, Hao B, Huang J, Leano EM, Li Q, Liang Y, Dall'occo A (2021) A 12-point checklist for surveillance of diseases of aquatic organisms: a novel approach to assist multidisciplinary teams in developing countries. Rev Aquac 1–19:1469. https://doi.org/10.1111/raq.12530

Bouwmeester MM, Goedknegt AM, Poulin R, Thieltges DW (2021) Collateral diseases: aquaculture impacts on wildlife infections. J Appl Ecol 58:453–464

Cai J, Lovatelli A, Aguilar-Manjarrez J, Cornish L, Dabbadie L, Desrochers A, Diffey S, Garrido Gamarro, E, Geehan J, Hurtado AQ, Lucente D, Mair G, Miao W, Potin P, Przybyla C, Reantaso M, Roubach R, Tauati M, Yuan X (2021) Seaweeds and microalgae: an overview for unlocking their potential in global aquaculture development. FAO Fisheries and Aquaculture Circular No. 1229. Rome, FAO. https://doi.org/10.4060/

Campbell I, Kambey CSB, Mateo JP, Rusekwa SB, Hurtado AQ, Msuya FE, Stentiford GD, Cottier-Cook EJ (2020) Biosecurity policy and legislation for the global seaweed aquaculture industry. J Appl Phycol 32:2133–2146. https://doi.org/10.1007/s10811-019-02010-5

Campbell I, Mateo JP, Rusekwa SB, Kambey CSB, Hurtado AQ, Msuya FE, Cottier-Cook EJ (2022) An international evaluation of biosecurity management capacity in the seaweed aquaculture industry. J Env Manag 304:114112. https://doi.org/10.1016/j.jenvman.2021.114112

Cottier-Cook EJ, Nagabhatla N, Badis Y, Campbell ML, Chopin T, Dai W, Fang J, He P, Hewitt CL, Kim GH, Huo Y, Jiang Z, Kema G, Li X, Liu F, Liu H, Liu Y, Lu Q, Luo Q, Mao Y, Msuya FE, Rebours C, Shen H, Stentiford GD, Yarish C, Wu H, Yang X, Zhang J, Zhou Y, Gachon CMM (2016) Safeguarding the future of the seaweed aquaculture industry. UN Univ Inst Water Environ Heal Scott Assoc Mar Sci 12

Cottier-Cook EJ, Nagabhatla N, Asri A (2021) Ensuring the sustainable future of the rapidly expanding global seaweed aquaculture industry – a vision. UNU-CRIS and Scottish Association for Marine Science Policy Brief. ISBN 978-92-808-9135-5. https://cris.unu.edu/gsstarpolicybrief

Cottier-Cook EJ, Cabarubias JP, Brakel J, Brodie J, Buschmann AH, Campbell I, Critchley AT, Hewitt CL, Huang J, Hurtado AQ, Kambey CSB, Lim PE, Liu T, Mateo JP, Msuya FE, Qi Z, Shaxson L, Stentiford GD, Bondad-Reantaso MG (2022) A new progressive management pathway for improving seaweed biosecurity. Nat Comm 13:7401

Department of Standards Malaysia (2012) Malaysian Standard-seaweed Cultivation No 2467–2012, SIRIM Berhad, Selangor, 6 pp

DOFM (2012) Manual pengkulturan rumpai laut (Kappaphycus/Eucheuma). Jabatan Perikanan Malaysia. 66 pp

DOFM (2018) Annual Report of Malaysian Seaweed Production, Portal Rasm. Jab. Perikan Malaysia, 2018. Accessed 5 Oct 2020. https://www.dof.gov.my/

Eraza DD, Bahron A, Alin J, Mahmud R, Malusirang S (2017) On going assessment of issues in the seaweed farming industry in Sabah. Malaysia J Asian Acad Appl Bus 4(2017):49–60

Faisan JP, Luhan MRJ, Sibonga RC, Mateo JP, Ferriols VMEN, Brakel J, Ward GM, Ross S, Bass D, Stentiford GD, Brodie J, Hurtado AQ (2021) Preliminary survey of pests and diseases of eucheumatoid seaweed farms in The Philippines. J Appl Phycol 33:2391–2405

FAO (2007) Biosecurity toolkit. FAO, Rome. http://www.fao.org/3/a1140e/a1140e00.htm

FAO (2020) The state of world fisheries and aquaculture. Sustainability in action. FAO, Rome. https://doi.org/10.4060/ca9229en

FAO FishStatJ (2021) Red seaweeds aquaculture production by species and country or area. Accessed 22 Jan 2022

Fasina FO, Ali AM, Yilma JM, Thieme O, Ankers P (2012) The cost–benefit of biosecurity measures on infectious diseases in the Egyptian household poultry. Prev Vet Med 103(2–3):178–191. https://doi.org/10.1016/j.prevetmed.2011.09.016

Foster R, Peeler E, Bojko J, Clark PF, Morritt D, Roy HE, Stebbing P, Tidbury HJ, Wood LE, Bass D (2021) Pathogens co-transported with invasive non-native aquatic species: implications for risk analysis and legislation. Neo Biota 69:79–102

Freitas J, Vaz-Pires P, Câmara JS (2020) From aquaculture production to consumption: freshness, safety, traceability and authentication, the four pillars of quality. Aquaculture 518:734857. https://doi.org/10.1016/j.aquaculture.2019.734857

GoT (1998) The United Republic of Tanzania, plant protection regulation. http://extwprlegs1.fao.org/docs/pdf/tan19459

.Hedberg N, von Schreeb K, Charisiadou S, Jiddawi NS, Tedengren M, Nordlund LM (2018) Habitat preference for seaweed farming – a case study from Zanzibar, Tanzania. Ocean Coast Manag 134:186–195

Henriksson PJG, Troell M, Banks LK, Belton B, Beveridge MCM, Klinger DH, Pelletier N, Phillips MJ, Tran T (2021) Interventions for improving the productivity and environmental performance of global aquaculture for future food security. One Earth 4:1220. https://doi.org/10.1016/j.oneear.2021.08.009

Hurtado AQ, Neish IC, Critchley AT (2019) Phyconomy: the extensive cultivation of seaweeds, their sustainability and economic value, with particular reference to important lessons to be learned and transferred from the practice of eucheumatoid farming. Phycologia 58:472–483. https://doi.org/10.1080/00318884.2019.1625632

Hurtado AQ, Luhan MRJ, Ferriols VMEN, Faisan JP, Mateo JP, Sibonga RM, Suyo JGB (2021) Towards a robust and resilient seaweed aquaculture in The Philippines. UPV -SEAFDEC Policy Brief #01

Indonesian Fisheries Statistics Year Book (2017) Statistics of marine and coastal resources. BPS Press, Jakarta. 708 pp

Jia B, St-Hilaire S, Singh K, Gardner IA (2017) Biosecurity knowledge, attitudes and practices of farmers culturing yellow catfish (*Pelteobagrus fulvidraco*) in Guangdong and Zhejiang provinces, China. Aquaculture 471:146–156

Kambey CSB, Campbell I, Sondak CFA, Nor ARM, Lim PE, Cottier-Cook EJ (2020) An analysis of the current status and future of biosecurity frameworks for the Indonesian seaweed industry. J Appl Phycol 32:2147. https://doi.org/10.1007/s10811-019-02020-3

Kambey CSB, Campbell I, Cottier-Cook EJ, Nor ARM, Kassim A, Sade A, Lim PE (2021a) Seaweed aquaculture: a preliminary assessment of biosecurity measures for controlling the ice-ice syndrome and pest outbreaks of a *Kappaphycus* farm. J Appl Phycol 33:3179. https://doi.org/10.1007/s10811-021-02530-z

Kambey CSB, Campbell I, Cottier-Cook EJ, Nor ARM, Kassim A, Sade A, Lim PE (2021b) Evaluating biosecurity policy implementation in the seaweed aquaculture industry of Malaysia, using the quantitative knowledge, attitude, and practices (KAP) survey technique. Mar Policy 134:104800. https://doi.org/10.1016/j.marpol.2021.104800

Kambey CSB, Lim PE, Cottier-Cook EJ, Campbell I, Poong ZW, Kassim A (2022) Standard operating procedure of eucheumatoid cultivation using biosecurity-based approach. Institute of Ocean and Earth Sciences, Universiti Malaya 40 pp

Largo DB, Msuya FE, Menezes A (2020) Understanding disease and control in seaweed farming in Zanzibar. FAO Fisheries and Aquacult Tech Paper No 662. FAO, Rome

Lim PE, Poong SW, Kambey CSB, Tan J, Asri A, Nagabhatla N, Nor AM, Kassim A, Le-Masson V, Shaxson L, Brodie J, Gachon CMM, Cottier-Cook EJ, De Lombaerde P (2021) Sustainable value chain for the seaweed industry in Malaysia and the ASEAN region: a roadmap for policy formulation. United Nations University Institute on Comparative Regional Integration Studies Policy Brief ISBN 978-92-808-9137-9. https://cris.unu.edu/seaweedindustry-malaysia-and-asean

Manap MGM, Pauzi MFA (2020) MyGAP: Malaysia's gift to seafood world. IOP Conf Ser Earth Environ Sci 414:012012. https://doi.org/10.1088/1755-1315/414/1/012012

Mantri VA, Kambey CSB, Cottier-Cook EJ, Usandizaga S, Buschmann AH, Chung IK, Lui T, Sondak CFA, Qi Z, Lim PE, Nguyen VN (2022) Overview of global *Gracilaria* production, the role of biosecurity policies and regulations in the sustainable development of this industry. Rev Aquac 15(2):801–819

Mariño M, Breckwoldt A, Teichberg M, Kasec A, Reuter H (2019) Livelihood aspects of seaweed farming in Rote Island. Indonesia Mar Policy 107:103600

Mateo JP, Campbell I, Cottier-Cook EJ, Luhan MRJ, Ferriols VMEN, Hurtado AQ (2020) Analysis of biosecurity-related policies governing the seaweed industry in The Philippines. J Appl Phycol 32:2009–2022

Mateo JP, Campbell I, Cottier-Cook EJ, Luhan MRJ, Ferriols VMEN, Hurtado AQ (2021) Understanding biosecurity: knowledge, attitudes and practices of seaweed farmers in The Philippines. J Appl Phycol 33:997–1010

Msuya FE (2006) The impact of seaweed farming on the social and economic structure of seaweed farming communities in Zanzibar. Tanzania, ETI Bioinformatics

Msuya FE (2020) Seaweed resources of Tanzania: status, potential species, challenges, and development potentials. Bot Mar 63(4):371–380

Msuya FE, Bolton J, Pascal F, Narrain K, Nyonje B, Cottier-Cook EJ (2022) Seaweed farming in Africa: current status and future potential. J Appl Phycol 34:985–1005. https://doi.org/10.1007/s10811-021-02676-w

Mulyaningrum SRH, Suwoyo HS, Paena M, Tampangallo BR (2019) Epiphyte identification on *Kappaphycus alvarezii* seaweed farming area in Arungkeke Waters, Jeneponto and the effect on carrageenan quality. Indo J Mar Sci 24(3):146–152

Naylor RL, Hardy RW, Buschmann AH, Bush SR, Cao L, Klinger DH, Little DC, Lubchenco J, Shumway SE, Troell M (2021) A 20-year retrospective review of global aquaculture. Nature 591:551–563. https://doi.org/10.1038/s41586-021-03308-6

Ndawala MA, Msuya FE, Cabarubias JP, Buriyo A, Cottier-Cook EJ (2021) Seaweed biosecurity in Tanzania: lessons to be learned from other major plant crops. Env Chall 5:100319. https://doi.org/10.1016/j.envc.2021.100319

Ndawala MA, Msuya FE, Cabarubias JP, Kambey CSB, Buriyo AS, Mvungi EF, Cottier-Cook EJ (2022) Effect of biosecurity practices and diseases on growth and carrageenan properties of *Kappaphycus alvarezii* and *Eucheuma denticulatum* cultivated in Zanzibar, Tanzania. J Appl Phycol 34:3069–3085. https://doi.org/10.1007/s10811-022-02835-7

Neish IC, Sepulveda M, Hurtado AQ, Critchley AT (2017) Reflection of commercial development of Eucheumatoid seaweed farming. In: Hurtado AQ, Critchley AT, Neish IC (eds) Tropical seaweed farming trends, problems and opportunities: focus on *Kappaphycus* and *Eucheuma*. Springer, pp 1–27

Nor AM, Gray TS, Caldwell GS, Stead SM (2017) Is a cooperative approach to seaweed farming effectual? An analysis of the seaweed cluster project (SCP), Malaysia. J Appl Phycol 29:2323–2337

Oidtmann BC, Thrush MA, Denham KL, Peeler EJ (2011) International and national biosecurity strategies in aquatic animal health. Aquacult 320:22–33

Paseka RE, White LA, Van de Waal DB, Strauss AT, González AL, Everett RA, Peace A, Eric W, Seabloom EW, Frenken T, Borer ET (2020) Disease-mediated ecosystem services: pathogens, plants, and people. Trends in Ecol and Evol 35(8): 731–743

Philippine Statistic Authority (2018) Fisheries Situationer January–December 2018. http://www.psa.gov.ph/sites/default/files/attachments/ird/specialrelease/SpecialRelease2018REV2.pdf. Accessed on 20 Jan 2020, 4 pp

Philippine Statistic Authority (PSA) (2013) Selected Statistics on Agriculture. ISSN: 2012-0362. http://www.psa.gov.ph. Accessed 15 Aug 2018, 67 pp

Philippine Statistic Authority (PSA) (2016) Fisheries Situationer January–December 2016. https://psa.gov.ph/sites/default/files/FisheriesSituationer2016_0.pdf. Accessed 15 Aug 2018, 39 pp

Porse H, Rudolph B (2017) The seaweed hydrocolloid industry: 2016 updates, requirements, and outlook. J Appl Phycol 29:2187–2200. https://doi.org/10.1007/s10811-017-1144-0

Quiaoit HAR, Uy WH, Bacaltos DGG, Chio PBR (2018) Seaweed area GIS-based mapping. production support system for sustainable seaweeds farming in The Philippines. Xavier University Press, Cagayan de Oro City, Philippines, 158 pp

Reed MS, Curzon R (2015) Stakeholder mapping for the governance of biosecurity: a literature review. J Integr Environ Sci 12:15–38

Rimmer MA, Larson S, Lapong I, Purnomo AH, Pong-Masak PR, Swanepoel L, Paul NA (2021) Seaweed aquaculture in Indonesia contributes to social and economic aspects of livelihoods and community wellbeing. Sustainability 13:10946. https://doi.org/10.3390/su131910946

Ritter C, Jansen J, Roche S, Kelton DF, Adams CL, Orsel K, Erskine RJ, Benedictus G, Lam TJGM, Barkema HW (2017) Invited review: determinants of farmers' adoption of management-based strategies for infectious disease prevention and control. J Dairy Sci 100:3329–3347

Rodgers C, Arzul I, Carrasco N, Furones Nozal DF (2019) A literature review as an aid to identify strategies for mitigating Ostreid herpesvirus 1 in *Crassostrea gigas* hatchery and nursery systems. Rev Aquac 11:565–585. https://doi.org/10.1111/raq.12246

Rusekwa SB, Campbell I, Msuya FE, Buriyo AS, Cottier-Cook EJ (2020) Biosecurity policy and legislation of the seaweed aquaculture industry in Tanzania. J Appl Phycol 32:4411–4422

Sekaran HR (2014) Biosecurity system in Malaysian fisheries: gearing up for safe and quality seafood. Fish People 12:33–37

Shannon C, Stebbing PD, Dunn AM, Quinn CH (2020) Getting on board with biosecurity: evaluating the effectiveness of marine invasive alien species biosecurity policy for England and Wales. Mar Policy 122:104275. https://doi.org/10.1016/j.marpol.2020.104275

Sikes BA, Bufford JL, Hulme PE, Cooper JA, Johnston PR, Duncan RP (2018) Import volumes and biosecurity interventions shape the arrival rate of fungal pathogens. PLoS Biol 16(5):e2006025. https://doi.org/10.1371/journal.pbio.2006025

Simatupang NF, Pong-Masak PR, Ratnawati P, Agusman PNA, Rimmer MA (2021) Growth and product quality of the seaweed *Kappaphycus alvarezii* from different farming locations in Indonesia. Aqua Rep 20:100685. https://doi.org/10.1016/j.aqrep.2021.100685

Soliman T, Inglis GJ (2018) Forecasting the economic impacts of two biofouling invaders on aquaculture production of green lipped mussels *Perna canaliculus* in New Zealand. Aquacult Environ Interact 10:1–12

Stentiford GD, Bateman IJ, Hinchliffe SJ, Bass D, Hartnell R, Santos EM, Devlin MJ, Feist SW, Taylor NGH, Verner-Jeffreys DW, van Aerle R, Peeler EJ, Hiqman WA, Smith L, Baines R, Behringer DC, Katsiadaki I, Froehlich HE, Tyler CR (2020) Sustainable aquaculture through the one health lens. Nature Food 1.8 (2020): 468-474

Suyo JGB, Le Masson V, Shaxson L, Luhan MRM, Hurtado AQ (2021) Navigating risks and uncertainties: risk perceptions and risk management strategies in the Philippine seaweed industry. Mar Policy 126:104408

Tano SA, Halling C, Lind E, Buriyo A, Wikström SA (2015) Extensive spread of farmed seaweeds causes a shift from native to non-native haplotypes in natural seaweed beds. Mar Biol 162:1983–1992

Tsiresy G, Preux J, Lavitra T, Dubois P, Lepoint G, Eeckhaut I (2016) Phenology of farmed seaweed *Kappaphycus alvarezii* infestation by the parasitic epiphyte *Polysiphonia* sp. in Madagascar. J Appl Phycol 28:2903–2914

Vairappan CS, Anangdan SP, Tan KL, Matsunaga S (2010) Role of secondary metabolites as defense chemicals against ice-ice disease bacteria in bio fouler at carrageenophyte farms. J Appl Phycol 22: 305–311

Ward GM, Faisan-Jr JP, Cottier-Cook EJ, Gachon C, Hurtado AQ, Lim PE, Matoju I, Msuya FE, Bass D, Brodie J (2020) A review of reported seaweed diseases and pests in aquaculture in Asia. J World Aquacult Soc 51:815–828

Ward GM, Kambey CSB, Faisan JP, Tan P, Daumich CC, Matoju I, Stentiford GD, Bass D, Lim PE, Brodie J, Poong SW (2021) Ice-Ice disease: an environmentally and microbiologically driven syndrome in tropical seaweed aquaculture. Rev Aquac 1–26:414. https://doi.org/10.1111/raq.12606

Watson DI, Shumway SE, Whitlatch RB (2009) Biofouling and the shellfish industry. In: Shumway SE, Rodrick GE (eds) Shellfish safety and quality. Woodhead Publishing, Cambridge, pp 317–337

Welsh MJ, Turner JA, Epanchin-Niell RS, Monge JJ, Soliman T, Robinson AP, Kean JM, Phillips C, Stringer LD, Vereijssen J, Liebhold AM, Kompas T, Ormsby M, Brockerhoff EG (2021) Approaches for estimating benefits and costs of interventions in plant biosecurity across invasion phases. Ecol Appl 31(5):e02319

Zamroni A (2021) Sustainable seaweed farming and its contribution to livelihoods in Eastern Indonesia. IOP Conf Series: Earth Env Sci 718:012099

Zamroni A, Yamao M (2011) Coastal resource management: fishermen's perception of seaweed farming in Indonesia. World Acad Sci Eng and Tech 5:2011–2020

The Bio Economic Seaweed Model (BESeM) for Modeling *Kappaphycus* Cultivation in Indonesia

8

P. A. J. van Oort, Nita Rukminasari, Gunarto Latama, Jan Verhagen, and A. van der Werf

Abstract

A simple model was presented for simulating production and economics of tropical seaweed cultivation. The model has a limited number of parameters, which makes it easily amenable to other seaweeds and other sites. Parameter estimates were presented and uncertainties in parameter estimation were discussed. A simulation example suggests farmers can increase income by using cuttings with a higher weight than they are currently using. A literature review suggests still much need for further experimentation.

Keywords

Simulation · Model · Production · Economics · Seaweed

8.1 Introduction

Seaweed farming is an important source of income for coastal communities in Indonesia and other tropical countries and a growing economic sector for decades (FAO 2018). Common tropical (red) seaweeds cultivated in Indonesia are *Gracilaria*, *Kappaphycus alvarezii* (cottonii), and *Eucheuma denticulatum* (spinosum). Important producers are India (Periyasamy and Rao 2017; Periyasamy et al. 2019), Indonesia (Setyawidati et al. 2017; Kasim et al. 2016), Malaysia (Phang et al. 2019), the Philippines (Dawes et al. 1994), and Vietnam (Bui et al. 2019). Seaweed growth in these tropical countries is markedly different from that in temperate climates. In temperate climates, for the example case, only one seaweed crop is grown per year, or two distinct seaweeds such as *Saccharina* in winter and *Ulva* in summer. Whereas in the tropics the same seaweed species is cultivated a number of times per year and year after year. In temperate climates, seaweed farms buy inoculated lines from companies specialized in seaweed propagation. In tropical climates cultivation is predominantly through on-farm, vegetative replanting part of the harvested seaweed. Most commonly in the tropics, seaweeds are cultivated in cycles of 45 days. A line is planted out and harvested after 45 days. The harvested seaweed is separated in two piles, one large pile to be dried and sold, a smaller part to be replanted. The replanted part then starts a new cycle of 45 days and so forth. Figure 8.1 provides an image of different steps in the cultivation cycle in Sulawesi, Indonesia. The FAO report by (McHugh 2003) provides a more elaborated overview of tropical seaweed cultivation in various countries.

It is unclear if, and how, current farm management practices can be improved. Consider a cultivation period of a year, 360 days. Is a farmer better off with 12 cycles of 30 days, 8 cycles of 45 days, or 6 cycles of 60 days? (note in all cases cultivation is for 360 days). While yield per harvest will normally be higher with a longer cycle, aggregate yield (over a year) might be higher with more, yet shorter cycles. A second important operational farm management decision is how to split up harvests into the fraction sold and the fraction replanted. Selling a large fraction of the harvest (say 90%) may seem attractive to the farmer. Imagine a harvest of 1 ton/ha at a selling fraction of 90%; in this scenario, the farmer sells 0.9 t/ha and replants 0.1 t/ha. A low replanting weight results in slow initial growth. Imagine harvest would be 1.6 t/ha at a selling fraction of 75%. In this second scenario, the farmer sells 1.6*0.75 = 1.2 t/ha and replants 0.25*1.6 = 0.4 t/ha. Total sales are then higher (1.2 t/ha vs. 0.9 t/ha), despite the fact that a smaller fraction of the harvest is sold (75% vs. 90%). Ultimately, farmers are more interested in net profit than in yield. The cultivation system is very labor intensive, but in a peaked manner. That is, almost no labor is needed while the seaweed is out in the sea growing

P. A. J. van Oort (✉) · J. Verhagen · A. van der Werf
Wageningen Plant Research (WPR), Wageningen, The Netherlands
e-mail: pepijn.vanoort@wur.nl; jan.verhagen@wur.nl; adrie.vanderwerf@wur.nl

N. Rukminasari · G. Latama
Faculty of Marine Science and Fisheries, Hasanuddin University, Makassar, Indonesia
e-mail: nita.r@unhas.ac.id

A. T. Critchley et al. (eds.), *Tropical Phyconomy Coalition Development*, Developments in Applied Phycology 11,
https://doi.org/10.1007/978-3-031-47806-2_8

Fig. 8.1 Pictures of typical seaweed cultivation in South West Sulawesi, Indonesia. Top left: seaweed is growing on lines in between wooden poles. Top right: seaweed is harvested by boat. Bottom left: seaweed to be sold is dried on platforms on the beach. Bottom right: seaweed sorting and tying up new lines is a labor-intensive process. Photos by Dr P.A.J. van Oort, 2019

on its line. A peak in labor demand occurs on the day of harvesting. That day, the seaweed is harvested, sorted into the two piles (sales and replants); the sales pile is placed on platforms for drying; the replant file is sorted into individual plants (in Indonesian Bahasa: bibit), which are attached to a new clean line; and these new lines are set up in the sea again. Depending on farm labor availability and costs, it may be more profitable for farmers to adopt a longer growing period (thus less harvests and less peak labor costs within 1 year), despite the lower aggregate yield associated with longer cycles.

The farm management decisions, which we discussed above, can be investigated purely experimentally. Describing these management decisions in a mathematical model offers significant benefits. Modeling can be useful for summarizing current knowledge and for identifying knowledge gaps. Modeling can be a way to integrate knowledge from different domains, such as biology and economy. Finally, once a model has been successfully validated, it can be used to compare management decisions such as those discussed above. We will present example comparisons in this chapter.

For temperate climates, recent advances have been made in seaweed modeling (Broch and Slagstad 2012; Duarte and Ferreira 1997; Jackson 1987; Lavaud et al. 2020; van der Molen et al. 2018; Venolia et al. 2020; Zhang et al. 2016). Modifying these models to a tropical context is quite impossible, because they require so many input parameters, while generally very limited experimental data are available from tropical seaweed cultivation systems. The most commonly used model in tropical seaweed studies in India, Indonesia, and the Philippines is the exponential growth model (Dawes et al. 1994; Dawes et al. 1993; Kasim et al. 2016; Periyasamy et al. 2019; Setyawidati et al. 2017; Hurtado et al. 2001), in which fresh weight $w_f(t) = w_{f,0} * exp(RGR * t)$ increases exponentially over time t since planting with relative growth rate RGR (g g^{-1} day^{-1}). An obvious limitation of this model is that it is only useful for describing the initial growth phase of a seaweed, which is indeed often exponential. Species do not continue growing exponentially forever. This is reflected in lower *RGR* values reported when a seaweed is grown for a longer period of time (Kasim et al. 2016; Periyasamy et al. 2019). Furthermore, all these biological studies lack an economic component. In turn, the few studies on tropical seaweed farm economics (Tahang et al. 2019; Limi et al. 2018) lack a biological component.

There is a need for a seaweed growth model positioned somewhere in between the complex biological models developed for the temperate climates and the simple exponential growth model, in combination with econ mics of seaweed cultivation. The objective of this chapter is to present such a model. The following part of this chapter contains two main sections. In Sect. 8.2, we present the BESeM model, and in Sect. 8.3, we present example simulations. We conclude the chapter with a discussion.

8.2 Bio Economic Seaweed Model (BESeM)

We start with listing model variables and parameters. The Bio Economic Seaweed Model (BESeM) consists of a biological module and an economic module. In the model, most variables and parameters are expressed per square meter; in Sect. 8.2.4, we show how the same variables can be expressed in other units, e.g., expressing yield per plant, per line or at the farm level.

8.2.1 Model Variables and Parameters

Model variables are defined in Table 8.1.

Intentionally, the model is kept as simple as possible, which makes it easily amenable for simulating different tropical seaweeds such as *Gracilaria*, *Kappaphycus alvarezii* (cottonii), and *Eucheuma denticulatum* (spinosum). The model has only 13 parameters, and for this study, we concentrate on cottonii, and carrageenan produced from it:

- Eight biochemical parameters ($rgrmax$, $w_{f,max}$, m_f, m_{sd}, cf_{min}, cf_{max}, cfk, and cft_{50}), of which two are for growth ($rgrmax$ and $w_{f,max}$), two are for converting from fresh weight to semidry weight (m_f and m_{sd}), and four are for calculating carrageenan concentration (cf_{min}, cf_{max}, cfk, and cft_{50}).
- Three economic parameters (FGP_c, PC_m, and PC_h).
- Two (farmer) decision parameters: hcl (harvest cycle length) and $w_{f,O}$ (replanting weight).

We present parameter estimates in Table 8.2. Additionally, in Table 8.2, we present parameters on farm configuration (number of lines per farm, line length, line spacing, plants per meter line), which can be useful for unit conversion but which are not critically needed as input for the model.

8.2.2 Seaweed Growth Model

Seaweed growth formulae cited in this section are listed in Fig. 8.2.

Time (t) in the model is expressed in days after planting. For a year with multiple plantings, t can be calculated as shown in Eq. (1) (Fig. 8.2), where doy is the Julian day of the year, doy_O is any planting day of the year, and hcl is the cycle length including the harvest day. The modulo—*mod*()—is a standard mathematical operator that returns the remainder after division of one number by another. For example $mod(44,45) = 44$, $mod(45,45) = 0$, $mod(46,45) = 1$, $mod(47,45) = 2$, $mod(89,45) = 44$, $mod(90,45) = 0$, $mod(91,45) = 1$, etc. It generates a sawtooth timeseries in which t increases by 1 every day until reaching the harvest date (after 45 days), drops back to 0, and then increases again by 1 every day until reaching the next harvest date. For a simulation for a year, we will simulate from $doy = 1$ to 360 and we set $doy_O = 1$.

We introduce a sigmoid biomass growth model (Fig. 8.3). Initially, the seaweed grows exponentially with a relative growth rate rgrmax. Then, growth becomes almost linear, due to various factors such as self-shading, erosion, predation by herbivores, etc. Finally, growth flattens off to a maximum weight $w_{f,max}$. Equation (2) is the standard differential equation used in ecophysiological models for describing sigmoid growth.

From Eq. (2), we can derive Eq. (3), showing gross biomass (kg m^{-2}) at any point in time t (days after planting) for a given start weight $w_{f,O}$.

Of the harvested gross yield $w_{f,g}(t)$, one-part $w_{f,O}$ is "replanted" (the same day, to avoid desiccation), and the

Table 8.1 Variables of the Bio Economic Seaweed Model

Variable	Unit	Description
t	days	Days after planting
$w_{f,g}(t)$	kg FW m^{-2}	Gross biomass fresh from sea, at time t after planting
$w_{f,n}(t)$	kg FW m^{-2}	Net biomass fresh from sea, at time t after planting
$W_{f,n}(hcl,w_{f,O})$	kg FW m^{-2} year^{-1}	Fresh (f) net (n) annual harvested biomass as a function of replanting weight $w_{f,O}$ and harvest cycle hcl
$W_{sd,n}(hcl,w_{f,O})$	kg SDW m^{-2} year^{-1}	Semidried (sd) net (n) annual harvested biomass
cf(t)		Kilogram carrageenan per kilogram of semidry seaweed
$CAR(hcl,w_{f,O})$	kg m^{-2} year^{-1}	Net annual harvested carrageenan
$FGP_{sd}(t)$	IDR kg^{-1} SDW	Farmgate price for semidry seaweed as a function of t, in Indonesian Rupiah
$I_g(hcl,w_{f,O})$	IDR m^{-2} year^{-1}	Gross annual income per m^2
$PC(hcl)$	IDR m^{-2} year^{-1}	Annual production cost per m^2
$I_n(hcl,w_{f,O})$	IDR m^{-2} year^{-1}	Net annual farm income per m^2

Table 8.2 Parameter estimates for species *Kappaphycus alvarezii* (cottonii)

Parameters	Unit[a]	Value	BED[b]	Description	Source
rgrmax	g g^{-1} day^{-1}	0.06	B	Maximum relative growth rate	Range reported in literature for *K. alvarezii*: 0.05–0.08 (Periyasamy et al. 2019; Dawes et al. 1994)
$w_{f,max}$	kg FW m^{-2}	5.26	B	Maximum fresh weight	Figure 8.3, author's estimate. This value of $w_{f,max}$ gives for the set $rgrmax = 0.06$, $w_{f,0} = 0.15$, $hcl = 45$ the (farmer) reported 40 kg FW per line = 1.6 kg FW m^{-2} (with $L_l = 25$ and $L_w = 1.0$)
$cf_{\min}$	kg carrageenan kg^{-1} SDW	0.234	B	Minimum chemical content at $t = -\infty$	Figure 8.2. Based on data in (Periyasamy and Rao 2017)
$cf_{\max}$	kg carrageenan kg^{-1} SDW	0.374	B	Maximum chemical content at $t = +\infty$	Figure 8.2
cfk	day^{-1}	0.162	B	Steepness of slope with which content increases with days after planting	Figure 8.2
cft_{50}	day	34.7	B	Days after planting at which chemical content is halfway between $cf_{\min}$ and $cf_{\max}$	Figure 8.2
m_f	g g^{-1}	0.90	B	Moisture content fresh from sea	Authors' interviews[c]
m_{sd}	g g^{-1}	0.343	B	Moisture content semidried (dried on platform on the beach by farmer)	Authors' interviews[c]
$w_{f,0}$	kg FW m^{-2}	0.15	D	Fresh biomass weight replanted	I. Neish pers. comm. (2021): 20–40 g/plant. Authors' calculation: 30 g/plant, with $PD = 5$ plants m^{-1}, line spacing $L_w = 1.0$ m–>0.001*30*5/(1.0) = 0.15
hcl	days	45	D	Harvest cycle length: days from planting to harvesting, including the harvest day	Authors' interviews[c]
FGP_c	IDR kg^{-1} carrageenan	65,341	E	Farmgate price carrageenan	Authors' interviews. Normal farmgate price in 2019 was 23,000 IDR kg^{-1} SDW. According to Eq. 10 in Fig. 8.2, carrageenan content at normal harvesting time (45d) is 0.343, thus $FGP_c = FGP_{sd}(45)/cf(45)$ = 23,000/0.352 = 65,341
PC_m	IDR m^{-2} dayear^{-1}	0.99	E	Production cost for maintenance (m). This includes depreciation of material plus daily maintenance of lines in sea	Authors' estimate based on (Tahang et al. 2019). Tahang et al. report "fixed costs" of in total 2,700,000 IDR farm^{-1} year^{-1}. Assuming a farm with 300 lines of 25 m at 1 m spacing we obtain 2,700,000/(300*25*1.0)/365 = 0.99. Note highly uncertain!
PC_h	IDR m^{-2} cycle^{-1}	580	E	Production cost at harvesting (h). This is the sum of costs of harvesting + drying + replanting	Authors' estimate based on (Tahang et al. 2019). Tahang et al. report "variable costs" of in total 21,750,000 IDR farm^{-1} year^{-1} and they report farmer harvesting 5× per year. Assuming a farm with 300 lines of 25 m at 1 m spacing, we obtain 21,750,000/(300*25*1.0)/5 = 580. Note highly uncertain!
L_N		300	E	Number of lines per farm	Authors' interviews[c]. Range 200–1000
L_l	m	25	E	Length of a line	Authors' observations[c]. Standard rope length
L_w	m	1	E	Width = spacing = distance between lines	Authors' observations[c]
L_{pd}	Plants m^{-1}	5		Plant density, number of plants per meter	Authors' observations[c]

[a] In the units, FW indicates fresh weight, SDW is semidry weight (dried on beach), and IDR is Indonesian rupiah
[b] BED: parameter types—B, biochemical; E, economic; D, decision
[c] Authors interviewed five farmers plus five traders in Sept 2019 in South West Sulawesi. Two of the authors are permanently based in South West Sulawesi

remaining part $w_{f,g}(t)$—$w_{f,0}$ is sold (after drying). Net fresh biomass production is therefore as shown in Eq. (4).

A number of growing cycles can be completed in 1 year. For simplicity, we will consider a year consisting of 360 days, and we will consider growing periods that give a round number of cycles. For example, harvest cycles of $hcl = 30$ and 45 days gives a round number of 12 and 8 cycles. Thus, we can compare cultivation systems over the same total period of 360 days. The annual number of cycles is $N_h = 360/hcl$ (Eq. 5).

(1) $t = mod(doy - doy_0, hcl)$

(2) $\frac{dw_{f,g}}{dt} = rgrmax * w_{f,g} * \left(1 - \frac{w_{f,g}}{w_{f,max}}\right)$

(3) $w_{f,g}(t) = \frac{w_{f,max}}{1+\left(\frac{w_{f,max}-w_{f,0}}{w_{f,0}}\right)*e^{-rgrmax*t}}$

(4) $w_{f,n}(t) = w_{f,g}(t) - w_{f,0}$

(5) $N_h = 360/hcl$

(6) $lgp = hcl - 1$

(7) $W_{f,n}(hcl, w_{f,0}) = N_h * w_{f,n}(lgp, w_{f,0})$

(8)

$W_{f,n}(hcl, w_{f,0}) = \left(\frac{360}{hcl}\right) * \left(\left(\frac{w_{f,max}}{1+\left(\frac{w_{f,max}-w_{f,0}}{w_{f,0}}\right)*e^{-rgrmax*(hcl-1)}}\right) - w_{f,0}\right)$

(9) $W_{sd,n}(hcl, w_{f,0}) = \left(\frac{1-m_f}{1-m_{sd}}\right) W_{f,n}(hcl, w_{f,0})$

(10) $cf(t) = cf_{min} + \frac{cf_{max}-cf_{min}}{1+e^{-cfk(t-cft_{50})}}$

(11) $CAR(hcl, w_{f,0}) = cf(hcl) * W_{n,sd}(hcl, w_{f,0})$

Fig. 8.2 Seaweed growth formulae cited in Sect. 8.2.2

Total net annual production can now be expressed as the annual number of harvests N_h times net biomass harvested per individual cycle. However, note that at the day of harvesting, the seaweed is harvested in the morning, sorted and replanted. Thus, in each cycle, 1 day is "lost," i.e., the seaweed is not growing while not in the sea. With cycles of *hcl* = 45 days, there are eight harvests per year. The seaweed is growing in the sea for 8**lgp* = 8 (*hcl* − 1) = 8* (45–1) = 352 days, and it is on land for sorting for 8 days. Although it matters little for annual production, a theoretically correct statement is that the gross yield at harvesting is

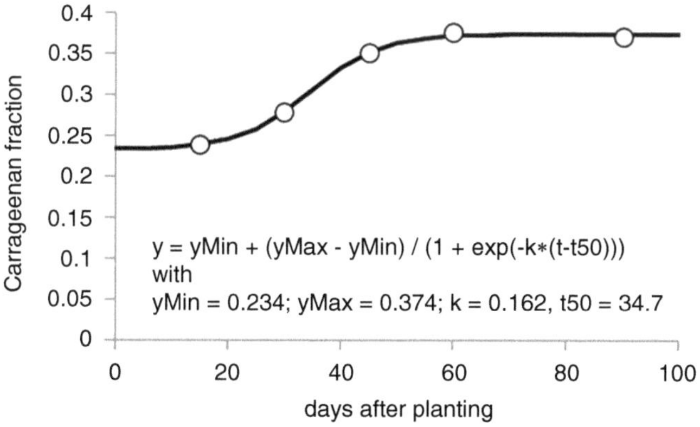

Fig. 8.3 Sigmoid function of carrageenan fraction in *Kappaphycus* as a function of plant age, fitted on data reported in (Periyasamy and Rao 2017)

not $w_{f,g}(hcl)$ but $w_{f,g}(hcl - 1)$. We define the length of the actual growing period *lgp* as *lgp* = *hcl* − 1 per Eq. (6).

Annual net fresh yield is calculated from the number of harvests and the net yield at harvesting per Eq. (7). We use a lower case "*w*" in Eq. (4) for biomass weight at time *t* and an uppercase "*W*" in Eq. (7) for annual total biomass.

Combining the above equations, we can express net annual fresh production $W_{f,n}$ as a function of the four parameters *hcl*, $w_{f,0}$, *rgrmax*, and $w_{f,max}$ as shown in Eq. (8).

After harvesting, the seaweed is dried on platforms on the beach. Annual net semidry weight $W_{sd,n}$ (kg m^{-2}) is calculated from moisture contents of the seaweed fresh from sea (m_f) and semidry (m_{sd}), i.e., after drying on platforms on the beach (Eq. 9).

Typical values for the moisture content fresh from sea (m_f) for the seaweed *Kappaphycus alvarezii* (cottonii) are m_f = 0.9 and for the moisture content after drying on the beach: m_{sd} = 0.35 (35% moisture content after drying; farmer's indication). The conversion factor for calculating semidry weight from fresh weight is then as follows: ((1-m_f)/(1-m_{sd})) = 0.15. For example, a harvest of 40 kg FW (Fresh weight) will contain 0.9*40 = 36 kg water and (1–0.9)*40 = 4 kg dry matter. After drying to m_{sd} = 0.35, the semidry weight is 0.15*40 = 6.15 kg SDW, of which 2.15 kg water and 4 kg dry matter. Check: 2.15/6.15 = 0.35.

Next, we calculate the carrageenan content, which is in most *Kappaphycus/Eucheuma* research expressed as kg carrageenan per kg semidry weight. Periyasamy et al. (2019) showed carrageenan content increases with plant age. From the data reported in (Periyasamy and Rao 2017), we fit the sigmoid function shown per Eq. (10).

This sigmoid function ranges from cf_{min} at $t = -\infty$ to cf_{max} at $t = \infty$; parameter cfk determines the steepness of the slope, and cft_{50} is the time (in days after planting). At 45 days after planting *cf*(45) = 0.352. The seaweed *Eucheuma denticulatum* (spinosum) has a lower carrageenan content than *Kappaphycus*. To make the same equation work for *Eucheuma*, probably the only thing we would need to do is replace the parameters of the *cf*(*t*) Eq. (11) to one that predicts lower *cf*(*t*) for *Eucheuma* than for *Kappaphycus*.

Net annual carrageenan production, which is what ultimately industry will extract from the semidried seaweed collected from farms, can be written as per Eq. (11).

The much-used exponential model (Dawes et al. 1993, 1994; Kasim et al. 2016; Periyasamy et al. 2019; Setyawidati et al. 2017; Hurtado et al. 2001) simulates seaweed biomass with one parameter, the relative growth rate. Here, we introduced a new (sigmoid) model with only two parameters, *rgrmax* and $w_{f,max}$, thus still simple. The second novelty of the model presented here is that we simulate multiple harvests in a cultivation system with replanting. Annual production depends on two (farmers') decision parameters *hcl* and $w_{f,0}$. In its simplest form of simulating fresh biomass that is all we need. We further added to the model two parameters mf and msd for converting fresh biomass to semidry biomass. And we added four parameters, cf_{min}, cf_{max}, cfk, and cft_{50}, for calculating the concentration of the chemical of interest; in the case of the seaweed *Kappaphycus*, we calculate the concentration of the chemical carrageenan.

8.2.3 Economic Model

Economic model formulae cited in this section are listed in Fig. 8.4.

Our interviews suggested that farmers receive a lower price per kg semi-dry seaweed when they harvested earlier than normal. This is consistent with Fig. 8.3, which shows lower carrageenan content for earlier harvest dates. In BESeM, farmgate price is calculated per Eq. (12) (Fig. 8.4), where FGP_c is a constant price per kg carrageenan. Table 8.2 shows how we estimated this parameter. Filling in FGP_c = 65,341 and the normal harvest cycle length of 45 days, we obtain $FGP_{sd}(45)$ = 65,341 * 0.352 = 23,000 IDR kg^{-1} semidry; that was the normal price in 2019 (pre-COVID). Harvesting at, for example, 30 days, with carrageenan content $cf(30)$ = 0.279 (Fig. 8.3), would result in FGP_{sd} (30) = 65,341 * 0.279 = 18,230 IDR kg^{-1} semidry.

Gross annual income I_g (IDR m^{-2} year^{-1}) is calculated per Eq. (13) (Fig. 8.4).

Combining Eqs. (10)–(13), we can also write gross income as a function of carrageenan production per Eq. (14) (Fig. 8.4).

Net income is calculated as gross income minus production costs. Production costs consist of two types of costs:

PC_m = "maintenance" costs, which include material depreciation (ropes, wood, boat, nets) and labor costs (checking growth while lines are in the sea, shaking the lines to get rid of epiphytes and sediment). We refer to these costs as maintenance (m) costs and express them in IDR per m^2 per day, since they occur on a continuous daily basis.

PC_h = "harvest" costs, which is the costs that occur only at the event of harvesting. This includes the full package of labor costs that occur from harvesting to replanting: harvesting, drying, tying vegetative parts to new lines for replanting, and placing the new lines back into the sea. We refer to these costs as harvest (h) costs and express them in IDR per m^2 per cycle, since they occur at the end of each cycle.

$$FGP_{sd}(t) = FGP_c * cf(t) \quad (12)$$

$$I_g(hcl, w_{f,0}) = W_{n,sd}(hcl, w_{f,0}) * FGP_{sd}(hcl) \quad (13)$$

$$I_g(hcl, w_{f,0}) = CAR(hcl, w_{f,0}) * FGP_c \quad (14)$$

$$PC(hcl) = 360 * \left(PC_m + \frac{1}{hcl} * PC_h\right) = 360 * PC_m + N_h * PC_h \quad (15)$$

$$I_n(hcl, w_{f,0}) = I_g(hcl, w_{f,0}) - PC(hcl) \quad (16)$$

Fig. 8.4 Economic model formulae cited in Sect. 8.2.3

Total production costs $PC(hcl)$ (IDR m^{-2} year^{-1}) summed over a period of 360 days for a seaweed farm can then be calculated per Eq. (15) (Fig. 8.4),

Where N_h is the number of harvests (Eq. 5, Fig. 8.2). We can see that a shorter harvest cycle (hcl) = more frequent harvesting (N_h) increases total costs through the right term $N_h * PC_h$.

Finally, net income I_n, expressed in IDR m^{-2} year^{-1}, is calculated per Eq. (16) (Fig. 8.4).

8.2.4 Unit Conversions

The BESeM expresses biomass in kg per square meter and profits in IDR m^{-2} per year. Expressing yields and profits per square meter is a scientific standardized method of reporting. Reporting in a standardized way (i.e., per m^2 per year) allows for objective comparisons of productivity of farms of different size or for comparisons between cases where plant density, standard line length, or line spacing is different from the values considered. The conversion from kg m^{-2} tot kg per line, gram per plant, or Indonesian rupiah per farm is straightforward and is illustrated below:

- The cultivation area for a line of length L_l = 25 m placed in parallel with L_w = 1.0 m between the lines is $L_l * L_w$ = 25*1.0 = 25 m^2.
- A biomass of 40 kg/line thus corresponds with 40/25 = 1.6 kg m^{-2}.
- Or the other way round: a simulated biomass of 2.0 kg m^{-2} corresponds with 2.0 * $L_l * L_w$ = 50 kg line^{-1}.

Conversion from kg m^{-2} to gram plant^{-1} is equally straightforward:

- If plants are spaced at 20 cm, then plant density on the line is L_{pd} = 1.0/0.20 = 5 plants m^{-1}.
- Plant density per square meter is $PD = L_{pd}/L_w$. At L_{pd} = 5 and parallel lines at 0.8 m, spacing between the lines is L_w = 0.8, PD = 5/0.8 = 6.25 plants m^{-2}. At a line spacing of L_w = 1.0 m, PD is 5/1.0 = 5 plants m^{-2}.
- Conversion from plant weight w kg m^{-2} to plant weight y in gram per plant is y = 1000 * w/PD = 1000 * $w/(L_{pd}/L_w)$ and, the other way round, w = 0.001 * y * PD. For the L_{pd} and L_w values in Table 8.2, a plant weight of 30 gram

per plant thus corresponds with $w = 0.001 * 30/(5/1.0) = 0.15$ kg m^{-2}

Conversion from Indonesian rupiah per square meter per year (IDR m^{-2} $year^{-1}$) to farm level (IDR $farm^{-1}$ $year^{-1}$) is by multiplying with farm area A (m^2):

- Farm size in square meters can be calculated from the number of lines that a farmer has (L_N), the length of lines (L_l), and the spacing between lines (L_w). For a farm with $L_N = 300$ lines, farm area A (m^2) can be calculated as $A = L_N * L_l * L_w = 300 * 25 * 1.0 = 7500$ m^2 (=0.75 ha).
- A net profit of $I_n = 40{,}000$ IDR m^{-2} $year^{-1}$ then corresponds with $I_n * A = 40{,}000 * 7500 = 300$ million IDR per farm per year.
- If the family household consists of four family members working in the seaweed enterprise, then income per head can be calculated by dividing the farm income by 4.

8.3 Model Illustration

8.3.1 Parameter Estimation

8.3.1.1 Biochemical Parameters

Farmers in a number of sites in South West Sulawesi (visited by the authors in 2019 and 2020) reported gross harvests of 40 kg FW per line, with a harvest cycle of *hcl* 45 days and a standard line length of $L_l = 25$ m, at $L_w = 1.0$ meter spacing. Farmers gross yield is thus $40/(1*25) = 1.6$ kg FW m^{-2}, and yield per plant at 45 dap is 320 gram fresh per plant (at 5 plants per m). Iain Neish (pers. comm., 2021) reported farmers' common practice replanting weights of 20–40 g. We observed similar values during our visits, but also quite large variation in farmers' response to our questions on planting density and planting weight. Assuming an average of 30 g $plant^{-1}$, this corresponds with $30/1000*125 = 3.7$ kg FW $line^{-1}$ and 0.15 kg FW m^{-2}. Parameter $rgrmax = 0.06$ was estimated from the literature (Table 8.2). With this, three out of four biological parameters ($rgrmax = 0.06$, $hcl = 45$ and $w_{f,0} = 0.15$) were known. The remaining unknown parameter $w_{f,max}$ was estimated, such that simulated $w_{f,g}(t = 45)$ (Eq. 3) equaled the observed gross yield at harvesting (1.6 kg FW m^{-2}). Figure 8.5 illustrates how we estimated wf,max on this minimal calibration set. Clearly, the only way to check if our estimate of wf,max is correct, one would have to monitor growth for a longer time than the common 45 days or less. Figure 8.5 also shows in the gray line the exponential model with infinite growth. Early growth is exponential, but we can see the exponential model overestimates biomass in later stages after planting (Fig. 8.5).

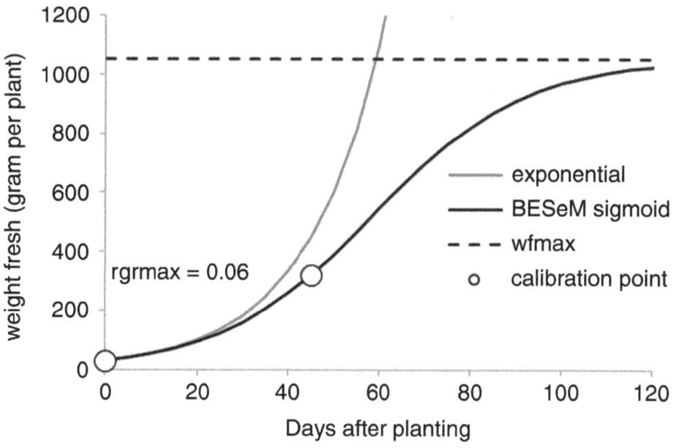

Fig. 8.5 Estimation of $w_{f,max}$ and illustration of the exponential and sigmoid model

8.3.1.2 Economic Parameters

At least for the biological parameters, we can estimate them through experiments. Economic parameters are the farmgate price FGPc and the production costs PCm ("maintenance" costs) and PCh ("harvest" costs). Farmgate prices are subject to economics of price and demand. Tourism in Indonesia has collapsed since the ongoing COVID-19 pandemic. Part of the people working in the tourism sector losing their jobs have moved into seaweed farming. This has led to increased supply, which has in turn had a negative effect on farmgate prices. According to Kospermindo staff in Makassar Indonesia, farmgate prices have recently dropped from 22,000–24,000 IDR kg^{-1} semidry in 2019 to 13,000–14,500 IDR kg^{-1} semidry in 2021. Kospermindo is a seaweed farmers' cooperative in Sulawesi. See also Langford (Langford et al. 2021) for effects of COVID-19 on cultivated area and prices.

Production costs PCm are aggregate values, consisting of depreciation of ropes, wooden poles, plastic bottles, boats, and more. We estimated this parameter based on Tahang et al. (2019); see Table 8.2. A major source of uncertainty is that (Tahang et al. 2019) did not report farm size, while we express production costs in IDR per m^2. In our calculation of PCm from Tahang's data, we assumed a farm size of $L_N = 300$ lines. Other sources of uncertainty are in the estimates of depreciation costs. For example, the number of years to replacement will depend on the quality of the material (boat, line, etc.), and how it is handled by the seaweed farmer. Replacement time may vary markedly. Also, prices at which material is bought will vary from time to time and from site to site.

Harvest costs PC_h are aggregate values, consisting of harvesting (normally done by the men), tying seaweed cuttings to new ropes (normally done by the women) and planting out the lines back into the sea (normally done by the men). We estimated this parameter based on Tahang et al.

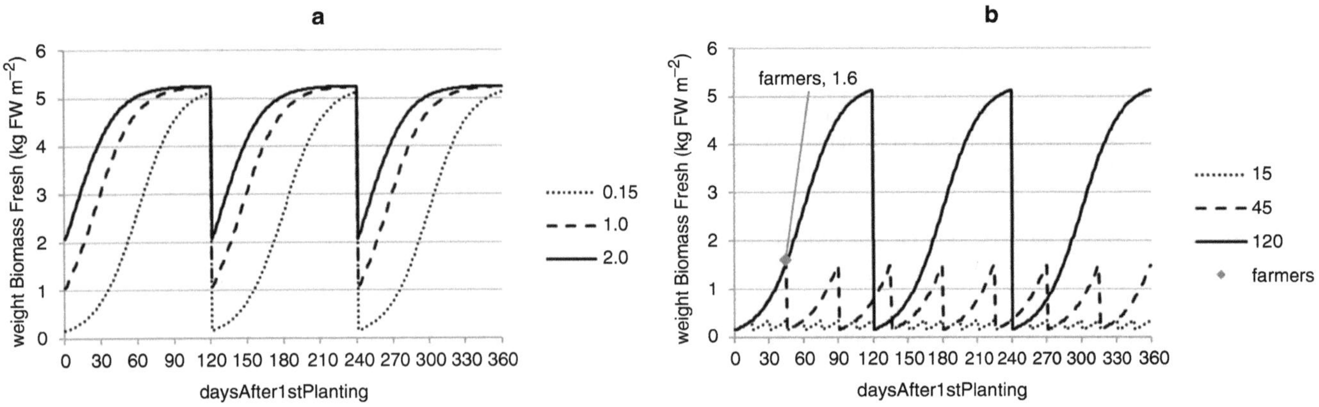

Fig. 8.6 Growth cycles for farming systems harvesting every $hcl = 15$, 45, and 90 days. (**a**) Fresh biomass for a harvest cycle length of $hcl = 120$ days, at three initial plant weights ($w_{f,0} = [0.15, 1.0, 2.0]$ kg FW m^{-2}); (**b**) fresh biomass (kg FW m^{-2}) with 3 harvest cycle lengths ($hcl = [15, 45, 120]$ days) at replanting weight 0.15 kg FW m^{-2}

(2019) (Table 8.2). Smaller farms will rely entirely on family labor; larger farms will also hire wage labor for the tying up of new ropes. The difficulty in estimating these labor costs is that seaweed is mainly a family business in which family members do not pay each other for their services. At best, these costs could be estimated from wages of laborers. The COVID-19 pandemic has led to increased unemployment in Indonesia, with a negative effect on wages. One may expect current PC_h to be lower than our (pre-COVID) estimate in Table 8.2. Both our farmgate price and the production costs are today probably lower than before COVID. Lower farmgate prices may partially cancel out against lower production costs, still leading to more or less realistic estimates of net income.

Evidently, our parameter estimates are quite uncertain. The following sections should be considered more as illustrations of the model than as predictions of the true values.

8.3.2 Harvest Cycles

Figure 8.6a shows three sigmoid curves for three different replanting weights. With a low replanting weight of 0.15 kg FW m^{-2} (farmers' practice), growth is initially exponential, changing to linear and then slowing down to the plateau of $w_{f,max}$. With this low replanting weight, the plateau of around 5 kg FW m^{-2} is reached only at approximately 120 days. A high replanting weight of 2.0 kg FW m^{-2} gives much stronger initial growth. Then the plateau of 5 kg FW m^{-2} is reached already at around 60 days. Figure 8.6b shows three cycles, of 15, 45, and 120 days, respectively. A short cycle of 15 days gives more (24) cycles per year but lower net yield at harvesting (0.33 kg FW m^{-2}). A long cycle of 120 days gives less (three) cycles per year and higher net yield at harvesting (5.12 kg FW m^{-2}).

8.3.3 Optimization

Farmers can influence their production and income through their decisions on replanting weight and harvest cycle length. As we will show below, the BESeM can be used to find the optimum replanting weight and optimum harvest cycle length. The BESeM was run 7*7 = 49 times, at seven different harvest cycle lengths ($hcl = [5, 15, 30, 45, 60, 90, 120]$ days) and seven different replanting weights ($w_{f,0} = [0.01, 0.05, 0.15, 0.5, 1.0, 2.0, 3.0]$ kg FW m^{-2}). Results are presented in Fig. 8.7. We discuss the results

1. Figure 8.7a shows gross yields per harvest cycle. At a start weight of 30 g/plant = 0.15 kg FW m^{-2} and for a cycle of 45 days, gross yield of 1.5 kg FW m^{-2} (Fig. 8.7a) corresponds with the gross yield of 1.6 kg FW m^{-2} shown in Fig. 8.6 (the difference between 1.6 in Fig. 8.6 and 1.5 in Fig. 8.7a is due to rounding).
2. Figure 8.7c shows net yield calculated as the gross yield minus the replanted weight. At $w_{f,0} = 0.15$ with $hcl = 45$ days, 1.5–0.15 = 1.35 kg FW m^{-2} is sold (Fig. 8.7c). At this particular combination of $w_{f,0}$ and hcl, the fraction sold is 1.35/1.5 = 0.90 (90%) and is shown in Fig. 8.7b.
3. A long cycle gives the highest net fresh yield per cycle (Fig. 8.7c), but one has less harvests per year. Figure 8.7d, e show annual net yields. Annual net yields are highest with a combination of a high replanting weight (2.0 kg FW m^{-2} = 400 g $plant^{-1}$) and a short cycle of only 15 days. The pattern in Figs. 8.5d and 8.7e is exactly the same; the only difference is in the numbers, as we are applying a standard conversion (Eq. 9) from fresh weight to semi-dry weight.
4. When considering carrageenan production (Fig. 8.7f), the optimum shifts a bit toward longer cycles (45 days) and a medium replanting weight (1.0 kg FW m^{-2} = 200 g

plant^{-1}), because according to Fig. 8.2 in our model, carrageenan concentration is lower shortly after planting.

5. Gross income (Fig. 8.5g) follows the same pattern as carrageenan (Fig. 8.7f); the only difference is in the numbers, as we are applying a standard conversion (Eq. 15) from carrageenan yield to gross income.
6. Finally, for net income (Fig. 8.7h), optimum farm management is similar as for the gross income. In other simulations (not shown) with higher harvesting costs (PC_h), we did find net income was maximized with less frequent harvesting.

An important message from Fig. 8.7a–h is that it is really the combination of start weight and cycle length that matters. For example, Fig. 8.7g shows that, at a low replanting weight, it is better to go for a longer cycle and, with a high replanting weight, it is better to go for a shorter cycle. It does not make sense recommending an optimal cycle length without considering replanting weight. And it does not make sense recommending an optimum replanting weight without considering cycle length. These two important farmers' decisions should be made as "package."

Table 8.3 presents a comparison of farmers' practice with the optimum, which we derived in Fig. 8.7h. The table suggests farmers can increase their income by a factor 2.5 (from 31,997 to 79,925 IDR m^{-2} year^{-1} with a similar cycle (45 days) and a much higher replanting weight (200 instead of 30 g fresh plant^{-1}). They would be selling a smaller part of their harvest (75% instead of 90%). The smaller fraction sold is more than offset by the much higher growth obtained at greater replanting weight.

From Fig. 8.7, we can calculate gross farmers' income per month as: 37.0 *1000 IDR m^{-2} year^{-1} (Fig. 8.7g) * 7500 m^2 (farm size)/12 (months per year)/4 (people per family) = 5.8 M IDR head^{-1} month^{-1}. This is about 2× the minimum wage of 3.2 mln IDR month^{-1} in provincial South West Sulawesi. This seems a reasonable model outcome; it seems reasonable that a seaweed farmer would earn more than an average wage worker. And it seems reasonable a seaweed farmer would not earn many orders of magnitude more than the average wage worker, e.g., it would be strange if our model had predicted

Harvesting cycles of 5, 10, 15, ..., 90 days. Simulation of 360 days.

(Re) planting fresh weight			(a) Gross fresh yield (kg FW m-2 cycle-1)							(b) fraction of gross yield sold						
g plant-1	kg line-1	kg m-2	5	15	30	45	60	90	120	5	15	30	45	60	90	120
2	0.25	0.01	0.0	0.0	0.1	0.1	0.3	1.4	3.6	0.21	0.56	0.81	0.92	0.97	0.99	1.00
10	1.25	0.05	0.1	0.1	0.3	0.6	1.2	3.4	4.8	0.21	0.55	0.81	0.91	0.96	0.99	0.99
30	3.75	0.15	0.2	0.3	0.7	1.5	2.6	4.5	5.1	0.20	0.54	0.79	0.90	0.94	0.97	0.97
100	12.5	0.5	0.6	1.0	1.9	3.1	4.1	5.0	5.2	0.19	0.51	0.74	0.84	0.88	0.90	0.90
200	25	1.0	1.2	1.8	3.0	4.0	4.7	5.2	5.2	0.17	0.46	0.67	0.75	0.79	0.81	0.81
400	50	2.0	2.3	3.1	4.1	4.7	5.0	5.2	5.3	0.13	0.35	0.51	0.58	0.60	0.62	0.62
600	75	3.0	3.3	4.0	4.7	5.0	5.2	5.2	5.3	0.09	0.25	0.36	0.40	0.42	0.43	0.43

(Re) planting fresh weight			(c) Net fresh yield (kg FW m-2 cycle-1)							(d) Net fresh yield (kg FW m-2 year-1)						
g plant-1	kg line-1	kg m-2	5	15	30	45	60	90	120	5	15	30	45	60	90	120
2	0.25	0.01	0.0	0.0	0.0	0.1	0.3	1.3	3.5	0.2	0.3	0.5	0.9	1.7	5.4	10.6
10	1.25	0.05	0.0	0.1	0.2	0.5	1.2	3.3	4.8	0.9	1.5	2.5	4.3	7.0	13.4	14.4
30	3.75	0.15	0.0	0.2	0.6	1.3	2.4	4.3	5.0	2.7	4.3	6.9	10.6	14.5	17.4	14.9
100	12.5	0.5	0.1	0.5	1.4	2.6	3.6	4.5	4.7	8.4	12.4	17.2	20.7	21.6	18.1	14.2
200	25	1.0	0.2	0.8	2.0	3.0	3.7	4.2	4.2	14.8	20.1	23.9	24.2	22.1	16.6	12.7
400	50	2.0	0.3	1.1	2.1	2.7	3.0	3.2	3.3	21.8	26.1	25.2	21.8	18.2	12.9	9.8
600	75	3.0	0.3	1.0	1.7	2.0	2.2	2.2	2.3	21.9	23.5	19.9	16.0	12.9	9.0	6.8

(Re) planting fresh weight			(e) Net semidried yield (kg SDW m-2 year-1)							(f) Carrageenan yield (kg m-2 year-1)						
g plant-1	kg line-1	kg m-2	5	15	30	45	60	90	120	5	15	30	45	60	90	120
2	0.25	0.01	0.0	0.0	0.1	0.1	0.3	0.8	1.6	0.01	0.01	0.02	0.05	0.10	0.31	0.61
10	1.25	0.05	0.1	0.2	0.4	0.7	1.1	2.1	2.2	0.03	0.05	0.11	0.23	0.40	0.77	0.83
30	3.75	0.15	0.4	0.7	1.1	1.6	2.2	2.7	2.3	0.10	0.16	0.29	0.57	0.83	1.00	0.86
100	12.5	0.5	1.3	1.9	2.6	3.2	3.3	2.8	2.2	0.30	0.45	0.72	1.11	1.23	1.04	0.82
200	25	1.0	2.3	3.1	3.7	3.7	3.4	2.6	2.0	0.53	0.74	1.01	1.30	1.26	0.96	0.73
400	50	2.0	3.4	4.0	3.9	3.4	2.8	2.0	1.5	0.79	0.96	1.06	1.17	1.04	0.74	0.56
600	75	3.0	3.4	3.6	3.1	2.5	2.0	1.4	1.0	0.79	0.86	0.84	0.86	0.74	0.52	0.39

(Re) planting fresh weight			(g) Gross Income (thousand IDR m-2 year-1)							(h) Net Income (thousand IDR m-2 year-1)						
g plant-1	kg line-1	kg m-2	5	15	30	45	60	90	120	5	15	30	45	60	90	120
2	0.25	0.01	0.4	0.7	1.4	3.3	6.4	20.2	40.0	-41.7	-13.6	-5.9	-1.7	2.6	17.5	37.9
10	1.25	0.05	2.2	3.6	7.0	15.0	26.3	50.3	54.0	-39.9	-10.7	-0.4	10.0	22.4	47.6	51.9
30	3.75	0.15	6.5	10.3	19.0	37.0	54.0	65.3	56.1	-35.6	-4.0	11.7	32.0	50.1	62.6	54.0
100	12.5	0.5	19.8	29.6	47.3	72.7	80.7	68.2	53.3	-22.3	15.4	40.0	67.7	76.9	65.5	51.2
200	25	1.0	34.9	48.3	65.8	84.9	82.6	62.6	47.9	-7.2	34.0	58.5	79.9	78.8	59.9	45.8
400	50	2.0	51.5	62.6	69.3	76.4	68.0	48.5	36.7	9.4	48.3	62.0	71.4	64.1	45.8	34.6
600	75	3.0	51.8	56.3	54.8	56.2	48.3	33.7	25.5	9.7	42.0	47.5	51.2	44.4	31.1	23.4

Fig. 8.7 Scenarios of production and income as a function of replanting weight (rows) and harvest cycle length (columns)

Table 8.3 Comparison between simulated farmers' practice and simulated optimum

	Farmers' practice	Optimum	References
Net income (IDR m^{-2} $year^{-1}$)	31,997	79,925	Figure 8.6h
Replanting fresh weight (kg FW m^{-2})	0.15	1.0	Figure 8.6h
Replanting fresh weight (g $plant^{-1}$)	30	200	Figure 8.6h
Harvest cycle length (days)	45	45	Figure 8.6h
Gross yield per harvest (kg FW m^{-2})	1.5	4.0	Figure 8.6a
Gross yield per harvest (g $plant^{-1}$)	294	806	Figure 8.6a
Fraction sold	90%	75%	Figure 8.6b
Fraction replanted	10%	25%	1—Fig. 8.6b

the seaweed farmer earning 10 or 100 times as much as the average wage worker.

8.4 Discussion

8.4.1 Scope for Improving Farm Income?

At the Phyconomy Coalition, online workshop July 7–8, 2021 (https://www.phyconomy.org/tpcd1/), Dr. Iain Neish responded to our presentation (van Oort et al. 2021) by stating "We did trials around Bantaeng and Bulukumba in the IFC-PENSA project (2004–2008) that showed much better return per unit cost/effort when bibit about 150–200 (or even as much as 500) g were used as starting material, compared to when small bibit about 20–40 g were used. Despite that, farmers persisted in using the very small bibit. The reason given, was that village ladies were paid on a piecework basis per planted line to attach bibit. Smaller bibit means more lines must be planted ... hence more income to the ladies. In fact, the size of harvested fronds in much of SulSel was smaller than the size of propagules typically planted around Zamboanga, Philippines."

Our optimization (Fig. 8.7h, Table 8.3) suggests a similar optimum replanting weight (100–200 g $plant^{-1}$), quite a remarkable outcome considering that our model was developed independently of the work by Dr. Neish. Our calculations suggested farmers can more than double their income by adopting this alternative management. Should we conclude farmers have great scope for increasing their income?

A word of caution is needed here. During our visit to seaweed farms in South West Sulawesi, farmers reported replanting weights ranged between 50 and 100 g per plant. In one case where we actually weighed a cutting, we measured 10 g. If actual replanting weight is 100 g per plant instead of the 30 g used in Table 8.3), then for these farmers the estimated net income would be 67,683 instead of 31,997 M IDR m^{-2} $year^{-1}$ (Fig. 8.7h), and thus for these 100 g farmers, there would be much less to gain toward the simulated maximum of 79,925 M IDR m^{-2} $year^{-1}$. We should note here there has been little published research on actual farmers' practice. This one caution against claiming high potential income gains. A second word of caution has to do with model uncertainty. Reality is more complex than captured here in our model, since a model is always a simplification of reality. In reality, farmers may need fast cash to repay loans and pay school fees, etc. This may lead them to opt for shorter cycles and sell a large percentage of their harvest (which as we have shown can lead to lower net income). A third caution is that it is easy to simulate high yields on our computer but much harder to achieve them in practice. Sometimes for various reasons, *Kappaphycus* simply will not grow at a specific geographical site. In that case, farmers opt for more robust seaweeds like *Gracilaria* or *Eucheuma denticulatum* (spinosum) for which farmgate prices are much lower and thus also income generated is lower. We might find income for the *Eucheuma* farmer is about 50% lower (compared with *Kappaphycus*), but this would apply both to the farmers' common practice and our optimized management. It does not necessarily change our conclusion that higher replanting weights generate more income.

This being said, the logic of our model seems to be clear and our model results do seem to be in agreement with previous experience quoted above. In the very least, it calls for more experimentation. Our model outcomes can be considered hypotheses which are easily enough tested in the field (or actually in the sea), simply by planting seaweed at different starting weights and monitoring its growth.

8.4.2 Effects of the Environment

The reader will have noted environmental effects, such as effects of nutrient concentrations, temperature, irradiance, turbidity, etc., are not incorporated in our biological model. We briefly discuss how such effects could be incorporated.

8.4.2.1 Temperature, Irradiance, Nutrients

Implicitly, our biological parameters rgrmax and wf,max are determined by the environment. Other sites with different environments will have different rgrmax and wf,max than those in Table 8.2, which were estimated for a specific site in this study. One may expect higher values for rgrmax and wf, max in more favorable environments. Consider spatial and temporal variation in environmental variables. More complex models are needed when environmental variables fluctuate strongly during the year. In temperate climates temperature and irradiance are low during winter and high during summer. In comparison, tropical climates have fairly constant temperature and irradiance throughout the year, which makes it less urgent to incorporate temporal variation into a tropical seaweed cultivation model. Light can be a limiting factor if cultivation is from seaweed growing upward from a deep seafloor. Light is hardly a limiting factor in tropical cultivation systems with seaweed cultivated on lines or in nets near the surface. Regardless of climate, nutrient concentrations are often higher near river mouths (which flush nutrients into the sea). Higher rgrmax and wf,max values may therefore be expected near river mouths. It is possible to determine from experiments functions of $w_{f,max}(X)$, i.e., $w_{f,max}$ as function of environmental variable X; for example, X could be phosphate (PO_4^{3-}) concentration. As a starting point of elaborating the BESeM with environmental effects, we propose to focus more on modeling spatial variation in *rgrmax* and $w_{f,max}$ in relation to spatial variation in environmental conditions and less on temporal fluctuations. It requires experiments with simultaneous monitoring of environmental conditions and biomass. We found no such studies in the literature. Clearly there is a need for further experimental research in this domain.

8.4.2.2 Monsoon

According to farmers interviewed, in South West Sulawesi, the December–January–February–March part of the year is unsuitable for seaweed cultivation because the sea is too rough during that (monsoon) time. Rather than explicitly modeling lower production during this period, we could simply limit our simulations to the shorter 8 months period that is suitable for cultivation. During this new reference period of now 8 months (8/12)*360 = 240 days, farmers can complete 8 cycles of 30 days, 5 cycles of 45 days, or 4 cycles of 60 days. Instead of calculating aggregate yields over a full year, carrageenan production and income would now be calculated over this period of 240 days. It would not fundamentally change our approach. For the incomes reported in Fig. 8.7, these could all be multiplied with (8/12), and we would still find income is maximized at the same combination of $w_{f,0} = 200$ gram per plant and $hcl = 45$ days.

8.4.3 Uncertainties, Need for Further Research

We already reflected on some of the uncertainties in our parameter estimates. The difficulty of estimating $w_{f,max}$ with limited data from interviews shows how very little is needed to make a first rough estimate of this parameter (Fig. 8.3). At the same time, it shows that without any observations closer to $w_{f,max}$ our estimate of $w_{f,max}$ remains very uncertain. Many of the parameters estimates listed in Table 8.2 are based on limited personal observation. For the biological parameters, we observed in the literature a general lack of experimentation, where growth was monitored longer than 45 days (see references quoted in the introduction). We hypothesized a sigmoid model. Although it is a biologically plausible model, to date there is not a single empirical tropical seaweed study that can confirm whether growth is indeed sigmoid. Perhaps future experiments will reveal biomass decreases again after certain time, due to processes of, e.g., herbivory. This remains to be tested. With lack of such longer time experiments, we had to improvise to estimate a key parameter, $w_{f,max}$, which can only really be determined from experiments, where growth is monitored for a longer period than 45 days. We also found no published studies where in the same growing environments growth was monitored for species planted at different initial plant weights, while our model suggests clear and strong effects of initial plant weights. There is a clear need for experiments in which growth is monitored for a longer period of time and experiments on effect of initial plant weight.

There is also much uncertainty about the economic parameters. Two studies provide the quantitative data we need (Tahang et al. 2019; Limi et al. 2018). One particular issue is that in BESeM economic production cost parameters are expressed in IDR square meter (IDR m^{-2}) while these two studies report production at the farm level without reporting farm size. A second issue is that our economic parameters were determined during pre-COVID-19 times. The COVID-19 pandemic seemed to drive many people into seaweed farming, leading to more and cheaper wage workers but also larger production volumes which in turn lead to reduced farmgate prices. Over time, it may be necessary to reestimate the economic parameters of the model.

8.5 Conclusions

A simple model was presented for simulating production and economics of tropical seaweed cultivation. The model has a limited number of parameters, which makes it easily amenable to other seaweeds and other sites. Parameter estimates were presented, and uncertainties in parameter estimation were discussed. A simulation example suggests farmers can increase income by using cuttings with a higher weight than they are currently using. A literature review suggests still much need for further experimentation.

Acknowledgments Funding: The authors would like to acknowledge funding from the Wageningen University and Research Program on "Food Security and Valuing Water" (project code KB-35-004-001) that is supported by the Dutch Ministry of Agriculture, Nature, and Food Quality.

References

Broch OJ, Slagstad D (2012) Modelling seasonal growth and composition of the kelp *Saccharina latissima*. J Appl Phycol 24(4):759–776. https://doi.org/10.1007/s10811-011-9695-y

Bui VTNT, Nguyen BT, Renou F, Nicolai T (2019) Structure and rheological properties of carrageenans extracted from different red algae species cultivated in Cam Ranh Bay, Vietnam. J Appl Phycol 31(3):1947–1953. https://doi.org/10.1007/s10811-018-1665-1

Dawes CJ, Trono GC Jr, Lluisma AO (1993) Clonal propagation *of Eucheuma denticulatum* and *Kappaphycus alvarezii* for Philippine seaweed farms. Hydrobiologia 260–261(1):379–383. https://doi.org/10.1007/BF00049044

Dawes CJ, Lluisma AO, Trono GC (1994) Laboratory and field growth studies of commercial strains of *Eucheuma denticulatum* and *Kappaphycus alvarezii* in The Philippines. J Appl Phycol 6(1): 21–24. https://doi.org/10.1007/BF02185899

Duarte P, Ferreira JG (1997) A model for the simulation of macroalgal population dynamics and productivity. Ecol Model 98(2–3): 199–214. https://doi.org/10.1016/S0304-3800(96)01915-1

FAO (2018) The global status of seaweed production, trade and utilization. FAO Globefish Research Programme, Rome

Hurtado AQ, Agbayani RF, Sanares R, de Castro-Mallare MTR (2001) The seasonality and economic feasibility of cultivating *Kappaphycus alvarezii* in Panagatan Cays, Caluya, Antique, Philippines. Aquaculture 199(3–4):295–310. https://doi.org/10.1016/S0044-8486(00)00553-6

Jackson GA (1987) Modeling the growth and harvest yield of the giant-kelp *Macrocystis pyrifera*. Mar Biol 95(4):611–624. https://doi.org/10.1007/Bf00393105

Kasim M, Mustafa A, Munier T (2016) The growth rate of seaweed (*Eucheuma denticulatum*) cultivated in longline and floating cage. AACL Bioflux 9(2):291–299

Langford A, Waldron S, Sulfahri SH (2021) Monitoring the COVID-19-affected Indonesian seaweed industry using remote sensing data. Mar Policy 127:104431. https://doi.org/10.1016/j.marpol.2021.104431

Lavaud R, Filgueira R, Nadeau A, Steeves L, Guyondet T (2020) A dynamic energy budget model for the macroalga *Ulva lactuca*. Ecol Model 418:108922. https://doi.org/10.1016/j.ecolmodel.2019.108922

Limi MA, Sara L, La Ola T, Yunus L, Suriana TSAA, Batoa H, Hamzah A, Fyka SA, Prapitasari M (2018) The production and income from seaweed farming after the sedimentation in Kendari bay. AACL Bioflux 11(6):1927–1936

McHugh DJ (2003) A guide to the seaweed industry. FAO Fish Tech Pap 441. http://www.fao.org/3/y4765e/y4765e00.htm

Periyasamy C, Rao PS (2017) Growth rate and carrageenan yield of cultivated *Kappaphycus alvarezii* (Doty) Doty in the coastal waters of Bay of Bengal at Chepala Timmapuram, Andhra Pradesh, east coast of India. J Appl Phycol 29(4):1977–1987. https://doi.org/10.1007/s10811-017-1099-1

Periyasamy C, Subba Rao PV, Anantharaman P (2019) Harvest optimization to assess sustainable growth and carrageenan yield of cultivated *Kappaphycus alvarezii* (Doty) Doty in Indian waters. J Appl Phycol 31(1):587–597. https://doi.org/10.1007/s10811-018-1562-7

Phang SM, Yeong HY, Lim PE (2019) The seaweed resources of Malaysia. Bot Mar 62(3):265–273. https://doi.org/10.1515/bot-2018-0067

Setyawidati N, Liabot PO, Perrot T, Radiarta N, Deslandes E, Bourgougnon N, Rossi N, Stiger-Pouvreau V (2017) In situ variability of carrageenan content and biomass in the cultivated red macroalga *Kappaphycus alvarezii* with an estimation of its carrageenan stock at the scale of the Malasoro Bay (Indonesia) using satellite image processing. J Appl Phycol 29(5):2307–2321. https://doi.org/10.1007/s10811-017-1200-9

Tahang H, Latama G, Kasri (2019) Development strategy and increased production of seaweed in Takalar District. In: IOP Conference Series: Earth and Environmental Science, 2019. https://doi.org/10.1088/1755-1315/370/1/012058

van der Molen J, Ruardij P, Mooney K, Kerrison P, O'Connor NE, Gorman E, Timmermans K, Wright S, Kelly M, Hughes AD, Capuzzo E (2018) Modelling potential production of macroalgae farms in UK and Dutch coastal waters. Biogeosciences 15(4): 1123–1147. https://doi.org/10.5194/bg-15-1123-2018

van Oort PAJ, Latama G, Rukminasari N, Verhagen J, van der Werf A (2021) The Bio-economic Seaweed Model (BESeM) for modelling *Kappaphycus* cultivation in Indonesia. Paper presented at the Phyconomy Coalition, online workshop 7–8 July 2021, Online

Venolia CT, Lavaud R, Green-Gavrielidis LA, Thornber C, Humphries AT (2020) Modeling the growth of sugar kelp (*Saccharina latissima*) in aquaculture systems using dynamic energy budget theory. Ecol Model 430:109151. https://doi.org/10.1016/j.ecolmodel.2020.109151

Zhang J, Wu W, Ren JS, Lin F (2016) A model for the growth of mariculture kelp *Saccharina japonica* in Sanggou Bay, China. Aquac Environ Interact 8:273–283. https://doi.org/10.3354/aei00171

Cultivation and Domestication of *Kappaphycus alvarezii* Strains at Ubatuba Bay, São Paulo State, Southeastern Brazil

9

Valéria C. Gelli, Estela M. Plastino, and Nair S. Yokoya

Abstract

The red seaweed *Kappaphycus alvarezii* (Doty) L.M. Liao was introduced in Brazil by Professor Édison José de Paula from São Paulo University, in January 1995. The initial tetrasporophytic segments came from Usa Bay, Shikoku Island, Japan, and were originated from commercial crops carried out in the Philippines. The process of domestication has been ongoing since 1995-1996 through the cultivation of brown tetrasporophytic strain into the Experimental Marine Farm of Fisheries Institute, Ubatuba Bay, São Paulo State, Brazil. The initial brown tetrasporophyte originated red and green tetrasporophytic strains, and, after selection of tetraspore progeny, a pale brown gametophytic strain (initially named G11, and later Edison de Paula strain) was also originated. During the domestication process, the occurrence of spontaneous color strains of both diploid and haploid phases was recorded, and environmental data of temperature, salinity, and water transparency at the cultivation site were monitored. Currently, 12 color strains are being cultivated in Ubatuba Bay. The present chapter provides a comprehensive review on cultivation methods, grow rates, productivity, carrageenan yield and characterization of *K. alvarezii* strains cultivated at Ubatuba Bay. Characterization of these new strains should encompass factors such as refined and semirefined carrageenan content, biofertilizer yield, temperature resilience, epibiont presence, growth rate, yield, and identification of economically valuable biomolecules. To facilitate the preservation of these strains, a biobank with all color strains is implemented in both natural environment and laboratory conditions. These initiatives ensure secure strain storage and supply of *K. alvarezii* strains for local farmers.

V. C. Gelli (✉)
Instituto de Pesca, Agência Paulista de Tecnologia dos Agronegócios, Secretaria da Agricultura e Abastecimento do Estado de São Paulo, São Paulo, Brazil
e-mail: valeria.gelli@sp.gov.br

E. M. Plastino
Departamento de Botânica, Instituto de Biociências, Universidade de São Paulo, São Paulo, Brazil

N. S. Yokoya
Environmental Research Institute, Biodiversity Conservation Center, São Paulo, Brazil

Keywords

Biobank · Color strains · Edison de Paula Strain · Mariculture · Seaweed

9.1 Historic

The red alga *Kappaphycus alvarezii* (Doty) L.M. Liao (Rhodophyta, Solieriaceae) was introduced in Brazil in January, 1995, by Professor Édison José de Paula of São Paulo University. The initial thallus segments came from an experimental cultivation maintained in Usa Bay, Kochi, Shikoku Island, Japan, and originated from commercial crops carried out in the Philippines (Paula et al. 1998). The material was provided by Professor Masao Ohno with whom Prof. Paula worked with during his postdoctoral degree at the Usa Marine Biological Institute (Kochi University) in the second semester of 1994 (Paula 2001). Back in Brazil, Prof. Paula began to cultivate a branch 2.5 cm in length (about 2.5 g) in the Laboratory of Marine Algae, LAM (today called LAM Édison José de Paula), Institute of Biosciences, São Paulo University, São Paulo City. A unialgal non-axenic culture of the brown tetrasporophyte was established to eliminate associated species, the presence of which is considered one of the greatest risks in the introduction of exotic species (Paula 2001). This procedure was decisive in ensuring the introduction of the species into the sea without risking the simultaneous introduction of undesirable organisms.

The first seedlings of *K. alvarezii* destined for seawater experiments were produced after 10 months of vegetative propagation of the brown tetrasporophytic strain in the laboratory (Paula and Pereira 1998; Paula et al. 1998). A total of

A. T. Critchley et al. (eds.), *Tropical Phyconomy Coalition Development*, Developments in Applied Phycology 11,
https://doi.org/10.1007/978-3-031-47806-2_9

Fig. 9.1 Original *Kappaphycus alvarezii* brown tetrasporophyte cultivated at the Experimental Marine Farm of the Fisheries Institute, Ubatuba Bay, São Paulo State. (Photo by Valéria Gelli)

20 branches (2.0–4.0 g) were produced and transferred to the subtropical waters of the southwestern Atlantic at the Experimental Marine Farm of the Fisheries Institute (EMFFI) (23° 26′90″S, 45°0′30″W), Paulista Technological Agribusiness Agency, Department of Agriculture and Supply, Ubatuba Bay, São Paulo State (Paula et al. 1999), in accordance with Brazilian legislation under the IBAMA Environmental License Process (IBAMA/MMA n° 02027.009179/1996-11) (Paula 2001; Paula et al. 2002). These *K. alvarezii* crops remain in cultivation to this day (Gelli 2019) (Fig. 9.1). Some studies reported on the introduction of a female gametophyte (Hayashi et al. 2007b), but it died after some years (Hayashi, L. pers. comm., 2022).

The initiative to introduce an exotic seaweed in Brazil and invest in techniques for its cultivation arose from the need to mitigate overexploitation of the natural banks of the native kappa carragenophyte *Hypnea pseudomusciformis* Nauer, Cassano and M.C. Oliveira (cited previously as *Hypnea musciformis*) and meet the demand for national importation of this product (Paula and Pereira 1998; Paula et al. 1999).

Kappaphycus alvarezii seedlings introduced in Ubatuba Bay were monitored and showed satisfactory growth (Paula et al. 1998, 2002). From some of these seedlings, originally brown in color, green and red branches appeared spontaneously and in low frequency (Fig. 9.2). These colored branches were then propagated and kept in cultivation at

Fig. 9.2 Brown, green, and red tetrasporophytic strains and pale brown EP gametophytic strain of *K. alvarezii* (from left to right) cultivated at the Experimental Marine Farm of the Fisheries Institute, Ubatuba Bay, São Paulo State. (Photos by Valéria Gelli)

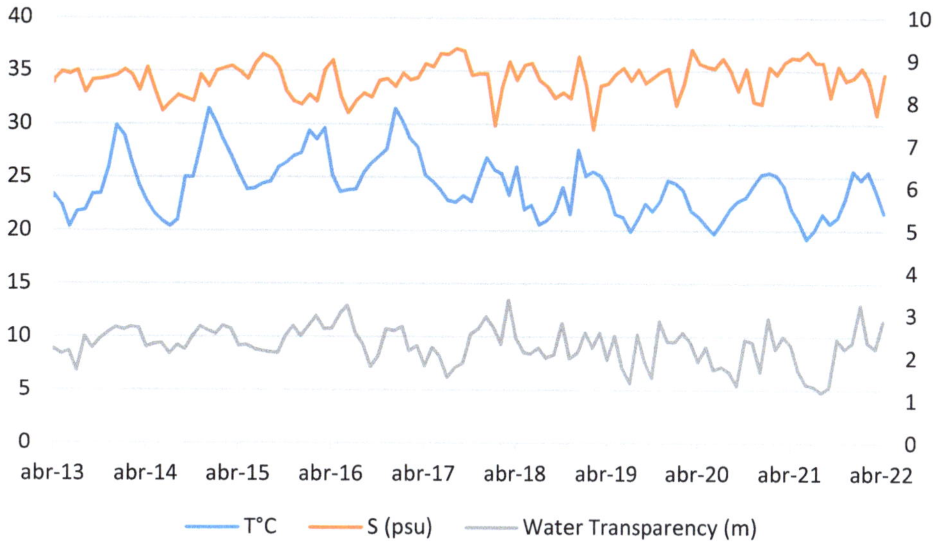

Fig. 9.3 Average variations of temperature, salinity, and water transparency from April 2013 to May 2022 in the Experimental Marine Farm of the Fisheries Institute, Ubatuba Bay, São Paulo State. Source: Fisheries Institute (2022)

sea, showing the stability of these characteristics (Paula 2001). These color strains were evaluated for their growth rates, carrageenan content (Hayashi et al. 2007a), and pigments (Paula 2001) compared to the original strain, and they also remain in the seawaters of EMFFI.

In the summer and autumn months, December 1995 to May 1996, tetraspores were produced in cultivation experiments at sea, quite unexpectedly, though, since the strain had been considered infertile (Paula et al. 1999). Experiments on the germination of these tetraspores showed low viability (Paula et al. 1999). They were produced from brown plants and both green and red color variants (Fig. 9.2) (Bulboa et al. 2008). Rare plantlets were finally obtained in the laboratory after great experimental effort. These plantlets were very delicate and sensitive (Paula 2001). After fertility observation, there was monitoring in the tidal region near the crops and in deeper places, through scuba diving. However, no fixed plants were found in any substrates. These results indicated that the establishment of *K. alvarezii* via spore production in the natural environment of the southeastern coast of Brazil is remote (Bulboa et al. 2008). These important conclusions provide scientific bases to the legislation for the sea cultivation of *K. alvarezii* along Brazilian coast.

After vegetative propagation, some *K. alvarezii* plantlets derived from tetraspores were transferred to the sea and showed lower growth rates than parental tetrasporophytes, though mostly unviable in winter (Paula et al. 1999). However, one plantlet, G11, was more promising. Although it presented lower growth rates, this new pale brown gametophytic strain had higher carrageenan content (Paula 2001; Hayashi et al. 2007a). Later, in honor of Prof. Édison José de Paula, this strain was called EP (Hayashi et al. 2008a). However, in the older papers mentioned in this review, it is still termed G11. The EP strain is still cultivated at EMFFI. The haploidy (n) of the EP gametophyte was confirmed by confocal fluorescence microscopy (Zitta et al. 2012).

9.2 Cultivation of *Kappaphycus alvarezii* Strains in the Natural Environment (Ubatuba Bay) and Laboratory

9.2.1 Cultivation

Since 1995, *K. alvarezii* has been cultivated in a floating raft system (Fig. 9.2) in Ubatuba Bay and periodically replanted using the tie-tie system (Paula et al. 1999; Hayashi et al. 2007a; Solorzano-Chavez et al. 2019). In this system, ten fragments of each strain weighing about 100 g were tied in ropes, 2 m in length, and spaced 20 cm apart (Fig. 9.2). Harvesting and planting were then carried out every 30 days according to monthly marine farm management records.

9.2.2 Monitoring of Environmental Parameters

Temperature (°C), salinity (psu), and seawater transparency (m) were recorded in situ near the *K. alvarezii* culture raft every day from April 2013 to May 2022 in Ubatuba Bay. Figure 9.3 shows the variation of abiotic parameters.

9.2.3 New Spontaneous Strains

Throughout these years of cultivation, other variants of *K. alvarezii* have arisen spontaneously from the four strains already described, and these strains also continue to be propagated and maintained in both seawater and laboratory. They have been photographed and catalogued. As it appeared, each new strain was cultivated in the natural environment on 2-m ropes and with 10 branches. Furthermore, apical segments (Fig. 9.4) of these strains (about 10 g) were cultivated at the Seaweed Laboratory of the Fisheries

Fig. 9.4 Thallus aspect at the beginning of culture (left) and more developed thallus (right) of the new green color variant of *Kappaphycus alvarezii* originated from a red tetrasporophytic branch cultivated at the Experimental Marine Farm of the Fisheries Institute, Ubatuba Bay, São Paulo State. (Photo by Valéria Gelli)

Institute (SLFI) with two replicates for each color strain. According to SLFI protocol, they were kept in a temperature-controlled room at 25 °C with a photoperiod of 12 h and maintained in seawater (35 psu) changed weekly (Fig. 9.5).

Since the introduction of the EP gametophytic strain to the sea in 1998 (Paula et al. 1999), its phenotypic plasticity has been observed, and two new color variants have arisen from it (Fig. 9.6). Moreover, six new variants arose from the tetrasporophytic strains (Fig. 9.7).

9.3 Brief Review of Publications on *Kappaphycus alvarezii* Cultivated in Ubatuba Bay

9.3.1 Physiological Responses

Experimental conditions and the main results on growth rates and viability of tetrasporophyte and gametophyte in both field and laboratory conditions are summarized in Table 9.1. In sea cultivation, seasonal variation of growth rates was clearly related to seawater temperature (Paula et al. 2002).

Bulboa et al. (2008) evaluated tetraspore viability and concluded that tetraspores had low survival rates, mostly dying after 20 days, and that the recruitment of tetraspores did not occur in the field, indicating that the establishment of *K. alvarezii* via spore production in the natural environment of the southeastern coast of Brazil is remote. These important conclusions provide scientific bases to support legislation aimed at cultivating *K. alvarezii* along Brazilian coastal seawaters.

9.3.2 Carrageenan Yields and Characterization

Carrageenan yields and properties of brown, red, and green tetrasporophytes, EP gametophyte, and red female gametophyte are summarized in Table 9.2. Although the EP gametophyte showed lower growth rates when compared to tetrasporophytic strains (see Table 9.1), carrageenan yields were higher in EP gametophyte than those of other strains (Table 9.2). Comparing cultivation period, depth, and planting density, the only parameter that influenced carrageenan quality was the cultivation period (Hayashi et al. 2007b).

9.3.3 Tissue Culture and Micropropagation

Tissue culture and micropropagation techniques were used in five *K. alvarezii* strains: brown (BR), green (GR), and red (RD) tetrasporophytes; pale brown female gametophyte (BFG); and a strain originating from tetraspore germination ("Edison de Paula," EP). Several experimental conditions were tested, such as culture media (half-strength of von Stosch (VS 50), Guillard and Ryther (F/2 50), and synthetic ASP 12-NTA medium) and the addition of glycerol or phytoregulators (indole-3-acetic acid (IAA), 2,4-dichlorophenoxyacetic acid (2,4-D), and benzylaminopurine (BA), either alone or in combination) and colchicine (0.01%). The EP strain showed higher potential for micropropagation than the other strains. Regeneration was stimulated by the addition of glycerol, IAA: BA (5: 1 mg L^{-1}), and colchicine, which proved useful for improving the micropropagation of *K. alvarezii* (Hayashi et al. 2008a).

Fig. 9.5 Different color strains of *Kappaphycus alvarezii* cultivated at the Experimental Marine Farm of the Fisheries Institute, Ubatuba Bay, São Paulo State. EP = EP gametophytic strain; Brown, Green, Red = brown, green, and red tetrasporophytic strains, respectively. (Photo by Valéria Gelli)

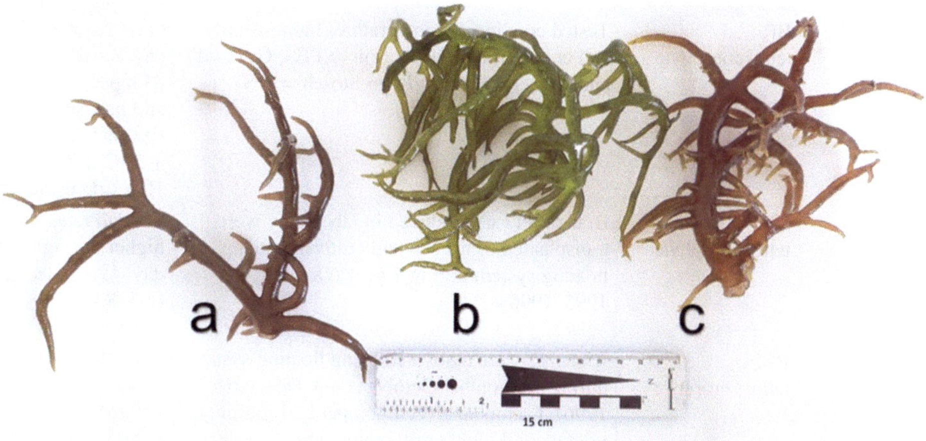

Fig. 9.6 EP gametophytic strain (**a**) which originated two spontaneous color strains (**b** and **c**) of *Kappaphycus alvarezii* cultivated at the Experimental Marine Farm of the Fisheries Institute, Ubatuba Bay, São Paulo State. (Photo by Valeria Gelli)

9.3.4 Ecological Traits

Kappaphycus alvarezii showed potential as a biofilter when cultivated with fish effluents (*Trachinotus carolinus*), removing nitrate (18.2%), nitrite (50.8%), ammonium (70.5%), and phosphate (26.8%) (Hayashi et al. 2008b). However, growth rates of algae cultivated with fish effluents were lower than those observed in sea cultivation as further claimed by the authors. Still, carrageenan yields did not differ between cultivation systems Therefore, *K. alvarezii* cultivation with fish effluents not only reduced eutrophication but also produced carrageenan.

The environmental monitoring of *K. alvarezii* cultivation in Ubatuba Bay was performed from November 2016 to

Fig. 9.7 Tetrasporophytic strains (**a**, **b**, and **c**) and its spontaneous color strains (**d**, **e**, **f**, **g**, **h**, and **i**) of *Kappaphycus alvarezii* cultivated at the Experimental Marine Farm of the Fisheries Institute, Ubatuba Bay, São Paulo State. (Photo by Valéria Gelli)

Table 9.1 Physiological responses of *Kappaphycus alvarezii* strains cultivated in the field (Ubatuba Bay, São Paulo State) or in laboratory (Institute of Biosciences, São Paulo University)

Cultivation location	Strain(s)	Cultivation conditions	Physiological responses	References
Laboratory	BR tetrasporophyte	Fertile branches (10 cm length, 50 g FW) were selected from the sea during the reproductive period (December 1995 to May 1996). After cleaning, these branches were cultured at 25 ± 2 °C, under 60–100 μmol photon m^{-2} s^{-1}, 14:10 h L:D cycle	Tetraspores were released and germinated but the majority died after 2 to 4 days after release. Only 20 plants derived from tetraspores were grown successfully for over a year in the laboratory. These haploid plants differed in morphology, color, size, and growth rates since the early developmental stages.	Paula et al. (1999)
Laboratory	BR tetrasporophyte	Tested conditions: temperature, light, salinity, and culture media (Provasoli = PES, Guillard and Ryther = F/2, and von Stosch = VS)	Growth rates of about 3% day^{-1} (2.8–3.6% day^{-1}) were obtained using PES, pulse-fed 24 h per week, or one-quarter strength 'F/2', and half-strength VS added continuously. Optimal conditions for growth were: 25 ± 2 °C, 170–210 μmol photons m^{-2} s^{-1}, 14: 10 h L:D cycle, and 32–35 psu	Paula et al. (2001)
Field	BR tetrasporophyte	Branches (3 g) produced in laboratory were transplanted to the sea and cultivated using floating system during 1 year (October 1995–1996)	Growth rates of transplanted branches were higher during the first 2 months (6.5–10.7% day^{-1}) and decreased in subsequent months (4.5–8.2% day^{-1}). Seasonal variation of GR was related to seawater temperature	Paula et al. (2002)
Field	BR tetrasporophyte	Branches were cultivated using floating system during 48 months (October 1995–December 1996). Factors analyzed: temperature, salinity, rainfall, and number of sunny hours	Growth rates of 3.6–8.9% day^{-1} correlated to seawater temperature: mean monthly values from 19.9 to 29.0 °C and extreme values from 17.0 to 33 °C low salinity and rainfall affected negatively the GR.	Paula and Pereira (2003)
Laboratory	BR tetrasporophyte	Experimental design: Temperatures 15, 18, 21, 24, 27, and 30 °C X PFDs 50, 100, and 150 μmol photon m^{-2} s^{-1}	*K. alvarezii* grew in temperatures from 18 to 30 °C but died at 15 °C, and the highest growth rate was 5.7% day^{-1} at 30 °C and 100 μmol photon m^{-2} s^{-1}	Bulboa and Paula (2005)
Field	BR, RE and GR tetrasporophytes, and EP	Cultivation from February to December 2001	Growth rates of brown, green, and red tetrasporophytes were similar (2.5%–6.6% day^{-1}) but higher than EP gametophytes (1.3–4.8% day^{-1}). For all strains, including EP strain, GR were higher from February to May and decreased from July to December	Hayashi et al. (2007a)

(continued)

Table 9.1 (continued)

Cultivation location	Strain(s)	Cultivation conditions	Physiological responses	References
Field	BR female gametophyte	Tested conditions: cultivation period (28, 44, and 59 days), depth (surface, 0.5 m and 1.0 m), and planting density (24, 12, and 8.4 plants m^{-2}) during the winter	Growth rates ranged 5.2–7.2% day^{-1}, and the optimal conditions for growth were as follows: cultivation for 28 days, at 0–0.5 m depth and planting density of 12 and 8.4 plants m^{-2}	Hayashi et al. (2007b)
Field and laboratory	BR, RE and GR tetrasporophytes, and EP	Laboratory experiments: temperatures (20, 25, and 30 °C) X PFDs (50, 100, 150 μmol photon m^{-2} s^{-1} photoperiods (LD cycles): 14:10 h, 12:12 h, and 10:14 h. Field experiments: substratum for tetraspore settlement was 25 cm^2 acrylic plates ($n = 84$). Forty-two cords were hung vertically from the culture raft, each with two settlement plates at 50 and 150 cm depth	Tetraspore survival (%) of brown (13%), red (3%), and green (5%) strains declined with time, showing high mortality after 20 days in all temperatures, PFDs, and photoperiods. Recruitment of tetraspores did not occur in the field; what indicated that the establishment of *K. alvarezii* via spore production in the natural environment is rather remote	Bulboa et al. (2008)
Field	BR tetrasporophyte	Cultivation in raft anchored in the bay, density of 6.7 plants m^{-2} for 30 days (May 2014)	Growth rate of 6.3% day^{-1} and productivity of 50.9 g m^{-2} day^{-1} (corresponding to 18.3 kg m^{-2} $year^{-1}$)	Roldán et al. (2017)
Field	BR, RE and GR tetrasporophytes, and EP	Cultivation in raft anchored in the bay, from May to June 2013	Growth rates ranged from 3.8 to 6.2% day^{-1}, and productivity ranged from 15.9 to 46.0 g m^{-2} day^{-1}	Masarin et al. (2016); Oliveira et al. (2019)
Field	BR, RE and GR tetrasporophytes, and EP	Cultivation in raft anchored in the bay, density of 6.7 plants m^{-2}, from August 2013 to July 2014	Growth rates differed among strains: 5.0–8.5% day^{-1} (RE), 5.3–9.0% day^{-1} (BR), 4.8–9.0% day^{-1} (GR), and 3.3–6.6% day^{-1} (G11)	Solorzano-Chavez et al. (2019)

BR, RE, and GR = brown, red, and green strains, respectively; EP—gametophyte previously cited as "G11," which was originated from tetraspore germination

January 2018, and vegetative dispersion and reproductive development were evaluated in four strains, three tetrasporophytes (brown, green, and red) and one gametophyte (EP) (Araújo et al. 2020b). No evidence of vegetative dispersion and/or establishment in Ubatuba Bay was observed. Also, reproductive structures were not found during the monitoring period, indicating the low invasive potential of *K. alvarezii* in this region (Araújo et al. 2020b).

9.3.5 Chemical Composition, Utilization, and Biotechnological Potential

The high and sustainable biomass production from *K. alvarezii* cultivation in Ubatuba Bay has provided raw material for chemical analysis and for biotechnological applications, such as biofuel production (Masarin et al. 2016; Roldán et al. 2017; Oliveira et al. 2019; Solorzano-Chavez et al. 2019), production of monomeric sugars for bioethanol or fine chemical production (Paz-Cedeno et al. 2019), hydrogen production (Rodrigues et al. 2019; Fonseca et al. 2020), antioxidants (Araújo et al. 2020a, 2022), and biofertilizers (Gelli et al. 2020).

An alternative found to leverage São Paulo's algiculture would be the new processing of fresh macroalgae by the producers themselves (Gelli and Barbieri 2015). Recent studies were carried out on the extraction of agricultural biostimulant using equal parts of the four cultivated strains (original brown, green, and red tetrasporophytes and the EP gametophyte (Gelli et al. 2020)). The results can be seen in Table 9.3.

9.4 Perspectives and Final Remarks

The commercial cultivation of *Kappaphycus alvarezii* was prohibited on the north coast of São Paulo. However, in June of 2022, commercial cultivation was finally authorized by environmental legislation in São Paulo State (Brazil 2020; São Paulo 2022).

In São Paulo State, algiculture activities can be expanded in a gradual and planned manner in the coming years (Gelli 2019). In this emerging phase, it will be necessary to develop new processing methods for the commercialization of fresh macroalgae directly by the mariculturist to stimulate the production chain at the artisanal level, followed by the development of the industrial production chain.

K. alvarezii color strains that originated spontaneously from the four initial strains need to be studied before they are made available to the production sector. More

Table 9.2 Carrageenan yield and properties of *Kappaphycus alvarezii* strains cultivated in Ubatuba Bay, São Paulo State, southeastern Brazil

Strain(s)	Extraction method	Carrageenan yields	Carrageenan characterization	References
BR, RE and GR tetrasporophytes and EP	Semi-refined and refined carrageenans were obtained from 20–50 g DW after alkaline transformation	Semi-refined carrageenan yields of EP gametophyte were higher (31–43%) than tetrasporophytic strains; all strains showed lower yields in May and higher from August to December Refined carrageenan yields were also higher in EP strain (15%–28%) than tetrasporophytic strains	Higher 3,6-anhydrogalactose contents were observed in the EP (30.62%) and green (24.51%) strains, while the brown strain showed the lowest values (7.45%). Sulfate contents varied from 23.08% to 33.48%, and gel strength varied from 688 to 926 g cm^{-2}, but these variations did not differ significantly among the strains	Hayashi et al. (2007a)
BR female gametophyte	Native carrageenan was extracted in distilled water, and alkali-modified carrageenan was produced by alkali transformation with isopropanol and KOH solution	Yields of native and alkali modified carrageenan varied, respectively, from 21 to 35% and from 20 to 32%, and the higher values were observed in samples cultivated for 28 days	The highest values of iota fraction, molecular weight, and gel strength of native and alkali-modified carrageenan were observed in samples cultivated for 59 days	Hayashi et al. (2007b)
BR and RE tetrasporophytes	Semi-refined carrageenan was extracted in distilled water using a biomass pretreated in cold alkaline solution	Yields varied between the strain: 63.5% (BR strain) and 60% (RE strain)	Carrageenans from BR and RE strains presented, respectively, 42.6 and 46.6% (galactan), 24.2 and 28.5% (ash), 0.3 and 0.3% (protein), 1.1 and 1.4% (insoluble aromatics), and 13.3 and 14.0% (sulfate groups).	Masarin et al. (2016)
BR tetrasporophyte	Semi-refined carrageenan was extracted in distilled water using a biomass pretreated in cold alkaline solution	Yield of 64.5 ± 1.2%	Galactan and sulfate contents were 44.3% and 13.8%, respectively	Roldán et al. (2017)
BR, RE and GR tetrasporophytes	Refined carrageenan was extracted in distilled water using a biomass pretreated in cold alkaline solution	The yields varied among strains: 53.2–69.1% (brown strain), 59.5–63.1% (red strain), and 56.1–63.2% (green strain). Higher yields in samples cultivated during the summer-autumn season	The main components found in the biomass: 26.9–33.1% (galactans), 12.7–16.6% (glucans), 14.4–17.1% (ashes), 1.8–5.6% (proteins), 2.5–4% (insoluble aromatics), 10.7–13.8% (sulfate groups), 1.5–8.6% (lipids), and 77.7–89.7% (sum of components) (w/w)	Solorzano-Chavez et al. (2019)

BR, RE and GR = brown, red and green strains, respectively; EP—gametophyte previously cited as "G11", which was originated from tetraspore germination

Table 9.3 Chemical characteristics of extract composed by equal parts of the brown, green, red, and EP strains of *Kappaphycus alvarezii* cultivated in Ubatuba Bay (Gelli et al. 2020)

	Values
pH	5.97
Density	0.95 g mL^{-1}
Organic matter	6.89 g L^{-1}
Soluble mineral residue	36.53 g L^{-1}
Total Carbon	3.83 g L^{-1}
Total mineral residue	36.57 g L^{-1}
Mineral insoluble residue	0.036 g L^{-1}
Potassium (K_2O)	20.17 g L^{-1}
Total Carbon	3.83 g L^{-1}
Total Nitrogen	0.42 g L^{-1}
Sulfur (S)	0.35 g L^{-1}
Calcium (Ca)	0.25 g L^{-1}
Magnesium (Mg)	0.2367 g L^{-1}
Phosphorus (P_2O_5)	0.090 g L^{-1}
Iron (Fe)	0.0037 g L^{-1}
Zinc (Zn)	0.0010 g L^{-1}
Manganese (Mn)	0.0003 g L^{-1}
Copper (Cu)	0.0 g L^{-1}

investigations will be needed to characterize these new strains, such as the amount of refined and semi-refined carrageenan, biofertilizer yield, temperature resistance, epibionts, growth rate and yield, as well as characterization of biomolecules of economic interest.

The implanted biobank (SLFI) and the strain bank in a natural environment (EMFFI) will guarantee the deposit of these variants and the supply of different strains adapted to the needs and conditions of each farmer.

References

Araújo PG, Nardelli AE, Duran R, Pereira MS, Gelli VC, Mandalka A, Eisner P, Fujii MT, Chow F (2022) Seasonal variation of nutritional and antioxidant properties of different *Kappaphycus alvarezii* strains (Rhodophyta) farmed in Brazil. J Appl Phycol 34:1677–1691. https://doi.org/10.1007/s10811-022-02739-6

Araújo PG, Nardelli AE, Fujii MT, Chow F (2020a) Antioxidant properties of different strains of *Kappaphycus alvarezii* (Rhodophyta) farmed on the Brazilian coast. Phycologia 59:272–282. https://doi.org/10.1080/00318884.2020.1736878

Araújo PG, Nardelli AE, Gelli VC, Fujii MT, Chow F (2020b) Monitoring environmental risk of the exotic species *Kappaphycus alvarezii* (Rhodophyta), after two decades of introduction in southeastern Brazil. Bot Mar 63:551–558. https://doi.org/10.1515/bot-2020-0052

Brazil (2020) Normative instruction No. 1, of January 21, 2020. It allows the cultivation of *Kappaphycus alvarezii* on the coast of Santa Catarina, Rio de Janeiro and São Paulo in the areas delimited in this standard. https://www.ibama.gov.br/component/legislacao/?view=legislacao&legislacao=138683. Accessed 2 Feb 2022

Bulboa CR, Paula EJ (2005) Introduction of non-native species of *Kappaphycus* (Rhodophyta, Gigartinales) in subtropical waters: comparative analysis of growth rates of *Kappaphycus alvarezii and Kappaphycus striatum* in vitro and in the sea in South-Eastern Brazil. Phycol Res 53:183–188. https://doi.org/10.1111/j.1440-183.2005.00385.x

Bulboa CR, Paula EJ, Chow F (2008) Germination and survival of tetraspores of *Kappaphycus alvarezii* var. *alvarezii* (Solieriaceae, Rhodophyta) introduced in subtropical waters of Brazil. Phycol Res 56:39–45. https://doi.org/10.1111/j.1440-1835.2008.00483.x

Fonseca BC, Dalbelo G, Gelli VC, Carli S, Meleiro LP, Zimbardi ALRL, Furriel RPM, Tapia DR, Reginatto V (2020) Use of algae biomass obtained by single-step mild acid hydrolysis in hydrogen production by the β-Glucosidase-producing *Clostridium beijerinckii* Br21. Waste Biomass Valorization 11:1393–1402. https://doi.org/10.1007/s12649-018-0430-7

Gelli VC (2019) Desenvolvimento ordenado e potencial da produção da macroalga *Kappaphycus alvarezii* no estado de São Paulo para extração de biofertilizante. Universidade Estadual de Campinas, SP 111 pp. https://doi.org/10.47749/T/UNICAMP.2019.1092073

Gelli VC, Barbieri E (2015) Cultivo e aproveitamento da macroalga *Kappaphycus alvarezii* para pequenos maricultores. In: Editores P& J (ed) Aquicultura no Brasil: Novas Perspectivas. São Carlos - SP, Brasil, pp 641–658

Gelli VC, Patino MTO, Rocha JV et al (2020) Production of the *Kappaphycus alvarezii* extract as a leaf biofertilizer: technical and economic analysis for the north coast of São Paulo-Brazil. Bol do Inst Pesca 46:1–12. https://doi.org/10.20950/1678-2305.2020.46.2.568

Hayashi L, Paula EJ, Chow F (2007a) Growth rate and carrageenan analyses in four strains of *Kappaphycus alvarezii* (Rhodophyta, Gigartinales) farmed in the subtropical waters of São Paulo state, Brazil. J Appl Phycol 19:393–399. https://doi.org/10.1007/s10811-006-9135-6

Hayashi L, Oliveira EC, Bleicher-Lhonneur G, Boulenguer P, Pereira RTL, von Seckendorff R, Shimoda VT, Leflamand A, Vallée P, Critchley AT (2007b) The effects of selected cultivation conditions on the carrageenan characteristics of *Kappaphycus alvarezii* (Rhodophyta, Solieriaceae) in Ubatuba Bay, São Paulo, Brazil. J Appl Phycol 19:505–511. https://doi.org/10.1007/s10811-007-9163-x

Hayashi L, Yokoya NS, Kikuchi DM, Oliveira EC (2008a) Callus induction and micropropagation improved by colchicine and phytoregulators in *Kappaphycus alvarezii* (Rhodophyta, Solieriaceae). J Appl Phycol 20:653–659. https://doi.org/10.1007/s10811-007-9234-z

Hayashi L, Yokoya NS, Ostini S, Pereira RTL, Braga ES, Oliveira EC (2008b) Nutrients removed by *Kappaphycus alvarezii* (Rhodophyta, Solieriaceae) in integrated cultivation with fishes in re-circulating water. Aquaculture 277:185–191. https://doi.org/10.1016/j.aquaculture.2008.02.024

Masarin F, Cedeno FRP, Chavez EGS, Oliveira LE, Gelli VC, Monti R (2016) Chemical analysis and biorefinery of red algae *Kappaphycus alvarezii* for efficient production of glucose from residue of carrageenan extraction process. Biotechnol Biofuels 9:122. https://doi.org/10.1186/s13068-016-0535-9

Oliveira LE, Cedeno RF, Chavez EG, Gelli VC, Masarin F (2019) Red macroalgae *Kappaphycus alvarezii* as feedstock for nutraceuticals, pharmaceuticals and fourth generation biofuel production. Renew Energy Power Qual J 17:546–549. https://doi.org/10.24084/repqj17.370

Paula EJ (2001) Marinomia da alga exótica, *Kappaphycus alvarezii* (Rhodophyta), para produção de carragenanas no Brasil. Universidade de São Paulo, SP, p 39

Paula EJ, Pereira RTL (1998) Da "marinomia" a maricultura da alga exótica *Kappaphycus alvarezii* para produção de carragenanas no Brasil. Panorama Aquicult 8:10–15

Paula EJ, Pereira RTL (2003) Factors affecting growth rates of *Kappaphycus alvarezii* (Doty) Doty ex P. Silva (Rhodophyta, Solieriaceae) in subtropical waters of São Paulo state, Brazil. In: Chapman ARO, Anderson RJ, Vreeland V, Davison I (eds) Proceedings of the XVII international seaweed symposium. Oxford University Press, New York, pp 381–388

Paula EJ, Pereira RTL, Ostini S (1998) Introdução de espécies exóticas de *Eucheuma* e *Kappaphycus* (Gigartinales, Rhodophyta) para fins de maricultura no litoral brasileiro: abordagem teórica e experimental. In Anais. São Paulo: Sociedade Ficológica da América Latina e Caribe/Sociedade Brasileira de Ficologia, pp 341–357

Paula EJ, Pereira RTL, Ohno M (1999) Strain selection in *Kappaphycus alvarezii* var. *alvarezii* (Solieriaceae, Rhodophyta) using tetraspore progeny. J Appl Phycol 11:111–121. https://doi.org/10.1007/978-94-011-4449-0_77

Paula EJ, Erbert C, Pereira RTL (2001) Growth rate of the carrageenophte *Kappaphycus alvarezii* (Rhodophyta, Gigartinales) in vitro. Phycol Res 49:155–161. https://doi.org/10.1046/j.1440-1835.2001.00235.x

Paula EJ, Pereira RTL, Ohno M (2002) Growth rate of the carrageenophyte *Kappaphycus alvarezii* (Rhodophyta, Gigartinales) introduced in subtropical waters of São Paulo state, Brazil. Phycol Res 50:1–9. https://doi.org/10.1046/j.1440-1835.2002.00248.x

Paz-Cedeno FR, Solórzano-Chávez EG, de Oliveira LE, Gelli VC, Monti R, Oliveira SC, Masarin F (2019) Sequential enzymatic and mild-acid hydrolysis of by-product of carrageenan process from *Kappaphycus alvarezii*. Bioenergetics 12:419–432. https://doi.org/10.1007/s12155-019-09968-7

Rodrigues EL, Fonseca BC, Gelli VC, Carli S, Meleiro LP, Furriel RPM, Reginatto V (2019) Enzymatically and/or thermally treated macroalgae biomass as feedstock for fermentative H_2 production.

Materia (Rio de Janeiro) 24:1–14. https://doi.org/10.1590/S1517-707620190002.0678

Roldán IUM, Mitsuhara AT, Desajacomo JPM, Oliveira LE, Gelli VC, Monti R, Sacramento LVS, Masarin F (2017) Chemical, structural, and ultrastructural analysis of waste from the carrageenan and sugar-bioethanol processes for future bioenergy generation. Biomass Bioenergy 107:233–243. https://doi.org/10.1016/j.biombioe.2017.10.008

São Paulo (2022) DECREE No. 66,823 OF JUNE 7, 2022 Approves the management plan for the North Coast Marine Environmental Protection Area, created by Decree No. 53,525, of October 8, 2008. https://www.al.sp.gov.br/repositorio/legislacao/decreto/2022/decreto-66823-07.06.2022.html. Accessed 29 Jun 2022

Solorzano-Chavez GE, Paz-Cedeno FR, Oliveira LE, Gelli VC, Monti R, Oliveira SC, Masarin F (2019) Evaluation of the *Kappaphycus alvarezii* growth under different environmental conditions and efficiency of the enzymatic hydrolysis of the residue generated in the carrageenan processing. Biomass Bioenerg 127:105254. https://doi.org/10.1016/j.biombioe.2019.105254

Zitta CS, Oliveira EM, Bouzon ZL, Hayashi L (2012) Ploidy determination of three *Kappaphycus alvarezii* strains (Rhodophyta, Gigartinales) by confocal fluorescence microscopy. J Appl Phycol 24:495–499. https://doi.org/10.1007/s10811-011-9704-1

Kappaphycus alvarezii Farming in Brazil: A Brief Summary and Current Trends

10

Leila Hayashi, Alex Alves dos Santos, Thallis Felipe Boa Ventura, Felipe Schwahofer Landuci, Valéria Cress Gelli, and Beatriz Castelar

Abstract

Since our last review in 2017, the main progress related to *Kappaphycus alvarezii* farming in Brazil came from southern Brazil. The region is a hot spot in marine aquaculture because of its tradition in mollusk farming. In 2020, after 12 years of studies and negotiation with the Brazilian government, the State of Santa Catarina had permission to cultivate the species commercially. However, only in 2021 local producers got the license and has started the commercial production. On the southeast coast, some marine farms were established in Rio de Janeiro, but according to the last report of the Secretary of Aquaculture and Fisheries (SAP) from the Ministry of Agriculture, Livestock, and Supply, there was no production until 2020. Nevertheless, the possibility of expanding the marine farms for *K. alvarezii* is higher since the SAP debureaucratized the system to apply for marine area grants. Sao Paulo State is still struggling to establish the areas because most of its coastline is protected. Biostimulants seem to be a new product of *K. alvarezii* production in Brazil. This market has a vast national demand thanks to the importance of agribusiness in the country. The price paid for seaweed is higher than that paid for the carrageenan industry. Moreover, producers can process their seaweeds and directly deliver liquid biostimulant to the industry without depending on the international market. Since the activity is starting fresh, there are many possibilities, and depending on where the species is cultivated in Brazil, the final product can be different.

L. Hayashi (✉) · T. F. B. Ventura
Department of Aquaculture, Federal University of Santa Catarina (UFSC), Florianópolis, Santa Catarina, Brazil
e-mail: leila.hayashi@ufsc.br

A. A. dos Santos
EPAGRI – Company of Agricultural Research and Rural Extension of Santa Catarina, Florianópolis, Santa Catarina, Brazil

F. S. Landuci
FIPERJ - Fisheries Institute Foundation of the State of Rio de Janeiro, Rio de Janeiro, Brazil

V. C. Gelli
Sao Paulo Fisheries Institute, São Paulo, Brazil

B. Castelar
ATMOO – the algae factory, Italy; D'Alga Urban Aquaculture, Rio de Janeiro, Brazil

Keywords

Biostimulant · Cotonii · Eucheumatoid species · Red seaweed · South America

10.1 Brief Summary of *Kappaphycus alvarezii* History in Brazil

Kappaphycus alvarezii was experimentally introduced in 1995 by Dr. Edison Jose de Paula, with authorization of the Brazilian Institute for the Environment and Renewable Natural Resources (IBAMA). The experiments were developed in the Sao Paulo Fisheries Institute at Ubatuba Bay, São Paulo, in a joint project with Sao Paulo University (Paula et al. 2002). Seedlings from Japan originating from the Philippines were kept in quarantine for 10 months at the Laboratory of Marine Algae "Edison Jose de Paula" at São Paulo University before transplantation to the sea. Some seedlings, which showed fertile structures in the sea farm, were separated and cultivated in the laboratory. Most of the tetraspores died 3 or 4 days after being released, and only one survived in laboratory conditions (Paula et al. 1999; Zitta et al. 2012). Later, *Kappaphycus striatus* was also introduced in the same place (Ubatuba Bay) (Bulboa and Paula 2005). However, because this strain became fertile and produced viable tetraspores, these last authors recommended keeping only *K. alvarezii* farms for environmental safety. Currently, the cultivation of *K. striatus* is forbidden in Brazil (BRAZIL 2020).

In 1998, informal commercial farming of *K. alvarezii* started at Ilha Grande Bay in Rio de Janeiro with seedlings

A. T. Critchley et al. (eds.), *Tropical Phyconomy Coalition Development*, Developments in Applied Phycology 11,
https://doi.org/10.1007/978-3-031-47806-2_10

imported from Venezuela, which underwent a quarantine period in Santa Catarina (Silva et al. 2010, R. Derner, personal communication). In 2003, other farms started in Sepetiba Bay (Castelar et al. 2009). Nevertheless, only in 2007, after the publication of the first regulation of *K. alvarezii* cultivation in Brazil (BRAZIL 2007), did the number of farms increase, reaching 30 farms. This regulation allowed the cultivation of this species, particularly between Sao Paulo and Rio de Janeiro states, for those with a term of commitment. In the subsequent year, another regulation was published (BRAZIL 2008), replacing this first one, now legalizing one small region between the north coast of Sao Paulo and the south coast of Rio de Janeiro (Fig. 10.1a). Despite all efforts, in 2012, the activity stopped in Rio de Janeiro, probably because of the lack of profitability since the farms were too far from the land (Hayashi et al. 2017). Some farms have been operating recently, and details will be described below.

Some propagules were also irregularly introduced on the Brazilian northeast coast, particularly in the states of Paraiba, Ceara, Pernambuco, and possibly Bahia in the 2000s (Silva et al. 2010). This zone was classified as being under high environmental risk for *K. alvarezii* cultivation (Castelar et al. 2015). However, none of the irregular introductions was commercially successful. In Paraíba State, in particular, 100-kg propagules of unknown origin were transferred to one farm without quarantine procedures and environmental monitoring (Araujo et al. 2013). These last authors identified these samples molecularly and compared them with samples from other regions of Brazil, in addition to Hawaii, Venezuela, Malaysia, and Tanzania, which have never shown invasive behavior, finding substantial similarities.

In South Brazil, particularly in Florianopolis, State of Santa Catarina, the first attempt to cultivate *K. alvarezii* took place in September 1998 as part of the Program of Aquaculture and Fisheries developed by the Brazilian Support Service for Micro and Small Enterprises (SEBRAE), a nonprofit private entity. Approximately 10 g of *Kappaphycus alvarezii* and *Gracilaria lemaneiformis*, imported from Venezuela, was kept in quarantine at the Marine Shrimp Laboratory (LCM) from the Federal University of Santa Catarina (UFSC). At the same time, IBAMA evaluated the authorization to cultivate these species in Brazil. However, since the answer from IBAMA never came, the project ended, and one part of the seedlings was transferred to Rio de Janeiro and another discarded (Silva et al. 2010, R. Derner, personal communication). In 2006, a second trial was made by the Agricultural Research and Rural Extension Company of Santa Catarina (EPAGRI), a Santa Catarina State Company, in partnership with UFSC. At this time, seedlings from the seedstock kept at the Laboratory of Marine Algae "Edison de Paula" from Sao Paulo University were transferred to the Plant Cell Biology Lab (UFSC) and kept in quarantine for 2 years until the release of the experimental license to cultivate this species, issued by IBAMA. In 2008, the seedlings were finally transplanted to the sea, and after 3 years of study, the environmental safety and economic viability were proven (Hayashi et al. 2011). However, it was only in 2020, 12 years after the beginning of the experimental cultivation, that the authorization to commercially cultivate the species in Santa Catarina was released (Brazil 2020) (Fig. 10.1a).

We will present the recent advances in *K. alvarezii* cultivation in the regions where commercial farming is approved. We considered only the activities developed regularly following the Brazilian Regulation for Aquaculture. We based this review on published data and an official database from the Secretary of Aquaculture and Fisheries from the Ministry of Agriculture, Livestock and Supply (MAPA).

10.2 *K. alvarezii* Farming in Sao Paulo

Even with all the studies focused on cultivation technologies, biology, and extraction of carrageenan since 1995 (Paula et al. 1999, 2001; Bulboa and Paula 2005; Bulboa et al. 2008; Hayashi et al. 2007a, b, 2008), Sao Paulo State did not develop commercial farms. This occurred because, since 2008, this state has had environmental restrictions for implementing commercial farming because of Decree N° 53.525, which creates environmental protection areas on all north coasts (Sao Paulo 2008). The good news is that the Decree on the Management Plan for the Marine Environmental Protection Area of the North Coast of São Paulo (Sao Paulo 2022) has recently been published, allowing small producers to apply for licenses to start commercial farming.

Meanwhile, waiting for the publication of this last Decree, Sao Paulo has thus contributed to research on introduced (tetrasporophytes) and cultivated (gametophyte) strains in the Experimental Marine Farm of the Fisheries Institute, focusing on developing subproducts for industrial or artisanal applications, such as biofuels (Masarin et al. 2016; Roldán et al. 2017; Cedeno et al. 2018; Paz-Cedeno et al. 2019; Oliveira et al. 2019; Solorzano-Chavez et al. 2019) and biohydrogen (Dalbelo et al. 2016; Rodrigues et al. 2019; Fonseca et al. 2020). In addition, studies on the extraction of glucose in hydrolysate form to produce several bioproducts (Masarin et al. 2018), agricultural stimulants (Gelli et al. 2020), and antioxidants (Araújo et al. 2020a, 2022) have been developed.

In addition, other studies have contributed to preparing the state for the responsible development of *Kappaphycus alvarezii* farming. Araújo et al. (2020b) concluded, after an environmental monitoring investigation, that this species has a low potential for bioinvasion in the region. Gelli et al. (2020) also demonstrated the economic feasibility of

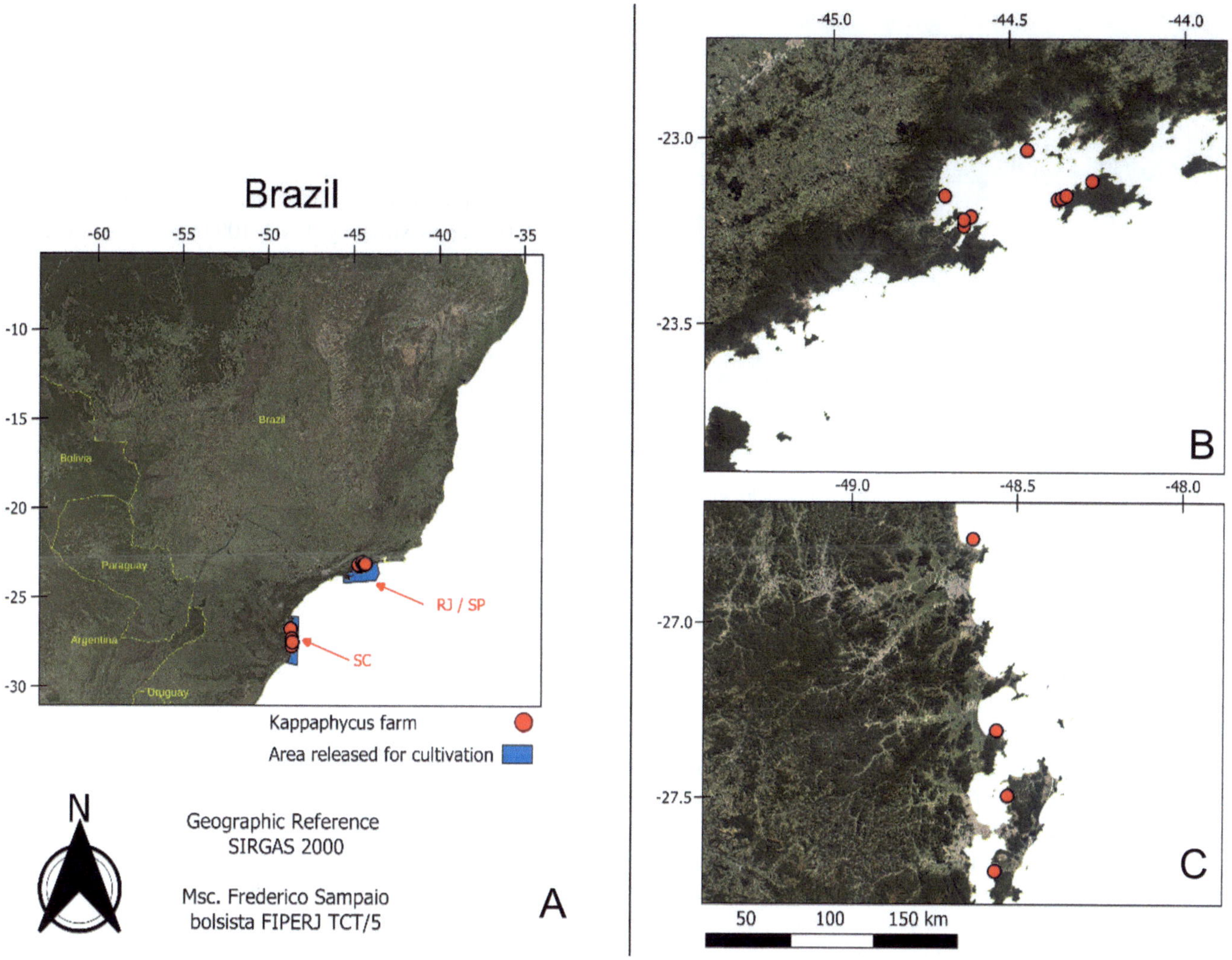

Fig. 10.1 *Kappaphycus alvarezii* marine farms in Brazil. (**a**) States where *Kappaphycus alvarezii* cultivation is authorized. (**b**) Commercial farms established on the south coast of Rio de Janeiro. (**c**) Experimental farms established between 2018 and 2019 to evaluate the viability of commercial cultivation in Santa Catarina, South Brazil. From the north to the South: Penha; Governador Celso Ramos; Sambaqui Beach; Ribeirão da Ilha. Images: Felipe S. Landuci

cultivating the species to produce agricultural biofertilizer on the coast of São Paulo on a small scale. The north coast of São Paulo already has a spatial planning map for implementing marine seaweed farms, and producers are patiently waiting for the regulation to start this new activity.

10.3 *K. alvarezii* Farming in Rio de Janeiro

Although experimental cultivation started in the State of Rio de Janeiro in mid-1998, the activity has not yet reached its full potential. Among the reasons are low profits, little financial support for research and development, the bureaucracy for regulating cultivation areas, the reduced number of people qualified in seaweed cultivation, and the lack of knowledge of the business potential (Suzart and Vendramini 2021; Reis et al. 2017).

Cultivation of *K. alvarezii* in Rio de Janeiro is allowed only in the southern stretch of the "Green Coast." The zone is considered one of the most suitable locations for mariculture in Rio de Janeiro due to its environmental and economic features (Landuci et al. 2021). At present, nine *K. alvarezii* farms are distributed in two municipalities, Angra dos Reis and Paraty, inside Ilha Grande Bay, occupying a total area of only 2.3 ha (Fig. 10.1b).

The most widespread production system is floating rafts using tubular nets (Fig. 10.2). There are five to eight harvests *per year* in production cycles of 45–65 days year-round. The Fisheries Institute of Rio de Janeiro (FIPERJ) estimates the annual production of *K. alvarezii* in Rio de Janeiro to be approximately 2250 tons per year, considering an average

Fig. 10.2 Commercial farm in Rio de Janeiro (Alexandre Feder Commercial Farm). Image: Jandyr Almeida Rodrigues

productivity of 15 kg fw m^{-2} per cycle, using remote sensing and updated satellite imaging (see methods in Landuci et al. 2020). However, incongruences are observed in the official databases. According to the Ministry of Agriculture, Livestock, and Supply (MAPA), in 2020, there were only four assignment contracts for using Brazilian waters for seaweed cultivation. Of these, only two assignees delivered the mandatory Annual Report of Production and declared that they had no production in 2020 (Brazil 2021).

These incongruences are mainly related to legal regulation since the permission to use Brazilian waters for aquaculture is the responsibility of the federal government. At the same time, the license for developing the activity is the attribution of the states. Therefore, despite two decades of activity, the first commercial seaweed company was granted in Rio de Janeiro only recently. Nevertheless, several other enterprises are still operative since the Rio de Janeiro government allows the continuity of operations as long as the environmental license and the licensing process for establishing marine farms have been protocolled (Rio de Janeiro 2015). In recent years, the environmental licensing of the activity has been conducted by the Fishery and Aquaculture Secretary of the municipalities in a simpler and less expensive way than.. and less expensively than the process asked by the State Environmental Agency.

Efforts of bureaucracy simplification have been made by the government in many instances, enhancing the interest in seaweed cultivation, a trend validated by the increasing number of solicitations for environmental licensing and assignments for the use of Brazilian waters for *K. alvarezii* cultivation. The biostimulant trade and cosmetic and other biotechnological applications, such as carbon credits, are the main reasons for the increased interest in the activity. Some promising activities have already been developed, and start-ups born in universities have been dedicated to cultivating and transforming this species in Rio de Janeiro (Suzart and Vendramini 2021).

New investments by the government in the sector also demonstrated the rising interest in seaweed aquaculture in Rio de Janeiro. In 2021, the Chamber of Deputies approved the first parliamentary amendment for financing R$ 530,000.00 (approximately USD 100,000.00) for research on alternative proteins. This project, led by the Federal University of Rio de Janeiro, will examine the utilization of *K. alvarezii* as a potential ingredient for plant-based products (more information in http://algicultura.com).

Another incentive was the publication of the "Law of the Sea Economy" (Rio de Janeiro 2021) by the Legislative Assembly of the State of Rio de Janeiro. This law, defined as "State Policy of Incentive to the Sea Economy as a strategy for the socioeconomic development of the State of Rio de Janeiro," lists the main economic activities in the region, including aquaculture and marine biotechnology. This law classified *Kappaphycus* farming as a strategic activity for sustainable development, creating expectations for this new phase in the State of Rio de Janeiro.

10.4 *K. alvarezii* Farming in Santa Catarina

Since the last publication in 2017 (Hayashi et al. 2017), several advances have been made in southern Brazil. From 2018 to 2019, experimental cultivation was expanded to three other locations, in addition to Sambaqui beach in northern Florianopolis Island, in response to a request from IBAMA to evaluate the viability of commercial cultivation in the region. Experimental farms were installed in the south of Florianopolis Island, in Ribeirão da Ilha, where mollusk farms are established; in Governador Celso Ramos, the mainland of Santa Catarina State, just in front of Florianopolis Island; and in the north of the state, where the water temperature is slightly higher in winter (Fig. 10.1c).

The environmental and commercial viability of the experimental farms was proven in all regions. Additionally, the environmental monitoring protocol was tested, and recently, this protocol has been incorporated into the regulation (Brazil 2020). In the mandatory annual report, producers need to monitor the surrounding coastline during low tide, considering 500 m north and south of the limits of the farm area. In addition, they need to take dated and georeferenced pictures from random points to demonstrate the presence or absence of *K. alvarezii*. The frequency of this monitoring protocol will depend on the location and the respective responsible environmental agency. The idea was to develop a straightforward protocol for producers to comply with using only their cell phones without the need for specific tools.

Rafts also needed to be adapted in the southern region. For example, the polyvinyl chloride (PVC) tubes used in Sao Paulo and Rio de Janeiro to promote floating were too fragile for Santa Catarina waters. They were replaced by buoys designed exclusively for seaweed cultivation (Santos et al. 2018; Santos 2022). These buoys were also designed to integrate the cultivation of seaweeds and mollusks since Santa Catarina is Brazil's main producer of oysters, mussels, and scallops. This structure will also facilitate the mechanization of the farms. A harvester prototype was developed and tested successfully; a planting prototype has recently been under development (Hayashi et al. 2019, 2019). Mechanization is necessary since Santa Catarina State has 612 marine farms totaling 1346.90 ha (Santos et al. 2018; Brazil 2022).

One of the main bottlenecks detected immediately after the first experimental cultivation was the production seasonality, as the subtropical weather provides a hot summer and a cold winter. Although the summer temperatures are ideal for the species, with growth rates close to 3.5% day^{-1} (Hayashi et al. 2017), the low temperatures in wintertime are a challenge for keeping the seedlings at sea. One alternative to solve this problem was to keep the seedlings in tanks, just for maintenance, until the environmental conditions improved (Fig. 10.3). Several experiments were carried out considering the low cost and economic viability of the activity, focusing on fertilization strategies and possible effects on the growth of seedlings after transportation and planting on sea farms. Fertilization methods consider time, duration, and concentration (Martino et al. 2021; Pires et al. 2021); in addition to the type of fertilizers, such as laboratory culture medium, the effluent of shrimp reared in the biofloc system (Pedra et al. 2017; Martino et al. 2021; Pires et al. 2021) or even supplements, such as inorganic carbon addition (Ventura et al. 2020; Baran 2021), were evaluated. One interesting result was that seaweed maintained in tanks under different fertilization methods and fertilizers presented positive responses regarding its biochemical profile, cell structure, and carrageenan yield and quality (Pedra et al. 2017; Ventura et al. 2020; Martino et al. 2021; Baran 2021; Pires et al. 2021).

Since experimental farming started, economic viability studies have been centralized in the carrageenan industry (Santos et al. 2018), mainly due to the Brazilian dependency on this commodity. Official data indicated that in 2021, Brazil imported 3185 of carrageenan, corresponding to USD 24.5 million (Ministry of Economy 2022). However, since the beginning of commercial activities in Santa Catarina State, producers have received business proposals from the agricultural biostimulant industry, which offers values higher than those of the carrageenan industry. Moreover, producers could process biostimulants and profit more than just selling dry seaweed for the carrageenan industry. They also considered that the drying process is still a bottleneck in production, so they started the activity focusing on the biostimulant industry. Agribusiness is one of the main economic activities in Brazil, so production could attend to the national market, avoiding dependence on the international market. Moreover, since the south of Brazil has four well-defined seasons, what we consider a significant challenge in the production chain can now be considered an advantage since seaweed can produce several bioactive compounds that could be interesting to agribusiness, such as resistance to desiccation and low temperatures. Currently, there are four formalized producers in the state, totaling 3.2 ha (Fig. 10.4). Between September 2021 and April 2022, the total production of fresh seaweed was 102.3 tons, all for the biostimulant industry.

Fig. 10.3 Tank cultivation of *K. alvarezii* at Seaweed Laboratory (LCM Macroalga) in the Federal University of Santa Catarina. Image: Thallis Felipe Boa Ventura

10.5 Future Prospects

Seaweed cultivation in Brazil remains incipient, but we have made more progress in this field in the last 5 years than in the last 20 years. One of the main problems was the complicated bureaucratic process of asking for or formalizing the marine farms' areas. Only Santa Catarina State has officially mapped aquaculture parks and areas, totaling almost 970 ha of formalized marine farms for mollusk production, making this state the main Brazilian producer. By 2020, the request for the use of Brazilian waters for aquaculture can now be made online; once approved, it is mandatory to send to the Secretariat of Aquaculture and Fisheries (SAP) of the Ministry of Agriculture, Livestock, and Supply (MAPA), an annual production report. The inclusion of seaweed production in this report was important to track the production of formal activities. Until now, no production of seaweeds was declared in the Annual Production Report, but this will

Fig. 10.4 Commercial farm in Santa Catarina (Algas Brasil Comercial Farm). Image: Ademir Dário dos Santos

probably change in 2022 because of the marine farms in Rio de Janeiro and Santa Catarina. This last state started *Kappaphycus* production in September 2021.

It is important to note that since Brazil is a federation, each state has the autonomy to add its regulation. Normative Instruction no. 1 authorized the cultivation of *K. alvarezii* in Santa Catarina, the north coast of Sao Paulo, and the south coast of Rio de Janeiro. To develop the activity, the producer should also have permission to use the Brazilian waters for aquaculture activities and the license to develop the activity. This last license is granted at the state or municipality level, while the first one is granted by federal regulation. This formalization of the activity is extremely important because the producers will have access, for example, to government funding programs or private initiative funding.

The commercial activity of *K. alvarezii* is showing huge potential. Unlike other aquaculture activities, which have mainly the food market as the main target, seaweed farming can serve different branches of the industry: food, biofertilizers, chemistry, biotechnology, and so on. However, there are still bottlenecks to solve, mainly related to postharvest processing, but we hope to bring solutions for that in the next chapter.

References

Araujo PG, Miranda GEC, Barros-Barreto MB, Fujii MT (2013) Molecular identification of the exotic lineage of *Kappaphycus alvarezii* (Rhodophyta, Solieriaceae) cultivated in the tropical region of Brazil. Phytotaxa 109:17–26. https://doi.org/10.11646/phytotaxa.109.1.2

Araújo PG, Nardelli AE, Fujii MT, Chow F (2020a) Antioxidant properties of different strains of *Kappaphycus alvarezii* (Rhodophyta) farmed on the Brazilian coast. Phycologia 59:272–279. https://doi.org/10.1080/00318884.2020.1736878

Araújo PG, Nardelli AE, Gelli VC, et al (2020b) Monitoring environmental risk of the exotic species *Kappaphycus alvarezii* (Rhodophyta), after two decades of introduction in southeastern. Bot Mar 1–8. https://doi.org/10.1515/bot-2020-0052

Araújo PG, Nardelli AE, Duran R et al (2022) Seasonal variation of nutritional and antioxidant properties of different *Kappaphycus alvarezii* strains (Rhodophyta) farmed in Brazil. J Appl Phycol 1: 23. https://doi.org/10.1007/s10811-022-02739-6

Baran TBM (2021) Sodium bicarbonate addition as strategy for indoor maintenance of *Kappaphycus alvarezii* strains. Master Dissertation, Programa de Pós-Graduação em Aquicultura. Universidade Federal de Santa Catarina, Florianópolis (In Portuguese)

Brazil (2007) Normative Instruction no.165 from Brazilian Institute for the Environment and Renewable Natural Resources (IBAMA), at July 17 2007. Brazilian Official Gazette. Brasília, 18 July 2007, n° 137, Section 1, p.103 (In Portuguese - revoked)

Brazil (2008) Normative Instruction no.185 from Brazilian Institute for the Environment and Renewable Natural Resources (IBAMA), at July 22 2008. Brazilian Official Gazette. Brasília, July 23 2008, n° 140, Section 1, p.60 (In Portuguese - partially revoked)

Brazil (2020) Normative Instruction no.01 from Brazilian Institute for the Environment and Renewable Natural Resources (IBAMA), at January 21 2020. Brazilian Official Gazette. Brasília, 23 January 2020, n° 19, Section 1, p.76 (In Portuguese - actual)

Brazil (2021) Ministry of Agriculture, Livestock and Supply. Bulletin Aquaculture in Brazilian Waters. Annual Report of Production. Secretary of Aquaculture and Fisheries. Brazilia MAPA/SAP 2021. boletim-aquicultura-em-aguas-da-uniao-2020.pdf (www.gov.br). Accessed 23 May 2022 (In Portuguese)

Brazil (2022) Ministry of Agriculture, Livestock and Supply. List of winners of the Aquaculture Parks of Santa Catarina. https://www.gov.br/agricultura/pt-br/assuntos/aquicultura-e-pesca/aquicultura-1/aquicultura-em-aguas-da-uniao. Accessed 6 July 2022 (In Portuguese)

Bulboa CR, Paula EJ (2005) Introduction of nonnative species of *Kappaphycus* (Rhodophyta, Gigartinales) in subtropical waters: comparative analysis of growth rates of *Kappaphycus alvarezii* and *Kappaphycus striatum* in vitro and in the sea in south-eastern Brazil. Phycol Res 53:183–188. https://doi.org/10.1111/j.1440-183.2005.00385.x

Bulboa C, de Paula EJ, Chow F (2008) Germination and survival of tetraspores of *Kappaphycus alvarezii* var. *alvarezii* (Solieriaceae, Rhodophyta) introduced in subtropical waters of Brazil. Phycol Res 56:39–45. https://doi.org/10.1111/j.1440-1835.2008.00483.x

Castelar B, Reis RP, Moura AL, Kirk R (2009) Invasive potential of *Kappaphycus alvarezii* off the south coast of Rio de Janeiro state, Brazil: a contribution to environmentally secure cultivation in the tropics. Bot Mar 52:283–289. https://doi.org/10.1515/BOT.2009.002

Castelar B, de Siqueira MF, Sánchez-Tapia A, Reis RP (2015) Risk analysis using species distribution modelling to support public policies for the alien alga *Kappaphycus alvarezii* aquaculture in Brazil. Aquaculture 446:217–226. https://doi.org/10.1016/j.aquaculture.2015.05.012

Cedeno FRP, Solarzano CEG, Oliveira LEVCG et al (2018) Perspectives for the production of bioethanol from macroalgae biomass. In: Brienzo M (ed) Bioethanol and beyond advances in production process and future directions, 1st edn, New York, pp 203–230

Dalbelo G, Zimbardi ALRR, Fonseca BC, et al (2016) Saccharification of algae biomass for fermentative hydrogen production. In: WHEC 2016 - 21st world hydrogen energy conference 2016, proceedings, pp 891–892

Fonseca BC, Dalbelo G, Gelli VC et al (2020) Use of algae biomass obtained by single-step mild acid hydrolysis in hydrogen production by the β-glucosidase-producing *Clostridium beijerinckii* Br21.

Waste Biomass Valorization 11:1393–1402. https://doi.org/10.1007/s12649-018-0430-7

Gelli VC, Patino MTO, Rocha JV et al (2020) Production of the *Kappaphycus alvarezii* extract as a leaf biofertilizer: technical and economic analysis for the north coast of São Paulo-Brazil. Bol do Inst Pesca 46:1–12. https://doi.org/10.20950/1678-2305.2020.46.2.568

Hayashi L (2019) Protótipo de colheitadeira mecânica de macroalgas/Seaweed mechanical harvester prototype. https://www.youtube.com/watch?v=nxaEdImn_1U. Accessed 04 May 2022

Hayashi L, de Paula EJ, Chow F (2007a) Growth rate and carrageenan analyses in four strains of *Kappaphycus alvarezii* (Rhodophyta, Gigartinales) farmed in the subtropical waters of São Paulo state, Brazil. J Appl Phycol 19:393–399. https://doi.org/10.1007/s10811-006-9135-6

Hayashi L, Oliveira EC, Bleicher-Lhonneur G et al (2007b) The effects of selected cultivation conditions on the carrageenan characteristics of *Kappaphycus alvarezii* (Rhodophyta, Solieriaceae) in Ubatuba Bay, São Paulo, Brazil. J Appl Phycol 19:505–511. https://doi.org/10.1007/s10811-007-9163-x

Hayashi L, Yokoya NS, Ostini S et al (2008) Nutrients removed by *Kappaphycus alvarezii* (Rhodophyta, Solieriaceae) in integrated cultivation with fishes in recirculating water. Aquaculture 277:185–191. https://doi.org/10.1016/j.aquaculture.2008.02.024

Hayashi L, Santos AA, Faria GS, Nunes BG, Souza MS, Fonseca AL, Barreto PLM, Oliveira EC, Bouzon ZL (2011) *Kappaphycus alvarezii* (Rhodophyta, Areschougiaceae) cultivated in subtropical waters in southern Brazil. J Appl Phycol 23(3):337–343. https://doi.org/10.1007/s10811-010-9543-5

Hayashi L, Reis RP, Santos AA, Castelar B, Robledo D, Vega GB, Msuya FE, Eswaran K, Yasir SM, Ali MKM, Hurtado AQ (2017) The cultivation of *Kappaphycus* and *Eucheuma* in tropical and subtropical waters. In: Hurtado AQ, Critchley AT, Neish IC (eds) Tropical seaweeds farming trends, problems and opportunities - focus on *Kappaphycus* and *Eucheuma* of commerce. Springer, Cham, pp 55–90. https://doi.org/10.1007/978-3-319-63498-2_4

Hayashi L, Novaes ALT, Santos AA (2019) Preliminary performance assessment of *Kappaphycus alvarezii* mechanical harvester prototype. In: 23rd international seaweed symposium - program and abstracts book. Jeju, p 151

Landuci FS, Rodrigues D, Fernandes AM, Scott PC, Poersch LHS (2020) Geographic information system as an instrument to determine suitable areas and identify suitable zones to the development of emerging marine finfish farming in Brazil. Aquac Res 51:3305–3322. https://doi.org/10.1111/are.14666

Landuci FS, Bez MF, Ritter PD, Costa S, Silvestri F, Zanette GB, Castelar B, Costa PMS (2021) Mariculture in a densely urbanized portion of the Brazilian coast: current diagnosis and directions for sustainable development. Ocean Coastal Manage 213:105889. https://doi.org/10.1016/j.ocecoaman.2021.105889

Martino R, Mariot LV, Silva FZ, Simioni C, Carneiro MAA, Oliveira ER, Maraschin M, Santos AA, Hayashi L (2021) Effects of biofloc effluent in different regimes as a fertilizer for *Kappaphycus alvarezii*: indoor maintenance and sea cultivation. J Appl Phycol 33:3225–3237. https://doi.org/10.1007/s10811-021-02539-4

Masarin F, Cedeno FRP, Chavez EGS et al (2016) Chemical analysis and biorefinery of red algae *Kappaphycus alvarezii* for efficient production of glucose from residue of carrageenan extraction process. Biotechnol Biofuels 9:122. https://doi.org/10.1186/s13068-016-0535-9

Masarin F, Cedeno FRP, de Oliveira LE, et al (2018) Processo de obtenção de hidrolisado de glicose, hidrolisado de glicose e uso do mesmo. Patente: BR 10 2017 004803 9

Ministry of Economy (2022) Statistical database of Brazilian Foreign Trade. http://comexstat.mdic.gov.br/pt/geral. Accessed July 5 2022

Oliveira LE, Cedeno RF, Chavez EG, et al (2019) Red macroalgae *Kappaphycus alvarezii* as feedstock for nutraceuticals, pharmaceuticals and fourth generation biofuel production. Int. Conf. Renew. Energy power Qual. Tenerife (Spain), 10th to 12th April. 2019 renew. Energy Power Qual. J. ISSN 2172-038 X, no.17 April 2019 12–15

Paula EJ, Erbert C, Toledo Lima Pereira R (2001) Growth rate of the carrageenophyte *Kappaphycus alvarezii* (Rhodophyta, Gigartinales) in vitro. Phycol Res 49:155–161. https://doi.org/10.1046/j.1440-1835.2001.00235.x

Paula EJ, Pereira RTL, Ohno M (2002) Growth rate of the carragenophyte *Kappaphycus alvarezii* (Rhodophyta, Gigartinales) introduced in subtropical waters of São Paulo state, Brazil. Phycol Res 50:1–9. https://doi.org/10.1046/j.1440-1835.2002.00248.x

Paz-Cedeno FR, Solórzano-Chávez EG, de Oliveira LE et al (2019) Sequential enzymatic and mild-acid hydrolysis of by-product of carrageenan process from *Kappaphycus alvarezii*. Bioenergy Res 12:419–432. https://doi.org/10.1007/s12155-019-09968-7

Pedra AGLM, Ramlov F, Maraschin M, Hayashi L (2017) Cultivation of the red seaweed *Kappaphycus alvarezii* with effluents from shrimp cultivation and brown seaweed extract: effects on growth and secondary metabolism. Aquaculture 479:297–303. https://doi.org/10.1016/j.aquaculture.2017.06.005

Pires CM, Bazzo GC, Barreto PLM, Espírito Santo CM, Boa Ventura TF, Pedra AGLM, Rover T, McGovern M, Hayashi L (2021) Cultivation of the red seaweed *Kappaphycus alvarezii* using biofloc effluent. J Appl Phycol 33:1047–1058. https://doi.org/10.1007/s10811-020-02335-6

Reis RP, Castelar B, Santos AA (2017) Why is algaculture still incipient in Brazil? J Appl Phycol 29:673–682. https://doi.org/10.1007/s10811-016-0890-8

Rio de Janeiro (2015) Resolution CONEMA N° 68 at August 11 2015. Rio de Janeiro Official Gazette. Rio de Janeiro, August 20 2015, n° 151, Part I, p. 12 (in Portuguese – actual)

Rio de Janeiro (2021) Law N° 9466 at November 25 2021. Rio de Janeiro Official Gazette. Rio de Janeiro, November 26 2021, n° 223, Part I, p. 1 (in Portuguese – actual)

Rodrigues EL, Fonseca BC, Gelli VC et al (2019) Enzymatically and/or thermally treated macroalgae biomass as feedstock for fermentative H2 production. Rev Mater 24. https://doi.org/10.1590/s1517-707620190002.0678

Roldán IUM, Mitsuhara AT, Munhoz Desajacomo JP et al (2017) Chemical, structural, and ultrastructural analysis of waste from the carrageenan and sugar-bioethanol processes for future bioenergy generation. Biomass Bioenergy 107:233–243. https://doi.org/10.1016/j.biombioe.2017.10.008

Santos AA (2022) Cultivation system of the seaweed *Kappaphycus alvarezii* in Santa Catarina. EPAGRI, Florianópolis, SC. 56 pp (In Portuguese)

Santos AA, Novaes ALT, Marchiori NC, Hayashi L (2018) Novel flotation model for the experimental culture of macroalgae *Kappaphycus alvarezii* in Florianópolis, Brazil. Agropecuária Catarinense 31:45–48. https://doi.org/10.22491/RAC.2018.v31n2.4

Sao Paulo (2008) Decree N° 53.525 at October 8 2008. Sao Paulo Official Gazette. Sao Paulo, October 9 2009, n° 191, vol 118, pp 1–5 (In Portuguese – actual)

Sao Paulo (2022) Decree N° 66.823, at June 7 2022. Sao Paulo Official Gazette. Sao Paulo, June 8 2022, n° 132, volume 112, p. 1–4 (In Portuguese – actual)

Silva BNT, Amancio CE, Oliveira-Filho EC (2010) Exotic marine macroalgae on the Brazilian coast: a revision. Oecolol Austral 14:403–414. https://doi.org/10.4257/oeco.2010.1402.05

Solorzano-Chavez EG, Paz-Cedeno FR, Ezequiel de Oliveira L et al (2019) Evaluation of the *Kappaphycus alvarezii* growth under different environmental conditions and efficiency of the enzymatic hydrolysis of the residue generated in the carrageenan processing.

Biomass Bioenergy 127:105254. https://doi.org/10.1016/j.biombioe.2019.105254

Suzart LGC, Vendramini ALA (2021) Biotechnological applications of the seaweed *Kappaphycus alvarezii*: a prospective study. Cadernos de Prospecção 14(4):1145–1158. https://doi.org/10.9771/cp.v14i4.42328. (In Portuguese)

Ventura TFB, Bruzinga CP, Santos AA, Simioni C, Hayashi L (2020) Addition of carbon dioxide, followed by irradiance increase, as optimization strategy for the cultivation of the red seaweed *Kappaphycus alvarezii*. J Appl Phycol 32:4113–4126. https://doi.org/10.1007/s10811-020-02210-4

Zitta CS, Oliveira EM, Bouzon ZL, Hayashi (2012) Ploidy determination of three *Kappaphycus alvarezii* strains (Rhodophyta, Gigartinales) by confocal fluorescence microscopy. J Appl Phycol 24:495–499. https://doi.org/10.1007/s10811-011-9704-1

Paula EJ, Pereira RTL, Ohno M (1999) Strain selection in *Kappaphycus alvarezii* var. *alvarezii* (Solieriaceae, Rhodophyta) using tetraspore progeny. J Appl Phycol 11:111–121. https://doi.org/10.1007/978-94-011-4449-0_77

11 Developing Cultivation Systems and Better Management Practices for Caribbean Tropical Seaweeds in US Waters

L. M. Roberson, G. S. Grebe, I. B. Arzeno-Soltero, D. Bailey, S. Chan, K. Davis, C. A. Goudey, H. Kite-Powell, S. Lindell, D. Manganelli, M. Marty-Rivera, C. Ng, F. Ticona Rollano, B. Saenz, A. M. Van Cise, T. Waters, Z. Yang, and C. Yarish

Abstract

At the time of writing, the Caribbean's small countries and island nations were experiencing a loss of resources resulting from climate change, nutrient pollution, ocean acidification, seagrass bed habitat loss, and overfishing and tourism revenue losses due to a global pandemic and *Sargassum* blooms. Sustainably managed development of a tropical seaweed industry in the region may help to assuage these issues while also providing a new seaweed biomass source for existing markets (e.g., carrageenan and agar) as well as newly emerging markets (e.g., bioplastics). Here we explore ways to support the development of the seaweed industry in tropical US waters of the Caribbean and Gulf of Mexico drawing on knowledge and expertise from our many partners and test sites in Puerto Rico, Florida, and Belize. In collaboration, we are prototyping seaweed farm systems and techniques that can be used in offshore areas, creating tools to automate seeding and harvesting of biomass, assessing the habitat provisioning and impacts of these farm systems and their risks from storms, identifying likely species of tropical algae with high growth and compounds of commercial interest, and conducting economic and life cycle analyses of these advanced aquaculture systems and processing. Although we describe systems optimized for the Caribbean and the Gulf of Mexico, they can easily be adapted for use in other regions, including temperate zones, thus advancing available technology for the US seaweed industry as a whole.

Keywords

Macroalgae · Aquaculture · Carrageenophytes · Rhodophyta · *Eucheuma* · *Gracilaria*

L. M. Roberson (✉) · G. S. Grebe
Marine Biological Laboratory, Woods Hole, MA, USA
e-mail: lroberson@mbl.edu

I. B. Arzeno-Soltero
Stanford University, Palo Alto, CA, USA

D. Bailey · H. Kite-Powell · S. Lindell
Woods Hole Oceanographic Institution, Woods Hole, MA, USA

S. Chan
The Nature Conservancy, Belize Program, Belmopan, Belize

K. Davis
University of California, Irvine, Irvine, CA, USA

C. A. Goudey
C.A. Goudey & Associates, Newburyport, MA, USA

D. Manganelli
TendOcean, LLC, Newburyport, MA, USA

M. Marty-Rivera · C. Ng
University of Connecticut, Stamford, CT, USA

F. T. Rollano · Z. Yang
Pacific Northwest National Laboratory, Seattle, WA, USA

B. Saenz
Stanford University, Palo Alto, CA, USA

University of California, Irvine, Irvine, CA, USA

A. M. Van Cise
University of Washington, School of Aquatic and Fisheries Sciences, Seattle, WA, USA

T. Waters
The Nature Conservancy, Belize Program, Belmopan, Belize

The Nature Conservancy, Washington, DC, USA

C. Yarish
Woods Hole Oceanographic Institution, Woods Hole, MA, USA

University of Connecticut, Stamford, CT, USA

11.1 Introduction

Like many small island nations around the world, Caribbean countries are highly dependent on marine resources for their economies and face growing threats from climate change, pollution, habitat loss, fishing pressure, and loss of tourism revenue from the COVID-19 pandemic (Diez et al. 2019; Clegg et al. 2020; Mulder 2020). Fostering the development of a tropical macroalgae farming industry in this region may simultaneously address the growing environmental crisis and

A. T. Critchley et al. (eds.), *Tropical Phyconomy Coalition Development*, Developments in Applied Phycology 11,
https://doi.org/10.1007/978-3-031-47806-2_11

create economic opportunities for Caribbean coastal communities. Well-managed macroalgal farms can reduce the impact of environmental threats by preventing coastal erosion, removing excess nutrients from coastal waters that fuel opportunistic blooms of micro- and macroalgae, providing habitat for marine life, and creating new marine livelihoods for local communities without the need for scarce land or freshwater resources or fertilizers (Walls et al. 2016; Waters et al. 2019; Tanaka et al. 2020; Theuerkauf et al. 2021). To be a successful long-term business enterprise in the region, however, macroalgal aquaculture must also be economically viable, environmentally sustainable, and socially accepted.

Small-scale cultivation of tropical seaweeds in the Caribbean and Gulf of Mexico developed sporadically at several locations from the 1970s to the present (FAO 1980; Smith 1992). However, farming operations currently remain limited to a few locations (e.g., Belize, St. Lucia, and Venezuela), where there is a culture of seaweed products such as blended sea moss drinks made from dried *Eucheuma isiforme*, spices, and milk (Rincones 2000; Hayashi et al. 2017). There are several initiatives in Belize, St. Lucia, and Mexico attempting larger-scale production (Cai et al. 2021), in addition to our test farms in Puerto Rico and Florida. There are also several startups in the region (e.g., *Sargassum* Ocean Sequestration Carbon (SOS Carbon), Carbon Wave, Salgax Biotecnologia Marina Aplicada) that are attempting to harvest large quantities of wild, floating *Sargassum* from coastal waters before the seaweed washes ashore and becomes a nuisance for nearby coastal communities (Thompson et al. 2020; Oxenford et al. 2021).

Despite the lack of historic production, the Caribbean and Gulf of Mexico region offers enormous potential and advantages for increased production of macroalgal biomass (Merrill and Waaland 1998; Bjerregaard et al. 2016). Warm water temperatures and ample available sunlight facilitate rapid algal growth. In addition, most if not all the countries in this region have large marine areas, and the offshore portions of these marine areas could be particularly good locations for aquaculture development if proper technologies can be developed to exploit these challenging environments. There are several native carrageenophytes and agarophytes of commercial interest, including *Eucheumatopsis isiformis* (syn. *Eucheuma isiforme*), *Hypnea musciformis*, *Soliera filiformis*, *Meristiella* spp., *Agardhiella* spp., and various species of *Gracilaria* (Zertuche-Gonzalez 1993, 1994; 1998). Other species of potential interest to emerging markets include *Laurencia* spp. (methane-reduction properties for animal feed), *Caulerpa* spp. (culinary intrigue), and *Botryocladia* spp. (culinary). In addition to these physical and biological assets, there are also market-based advantages to farming seaweeds in the Caribbean and Gulf of Mexico. Firstly, the development of local, circular economies is a critical need for small island nations (e.g., Kerr 2005), and given the many existing and potential uses of seaweed biomass, it can likely be incorporated into these larger economic initiatives. Secondly, seaweed producers in the Caribbean and Gulf of Mexico region have direct access to the large North American market (and manufacturers in Mexico/Central America), which will reduce the transport cost and time associated with delivering seaweed biomass to processors. Additionally, when algal processing streams are further established in North America, there is a possibility that cultivated and nuisance algae (i.e., *Sargassum* spp.) from the region could be aggregated into the same supply chain.

The United States (US) Department of Energy Advanced Research Projects Agency-Energy (ARPA-E) Macroalgae Research Inspiring Novel Energy Resources (MARINER) Program was developed in 2017 to advance the development of tools to support large-scale production of renewable marine biomass in US waters for use as feedstock for biofuel, valuable bioproducts, and animal feed (Kim et al. 2019). MARINER technology development includes large-scale farm systems, computational modelling tools, farm and water quality monitoring tools, and advanced breeding and genetic tools. In the USA, marine macroalgal farming is limited by a lack of knowledge of uses and a ready local market, a burdensome permitting process and limited social license, as well as the labor-intensive, and therefore high-cost, farming methods commonly used at present (Grebe et al. 2019; Duarte et al. 2022). Additionally, while there is a large body of work on marine macroalgae ecology and biology and responses to environmental factors (e.g., Oyieke 1994; Roberson and Coyer 2004), few rigorous studies have been done in the field with species from the Caribbean and Gulf of Mexico and at a scale relevant to even a small farm system (but see Pereira and Yarish 2008; Abreu et al. 2009; Valderrama et al. 2013; Buschmann et al. 2014). Similarly, much work has been done assessing the impact of kelps and flow on nutrient transport in kelp beds (e.g., Gaylord et al. 2007), but very little has been done on red algae, and nothing has been done on carrageenophytes or agarophytes from the tropical USA or Caribbean. This information is crucial for understanding nutrient dynamics in and around large-scale farms where upstream propagules might limit nutrient availability to downstream propagules and the forces that will impact the farm system and the seaweed itself, particularly under more harsh offshore conditions.

The MARINER program presented a clear opportunity to develop farming methods for the large US exclusive economic zone (EEZ), the second largest in the world, most of which is in warm, tropical waters. In total, 21 projects were funded, with approximately $62 million allocated to the MARINER program. Our funded project, Techniques for Tropical Seaweed Cultivation and Harvesting, proposed to investigate and develop cost-effective opportunities to produce biomass in underutilized areas of the Gulf of Mexico, Caribbean, and tropical US EEZ. In this region, for example, Puerto Rico and the US Virgin Islands share 211,429 km^2 of unexploited waters with minimal conflicts of use (Jossart et al. 2019). With this project, we sought to develop six critical areas to jump start the tropical seaweed industry in the USA: (1) optimization of farming practices; (2) development of a farming platform with mechanized cultivation and harvesting of climate-resistant macroalgae; (3) characterization of nutrient and hydrodynamic loads at the farm scale; (4) chemical and growth characteristics of the tropical macroalgal biomass; (5) identification of the social and environmental impacts of macroalgal farms in the Caribbean and Gulf of Mexico; and (6) economic modelling and life cycle assessment of tropical macroalgal biomass production incorporating technological, social, and environmental factors. Our approach involved scientific investigations across laboratories, in silico, and at nearshore research sites in southern Belize, western Florida, and southwestern Puerto Rico.

In this chapter, we outline our work and progress in the Caribbean and Gulf of Mexico to date and provide recommendations for better management practices to support the nascent macroalgae industry in these regions. We group this information and discussion within the following subsections: (1) farm design, outplanting, operating, and harvesting efficiency; (2) onshore nursery development and use; (3) farm monitoring and maintenance in response to epiphytes and grazers; (4) marine mammal interactions; (5) habitat provisioning; (6) farm site suitability under different nutrient loads; (7) storm surge and extreme wave conditions; (8) technoeconomic forecasts and industry development; and (9) needs for additional research and development.

11.2 Methods, Results, and Discussion

11.2.1 Farm Design, Outplanting, Operating, and Harvesting Efficiency

Increasing seaweed farm scale requires new farm design, the mechanization of outplanting, maintenance, and harvesting, and specialized work vessels. In this subsection, we describe our developments and recommendations for improvements in each of these categories. Due to the proprietary nature of these developments, some of the design and operation details are intentionally limited.

The Techniques for Tropical Seaweed Cultivation and Harvesting project has designed and tested three seaweed cultivation arrays. Each of the cultivation systems comprises a Hydropro or Colmega braid frame (≤23-mm diameter), hollow polypropylene growth lines (≤19-mm diameter), drag embedment or plow anchors (≤91 kg) with nylon or Hydropro anchor lines (≤23-mm diameter), and inflatable buoys. The first system, a five-line micro-farm array, has a total footprint of 92 × 3 m, with five parallel, submerged grow lines that are each 23 m long (Fig. 11.1). This system is better suited for shallow depths (<10 m) and for deployment without the need for specialized hauling equipment or large vessels. The second system, a five-line mini-farm array, has a 132 × 3 m footprint, with five grow lines that are each 61 m long (Fig. 11.1). This system operates in deeper water (>10 m) and is a useful platform for testing at the farm site

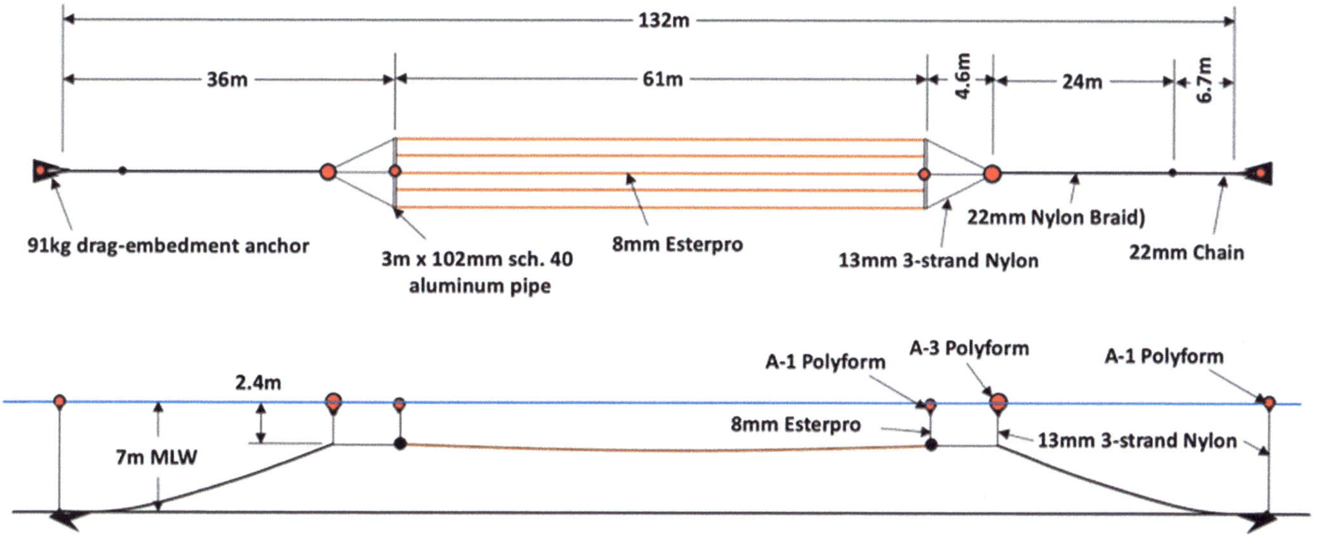

Fig. 11.1 Bird's-eye and side view of the five-line mini-farm array. The five-line micro-farm array shares the same basic design but with shorter and lighter components for ease of deployment. Source: C. A Goudey & Associates

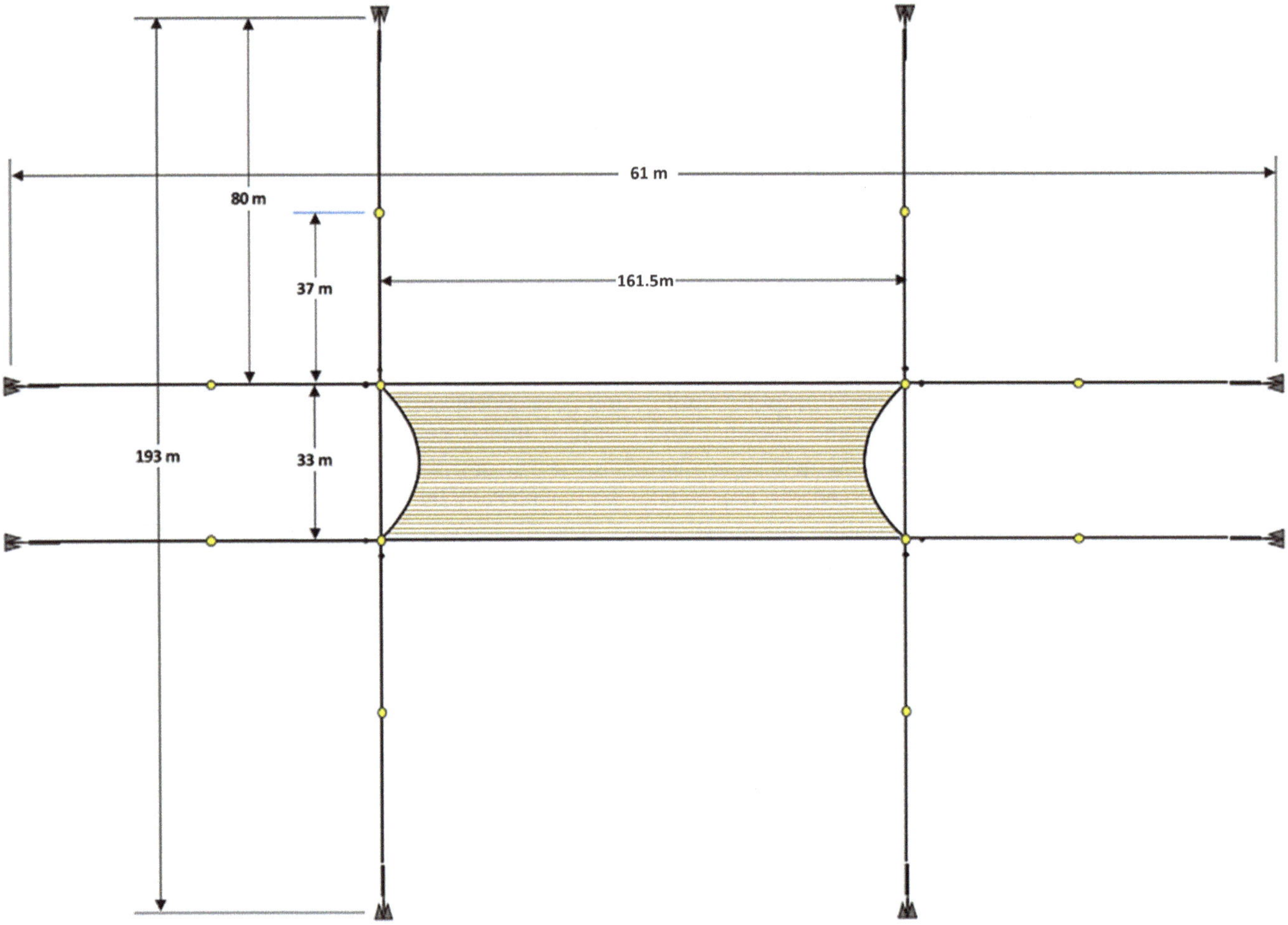

Fig. 11.2 Bird's-eye view of the catenary array and anchor system for tropical and temperate seaweed farming. Source: C. A Goudey & Associates

before committing to the deployment of a large-scale system that requires specialized equipment or vessels to deploy and maintain. The third system, a catenary array, has a footprint 161.5 m long × 33 m wide, with up to 60 161.5-m-long grow lines tensioned between two catenaries (Fig. 11.2; Goudey 2019). This system is modular and scalable; multiple catenary arrays may be anchored adjacent to each other to form a larger module (Fig. 11.3). The horizontal components of all arrays are submerged at least 3 m below the water surface.

The development of farming arrays is based on engineering principles and the results from numerical simulations using anticipated site conditions and an understanding of the drag characteristics of seaweed and farm components in currents and waves. The development of these novel systems is also informed by over 2 years of experience in designing, fabricating, installing, and managing similar large-scale systems in New England and Alaska (Kite-Powell et al. 2022). The experience of many members of this team with other seaweed cultivation approaches has been key to anticipating the advantages that catenary systems offer the emerging seaweed farming sector, and these potential advantages will be validated with precise measurements, methodical installations, and robust monitoring and intervention at our project field site in Puerto Rico.

In addition to new farm designs, methods that greatly reduce the manual labor associated with outplanting must be developed to be economically competitive with macroalgal biomass produced in developing or low-wage countries. Traditionally, tropical seaweed farming begins with the outplanting of vegetative material on single lines or small rafts (Doty et al. 1987; Ask 2003). The "seed" material is collected and portioned into propagules, a process usually done from a recently harvested crop. Large propagules must be cut into suitable pieces of approximately 50–100 g, and then all propagules are attached to, or supported, by a growth substrate. Traditionally, propagules are attached to a rope that is supported horizontally below the sea surface (Pereira and Yarish 2008). Although there are several options available for attaching propagules to a grow line, they all involve time-consuming manual labor. The most popular attachment methods are the tie-tie or "Made Loop" where a loop of twine is attached to the grow line, and then

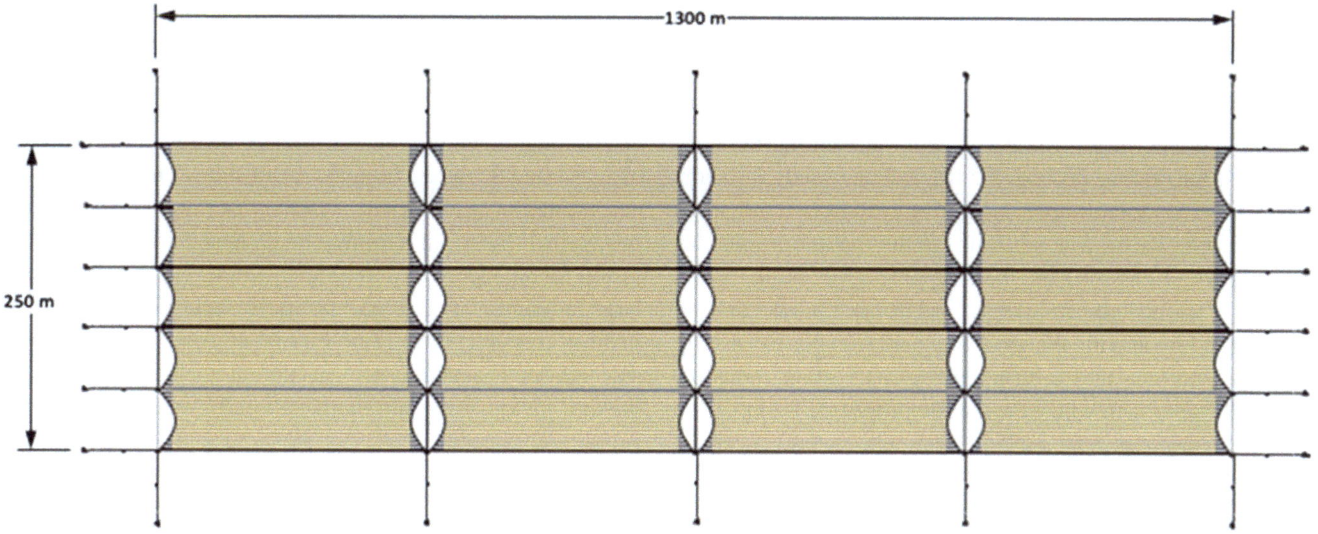

Fig. 11.3 Bird's-eye view of a module made from 20 catenary arrays that share anchors and framing lines. Source: C. A Goudey & Associates

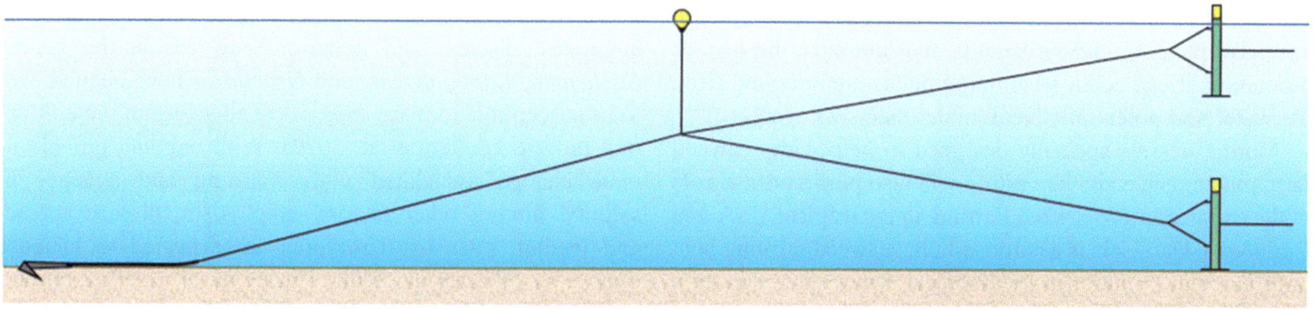

Fig. 11.4 Variable displacement spar buoy in floating and sunk positions. Source: C.A. Goudey and Associates

each propagule is manually inserted into and captured by the twine or vice versa (Doty et al. 1987; Ask et al. 2003). Alternatively, propagules can be placed inside a section of flexible tube netting where growth occurs both inside and outside the net (Zertuche-González 1996; Zertuche-González et al. 1999, 2001). One possibility for mechanizing the outplanting of tropical seaweed biomass is the use of a clip to attach the seaweed propagule to the growth line. We have devised several clip prototypes (Goudey 2022) and are experimenting with permanently attaching them to the line or inserting the clip into the line at the time of propagule attachment. Either approach seems amenable to mechanization, although the physical handling of the propagules is not trivial and will likely require machine vision and custom robotics.

We are also experimenting with techniques to mechanize the operation and maintenance of the cultivation array once it is deployed and seeded. Our experience to date suggests that seaweed farms deployed in moderately deep (10–100 m) and protected locations in the Caribbean (i.e., in the lee of a caye or near a deeper channel) may incur more sedimentation on the array lines and cultivated seaweed than what is currently observed on farms located in the intertidal zones in Southeast Asia and East Africa where tidally driven flow of water across the seaweed aids in minimizing sedimentation. Thus, methods to introduce minor movement into the array to dislodge this sediment or low-effort techniques for cleaning the seaweeds are being considered and further developed.

Seasonal tropical storms and hurricanes in our project region have also inspired smart design and automated operation and maintenance processes on the seaweed cultivation array. To reduce the risk of storm-related damage to the seaweed crop, we have developed a variable displacement spar buoy (VDSB) that sits at the sea surface when filled with air but can be sunk to the bottom while preserving its upright stability (Fig. 11.4). Deeper water has less kinetic energy during storms (Aleman and Constantin 2020), so by sinking the farming array and seaweed crop prior to a tropical storm or hurricane, the crop is more protected from strong waves and currents than it would be on the water surface. Once the storm passes, the buoys can be refilled with air, and the farm returns to its original position in the water column. Automation of the sinking process so that it can be done remotely from shore is currently in development.

Like the challenges associated with outplanting, the conventional harvesting process for tropical seaweeds is labor-intensive. The default approach is the complete removal of the crop with its growth substrate, or the mature propagules are simply cut or torn from their support (Doty et al. 1987). Efficiency in the harvesting of farmed tropical seaweeds may be achieved by leaving the grow line in place after harvesting the algal biomass and utilizing partial harvesting methods. Again, the approach where the mature propagule is clipped to the grow line may lend itself to such automated processes. Ideally, harvesting activities can be immediately followed by the seeding methods described above, so the ideal harvesting method for a particular array design or macroalgal species will vary with different seeding methods (i.e., using growing tips vs. mature propagules). We also note that accomplishing both seeding and harvesting underwater, resulting in minimal disruption to the grow line and the associated farm system, may also have efficiency benefits and minimize the loss of biomass that can occur when large plants are removed from the water and potentially break under their own weight.

Marine vessels specially designed to help with seaweed farm maintenance and harvesting will also play a critical role in the expansion of seaweed farming in the tropical USA and Caribbean. Damisela is a purpose-built seaweed farming boat built for the Techniques for Tropical Seaweed Cultivation and Harvesting (Fig. 11.5). We use the vessel for installation and moving the farm array, attaching new seaweed for cultivation, harvesting mature seaweed, and routine maintenance of the farm structure. Damisela was designed to fit in a shipping container for easy transport and is therefore less than 6 m long. We are continuing to evaluate the performance of the vessel throughout the course of the project, but to date, it has performed well under farm conditions and believe that the design will be of great use for the emerging industry.

Ultimately, achieving economies of scale in seaweed farming in the USA and Caribbean will be greatly facilitated by reducing labor and fuel costs associated with the transport of harvested seaweed biomass from the farming site to the nearest aggregation facility, landing, or port. The drone tug, prototyped by C.A. Goudey and Associates, provides a vision for an uncrewed and efficient vessel capable of moving large amounts of seaweed biomass secured on barges or floating transport bags (Kite-Powell et al. 2022). Additional research and development of complementary equipment, on-site processing alternatives, and renewable power supplies are needed.

11.2.2 Onshore Nursery Development and Use

While developments in outplanting and harvesting practices will substantially increase farm output, tropical seaweed farms will still have to contend with stressors such as epiphytes, disease, and grazing. Seaweeds in the genera *Eucheuma*, *Kappaphycus*, and *Gracilaria* have been shown to be susceptible to these stressors at all stages of the cultivation process (Behera et al. 2022). With ongoing global climate change associated with warming and changes in salinity, among other factors, epiphytism, disease spread, and feeding rates by herbivores are expected to increase over time (Vairappan 2006; Harley et al. 2012; Loureiro et al. 2017; Egan and Gardiner 2016). Thus, it is imperative for seaweed farmers in the tropical USA and Caribbean to address these issues by incorporating the use of bio-secure nurseries or risk major economic losses due to lower yield. Nursery-tank culturing of *Eucheuma* spp. and other red algal species will benefit the industry by providing marine seaweed farms with a consistent source of seed material rather than

Fig. 11.5 The seaweed farm service vessel Damisela underway. Source: C.A. Goudey and Associates

relying solely on material from existing farms or wild populations (Redmond et al. 2014; Reddy et al. 2017).

Two of the challenges associated with nursery and tank maintenance for *Eucheuma* spp. are obtaining higher specific growth rates in the tank environment and preventing epiphytes. Previous studies indicate that *E. isiforme* grown in tanks has a low daily growth rate of 1–3% per day (Moon and Dawes 1976), but the application of biostimulants may help to increase *E. isiformis* growth rates. Biostimulants derived from *Ecklonia maxima* and *Ascophyllum nodosum* have been shown to increase the number of shoots per *Eucheuma* propagule after 28 days of constant exposure at 5 ppm (Umanzor et al. 2020). Additionally, biostimulants may reduce the incidence of epiphytes after transplanting material to the sea. In a study with four strains of *Kappaphycus* from Malaysia, after 45 days of exposure to an *Ascophyllum*-derived biostimulant, segments bearing shoots were transplanted to a sea-based nursery. Segments that had been exposed to the biostimulant had less epiphytic growth than those that were not exposed (Ali et al. 2020). Therefore, the use of biostimulants seems promising for nurseries to adopt, although these previous studies indicate that the dose will depend on the species being cultivated.

Nursery-tank culture also requires movement of seed material from the field to the lab. Our nursery work with *E. isiforme* has revealed that this species can be sensitive to changes in conditions, especially if the material is moved or shipped. Within 2–4 days, it has been common to see shipments turn white, possibly evidence of ice-ice syndrome, which is thought to be caused by infectious microbes (Vairappan et al. 2010; Ward et al. 2021). Even though material was free of epiphytes and symptoms of disease prior to shipment, nursery conditions that provide ample light, nutrients, and appropriate temperature conditions likely facilitate the growth of nuisance species. Sterilization techniques such as iodine rinses and treatment with antibiotics can reduce losses (Roberson L, unpub. data), but further work to develop the best approach is still needed. Appropriate acclimation strategies for material transferred from the field to the lab will be necessary for the success of outplanted material on marine seaweed farms, and thus, this is a continued focal point for our team.

There are substantial capital and operating costs associated with the inclusion of onshore nursery tanks in the production of tropical seaweeds. Nonetheless, these tanks will also provide a critical, safe reservoir of high-quality plant material to restart the farm after an intentional or accidental shutdown or loss of crop. If we assume that the farm requires starting material of 0.5 kg wet weight (WW) per meter per line, the amount of starting biomass available when the farm is restarted determines how long it takes to bring the farm to full production. Optimizing the amount of onshore tank capacity therefore involves a tradeoff between the higher cost of maintaining a larger amount of biomass onshore and the higher cost imposed by a longer ramp-up (reduced harvest biomass) when there is less starting biomass available. A more in-depth cost analysis associated with the inclusion of a nursery system is presented in the technoeconomic forecasts and industry development subsection of this chapter.

11.2.3 Farm Monitoring and Maintenance in Response to Epiphytes and Grazers

Epiphytes, grazers, and disease can cause huge losses on marine macroalgal farms (Valderrama et al. 2013; Behera et al. 2022). Common organisms found on or in the tissue of seaweeds include microalgae, fungi, bacteria, viruses, and other seaweeds. Grazers commonly found on tropical seaweed farms include small gastropods, amphipods, polychaetae worms, and herbivorous fish (Kasim et al. 2020; Behera et al. 2022). These epiphytes, including disease-causing bacteria and fungi, are one of the most prevalent causes of lowered yield, and high seaweed stocking density ($\geq$500 g m^{-1}) is one of the driving factors of epiphytism on cultivated seaweed because it allows epiphytes to spread quickly and leads to massive crop collapse in a short timespan (Sahu et al. 2020). To reduce these losses, the attentive maintenance of healthy propagules is critical for the development of highly productive seaweed farms in the tropical USA and Caribbean.

The health status of new propagules and routine maintenance are critical for the growth of *Eucheuma* and other carrageenophytes (Valderrama et al. 2013; Hurtado et al. 2019). In a recent study looking at biosecurity measures on *Kappaphycus* farms in Malaysia, the incidence of ice-ice, biofilms, and epiphytes was reduced using a variety of cleaning treatments and monitoring practices. For example, propagules were checked and cleaned prior to outplanting and were routinely cleaned every 2 days using a tissue or soft cloth (Kambey et al. 2021). In Madagascar, the relationships between environmental factors and the growth and health of *Kappaphycus* were studied across various substrate types and cover (i.e., sand, rocky reef, fleshy algae, and seagrass beds), and a significantly positive correlation was observed between sediment cover on the seaweed and the health factors of epiphytes, fish grazing, and disease (Ateweberhan et al. 2015). However, even with attentive monitoring and cleaning, studies suggest that farm siting and design, water temperature, and density of outplanted seedlings are also important factors influencing the health of and degree of fouling on farmed tropical seaweeds (Hurtado et al. 2019; Behera et al. 2022; Ward et al. 2021). Furthermore, there is regional variation in the extent and frequency of cleaning needed on tropical seaweed farms. In contrast to the 2-day cleaning regimen applied by Kambey et al. (2021) in Malaysia, broader recommendations for Eucheumatoid farm cleaning in Asia are twice weekly (Cottier-Cook et al. 2021).

Preliminary assessments from our research farms in Belize, Florida, and Puerto Rico suggest that frequent maintenance of the algal biomass will play a key role in maintaining algal health and high growth rates. However, we are still investigating the best methods and frequency for cleaning. We have noticed that most of our losses occurred within 1 week of transplantation, which highlights both the importance of the initial sanitation procedures and the need for improved acclimation techniques. Additionally, at our sites in Belize and Puerto Rico, we hypothesize that high sedimentation on the algae plays a large role in decreasing photosynthetic rates. Maintenance and evaluation of seaweed every couple of days are labor-intensive and increasingly expensive when farming sites are located farther from shore. Therefore, as previously mentioned, we are currently testing several cleaning methods that can be automated.

11.2.4 Marine Mammal Interactions

Fishing gear and marine debris have been shown to present entanglement risks to marine mammals, which can be a significant threat to many marine mammals and other protected species (e.g., turtles and sea birds) globally (Würsig and Gailey 2002; Moore and van der Hoop 2012; Reeves et al. 2013; Poonian and Lopez 2016; Nelms et al. 2021). A less fatal effect of such underwater structures is the potential for long-term disruption of important ecological behaviors (e.g., foraging, resting, mating) if the selected sites were previously important to local populations for specific activities and are subject to habitat degradation or loss of prey resources (Würsig and Gailey 2002; Kemper et al. 2003; Markowitz et al. 2004). Although the farm structures deployed in Belize, Florida, and Puerto Rico for the Techniques for Tropical Seaweed Cultivation and Harvesting project are highly tensioned to minimize entanglement risk and have had no known negative interactions, it is still an important practice to monitor marine mammal interactions with the farm site and is an area of particular concern for permitting agencies, as very little is known about seaweed farming gear in US waters and how protected species will respond to the gear and seaweed. Here, we present a method amenable to remote monitoring that can also improve our understanding of protected species distribution and habitat use. We are using acoustic monitoring as a noninvasive method to document marine mammal presence at the farm site both before and after the installation of underwater farming structures, and once a real-time system can be developed, it can be an effective tool for remotely monitoring activity at the farm site. As these data provide valuable information to resource managers and can greatly improve the design and sustainable deployment of seaweed farms, we have included details on the best practices of bioacoustic monitoring and analysis below.

There are several considerations to make to ensure that bioacoustic data are recorded, analyzed, and interpreted correctly; some preliminary guidelines and best practices are reviewed in Oswald et al. (2022). In determining the experimental design, it is important to consider the size of the farm, recorder sensitivity, and vocalizing range of the animals to be monitored to determine how many recorders are needed to cover the entire site. The nature of the monitoring goals will also affect the number of recorders: a single recorder is sufficient for presence/absence monitoring, two recordings can be used to determine the bearing and distance of a vocalizing animal relative to the recorder positions, and three or more are needed to accurately triangulate an animal's position.

Many recorders suitable for passive acoustic monitoring are available commercially. When choosing which recorders to use, there are several considerations to make to ensure data are accurate and interpreted correctly. The sampling rate and flat-response frequency range of the recording package should be sufficient to capture signals produced by the species of interest without aliasing the resulting data, i.e., the sampling rate should be two times the highest frequency of interest (Grenander 1959; Stiltz 1961). To accurately capture delphinid echolocation signals, we used SoundTrap recorders (Ocean Instruments, Inc.) with a working frequency range of 20 Hz–60 kHz. The recorders were duty cycled to record 1 min out of every 5 min to balance storage capacity and battery life during the study period. Manufacturers will often have detailed information on the number of days a recorder will function at a given duty cycle and whether the maximum recording period is limited by storage capacity or battery; this information should be consulted before choosing a recording package. We chose to deploy our recorders for 30-day periods, during which time recorders were duty cycled to turn on for one out of every 5 min to ensure that the batteries and data storage were sufficient for the entire 30-day deployment.

To detect the mammals' response to the new presence of a farm underwater, it is important to collect data on the proposed farm site before any equipment is installed. Baseline data should be collected with the same recorder and in the same general location as will be used to monitor the farm once installed. Because marine mammals exhibit a large amount of diel, seasonal, and environmentally driven variability in their movements and vocalizing behaviors (Carlström 2005; Lin et al. 2015; Lewis and Širović 2018; Chou et al. 2020; Van Opzeeland and Hillebrand 2020), it is important to collect baseline data that are representative of this variability. For example, if the farm is installed year round, the best practice method to compare marine mammal behavior pre- and post-farm is to collect baseline data for at least one full year preceding the installation of the farm to understand baseline variability in marine mammal activity on the farm site. Once the farm is installed, acoustic monitoring should be conducted consistently while the structure is under water to allow for downstream analyses of trends in habitat use and interaction with the farm.

During the preinstallation phase of the project, frequent site visits are common. These can be used as opportunities to visually observe marine mammal presence on the farm and confirm the species ID of animals that are present. Data collected during these observations should include the date and start and end time of the encounter (i.e., when the animals arrived and left the area), group behavior during the encounter (e.g., travelling, resting, socializing at the surface, feeding, etc.), and species ID for all species present. These visual observations can be used in downstream acoustic analyses to generate visually validated acoustic datasets that will serve to train species classifiers (Oswald et al. 2003; Rankin et al. 2008, 2017; Keating et al. 2016).

Once acoustic data are collected, adhering to best practices in the downstream processing of the data is important to ensure that the results are accurate and interpreted correctly. Marine mammal signals are processed in two broad steps prior to data analysis. In the first step, acoustic signals are detected and extracted from the recordings. This can be done manually by representing the acoustic data in images and visually identifying marine mammal signals, or signal detection can be automated using a number of available automated detection algorithms. The decision of whether to use manual or automated detection methods will depend primarily on the complexity of the signals produced, the amount of background noise in the recordings, and the study-specific needs for precision and accuracy. Simple signals (e.g., fin whale 20 Hz pulses) with low background noise and a moderate tolerance for precision and accuracy (i.e., it is not important to capture every single signal in the recording) are ideal for automated detectors. To detect animal presence on or near a seaweed farm site, it is not necessary to detect 100% of signals, which allows some flexibility in the accuracy and precision of the automated detector. However, as signals grow in complexity and background noise increases, more signals will fail to be detected automatically. To a certain extent, this can be dealt with by increasing the sensitivity of the detector, but that will also increase the number of false positives in the dataset. We chose to use PAMGUARD (Gillespie et al. 2008), an open-source, freely available program for visualizing and processing acoustic data and combined an automated detection step with manual filtering of the detections. We recommend using open-source programs such as PAMGUARD for data processing because of the benefits associated with their availability to the whole scientific community (e.g., increased collaboration, increased replicability). Most automated detectors can be trained using a set of verified data or parameterized based on knowledge of the species of interest (e.g., peak frequency and frequency range of interest). Training and parameterizing detectors in this way will decrease the number of false positives and may eliminate the need for manually verifying detections. In our case, a large amount of biological background noise (e.g., from snapping shrimp) interfered with the accurate detection of delphinid echolocation clicks, requiring manual post-processing to remove false positives from the dataset.

Acoustic detections are then classified to the lowest taxonomic category possible based on the characteristics of the signals. In the case of some delphinids, species identification may not be possible due to high similarity in acoustic signals; increasing the quality and quantity of training data used to develop the classifier will improve the taxonomic resolution of the classifier. Developing and training a classifier is a complex and sensitive process that should be preceded by a thorough review of the relevant literature (e.g., Mellinger and Heimlich 2013; Nanaware et al. 2014; González-Hernández et al. 2017; Seger et al. 2018; Shiu et al. 2020; Thomas et al. 2020; Hildebrand et al. 2022) and informational interviews with subject experts. Classifiers for monitoring marine mammals are trained using visually validated, manually filtered datasets of acoustic signals that represent the full vocal repertoire of every marine mammal species that inhabits the region (not just the species of interest). Best practices include using training data only from acoustic encounters where a species was visually verified, with visual verification that no other marine mammals are in the area at the time. Recordings should come from the same geographic region and the same recorder as the data of interest, using the same recording parameters (e.g., sampling rate, duty cycle). When including vocally diverse species such as most delphinids, it is important to collect a temporally diverse training dataset (e.g., multiple seasons and years when possible) to ensure that the dataset is representative of the signals produced in all behavioral states, seasons, etc. Processing data for training datasets requires manual verification of all detections; this processing should use the same spectrogram parameters (e.g., fast Fourier transform (FFT) sample size, window function, bandwidth) as will be used for processing monitoring data. Building a training dataset for the classification of acoustic signals is a time-intensive process involving not only collecting a large amount of acoustic data from visually verified species but also a large amount of manual processing and verification of datasets. This step should not be underestimated if classification is desired, and building of the training dataset should begin as early as possible in the process.

Acoustic data collection has been focused on the farm site in Puerto Rico, where recorders have been deployed on a semi-regular basis since the first farm prototype was installed. To date, only preliminary data analysis has been conducted, and all results are subject to change upon refinement of the data analysis. Preliminary analysis of the echolocation data only indicates that delphinids were present at the farm site significantly more often in the first couple of days immediately following deployment of the pilot equipment – approximately 20% of the time over a 3-day period following deployment, compared to 1–6% of the time in the following months. This may indicate an initial attraction to the equipment followed by acclimation to its presence. However, because reference data were not collected at this site before the farm equipment was installed, we cannot assess the extent to which this habitat was used by delphinids prior to the installation of the farm—it is possible that this habitat was previously used by animals that now avoid the area or that the pattern we observed is an uncorrelated response to environmental or temporal variability in the area. Further analysis indicates that the majority of delphinid interactions with the farm had a duration <5 min; therefore, it is likely that these animals are simply passing through the area without interacting extensively with the farm itself. Finally, there is some evidence of a diel pattern in acoustic behavior around the farm, with a midday lull and two activity peaks in the morning and evening. Given the transitory nature of delphinid interaction with the farm, this likely reflects the animal's diel behavioral patterns and is not driven by the farm itself. However, based on this behavior, we expect that morning and evening would be the greatest risk periods for delphinid entanglement in the farm structure.

11.2.5 Habitat Provisioning

In recent years, considerable interest has been given to the potential for seaweed farming activities to provide valuable habitat provisioning for pelagic and benthic marine organisms (Walls et al. 2016; Teagle et al. 2017; Theuerkauf et al. 2021). The appeal of tangible habitat benefits from seaweed farming exists both for environmental managers tasked with marine conservation and restoration and for seaweed farmers and processors who are seeking additional public support for cultivated seaweed (i.e., social licensing) (Grebe et al. 2019). A need for more rigorous assessment and documentation of the potential habitat provisioning of seaweed farms has been identified because of this recent interest, and numerous studies around the world are currently being conducted with this objective in mind. In the meantime, meta-analyses such as Theuerkauf et al. (2021) have found that although there is a high density (0.3–1.15 individuals m^{-2}) of larger and mobile fish species within the boundaries of tropical seaweed farms, densities at reference sites were generally similar or higher (0.37–1.4 m^{-2}).

Thus, in tropical waters where marine species diversity and abundance are naturally very high, frequent and robust sampling methods are necessary to accurately determine the habitat provisioning provided by the installation of a new seaweed farm. As previously described for marine mammal detection, fish and benthic surveys should be conducted prior to deployment of the array, and they should be conducted across different seasons and life cycles of key species of interest (i.e., lobster, conch). When possible, staged deployment of a new farming array paired with intensive surveying could facilitate further elucidation of the relationships between array design or seaweed density and pelagic or benthic species presence, abundance, and diversity. Similarly, reassessment of these relationships during the maximum suspected impact and before and after notable changes in density, species, and farming period is warranted. The Techniques for Tropical Seaweed Cultivation and Harvesting project has conducted pre-deployment diver surveys and surveys during deployment but will employ autonomous surface vehicles (ASVs) and side scan sonar to assess and document habitat provisioning during peak seaweed density at our research sites. We are also testing the feasibility of hyperspectral imaging by aerial drones to assess the nutrient content of the seaweed crop (Cavanaugh et al. 2021) and have equipped cultivation arrays with passive loggers and sondes to document relevant fluctuations in environmental conditions potentially affecting the growth of the seaweed or interactions between the crop and other marine organisms. If successful, these autonomous survey methods will facilitate the monitoring of large-scale offshore farms and can aid in the validation of ecosystem services provided by seaweed farms (see Eklöf et al. 2005 and Theuerkauf et al. 2021), which can provide another income stream for farmers and local communities.

11.2.6 Evaluating Farm Site Suitability Under Different Nutrient Loads

One of the challenges in establishing a seaweed mariculture industry is selecting suitable farming sites. While doing so, farmers need to consider the physiochemical properties of potential locations (Kerrison et al. 2015; Visch et al. 2020) and how these pertain to the seaweed species under consideration. Nutrient availability, specifically, is a critical element limiting seaweed growth (He et al. 2008; Zhang et al. 2016; Froehlich et al. 2019; Arzeno-Soltero et al. preprint). Coupled biophysical numerical models are an appropriate tool for site suitability assessments, given the variety of environmental factors that must be simultaneously examined to evaluate the mariculture potential of different regions.

The Caribbean and the Gulf of Mexico offer a potentially vast area of opportunity for macroalgae cultivation, within which heterogeneous biogeochemical water qualities will determine the cultivation potential at local scales. Coupled biophysical numerical models are useful tools for assessing which locations within these regions are more suitable for establishing seaweed farms, leveraging ocean observations to estimate macroalgae yields. In addition, these models can help evaluate farm designs, assess cultivation operational procedures, and predict the impacts of cultivation on coastal ecosystems. Although many regional seaweed models exist (e.g., Solidoro et al. 1997; Trancoso et al. 2005; Broch and Slagstad 2012; Ren et al. 2014; Hadley et al. 2015; Zhang et al. 2016; Van der Molen et al. 2018; Jouanno et al. 2021; Frieder et al. 2022), none currently focuses on the Caribbean or Gulf of Mexico. Here, we apply the recently developed Global MacroAlgae Cultivation MODelling System (G-MACMODS; Arzeno-Soltero et al., preprint) to gauge the site suitability for seaweed farming in this region of interest and examine the effects of the harvest period on seaweed biomass.

In general, G-MACMODS estimates the depth-integrated global potential for offshore macroalgae cultivation based on hydrodynamics, macronutrient (here, nitrate) availability, and internal seaweed-specific parameters for a variety of seaweed groups at a 1/12th° horizontal resolution (Arzeno-Soltero et al., preprint). Environmental inputs to G-MACMODS include current velocities, sea surface temperature (SST), surface photosynthetically active radiation (PAR), chlorophyll, significant wave height, mean wave period, and nitrate concentrations. For the site suitability assessment presented in this chapter, we use environmental inputs from 2017, the most recent year of available data without a significant El Niño–Southern Oscillation (ENSO) event. Current velocities were extracted at the surface level from the HYbrid-Coordinate Ocean Model (HYCOM; Wallcraft et al. 2009) Global Ocean Forecasting System (GOFS) 3.1. Sea surface temperature and PAR were taken from the MODerate resolution Imaging Spectroradiometer (MODIS; R2018) from the NASA Earth Observing System. Net oceanic primary productivity (NPP) was estimated from MODIS chlorophyll measurements using the vertically generalized production model (VGPM; Behrenfeld and Falkowski 1997). Sea surface temperature, PAR, and NPP were downloaded from the Ocean Productivity website. Wave information stems from the European Centre for Medium-Range Weather Forecasts (ECMWF) Reanalysis V5 product (ERA5; Hersbach et al. 2020) distributed by the Copernicus Climate Change Service (C3S 2017).

Nitrate concentrations were extracted from a biogeochemical hindcast model produced at Mercator-Ocean, distributed by E.U. Copernicus Marine Service Information (DOI:10.48670/moi-00019). We carried out a Caribbean and Gulf of Mexico site suitability assessment using nitrate concentrations averaged over the (1) top 20 m and (2) top 100 m of the water column to evaluate the productivity of the region under the currently available surface nitrogen supply, as well as any enhanced nitrogen supply that may result from artificial upwelling (Fan et al. 2020; Wu et al. 2023) or depth cycling (Navarrete et al. 2021). All environmental inputs to G-MACMODS are 8-day averages adjusted to be on a 1/12th ° horizontal grid.

We conducted simulations to evaluate the site suitability of the Gulf of Mexico and Caribbean if farmers cultivate tropical, red seaweed species and if harvesting takes place over 45-day and 60-day cycles throughout the year. We presume that seaweed is fully harvested and reseeded (using some of the harvested biomass) within each harvest cycle. Our "standard runs" use the biological parameters most supported by the literature (Table 11.1 of Arzeno-Soltero et al., preprint). However, we assess the uncertainty of our results by performing a suite of ~350 simulations in which we vary the seaweed mortality rate and the nitrogen exudation rate using Monte Carlo methods; these were shown to be two of the model parameters that most influence annual harvest yields (Arzeno-Soltero et al., preprint). Specifically, the seaweed mortality rate and the nitrogen exudation rate are allowed to take on values between 0.003–0.03 day^{-1} and 0.001–0.1 day^{-1} (Table 11.1), respectively, and we assume that these parameters have a uniform distribution within their respective ranges.

Generally, G-MACMODS shows that the most productive region within the Gulf of Mexico and Caribbean system is near the mouth of the Mississippi River, where the annual harvest yield can reach over 2 kg dry weight (DW) m^{-2} (Fig. 11.6a, b). Some productive pockets exist within the Caribbean (e.g., near the mouth of the Orinoco River; Fig. 11.6e, f), but most of this region remains barren when macroalgae rely only on surface nutrients (Fig. 11.6a, d). Higher nitrate concentrations, such as those that might be present under artificial upwelling (Fan et al. 2020; Wu et al. 2023) or depth cycling (Navarrete et al. 2021), expand the range of suitable areas for seaweed farming (Fig. 11.6b). Regions farther away from the mouth of the Mississippi River become productive once deeper nutrients are considered (Fig. 11.6g). Some regions that were productive under the surface nutrient scenario (e.g., near the Orinoco River) generated even larger yields when deeper nutrients were considered (Fig. 11.6i). However, many areas in the Caribbean are predicted to not support red seaweed cultivation, even when exposed to nitrogen concentrations averaged over the top 100 m of the water column (Fig. 11.6h). The large uncertainty presented in Fig. 11.6c–j suggests the need for more research into tropical red seaweed mortality and exudation rates to improve model estimates.

The G-MACMODS also serves to evaluate different harvest operations. Generally, harvesting every 60 days, as opposed to every 45 days, allows the seaweed in the model

Table 11.1 Biological parameters implemented in G-MACMODS runs in the Gulf of Mexico and Caribbean Sea, adapted from Arzeno-Soltero et al. (preprint)

Parameter	Value	References
Maximum uptake rate (V_{max}) μmol-N/(g DW h^{-1})	9.7	Nishihara and Terada (2010), Cabello-Pasini et al. (2003)
Half-saturation constant (K_m) μM	5.6	–
Maximum growth rate (μmax) 1/day	0.2	Wakibia et al. (2006), Msuya and Salum (2011), Montoya Rosas et al. (2020)
Saturation irradiance (PAR_s) μmol photon/(m^2 s^{-1})	125.9	Dawes (1992), Nishihara et al. (2011), Dangan-Galon et al. (2013)
Compensation irradiance (PAR_c) μmol photon/(m^2 s^{-1})	13.5	Dawes and Kovach (1992), Nishihara et al. 2011), Dangan-Galon et al. (2013), Borlongan et al. (2017a), Borlongan et al. 2017b)
Minimum nitrogen cell quota (Q_{min}) mg-N/g-DW	5.8	Angell et al. (2016)
Maximum nitrogen cell quota (Q_{max}) mg-N/g-DW	44	Angell et al. (2016)
Drag coefficient (C_D) unitless	0.5	–
Ratio of biomass to surface area g-DW/m^2	94.8	Lab measurements (L. Roberson)
Nitrogen exudation rate (E) 1/day	0.001–0.1	Martin et al. (2018), Rassweiler et al. (2018), Chen et al. (2020)
Mortality rate (d) 1/day	0.003–0.03	Duarte and Ferreira (1993), Hadley et al. (2015), Zhang et al. (2016), Martin et al. (2018), Rassweiler et al. (2018)
Optimal temperature (T_{opt}) °C	22.5–27.5	Hayashi et al. (2017), Araújo et al. 2014), Terada et al. (2016)

to accumulate more biomass. For example, near the Orinoco River, 45-day and 60-day harvests render 405 g DW m^{-2} and 495 g DW m^{-2}, respectively, when only surface nitrate concentrations are considered (Fig. 11.6f). When deeper nutrients are present, the annual harvest yields are closer to 796 g DW m^{-2} and 1090 g DW m^{-2} for the 45-day and 60-day harvest cycles, respectively (Fig. 11.6j). These assumptions will be ground truthed in the field by the Techniques for Tropical Seaweed Cultivation and Harvesting project.

11.2.7 Storm Surge and Extreme Wave Conditions

Macroalgal farms located on the coasts of Puerto Rico and West Florida are subject to the risk induced by active tropical cyclones that propagate from the tropical Atlantic Ocean to the Gulf of Mexico and Caribbean Sea. To support the risk analysis and system design of macroalgae farms under extreme sea state conditions, such as large waves and storm surges, detailed and accurate wave climate and hydrodynamic information are needed, including the three-dimensional (3D) current distribution, extreme wave height and return period, and hydrodynamic load on the farm system structure.

To better understand extreme storm conditions in our region of focus, we developed a high-resolution coupled storm surge and wave model using the coastal ocean model (ADvanced CIRCulation (ADCIRC)) and wave model (Simulation WAves Nearshore model (SWAN)) for the project sites in Puerto Rico and West Florida. The coupled ADCIRC +SWAN modelling system has been used for modelling storm surges and extreme waves induced by historical hurricanes (Dietrich et al. 2011; Joyce et al. 2019; Yang et al. 2020). A recent study by Holzenthal et al. (2022) used the coupled ADCIRC+SWAN model to investigate the influence of buoyant submerged aquatic vegetation, such as seagrasses and kelps, on estuarine hydrodynamics. In our study, the model domain encompasses the US East Coast, the Gulf of Mexico, and the Caribbean with an open boundary extended to the deep basin of the North Atlantic Ocean to account for the widespread of tropical cyclone trajectories in the region. The unstructured-grid framework of ADCIRC +SWAN allowed us to specify high spatial grid resolution (~100 m) around the project sites to accurately represent the seaweed farming systems. The coupled model was forced with tides along the open boundary and wind and atmospheric pressure at the sea surface. Eight tidal constituents (M2, O2, S2, N2, K2, K1, Q1, P1) were used to define the tidal open boundary conditions. Hourly sea surface wind and atmospheric pressure fields were retrieved from the CFSv2 database (Saha et al. 2014) for model validation runs. The coupled model was run in 3D mode to provide detailed and accurate current and wave information.

The model was validated with available observed water level and wave height data for day-to-day normal operation conditions and historical hurricanes (Ticona Rollano et al. 2022). For this purpose, we simulated storm surge and wave climate under two historical hurricane events that severely impacted Puerto Rico within weeks of one another in September 2017: Hurricanes Irma and Maria. Snapshots of modelled significant wave height and depth-averaged current magnitudes in the region of the cultivation sites in southwestern Puerto Rico at the time of greatest local intensity of Hurricane Maria are shown in Fig. 11.7. The model shows the sheltering effects of nearby reefs on waves and currents; the potential macroalgae array sites are exposed to diminished waves compared to their surroundings but are also more exposed to local currents influenced by these features.

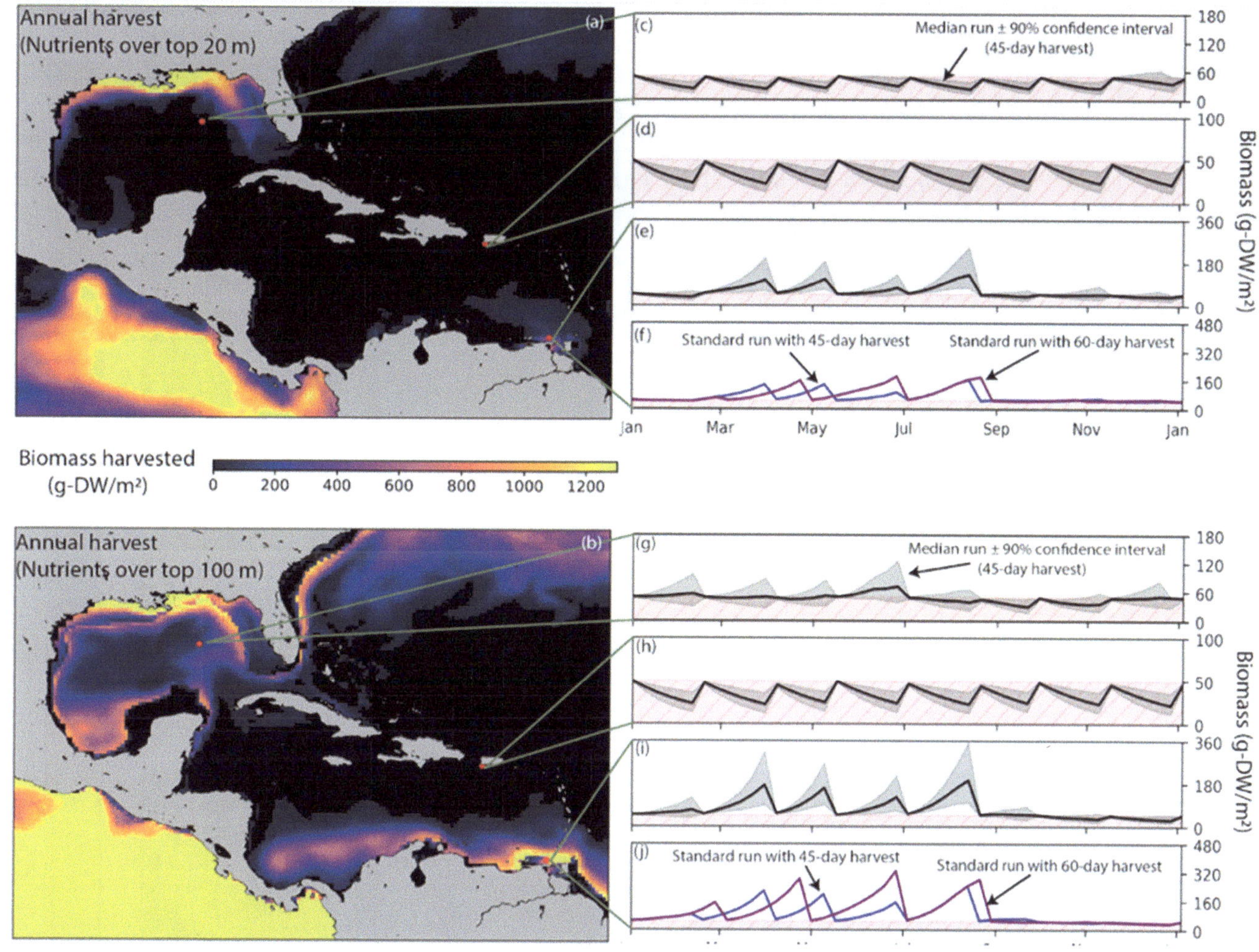

Fig. 11.6 Annual harvest yields from G-MACMODS standard runs using (**a**) only surface nitrate concentrations (averaged over the top 20 m of the water column) and (**b**) nitrate concentrations averaged over the top 100 m of the water column. Time series of select locations show the median seaweed biomass and its corresponding uncertainty throughout the year as estimated by ~350 Monte Carlo simulations (**c**–**e** and **g**–**i**). The red hatched area denotes biomass below the seeding weight (no net growth in the farm). The sawtooth pattern is a consequence of periodic seeding and harvesting. The impact of varying the harvest cycle period is shown in (**f**) and (**j**). Please note the variable y-axes. Source: University of California, Irvine

Once the coupled model was validated, it was applied to simulate storm surges and extreme waves under various extreme storm events. Hydrodynamic load analyses are critical to assess the growth and survivability of macroalgae due to wave- and current-induced disturbances (Utter and Denny 1996; Jones et al. 2015); thus, these load analyses help optimize macroalgae array designs. The probability distribution of velocity, storm surge, and wave height under extreme conditions was generated for the project sites based on large ensemble simulations of synthetic hurricane events of various intensities based on historical meteorological data patterns. For this purpose, we obtained a dataset of 3900 synthetic hurricane tracks that impact the main island of Puerto Rico, generated by Dr. Kerry Emanuel of the Massachusetts Institute of Technology using the methodology described in Emanuel et al. (2006). This dataset was subsampled down to 149 synthetic tracks that reach hurricane intensities of Categories 1–5 near the project sites in southwestern Puerto Rico. Hurricane wind and pressure fields derived from the 149 hurricane tracks were used to drive the coupled ADCIRC + SWAN model to generate large datasets of storm surge and extreme waves around the Puerto Rico region. The simulated water level, 3D velocity, and wave climate driven by synthetic hurricanes can be analyzed to calculate the hydrodynamic load through the water column and assess the risk induced by extreme storm events. Model results from storm surge and wave simulations, together with the Caribbean Coastal Ocean Observing System (CariCOOS) hydrodynamic model, can be used to support hydrodynamic load modelling and fine-scale LES/macroalgal simulations at project sites.

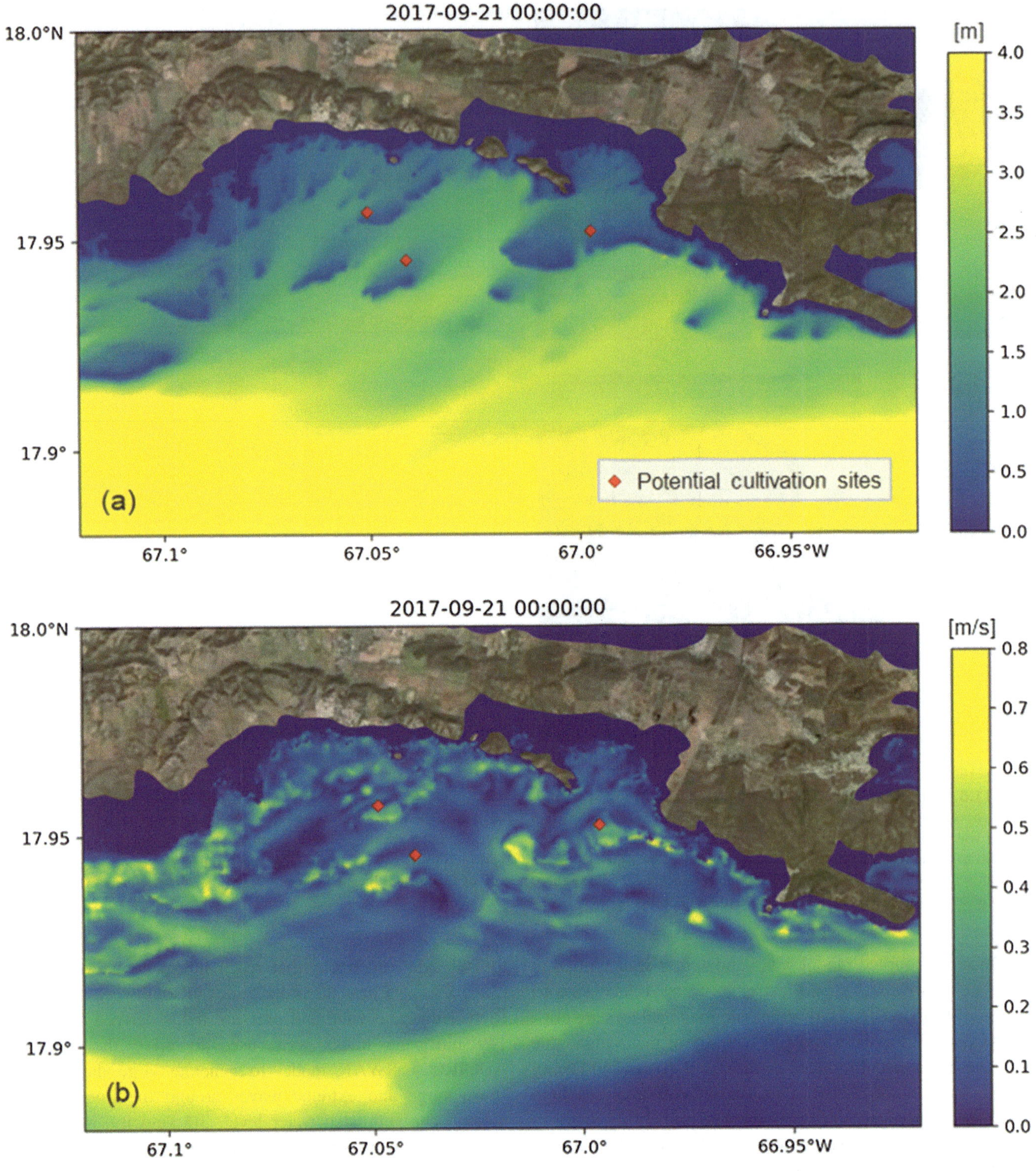

Fig. 11.7 Modelled significant wave height (**a**) and depth-average current magnitude (**b**) in potential macroalgae farm sites in southwestern Puerto Rico under the influence of Hurricane Maria. Source: Pacific Northwest National Laboratory

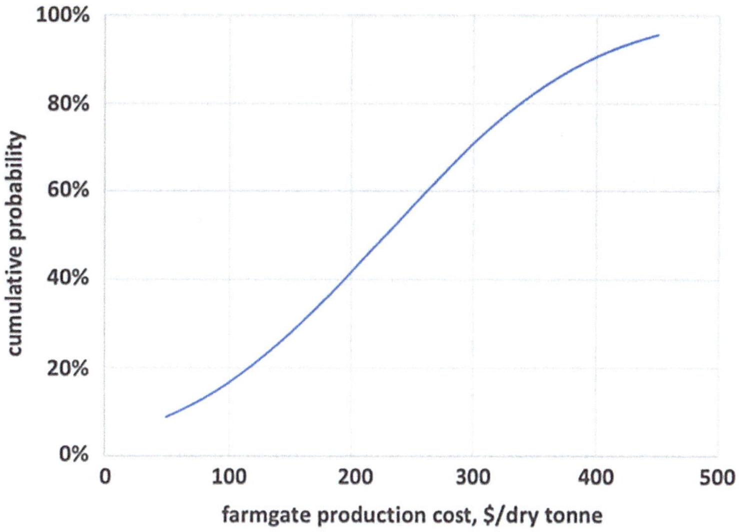

Fig. 11.8 Predicted farm production cost (1000 ha scale) in US dollars per DW tonne. Source: Woods Hole Oceanographic Institution

Table 11.2 Quantity of onshore tanks needed to restart grow-out operations as a function of likelihood of annual storm loss. Source: Woods Hole Oceanographic Institution

Likelihood of loss/year (%)	Optimal # tanks
10	100
20	200
30	300
40	400

11.2.8 Technoeconomic Forecasts and Industry Development

In the last decade, interest and investment in seaweed aquaculture have grown rapidly throughout the Western hemisphere. While most existing installations and operations are currently small in scale, many of the ecosystem services (i.e., carbon capture) and most exciting substitution prospects (i.e., plastic alternatives or biofuel feedstocks) associated with seaweed aquaculture will require a vastly larger scale of production. We have been conducting a technoeconomic assessment to model the production costs of these future farms, which will need to be designed for the exposed, open ocean and a scale of at least 1000 ha. At this scale, our model suggests that farm gate production costs of \$100–\$250 DW tonne $(t)^{-1}$ are likely when farms are sited up to 200 km from the nearest onshore support base (Kite-Powell et al. 2022). In a Monte Carlo analysis, production costs < \$100 DW t^{-1} are predicted to be achievable in some settings (Ibid; Fig. 11.3). The predicted 1000-ha farm production costs shown in Fig. 11.8 are derived from numerous hypotheses about the nursery operations, grow-out logistics, and profit margins on the harvested seaweed biomass.

Our modelling of nursery operations assumes onshore tanks made of concrete and sunk in the ground with a depth of 1 m and a volume of 10,000 L per tank. The tanks are connected to a seawater supply system. Biomass is started in the tanks at 3 g L^{-1} and grows at 5% day^{-1}, and then it is thinned back to 3 g L^{-1} when it reaches 7 g L^{-1}. We assume capital costs of \$2000 $tank^{-1}$, including plumbing, and a life of 40 years. The land lease cost is assumed to be \$20,000 ha^{-1} $year^{-1}$. Labor requirements are 10 h $tank^{-1}$ $year^{-1}$, and labor cost is assumed to be \$15 h^{-1}. Power and expendables are assumed to cost \$50 $tank^{-1}$ $year^{-1}$. As a result, the anticipated cost of maintaining biomass in the onshore tanks is \$9800 WW t^{-1} $year^{-1}$, and seeding the whole farm at once requires 83 WW t. The economically optimal amount of biomass to maintain onshore for annual restarts is 25–30 WW t, which requires approximately 900 tanks and 2 ha of land area. The economically optimal number of tanks to support storm loss restarts depends on the annual likelihood of storm loss (Table 11.2). For a farm that anticipates annual restarts, the annualized cost of the onshore tank operation is approximately \$250,000 $year^{-1}$. This adds \$5–6 to the total production cost of 1 DW t of biomass on the 1000-ha farm.

Our modelling of grow-out logistics includes a fleet of small, custom-built farm boats (capacity ≤ 4 people) that are used full time for outplanting, harvesting, and maintenance. A rotating crew completes multiple, 8-h shifts day^{-1} (weather permitting), and these workers are either shuttled from a common location onshore or they have living accommodations on a tethered platform at the farming site. The crew conducts harvest and replanting activities continuously, with grow-out cycles from seed to harvest spanning 1–2 months. The core farm infrastructure (mooring anchors, catenaries, and other structural lines) is permanently on site, but the grow ropes are potentially removed or replaced at harvest.

We assume a 15% profit margin on biomass produced by the farm or approximately \$5 WW t^{-1} sold. (This is equivalent to an annual profit of approximately $1.1 million for the 1000-ha farm). We then model two scenarios: (1) the annual restart, where the farm is routinely shut down for one or more months each year due to unfavorable growing conditions, and (2) the storm loss restart, where operations are continuous year round, but there is a certain probability of a storm event that will result in the loss of all seaweed biomass from the farm and therefore require a restart.

11.2.9 Needs for Additional Research and Development

For seaweed aquaculture to be a successful, long-term business enterprise in the tropical USA and Caribbean, we must show that the activity can be economically viable, environmentally sustainable, and socially accepted. Despite the technical advances we present above and those made by other initiatives, there are still several fundamental social, economic, and policy challenges that must be addressed for the seaweed industry to expand in these regions effectively and safely. Here, we briefly discuss three of these challenges, namely, (1) lack of existing policy and regulations for seaweed aquaculture, (2) limited social license for aquaculture, and (3) restricted participation in the global seaweed market.

Policy and regulations informing sustainable aquaculture development in the tropical USA and Caribbean are not currently strong enough to meet the enormous potential for seaweed aquaculture in the area. There is an immediate need for additional government engagement at national, regional, and local levels to create and enforce best practices related to farm siting and permitting, biosecurity measures, and diverting funds toward key infrastructure such as nursery or processing facilities. However, in recent years, many Caribbean governments have been much more focused on tourism and green tourism to mitigate and rebound from the economic impacts of COVID-19. Additionally, in most places in the Caribbean, the few artisanal seaweed farmers that are currently engaged in farming activities are not registered; there is minimal documentation of the practices they are currently using and any production, environmental, or social issues that they might be confronting. This lack of documented information presents an additional hurdle for engaging these farmers in innovative and responsible practices.

Of course, there are several exceptions to this general trend. For example, around St. Vincent Island, the government has designated a specific zone for seaweed cultivation (CANARI 1997). In Belize, the national government has already passed into law a Mariculture Policy (developed with funding from The Nature Conservancy, Belize). Mariculture specialists, seaweed farming cooperatives, and other promoters of seaweed farming in Belize are now hoping to secure funding to (1) draft national regulations (including a cabinet paper and presentation on the draft regulations); (2) conduct needed stakeholder consultations; (3) complete an industry analysis and feasibility study; (4) establish a baseline for the seaweed mariculture industry (i.e., number of farms, location and size of farms, etc.); (5) develop an investment module to guide investors; and (6) launch a socialization campaign on the new mariculture regulations. The engagement and diligence of the Belizean government, environmental organizations, and seaweed farmers to encourage the development and regulation of seaweed farming in Belize should serve as an example of what is possible, and necessary, across the region.

As a complement to efforts such as the one in Belize, we envision a region-wide Caribbean sustainable seaweed working group to help address critical gaps in the scalability, marketability, and ecological knowledge of farmed seaweed from the Caribbean. Activities such as developing a platform for Caribbean governments to collaboratively evaluate and grow regional market opportunities for seaweed farming and building regional standards for environmental, food and consumer, and occupational safety associated with seaweed farming are achievable yet critical first steps to supporting effective and safe seaweed industry expansion in the tropical USA and Caribbean.

The social license to operate also needs a strong boost before a productive seaweed aquaculture industry in this region of the world. Perceived conflicts between seaweed aquaculture, fishing, and tourism and/or concerns about the ecological impacts of seaweed farming on sensitive habitats or protected species affect are likely to influence the degree of acceptance for new aquaculture activities in the region, especially if outreach and education efforts are minimal. Unfortunately, negative experiences with *Sargassum* blooms impacting economically important beaches and other algal infestations smothering coral reefs add to the public's general wariness about intentional macroalgal cultivation. Here lies the vital work of environmental organizations, scientists, and governments to engage the public in educational activities and demonstrations to elevate public literacy about the potential benefits and risks of unregulated seaweed aquaculture expansion in the tropical USA and Caribbean. We hope that site suitability analyses, pilot farms, and extensive and robust monitoring efforts such as those conducted by the Techniques for Tropical Seaweed Cultivation and Harvesting project will also help to build social licenses for aquaculture in the region.

Ultimately, access to global seaweed markets must also be established for seaweed farming to be economically viable in the tropical USA and Caribbean. Currently, there are no entities in the region producing enough seaweed biomass to attract seaweed buyers in the growing global market. As such, a strategy for aggregating seaweed biomass from numerous producers or locations will be necessary to meet minimum order requirements by major buyers such as Cargill Inc. and DuPont. However, such an approach may also be costly (affecting the competitiveness of this biomass when compared to material from Asia), and it has the potential to also pull money away from microenterprises. Increased financing mechanisms for new farming operations (i.e., micro-funding and microloans) and additional education and support in business development are essential to bringing more people into the industry, either through direct seaweed biomass production or value-added processing of seaweeds.

11.3 Conclusion

The Techniques for Tropical Seaweed Cultivation and Harvesting project is poised to help reduce the risks of seaweed farming in the Gulf of Mexico and the Caribbean by testing farm systems and species that can be grown on a large scale in a sustainable, economically viable manner. The tools developed by the MARINER program aim to provide farmers with data needed to optimize operations and reduce costs. We believe that this will allow the nascent US seaweed farming industry to be more equitable through mechanization and automation, which leads to a more livable wage, and more sustainable by frequent monitoring of the biology and ecology of the farm using autonomous vehicles and growth models (i.e., GMACMODS). Bespoke farm vessels can be multipurposed for both farm maintenance (including installation) and seeding and harvesting. The farm system itself can be used to enhance flow (thereby increasing flux and reducing epiphyte load), increase exposure to nutrients and cooler water, and avoid storms through automated submergence and resurfacing. Finally, the detailed economic model can inform design and processing decisions. The project farms will provide the baseline needed for future farm permitting and support local stakeholders by providing an opportunity to learn about cutting-edge tropical seaweed cultivation systems, operations, and best practices. While we have focused on the Caribbean/Gulf of Mexico, these principles can be applied and adapted to other regions of the world.

Acknowledgments This activity was supported by United States Department of Energy Advanced Research Projects Agency-Energy (ARPA-E) Macroalgae Research Inspiring Novel Energy Resources (MARINER) award #DE-AR0000912 to the Marine Biological Laboratory.

Declarations Funding: This activity was supported by the US Department of Energy Advanced Research Projects Agency-Energy (ARPA-E) Macroalgae Research Inspiring Novel Energy Resources (MARINER) award #DE-AR0000912 to the Marine Biological Laboratory.

Competing interests: None.

References

Abreu MH, Varela DA, Henríquez L, Villarroel A, Yarish C, Sousa-Pinto I, Buschmann AH (2009) Traditional vs. integrated multi-trophic aquaculture of *Gracilaria chilensis* CJ Bird, J. McLachlan & EC Oliveira: productivity and physiological performance. Aquaculture 293(3–4):211–220

Aleman A, Constantin A (2020) On the decrease of kinetic energy with depth in wave-current interactions. Math Ann 378:853–872. https://doi.org/10.1007/s00208-019-01910-8

Ali MKM, Critchley AT, Hurtado AQ (2020) The impacts of AMPEP K + (*Ascophyllum* marine plant extract, enhanced with potassium) on the growth rate, carrageenan quality, and percentage incidence of the damaging epiphyte *Neosiphonia apiculata* on four strains of the commercially important carrageenophyte *Kappaphycus*, as developed by micropropagation techniques. J Appl Phycol 32:1907–1916

Angell AR, Mata L, de Nys R, Paul NA (2016) The protein content of seaweeds: a universal nitrogen-to-protein conversion factor of five. J Appl Phycol 28(1):511–524

Araújo PG, Ribeiro ALN, Yokoya NS, Fujii MT (2014) Temperature and salinity responses of drifting specimens of *Kappaphycus alvarezii* (Gigartinales, Rhodophyta) farmed on the Brazilian tropical coast. J Appl Phycol 26(5):1979–1988

Arzeno-Soltero, IB, Frieder C, Saenz BT, Long MC, DeAngelo J, Davis SJ, Davis KA (n.d.) Biophysical potential and uncertainties of global seaweed farming. Submitted to Nature Geoscience and available as pre-print. https://eartharxiv.org/repository/view/3094/. https://doi.org/10.31223/X52P8Z

Ask EI (2003) Creating a sustainable commercial *Eucheuma* cultivation industry: the importance and necessity of the human factor. In: Chapman ARO, Anderson RJ, Vreeland VJ, Davison IR (eds) . Proceedings of XVIIth International Seaweed Symposium, Oxford University Press, pp 13–18

Ask E, Azanza R, Simbik M, Cay-An R, Lagahid J (2003) Technological improvements in commercial *Eucheuma* cultivation (a short communication). Sci Diliman 15:47–51

Ateweberhan M, Rougier A, Rakotomahazo C (2015) Influence of environmental factors and farming technique on growth and health of farmed *Kappaphycus alvarezii* (cottonii) in south-West Madagascar. J Appl Phycol 27:923–934

Behera DP, Ingle KN, Mathew DE, Dhimmar A, Sahastrabudhe H, Sahu SK, Krishnan MG, Shinde PB, Ganesan M, Mantri VA (2022) Epiphytism, diseases and grazing in seaweed aquaculture: a comprehensive review. Rev Aquac 14:1345–1370

Behrenfeld MJ, Falkowski PG (1997) Photosynthetic rates derived from satellite-based chlorophyll concentration. Limnol Oceanogr 42(1): 1–20. https://doi.org/10.4319/lo.1997.42.1.0001

Bjerregaard R, Valderrama D, Radulovich R, Diana J, Capron M, Mckinnie CA, Cedric M, Hopkins K, Yarish C, Goudey C, Forster J (2016) Seaweed aquaculture for food security, income generation and environmental health in tropical developing countries. World Bank Group, Washington, DC. http://documents.worldbank.org/curated/en/947831469090666344/Seaweed-aquaculture-for-food-security-income-generation-and-environmental-health-in-Tropical-Developing-Countries;jsessionid=4sLY8b149Hwa-8ramT5do35G

Borlongan IA, Gerung GS, Kawaguchi S, Nishihara GN, Terada R (2017a) Thermal and PAR effects on the photosynthesis of *Eucheuma denticulatum* and *Kappaphycus striatus* (so-called Sacol strain) cultivated in shallow bottom of Bali, Indonesia. J Appl Phycol 29(1):395–404

Borlongan IA, Nishihara GN, Shimada S, Terada R (2017b) Effects of temperature and PAR on the photosynthesis of *Kappaphycus* sp. (Solieriaceae, Rhodophyta) from Okinawa, Japan, at the northern limit of native *Kappaphycus* distribution in the western Pacific. Phycologia 56(4):444–453

Broch OJ, Slagstad D (2012) Modelling seasonal growth and composition of the kelp *Saccharina latissima*. J Appl Phycol 24(4):759–776

Buschmann AH, Pereda SV, Varela DA, Rodríguez-Maulén J, López A, González-Carvajal L, Schilling M, Henríquez-Tejo EA, Hernández-González MC (2014) Ecophysiological plasticity of annual populations of giant kelp (*Macrocystis pyrifera*) in a seasonally variable coastal environment in the northern Patagonian inner seas of southern Chile. J Appl Phycol 26(2):837–847

Cabello-Pasini A, Zertuche-González JA, Pacheco-Ruíz I (2003) Photosynthesis, growth and nitrogen uptake of competing marine macrophytes in the Gulf of California. Bot Mar 46:503–512

Cai J, Lovatelli A, Aguilar-Manjarrez J, Cornish L, Dabbadie L, Desrochers A, Diffey S, Garrido Gamarro E, Geehan J, Hurtado A, Lucente D (2021) Seaweeds and microalgae: an overview for unlocking their potential in global aquaculture development. FAO Fish Aquacult Circular 1229. 48 pp. https://doi.org/10.4060/cb5670en

CANARI (1997) Guidelines Series 1. Smith AH. Seamoss cultivation in the West Indies

Carlström J (2005) Diel variation in echolocation behavior of wild harbor porpoises. Mar Mamm Sci 21(1):1–12. https://doi.org/10.1111/J.1748-7692.2005.TB01204.X

Cavanaugh KC, Cavanaugh KC, Bell TW, Hockridge EG (2021) An automated method for mapping giant kelp canopy dynamics from UAV, vol 8. Front Enviro Sci

Chen S, Xu K, Ji D, Wang W, Xu Y, Chen C, Xie C (2020) Release of dissolved and particulate organic matter by marine macroalgae and its biogeochemical implications. Algal Res 52:102096

Chou E, Antunes R, Sardelis S, Stafford KM, West L, Spagnoli C, Southall BL, Robards M, Rosenbaum HC (2020) Seasonal variation in Arctic marine mammal acoustic detection in the northern Bering Sea. Mar Mamm Sci 36(2):522–547. https://doi.org/10.1111/MMS.12658

Clegg P, Mahon R, McConney P, Oxenford HA (eds) (2020) The Caribbean blue economy. Routledge, New York

Copernicus Climate Change Service (C3S) (2017) ERA5: fifth generation of ECMWF atmospheric reanalyses of the global climate. Copernicus Climate Change Service Climate Data Store (CDS). https://cds.climate.copernicus.eu/cdsapp#!/home

Cottier-Cook EJ, Nagabhatla N, Asri A, Beveridge M, Bianchi P, Bolton J, Gondad-Reantaso MG, Brodie J, Buschmann A, Cabarubias J, Campbell I, Chopin T, Critchley A, De Lombaerde P, Doumeizel V, Gachon CMM, Hayashi L, Hewitt CL, Huang J, Hurtado A, Kambey C, Kim GH, Le Masson V, Lim PE, Liu T, Malin G, Matoju I, Montalescot V, Msuya FE, Potin P, Puspita M, Qi Z, Shaxson L, Sousa Pinto I, Stentiford GD, Suyo J, Yarish C (2021) Ensuring the sustainable future of the rapidly expanding global seaweed aquaculture industry – a vision. UNU institute on comparative regional integration studies.15pp. https://researchrepository.murdoch.edu.au/id/eprint/64744/1/Seaweed.pdf

Dangan-Galon FD, Dumilag RV, Ganzon-Fortes E (2013) Photosynthesis-irradiance curves of four marine macroalgae from Bolinao, Pangasinan and Calatagan, Batangas, Philippines. Scholarly J Sci Res Essay 2(8):134–138

Dawes CJ (1992) Irradiance acclimation of the cultured Philippine seaweeds, *Kappaphycus alvarezii* and *Eucheuma denticulatum*. https://doi.org/10.1515/botm.1992.35.3.189

Dawes CJ, Kovach CW (1992) Ecology of the algae of a Florida key II. Effects of irradiance, salinity and desiccation on intertidal and subtidal populations of seven macroalgae. Bull Mar Sci 50(1):165–170

Dietrich JC, Zijlema M, Westerink JJ, Holthuijsen LH, Dawson C, Luettich RA, Jensen RE, Smith JM, Stelling GS, Stone GW (2011) Hurricane waves and storm surge using integrally-coupled. Scalable Computations Coast Eng 58(1):45–65. https://doi.org/10.1016/j.coastaleng.2010.08.001

Diez SM, Patil PG, Morton J, Rodriguez DJ, Vanzella A, Robin D, Maes T, Corbin C (2019) Marine pollution in the Caribbean: not a minute to waste. World Bank Group. Retrieved from https://policycommons.net/artifacts/1507616/marine-pollution-in-the-caribbean/2173404/ on 09 Aug 2022. CID: 20.500.12592/5xpk3z

Doty MS, Caddy JF, Santelices B (1987) The production and use of *Eucheuma*. FAO Fish Tech Pap 281:123–161

Duarte P, Ferreira JG (1993) A methodology for parameter estimation in seaweed productivity modelling. In: Fourteenth international seaweed symposium. Springer, Dordrecht, pp 183–189

Duarte CM, Bruhn A, Krause-Jensen D (2022) A seaweed aquaculture imperative to meet global sustainability targets. Nat Sustain 5(3):185–193

Egan S, Gardiner M (2016) Microbial dysbiosis: rethinking disease in marine ecosystems. Front Microbiol 7:991

Eklöf JS, de la Torre Castro M, Adelsköld L, Jiddawi NS, Kautsky N (2005) Differences in macrofaunal and seagrass assemblages in seagrass beds with and without seaweed farms. Est Coast Shelf Sci 63(3):385–396

Emanuel K, Ravela S, Vivant E, Risi C (2006) A statistical deterministic approach to hurricane risk assessment. Bull Am Meteorol Soc 87(3):299–314. https://doi.org/10.1175/BAMS-87-3-299

Fan W, Zhang Z, Yao Z, Xiao C, Zhang Y, Zhang Y, Liu J, Di Y, Chen Y, Pan Y (2020) A sea trial of enhancing carbon removal from Chinese coastal waters by stimulating seaweed cultivation through artificial upwelling. Appl Ocean Res 101:102260

FAO (Food and Agriculture Organization) (1980) Aquaculture development in the Caribbean – Report of a mission to Antigua, Jamaica, Montserrat, and St. Lucia. Retrieved from https://www.fao.org/3/p4495e/P4495E00.htm#TOC on 09 Aug 2022

Frieder C, Yan C, Chamecki M, Dauhajre D, McWilliams JC, Infante J, McPherson M, Kudela R, Kessouri F, Sutula M, Arzeno-Soltero IB, Davis K (2022) A macroalgal cultivation modeling system (MACMODS): evaluating the role of physical-biological coupling on nutrients and farm yield. Front Mar Sci 9:752951

Froehlich HE, Afflerbach JC, Frazier M, Halpern BS (2019) Blue growth potential to mitigate climate change through seaweed offsetting. Curr Biol 29(18):3087–3093

Gaylord B, Rosman JH, Reed DC, Koseff JR, Fram J, MacIntyre S, Arkema K, McDonald C, Brzezinski MA, Largier JL, Monismith SG (2007) Spatial patterns of flow and their modification within and around a giant kelp forest. Limnol Oceanogr 52(5):1838–1852

Gillespie D, Gordon J, Mchugh R, Mclaren D, Mellinger D, Redmond P, McHugh R, Deng XY, Thode A (2008) PAMGUARD: semiautomated, open source software for real-time acoustic detection and localisation of cetaceans. Proc Inst Acoust 30:2547

González-Hernández FR, Sánchez-Fernández LP, Suárez-Guerra S, Sánchez-Pérez LA (2017) Marine mammal sound classification based on a parallel recognition model and octave analysis. Appl Acoust 119:17–28

Goudey CA (2019) Growing macroalgae at sea USA Patent 10,257,990

Goudey CA (2022) Attaching seaweed to a line. Notice of allowance, 06/01/2022, Application No. 16/558.294

Grebe GS, Byron CJ, Gelais AS, Kotowicz DM, Olson TK (2019) An ecosystem approach to kelp aquaculture in the Americas and Europe. Aquac Rep 15:100215

Grenander U (1959) Probability and statistics: the Harald Cramér volume. J R Statist Soc Ser A (General:123). Wiley. https://doi.org/10.2307/2342476

Hadley S, Wild-Allen K, Johnson C, Macleod C (2015) Modeling macroalgae growth and nutrient dynamics for integrated multi-trophic aquaculture. J Appl Phycol 27(2):901–916

Harley CD, Anderson KM, Demes KW, Jorve JP, Kordas RL, Coyle TA, Graham MH (2012) Effects of climate change on global seaweed communities. J Phycol 48:1064–1078

Hayashi L, Reis RP, dos Santos AA, Castelar B, Robledo D, de Vega GB, Msuya FE, Eswaran K, Yasir SM, Ali MK, Hurtado AQ (2017) The cultivation of *Kappaphycus* and *Eucheuma* in tropical and sub-tropical waters. In: Hurtado AQ et al (eds) Tropical Seaweed farming trends, problems and opportunities, developments in applied phycology 9. https://doi.org/10.1007/978-3-319-63498-2_4

He P, Xu S, Zhang H, Wen S, Dai Y, Lin S, Yarish C (2008) Bioremediation efficiency in the removal of dissolved inorganic nutrients by the red seaweed, *Porphyra yezoensis*, cultivated in the open sea. Water Res 42(4–5):1281–1289

Hersbach H, Bell B, Berrisford P, Hirahara S, Horányi A, Muñoz-Sabater J, Nicolas J, Peubey C, Radu R, Schepers D, Simmons A (2020) The ERA5 global reanalysis. Q J R Meteorol Soc 146(730): 1999–2049

Hildebrand JA, Frasier KE, Helble TA, Roch MA (2022) Performance metrics for marine mammal signal detection and classification. J Acoust Soc Am 151(1):414–427

Holzenthal E, Hill D, Wengrove ME (2022) Multi-scale influence of flexible submerged aquatic vegetation (SAV) on estuarine hydrodynamics. J Mar Sci Eng 10(4):554. https://doi.org/10.3390/jmse10040554

Hurtado AQ, Neish IC, Critchley IT (2019) Phyconomy: the extensive cultivation of seaweeds, their sustainability and economic value, with particular reference to important lessons to be learned and transferred from the practice of eucheumatoid farming. Phycologia 58(5):472–483. https://doi.org/10.1080/00318884.2019.1625632

Jones T, Gardner JPA, Bell JJ (2015) Modelling the effect of wave forces on subtidal macroalgae: a spatial evaluation of predicted disturbance for two habitat-forming species. Ecol Model 313:149–161. https://doi.org/10.1016/j.ecolmodel.2015.06.026

Jossart J, Wickliffe LC, Riley KL, Morris Jr. JA (2019) Scalability analysis for US Caribbean, Gulf of Mexico, southeast, and Tampa Bay. CASS Technical Report. CSS, Inc. for NOAA NOS/NCCOS, Beaufort, NC

Jouanno J, Benshila R, Berline L, Soulié A, Radenac MH, Morvan G, Diaz F, Sheinbaum J, Chevalier C, Thibaut T, Changeux T (2021) A NEMO-based model of *Sargassum* distribution in the tropical Atlantic: description of the model and sensitivity analysis (NEMO-Sarg1.0). Geosci Model Dev 14(6):4069–4086

Joyce BR, Gonzalez-Lopez J, Van der Westhuysen AJ, Yang D, Pringle WJ, Westerink JJ, Cox AT (2019) U.S. IOOS coastal and ocean modeling testbed: hurricane-induced winds, waves, and surge for Deep Ocean, reef-Fringed Islands in the Caribbean. J Geophys Res: Oceans 124(4):2876–2907. https://doi.org/10.1029/2018JC014687

Kambey CS, Campbell I, Cottier-Cook EJ, Nor AR, Kassim A, Sade A, Lim PE (2021) Seaweed aquaculture: a preliminary assessment of biosecurity measures for controlling the ice-ice syndrome and pest outbreaks of a *Kappaphycus* farm. J Appl Phycol 33:3179–3197

Kasim M, Balubi AM, Mustafa A, Nurdin R, Patadjai RS, Jalil W (2020) Floating cage: a new innovation of seaweed culture. In: Lu Q, Serajuddin M (eds) Emerging technologies, environment and research for sustainable aquaculture. IntechOpen, London, p 103. https://doi.org/10.5772/intechopen.90035

Keating JL, Oswald JN, Rankin S, Barlow J (2016) Whistle classification in the California current: a complete whistle classifier for a large geographic region with high species diversity. NOAA-TM-NMFS-SWFSC-552. La Jolla, CA. https://doi.org/10.7289/V5/TM-SWFSC-552

Kemper CM, Pemberton D, Cawthorn M, Heinrich S, Mann J, Würsig B, Shaughnessy P, Gales R (2003) Aquaculture and marine mammals: coexistence or conflict? In: Gales R, Hindell N, Kirkwood M (eds) Marine mammals: fisheries, tourism and management issues. Csiro, Melbourne, pp 208–226

Kerr SA (2005) What is small Island sustainable development about? Ocean Coast Manag 48(7–8):503–524

Kerrison PD, Stanley MS, Edwards MD, Black KD, Hughes AD (2015) The cultivation of European kelp for bioenergy: site and species selection. Biomass Bioenergy 80:229–242

Kim JK, Stekoll M, Yarish C (2019) Opportunities, challenges and future directions of open water seaweed aquaculture in the United States. Phycologia 58(5):446–461. https://doi.org/10.1080/00318884.2019.1625611

Kite-Powell H, Ask E, Augyte S, Bailey D, Decker J, Goudey CA, Grebe G, Li Y, Lindell S, Manganelli D, Ng C, Roberson L, Stekoll M, Umanzor S, Yarish C (2022) Estimating production cost for large-scale seaweed farms. Appl Phycol 3(1):435–445

Lewis LA, Širović A (2018) Variability in blue whale acoustic behavior off southern California. Mar Mamm Sci 34(2):311–329. https://doi.org/10.1111/MMS.12458

Lin TH, Yu HY, Chen CF, Chou LS (2015) Passive acoustic monitoring of the temporal variability of Odontocete tonal sounds from a Long-term marine observatory. PLoS One 10(4):e0123943. https://doi.org/10.1371/JOURNAL.PONE.0123943

Loureiro RR, Hurtado AQ, Critchley AT (2017) Impacts of AMPEP on epiphytes and diseases in *Kappaphycus* and *Eucheuma* cultivation. In: Hurtado AQ et al (eds) Tropical seaweed farming trends, problems and opportunities. Developments in applied phycology 9. Springer, Cham, pp 111–119. https://doi.org/10.1007/978-3-319-63498-2_6

Markowitz TM, Harlin AD, Würsig B, Mcfadden CJ (2004) Dusky dolphin foraging habitat: overlap with aquaculture in New Zealand. Aquat Conserv 14(2):133–149. https://doi.org/10.1002/AQC.602

Martin J, Ambrose R, Wool T (2018) WASP8 Macro Algae: Model Theory and User's Guide, https://www.epa.gov/sites/default/files/2018-05/documents/wasp-macroalgae_manual-v3.pdf

Mellinger DK, Heimlich SL (2013) Introduction to the special issue on methods for marine mammal passive acoustics. J Acoust Soc Am 134(3):2381–2382

Merrill JE, Waaland JR (1998) The seaweed resources of The United States of America. Culture 23:45–53

Montoya Rosas EDV, Rosas Cabrera JA, Rincones León RE, Narváez Rodríguez JL (2020) Evaluación del crecimiento de *Eucheuma isiforme* (Rhodophyta, Gigartinales) en sistemas de cultivo suspendidos en la isla de Cubagua, Venezuela (sureste del Mar Caribe). AquaTechnica: Revista Iberoamericana de Acuicultura 2(2):86–98

Moon RE, Dawes CJ (1976) Pigment changes and photosynthetic rates under selected wavelengths in the growing tips of *Eucheuma isiforme* (C. Agardh) J. Agardh var deudatum Cheney during vegetative growth. Br Phycol J 11:165–174

Moore MJ, van der Hoop JM (2012) The painful side of trap and fixed net fisheries: chronic entanglement of large whales. J Mar Biol 1–4 https://doi.org/10.1155/2012/230653

Msuya FE, Salum D (2011) Effect of the presence of seagrass and nutrients on growth rates of farmed *Kappaphycus alvarezii* and *Eucheuma denticulatum* (Rhodophyta). West Indian Ocean J Mar Sci 10(2):129–135

Mulder N (coord.) (2020) The impact of the COVID-19 pandemic on the tourism sector in Latin America and the Caribbean, and options for a sustainable and resilient recovery. International Trade series, No. 157 (LC/TS.2020/147), Santiago, Economic Commission for Latin America and the Caribbean (ECLAC). https://repositorio.cepal.org/bitstream/handle/11362/46502/S2000751_en.pdf?sequence=3

Nanaware S, Shastri R, Joshi Y, Das A (2014) Passive acoustic detection and classification of marine mammal vocalizations. In: 2014 international conference on communication and signal processing. IEEE, pp 493–497

Navarrete IA, Kim DY, Wilcox C, Reed DC, Ginsburg DW, Dutton JM, Heidelberg J, Raut Y, Wilcox BH (2021) Effects of depth-cycling on nutrient uptake and biomass production in the giant kelp *Macrocystis pyrifera*. Renew Sust Energy Rev 141:110747

Nelms SE, Alfaro-Shigueto J, Arnould JPY, Avila IC, Nash SB, Campbell E, Carter MI, Collins T, Currey RJ, Domit C, Franco-Trecu V (2021) Marine mammal conservation: over the horizon. Endanger Species Res 44:291–325. https://doi.org/10.3354/ESR01115

Nishihara GN, Terada R (2010) Spatial variations in nutrient supply to the red algae *Eucheuma serra* (J. Agardh). J Agardh Phycol Res 58(1):29–34

Nishihara GN, Noro T, Terada R (2011) In vitro growth and photosynthesis of three edible seaweeds, *Betaphycus gelatinus*, *Eucheuma serra* and *Meristotheca papulosa* (Solieriaceae, Rhodophyta). J Aquacult Sci 59(4):563–571

Oswald JN, Barlow J, Norris TF (2003) Acoustic identification of nine delphinid species in the eastern tropical Pacific Ocean. Mar Mamm Sci 19(1):20–037

Oswald JN, Van Cise AM, Dassow A, Elliott T, Johnson MT, Ravignani A, Podos J (2022) A collection of best practices for the collection and analysis of bioacoustic data. Appl Sci 12(23):12046

Oxenford HA, Cox SA, van Tussenbroek BI, Desrochers A (2021) Challenges of turning the *Sargassum* crisis into gold: current constraints and implications for the Caribbean. Phycology 1(1): 27–48

Oyieke HA (1994) The effect of phenotypic plasticity on agar from *Gracilaria salicornia* (J. Ag.) Dawson (Gracilariales, Rhodophyta) in Kenya. Bioresour Technol 49(3):267–271

Pereira R, Yarish C (2008) Mass production of marine macroalgae. In: Jørgensen SE, Fath BD (eds) Ecological engineering. Vol. [3] of Encyclopedia of ecology, 5 vols. Elsevier, Oxford, pp 2236–2247

Poonian CNS, Lopez DD (2016) Small-scale mariculture: a potentially significant threat to dugongs (Dugong dugon) through incidental entanglement. Aquat Mamm 42(1):56–59. https://doi.org/10.1578/AM.42.1.2016.56

Rankin S, Barlow J, Oswald JN, Ballance L (2008) Acoustic studies of marine mammals during seven years of combined visual and acoustic line-transect surveys for cetaceans in the eastern and Central Pacific Ocean. NOAA-TM-NMFS-SWFSC-429

Rankin S, Archer F, Keating JL, Oswald JN, Oswald M, Curtis A, Barlow J (2017) Acoustic classification of dolphins in the California current using whistles, echolocation clicks, and burst pulses. Mar Mamm Sci 33(2):520–540

Rassweiler A, Reed DC, Harrer SL, Nelson JC (2018) Improved estimates of net primary production, growth, and standing crop of *Macrocystis pyrifera* in Southern California. Ecology 7:2068. https://doi.org/10.1002/ecy.2440

Reddy C, Yokoya NS, Yong WTL, Luhan MRJ, Hurtado AQ (2017) Micro-propagation of *Kappaphycu*s and *Eucheum*a: trends and prospects. In: Hurtado AQ, Critchley AT, Neish I (eds) Tropical seaweed farming trends, problems and opportunities. Springer, Amsterdam, pp 91–109

Redmond S, Green L, Yarish C, Kim JK, Neefus C (2014) New England Seaweed culture handbook. https://opencommons.uconn.edu/seagrant_weedcult/1/

Reeves RR, McClellan K, Werner TB (2013) Marine mammal bycatch in gillnet and other entangling net fisheries, 1990 to 2011. Endanger Species Res 20(1):71–97. https://doi.org/10.3354/ESR00481

Ren JS, Barr NG, Scheuer K, Schiel DR, Zeldis J (2014) A dynamic growth model of macroalgae: application in an estuary recovering from treated wastewater and earthquake-driven eutrophication. Estuar Coast Shelf Sci 148:59–69

Rincones R (2000) Marine agronomy: a sustainable alternative for coastal communities in developing countries. Global Aquaculture Advocate. https://www.globalseafood.org/advocate/marine-agronomy-a-sustainable-alternative-for-coastal-communities-in-developing-countries/

Roberson L, Coyer J (2004) Variation in blade morphology of the kelp *Eisenia arborea*: incipient speciation due to local water motion? Mar Ecol Prog Ser 282:115–128

Saha S, Moorthi S, Wu X, Wang J, Nadiga S, Tripp P, Behringer D, Hou YT, Chuang HY, Iredell M, Ek M (2014) The NCEP climate forecast system version 2. J Clim 27(6):2185–2208

Sahu SK, Ingle KN, Mantri VA (2020) Epiphytism in seaweed farming: causes, status, and implications. In: Gothandam K, Ranjan S, Dasgupta N, Lichtfouse E (eds) Environmental biotechnology, vol 1. Environmental chemistry for a sustainable world 44. Springer, Cham, pp 227–242. https://doi.org/10.1007/978-3-030-38192-9_9

Seger KD, Al-Badrawi MH, Miksis-Olds JL, Kirsch NJ, Lyons AP (2018) An empirical mode decomposition-based detection and classification approach for marine mammal vocal signals. J Acoust Soc Am 144(6):3181–3190

Shiu Y, Palmer KJ, Roch MA, Fleishman E, Liu X, Nosal EM, Helble T, Cholewiak D, Gillespie D, Klinck H (2020) Deep neural networks for automated detection of marine mammal species. Sci Rep 10(1): 607

Smith AH (1992) Seaweed cultivation in the West Indies. In: Mshigeni KE, Bolton J, Critchley A, Kiangi G (eds) Proceedings of the first international workshop on sustainable seaweed resource development in sub-Saharan Africa, Windhoek, Namibia, pp 337–351

Solidoro C, Pecenik G, Pastres R, Franco D, Dejak C (1997) Modelling macroalgae (*Ulva rigida*) in the Venice lagoon: model structure identification and first parameters estimation. Ecol Model 94(2–3): 191–206

Stiltz HL (1961) Aerospace telemetry, vol 1. Prentice Hall. https://doi.org/10.1016/0016-0032(68)90490-0

Tanaka Y, Ashaari A, Mohamad FS, Lamit N (2020) Bioremediation potential of tropical seaweeds in aquaculture: low-salinity tolerance, phosphorus content, and production of UV-absorbing compounds. Aquaculture 518:734853

Teagle H, Hawkins SJ, Moore PJ, Smale DA (2017) The role of kelp species as biogenic habitat formers in coastal marine ecosystems. J Exp Mar Biol Ecol 492:81–98

Terada R, Vo TD, Nishihara GN, Shioya K, Shimada S, Kawaguchi S (2016) The effect of irradiance and temperature on the photosynthesis and growth of a cultivated red alga *Kappaphycus alvarezii* (Solieriaceae) from Vietnam, based on in situ and in vitro measurements. J Appl Phycol 28(1):457–467

Theuerkauf SJ, Barrett LT, Alleway HK, Costa-Pierce BA, St. Gelais A, Jones RC (2021) Habitat value of bivalve shellfish and seaweed aquaculture for fish and invertebrates: pathways, synthesis and next steps. Rev Aquac 14(1):54–72

Thomas M, Martin B, Kowarski K, Gaudet B, Matwin S (2020) Marine mammal species classification using convolutional neural networks and a novel acoustic representation. Lecture notes in computer science (including subseries lecture notes in artificial intelligence and lecture notes in bioinformatics), 11908 LNAI, pp 290–305. https://doi.org/10.1007/978-3-030-46133-1_18/COVER/

Thompson TM, Young BR, Baroutian S (2020) Pelagic *Sargassum* for energy and fertiliser production in the Caribbean: a case study on Barbados. Renew Sust Energ Rev 118:109564

Ticona Rollano F, Yang Z, Wang T (2022) Coupled wave-current modeling for hydrodynamic load analysis of macroalgae cultivation farms (In prep)

Trancoso AR, Saraiva S, Fernandes L, Pina P, Leitão P, Neves R (2005) Modelling macroalgae using a 3D hydrodynamic-ecological model in a shallow, temperate estuary. Ecol Model 187(2–3):232–246

Umanzor S, Jang S, Antosca R, Critchley AT, Yarish C, Kim JK (2020) Optimizing the application of selected biostimulants to enhance the growth of *Eucheumatopsis isiformis*, a carrageenophyte with commercial value, as grown in land-based nursery systems. J Appl Phycol 32:1917–1922

Utter B, Denny M (1996) Wave-induced forces on the giant kelp *Macrocystis Pyrifera* (Agardh): field test of a computational model. J Exp Biol 199(12):2645–2654. https://doi.org/10.1242/jeb.199.12.2645

Vairappan CS (2006) Seasonal occurrences of epiphytic algae on the commercially cultivated red alga *Kappaphycus alvarezii* (Solieriaceae, Gigartinales, Rhodophyta). J Appl Phycol 18:611–617

Vairappan CS, Anangdan SP, Tan KL, Matsunaga S (2010) Role of secondary metabolites as defense chemicals against ice-ice disease bacteria in biofouler at carrageenophyte farms. J Appl Phycol 22: 305–311

Valderrama D, Cai J, Hishamunda N, Ridler N (eds) (2013) Social and economic dimensions of carrageenan seaweed farming. Fisheries and aquaculture technical paper no. 580. Rome, FAO. 204 pp

Van der Molen J, Ruardij P, Mooney K, Kerrison P, O'Connor NE, Gorman E, Timmermans K, Wright S, Kelly M, Hughes AD, Capuzzo E (2018) Modelling potential production of macroalgae farms in UK and Dutch coastal waters. Biogeosciences 15(4): 1123–1147

Van Opzeeland I, Hillebrand H (2020) Year-round passive acoustic data reveal spatio-temporal patterns in marine mammal community composition in the Weddell Sea, Antarctica. Mar Ecol Prog Ser 638:191–206. https://doi.org/10.3354/MEPS13258

Visch W, Nylund GM, Pavia H (2020) Growth and biofouling in kelp aquaculture (*Saccharina latissima*): the effect of location and wave exposure. J Appl Phycol 32(5):3199–3209

Wakibia JG, Bolton JJ, Keats DW, Raitt LM (2006) Factors influencing the growth rates of three commercial Eucheumoids at coastal sites in southern Kenya. J Appl Phycol 18(3):565–573

Wallcraft AJ, Metzger EJ, Carroll SN (2009) Software design description for the hybrid coordinate ocean model (HYCOM), Version 2.2. Naval Research Lab Stennis Space Center Ms Oceanography Div

Walls AM, Kennedy R, Fitzgerald RD, Blight AJ, Johnson MP, Edwards MD (2016) Potential novel habitat created by holdfasts from cultivated Laminaria digitata: assessing the macroinvertebrate assemblages. Aquac Environ Interact 8:157–169

Ward GM, Kambey CS, Faisan JP Jr, Tan PL, Daumich CC, Matoju I, Stentiford GD, Bass D, Lim PE, Brodie J (2021) Ice-ice disease: an environmentally and microbiologically driven syndrome in tropical seaweed aquaculture. Rev Aquac 14:414–439

Waters TJ, Lionata H, Prasetyo Wibowo T, Jones R, Theuerkauf S, Usman S, Amin I, Ilman M (2019) Coastal conservation and sustainable livelihoods through seaweed aquaculture in Indonesia: a guide for buyers, conservation practitioners, and farmers. The Nature Conservancy, Arlington

Wu J, Keller DP, Oschlies A (2023) Carbon dioxide removal via macroalgae open-ocean mariculture and sinking: an Earth system modeling study. Earth Syst Dynam 14(1):185–221. https://doi.org/10.5194/esd-2021-104

Würsig B, Gailey GA (2002) Marine mammals and aquaculture: conflicts and potential resolutions. In: McVey JP (ed) Stickney RR. CABI, Responsible marine aquaculture, pp 45–59. https://doi.org/10.1079/9780851996042.0045

Yang Z, Gorton AM, Wang T, Whiting JM, Copping AE, Haas K, Wolfram PJ, Yim S (2020) Multi-resolution, multi-scale modeling for scalable macroalgae production. No. PNNL-29881. Pacific Northwest National Lab (PNNL), Richland, WA (United States)

Zertuche-Gonzalez JA (1993) Situación actual de la industria de macroalgas productoras ficocoloides en America Latina y el Caribe (No. F009. 038). FAO

Zertuche-González JA (1994) Situación actual del cultivo de algas agarofitas en América Latina y el Caribe. Programa Cooperativo Gubernamental FAO-Italia. Proyecto AQUILA II. Documento de Campo 13:5–15

Zertuche-González JA (1996) Feasibility for establishing a carrageenan industry in Mexico. Hydrobiologia 326(327):381–386

Zertuche-González JA (1998) Macroalgal culture as a sustainable coastal livelihood in coral reef areas. Proceedings of the coral reef conference. World Bank, Washington DC, October 9 & 10, 1997, pp 53–54

Zertuche-González JA, García-Lepe G, Pacheco-Ruíz I, Chee-Barragán A, Gendrop-Funes V (1999) New approach for the seaweed culturing in Mexico. J World Aquac Soc 30(2):50–53

Zertuche-González JA, García-Lepe G, Pacheco-Ruíz I, Chee-Barragán A, Gendrop-Funes V, Guzmán JM (2001) Open water Chondrus crispus Stackhourse cultivation. J Appl Phycol 13:249–253

Zhang J, Wu W, Ren JS, Lin F (2016) A model for the growth of mariculture kelp *Saccharina japonica* in Sanggou Bay, China. Aquac Environ Interact 8:273–283

12 Diverse Seaweed Farming Livelihoods in Two Indonesian Villages

Zannie Langford, Scott Waldron, Jing Zhang, Radhiyah Ruhon, Zulung Zach Walyandra, Risya Arsyi Armis, Imran Lapong, Boedi Julianto, Irsyadi Siradjuddin, Syamsul Pasaribu, and Nunung Nuryartono

Abstract

The Indonesian marine seaweed industry has expanded dramatically over the past two decades and now supports the livelihoods of approximately 62,000 households. The production sector is dominated by smallholder seaweed farmers that operate in diverse agroclimatic and social landscapes. Understanding their production and marketing systems is fundamental to both industry development efforts and understanding associated livelihood impacts. This chapter provides an in-depth description of seaweed production, marketing, and livelihoods in two Indonesian villages, drawing on a household survey of 273 seaweed farmers and extensive qualitative research. It extends previous studies through a focus on heterogeneity in production and marketing systems derived from a rich set of quantitative and qualitative data. We show that there are substantial differences in seaweed livelihoods both between and within study locations, which is reflected in different forms of integration into broader seaweed value chains.

Keywords

Smallholder · Livelihood · Seaweed · Farming · Indonesia

Z. Langford (✉)
School of Agriculture and Food Sustainability, University of Queensland, Brisbane, QLD, Australia

The Australia-Indonesia Centre, Universitas Hasanuddin, Makassar, Indonesia

Griffith Asia Institute, Griffith University, Brisbane, Australia
e-mail: z.langford@griffith.edu.au

S. Waldron
School of Agriculture and Food Sustainability, University of Queensland, Brisbane, QLD, Australia

The Australia-Indonesia Centre, Universitas Hasanuddin, Makassar, Indonesia

J. Zhang
School of Agriculture and Food Sustainability, University of Queensland, Brisbane, QLD, Australia

R. Ruhon · Z. Z. Walyandra · R. A. Armis · I. Lapong · B. Julianto
PT Jaringan Sumber Daya, Makassar, Indonesia

I. Siradjuddin
PT Jaringan Sumber Daya, Makassar, Indonesia

Faculty of Science and Technology, UIN Alauddin University, Makasar, Indonesia

S. Pasaribu · N. Nuryartono
The Australia-Indonesia Centre, Universitas Hasanuddin, Makassar, Indonesia

Faculty of Economics and Management, IPB University, Bogor, Indonesia

12.1 Introduction

The Indonesian seaweed industry has grown rapidly over the last two decades and is now estimated to support the livelihoods of approximately 62,000 coastal households (BPS 2022a, b, p. 47). Indonesian seaweed production is almost completely dominated by coastal smallholder farmers, who work around seasonal and tidal variations and typically employ limited capital but high labor inputs. Understanding household seaweed production is important for several reasons. First, the scale of the development issue in terms of the number of livelihoods it supports is significant. Second, unlike most other agricultural sectors, seaweed is almost completely dominated by smallholders, with only a few cases of corporate production. Thus, households supply virtually all the raw materials required for the functioning of the international industry and downstream actors. Third, the household sector is a central component in the pursuit of social and policy objectives, including employment, economic growth, social stability, and government tax receipts.

Seaweed farming brings a range of social and economic benefits to a large number of farmers in diverse ways (Aslan et al. 2015, 2018; de La Torre-Castro et al. 2017; Langford 2024; Larson et al. 2020; Mariño et al. 2019; Neish 2013, 2015; Nuryartono et al. 2021; Salayo et al. 2012; Suyo et al. 2020), although the economic benefits of seaweed

A. T. Critchley et al. (eds.), *Tropical Phyconomy Coalition Development*, Developments in Applied Phycology 11,
https://doi.org/10.1007/978-3-031-47806-2_12

livelihoods are variable (Rimmer et al. 2021; Langford et al. 2023b) and linked to broader relations in national industries and global value chains (Komarek et al. 2023; Permani et al. 2023; Zhang et al. 2023). Households in the sector range from those with small plots, limited capital, and marginal profits to very large, high performing, and well capitalized farmers who enjoy bargaining power in the sale of their products. The diversity of farmers must be accounted for in understanding and improving the livelihood aspects of seaweed farming and their incorporation into different seaweed value chains.

This chapter provides insights into seaweed growth, post-harvest handling, and marketing practices in two seaweed farming locations to explore the diverse livelihoods in these areas and the implications for their incorporation into the Indonesian seaweed value chain. It combines a detailed survey of 273 seaweed farming households in two villages with insights gained through 16 months of ethnographic observation and semistructured interviews by four of the authors of this study. This analysis reveals the diversity of farmers and farming practices in these locations with implications for policy and technology development targeting the Indonesian seaweed value chain.

12.2 Methods

12.2.1 Study Location

The study villages are in South Sulawesi, the largest seaweed-producing province in Indonesia, which produces approximately 20% of the global supply of carrageenan seaweed (Ministry of Marine Affairs and Fisheries, Republic of Indonesia 2019, p. 320–321; FAO 2023). South Sulawesi has recently been the site of a major foreign investment in seaweed processing capacity by the Chinese carrageenan producer Shanghai Brilliant Gum Company (BLG), and its capital, Makassar, is undergoing major new international port development to facilitate the direct export of products such as seaweed from Eastern Indonesia. As such, it is a province in transition to become an increasingly commercialized exporting region. Within South Sulawesi, the regency of Takalar is the largest seaweed producer and is also widely considered the most established, having a long history of seaweed production, intensive production, and established production methods. These features make Takalar an important site for understanding the development of the carrageenan seaweed industry. The regency of Pangkajene dan Kepulauan (Pangkep) is a relatively less established seaweed farming area. While Takalar lies to the south of Makassar, Pangkep lies a similar distance to the north and exhibits different agroclimatic conditions. Pangkep has a long history of export-oriented aquaculture but has only recently transitioned to seaweed farming. The province also includes many small islands off the coast, some of which are quite distant from the mainland. This location provides an interesting contrast to Takalar and insight into more recent and ongoing seaweed farming transitions (Fig. 12.1).

12.2.2 Methodology

This study combines mixed qualitative and quantitative research methods in two case study locations (Fig. 12.2). The field research team consisted of four of the authors of this article, who relocated to the study villages lived in the village undertaking participant observation for 16 months. Two of the researchers worked in the regency of Pangkep and two in Takalar in South Sulawesi. The first step was a study of the village characteristics in the two regencies of Takalar and Pangkep to assess the seaweed production characteristics of these areas and to select villages for detailed analysis. Second, a detailed quantitative survey was conducted on 273 seaweed farming households in two villages. Third, this information was supplemented and cross-verified with ethnographic data collected by four researchers living in the two villages over a period of 16 months. Fourth, detailed interviews were conducted with value chain actors (including farmers, local and regional traders, and government officials). The combination of these quantitative and qualitative methods allows for a detailed and rigorous analysis of patterns in seaweed production practices and differences between households and between villages. The qualitative data provide in-depth insights into the reasons for these patterns and the way that these characteristics are experienced and negotiated in practice. This provides an understanding of the social and economic factors affecting the management of seaweed farms in these two villages and the way that they translate into broader economic relations in the Indonesian seaweed industry.

12.2.3 Village Survey

To select the case study villages, mainland coastal villages in both provinces (17 villages in Pangkep and 15 villages in Takalar) were included in an initial survey of village characteristics. This involved visiting village offices, consulting with village leaders, consulting with local seaweed industry experts (facilitated by the existing connections of one of the authors of the study with long-term experience in the seaweed industry in this region), collecting data from local government, and reviewing satellite imagery of seaweed production in the region. Island villages were excluded due to practical limitations and because analysis of satellite

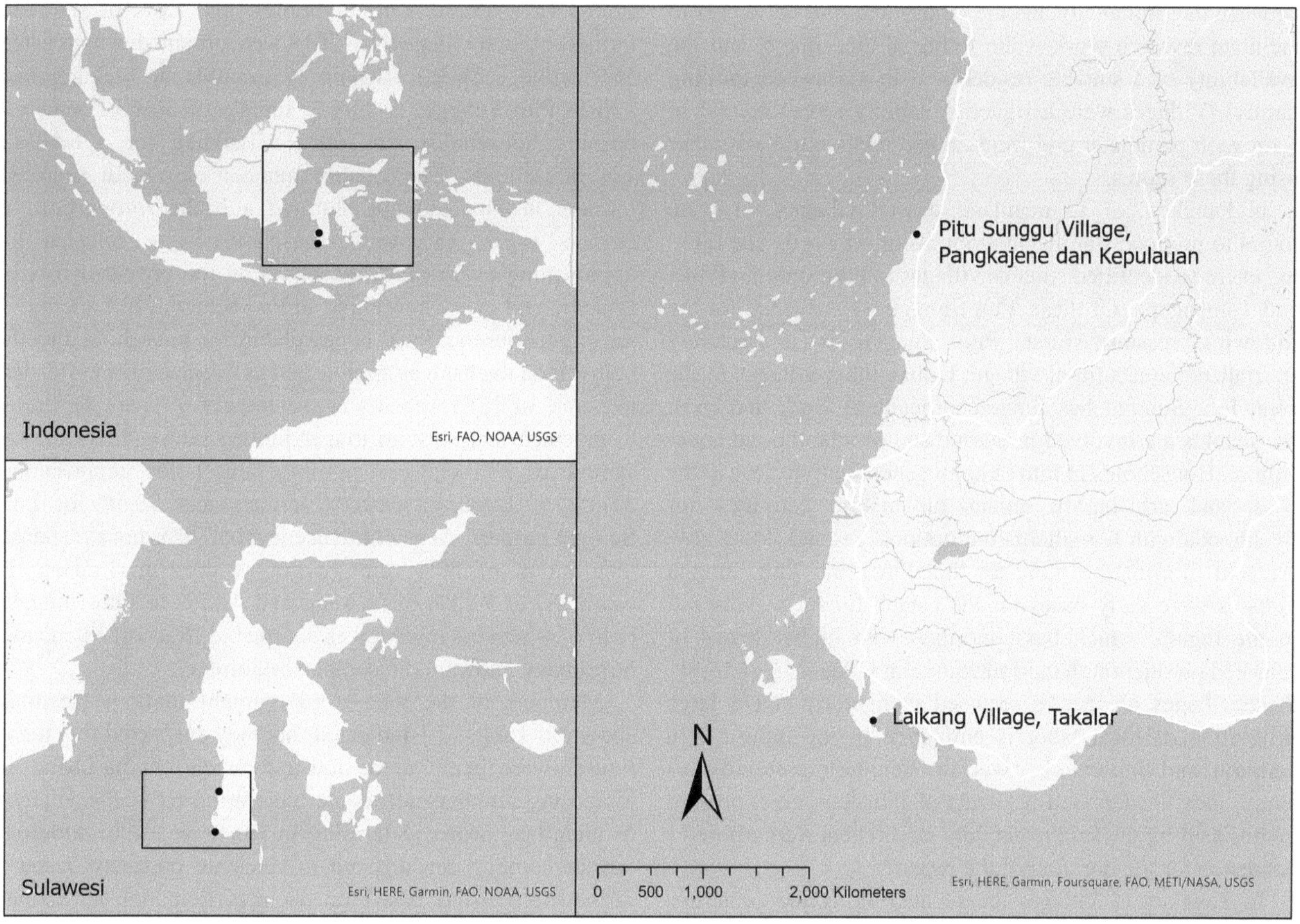

Fig. 12.1 Study site locations

imagery suggests that seaweed farming is concentrated around the mainland (Langford et al. 2021).

The villages were assessed based on criteria that fall into two categories. The first is the characteristics of the villages (the number of seaweed farming households, the proportion of households in the village that engage in seaweed production, the prevalence of other income sources, the number of fishing households, the volume of seaweed production, and the year that seaweed farming commenced). The second was the suitability of the village as a long-term study site and residence for the four researchers embedded in the villages (interest of the village leader and households in the study, the availability and willingness of local leaders to participate in the study, the condition of roads to facilitate regular travel

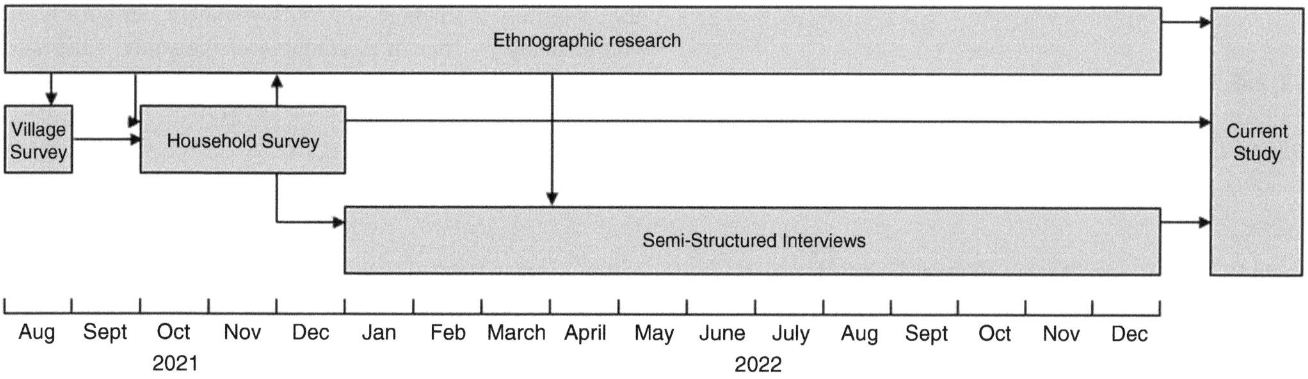

Fig. 12.2 Research methodology schematic

between the capital city and the village and Internet access to facilitate research work while living in the village, and the availability of a suitable residence with a seaweed farming family). Villages were assigned suitability scores from 1 to 5 for each parameter and were ranked relative to each other using these scores.

In Pangkep, of 17 mainland coastal villages, 11 were found to produce significant amounts of seaweed. The ranking exercise identified suitable villages as Pitu Sunggu, Pitue, and Tamangapa. Of these, Pitu Sunggu was selected for the following reasons. First, Pitu Sunggu is a relatively specialized aquacultural village. Unlike other villages in the area, Pitu Sunggu has limited agricultural land, and most households are involved in marine and/or inland pond aquaculture. Households in Pitu Sunggu generally live close to the coast and are highly reliant on marine activities for livelihoods with few alternative options. Second, local government information indicates that Pitue and Pitu Sunggu villages were early bases for the establishment of seaweed in the region, which has continued with higher levels of seaweed production than in surrounding villages. As a result, these villages are active seaweed cultivators, and a large proportion of local labor is employed in the industry. In addition, and importantly given the long-term nature of this study, local leaders and residents of Pitu Sunggu expressed enthusiasm for the study, and field researchers were offered a residence for the duration of the research.

In Takalar, 15 villages are located along the coastline; however, only 4 of these have significant seaweed production areas. The ranking exercise identified suitable villages of Laikang and Ujung Baji. Of these, Laikang was selected for a number of reasons. First, it is located in a large bay in which large areas of seaweed are intensively cultivated, alongside a range of other marine aquacultural activities. It is also an established seaweed production area where seaweed has been cultivated for several decades and farming techniques are well established. In addition, the local government responded favorably to the proposed study, and field researchers were offered a residence for the duration of the research.

12.2.4 Household Survey

During the village survey, basic data on Pitu Sunggu and Laikang were collected (Table 12.1) and used to design the survey of seaweed farming households. Lists of seaweed farmers in each village from 2018 were obtained, which listed 561 active seaweed farming households in Laikang and 136 in Pitu Sunggu. A survey sample of 30% of seaweed farming households was set for Laikang for a total of 181 households. The smaller number of overall seaweed farmers in Pitu Sunggu required a higher proportion of farmers to be surveyed. Households were selected by approaching every third household on the provided lists in Laikang and every household in Pitu Sunggu. In Laikang, if the target household was not available, the household directly below it on the list was approached as a replacement. This led to a total of 281 responses across the two villages. Of these, 8 indicated that they no longer farmed seaweed, for a total sample of 273 seaweed farming households, representing 31.6% of Laikang seaweed farmers and 73.5% of Pitu Sunggu farmers. At a confidence level of 90% and a response distribution of 50%, the margin of error in Laikang is estimated at 5.12% in Laikang and 4.53% in Pitu Sunggu, both of which are regarded as acceptable (Raosoft calculator, http://www.raosoft.com/samplesize.html).

Members of the village government in both locations shown in Table 12.1 indicated that they suspected that these figures were likely to be lower than current numbers, as increasing numbers of people have returned to the villages to undertake seaweed farming in recent years. In addition, farmer numbers are difficult to calculate precisely because farmers are listed based on their participation in farmer groups (which newer farmers may not be part of), and there is no clear differentiation between different categories of sea users at all levels of government data collection.

The results from the household survey were organized into descriptive statistics split by location. Most survey questions used discrete multiple choice and Likert scale questions, but provision was made for other "other" options and open text response. For questions with large numbers of "other" responses, these responses were organized into groups of additional responses and added to the data. This enabled us to record any unexpected options that in several cases yielded important insights. The survey was designed in the program Alchemer (https://www.alchemer.com/) and administered by four of the authors of the study. Data were organized and analyzed using Microsoft Excel.

Table 12.1 Sample numbers in Laikang and Pitu Sunggu

	Laikang	Pitu Sunggu
Total number of seaweed farming households (2018 data)	561	136
Number of seaweed farming households interviewed	177	96
Proportion of seaweed farming households interviewed	31.6%	73.5%

12.2.5 Participant Observation

The field research team undertook participant observation for 16 months. Two researchers were located in Pitu Sunggu village and two in Laikang. The field research team is gender balanced to provide male and female perspectives and support gender sensitivity in interviewing. The field researchers undertook ethnographic observation, prepared field notes documenting their observations, undertook interviews, and discussed their observations with members of the wider research group. These qualitative observations are used to interpret the results of the household survey.

12.2.6 Semistructured Interviews

Semistructured interviews were conducted from January to December 2022 with value chain participants such as farmers and traders. These interviews involve open-ended questions on a series of topics. They were undertaken in either Indonesian, Buginese, or Makassar language, according to participant preference, and were transcribed in Indonesian (Buginese and Makassar language interviews were translated and transcribed simultaneously). Repeat interviews were undertaken where necessary. A total of 171 interviews were undertaken with an average duration of 40 min. Respondents who were uncomfortable with the recorded format were instead invited to discuss topics informally, and notes taken during these informal interviews were presented in fieldnotes.

12.3 Results

12.3.1 Seaweed Aquaculture in Two Indonesian Villages

Respondents in both villages were similar in age and gender, with 60% of respondents between the ages of 34 and 54. Approximately 22% were older than this, and 18% were younger, indicating lower participation in the industry in these demographic brackets. Notably, just 1% of respondents were aged 18–24, indicating low participation of youth in the industry. Survey participants were 90% male, as researchers asked for a household representative of the household's choice to complete the survey, and most households offered men to complete the survey, often with assistance from other household members. Women were deliberately included in ethnographic research and semistructured interviewing to ensure incorporation of their views.

The most common species of seaweed grown were *Kappaphycus alvarezii* (known colloquially as "cottoni") and *Kappaphycus striatus* ("sacol"), with the majority of farmers in both villages growing both species. Cottoni was grown by 85% of Laikang farmers and 100% of Pitu Sunggu farmers. Sacol was grown by 93% of Laikang farmers and 81% of Pitu Sunggu farmers. In Laikang, a notable group of farmers (40%) also grew *Eucheuma denticulatum* ("spinosum"). No farmers in Pitu Sunggu reported growing spinosum. Our respondents indicated that the differences in the choice of seaweed depend on local oceanic conditions (temperature and salinity levels), rainfall patterns, price incentives, and other productive factors, such as growing season lengths and ease of propagule attachments to ropes. Along the western coast of South Sulawesi, it is common to grow cottoni in the wet season and sacol in the dry season. Both types of carrageenan seaweeds are used to produce the kappa-type of carrageenan, have the same price, and are often mixed by local traders. Spinosum, on the other hand, produces the iota-carrageenan type and cannot be mixed with these species. It is a significantly lower value species of seaweed. In Laikang, 2% of households also produce *Caulerpa* ("sea grapes"). This type of seaweed is grown in coastal ponds rather than on ropes in the sea and is sold primarily fresh into domestic markets where it is eaten raw. This is a much smaller commodity that enters different value chains, is not a eucheumatoid seaweed, and is not the focus of this study. Across the two villages, 83% of seaweed farmers indicated that seaweed was their main source of income, and most of these indicated that it accounted for more than half of their total household income. The findings suggest that seaweed farming is a major contributor to household income, and most seaweed farming households are relatively specialized in that activity.

However, the two villages differ in important ways. In Laikang, seaweed farming has a long history. Most farmers (65%) have been cultivating seaweed for over two decades, with very few (7%) commencing seaweed farming less than a decade ago. In contrast, in Pitu Sunggu, only 6% of farmers have cultivated seaweed for more than two decades, and nearly half (48%) commenced less than a decade ago. In addition, the locations are also geographically and environmentally different, which influences the choice of farming activities. Figure 12.3 shows the main production activities of households before and after commencing seaweed farming.

In Pitu Sunggu, seaweed farmers are highly specialized in marine fishing (including a range of species of fish and crab) and pond aquaculture. A very low proportion produce a staple food such as corn or rice. In this village, widespread conversion of rice paddies to brackish water aquaculture was undertaken from approximately 1980 to 2000 to capitalize on the high price of tiger prawns at the time. This industry lasted only a few years before it collapsed as a result of white spot syndrome virus and ceased completely in the region. Vannamei shrimp (*Litopenaeus vannamei*) were released by the government to farmers in 2001 and distributed around the

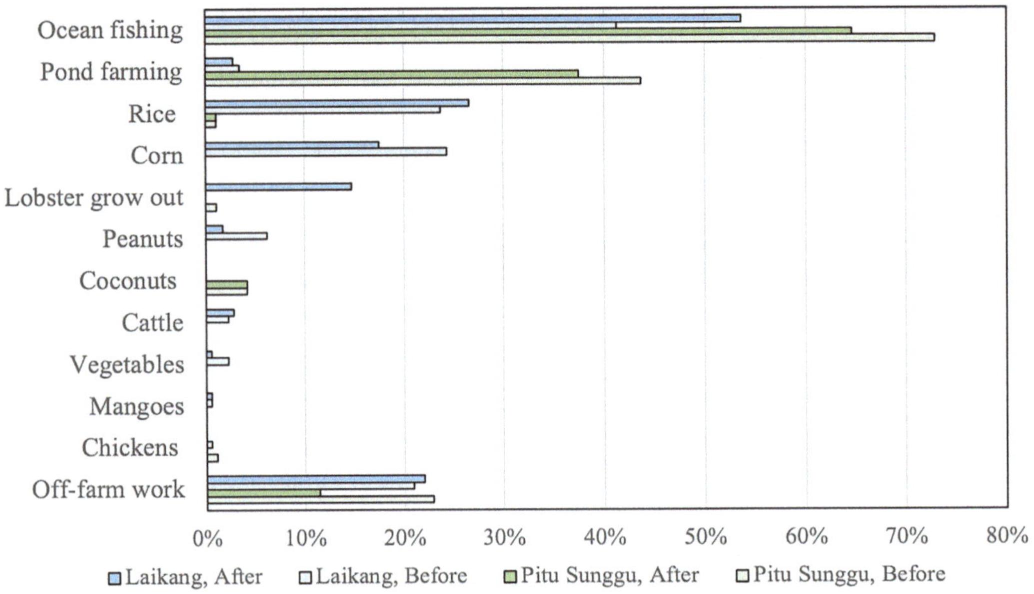

Fig. 12.3 Other major household activities before and after commencing seaweed farming

country, and it is this variety of shrimp that now dominates ponds in Pitu Sunggu, which with a few inland exceptions did not return to rice farming.

Prior to commencing seaweed farming, approximately 23% of seaweed farmers reported that they had earned a significant proportion of their household income from off-farm work, but this has now decreased by more than half. Many older village residents commented on this, noting that during the rice farming era in Pitu Sunggu, many people struggled to survive on the agricultural activities available in the village. With a distinct dry season and a lack of irrigation, only one rice crop was planted per year, which limited food and income and induced farmers to seek off-farm work-off farms in other seasons. As a result, Pitu Sunggu farmers pursued an agricultural development pathway based on specialization in marine capture and marine and brackish water aquaculture. Today, just 1% of seaweed farmers consider the production of a staple food such as rice to be a major household activity.

The mix of agricultural activities differs considerably in Laikang. Very few participate in land-based pond farming. Their land-based activities focus on food production, where 31% of seaweed farmers also grow a staple food such as corn or rice. Just over half also engage in ocean fishing for both food and sale. Another significant activity is the growth out of lobsters, which involves purchasing juvenile lobsters from nearby cities and growing them out in marine cages. Of the seaweed farmers interviewed, 15% reported participation in this activity, which appears to be a significant increase over recent years. Lobster grow-out is a lucrative activity but has higher capital costs than seaweed farmers, so it is being taken up by higher-income, entrepreneurial farmers or those who receive support from the government. Like Pitu Sunggu, it appears that many rice farmers in Laikang had also struggled to make a decent living from a single annual rice crop and sought to fill the gap through off-farm work. The production of commercial coastal products appears to have enabled farmers to pursue year-round, on-farm livelihood strategies.

The scale of production is an important indicator in the analysis of production and livelihood aspects of seaweed systems. The average scales of production were similar in the two villages (1.63 tonnes per household year in Laikang and 1.66 tonnes in Pitu Sunggu). There were, however, major differences in range. Laikang has an overall lower median level of production but several very large outlying producers (Fig. 12.4). This suggests the presence of highly entrepreneurial farmers who have managed to claim large areas of sea space for production.

The ability to farm seaweed is closely linked to ownership of a boat and engine, as well as rights to use marine space. Government aid in the form of boats, boat engines, and mariculture apparatus is common in both villages and indeed is a contentious topic of discussion in the villages. Some farmers appear to be regular recipients of aid, while others feel that they have missed out. In addition, in Pitu Sunggu, several farmers had dropped out of seaweed farming after their boat or engine broke, and they lacked the capital to replace it and hoped that the government would provide aid to enable them to return to seaweed farming.

Most farmers in both locations (91% overall) indicated that they plan to expand seaweed production in the next 5 years. The most commonly cited reason was good prices

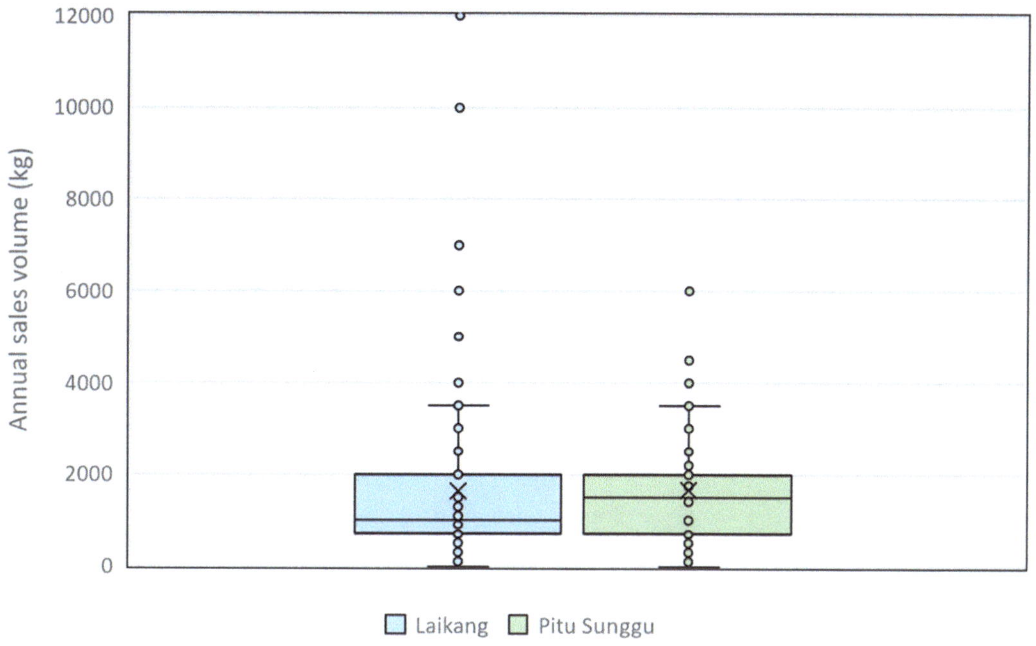

Fig. 12.4 Seaweed sales quantity over the last year

of seaweed at the time of survey in late 2021 (70%), followed by the potential to earn higher incomes from increased production (25%). However, in both locations, rising prices from June 2021 to August 2022 drove extensification of farming areas and led to shortages of sea space that enforced hard limits to expansion. Favorable seaweed cultivation areas were widely considered "full" in both villages by mid 2022. As one farmer put it, "the sea looks wide and empty, but actually, it is already full of other people's land stakes." Farmer understandings of marine tenure are based on local histories of seaweed farming establishment and, in many cases, are marked by conflict between seaweed farmers and other ocean users, such as fishermen and lobster growers. In both villages, sea space was traditionally a public good and could be used by anyone. However, with the introduction of seaweed farming, farmers needed to claim exclusive rights to areas and prevent those areas from being used for other activities, such as boat travel or crab fishing. A small number of farmers began claiming portions of the sea for seaweed farms and installing permanent anchors to mark and maintain their sites. This often occurred alongside conflict with other sea users, but over time, the conflict dissipated, and the right of individuals to claim parts of the sea in this way began to be acknowledged.

Today, the rights of seaweed farming plot owners to exclusively use the sea are widely acknowledged by convention within the village. Seaweed farming plots are defined by anchors installed by farmers, and local governance systems of different degrees of formality protect an informal form of property rights over these areas. The right to use an area may be sold or leased to another farmer, but in both villages, residents do not consider it possible to "own" the sea. Rather, they conceptualize these rights as being based on capital investment in setting up the anchors needed to farm seaweed in any given location. As one farmer put it, "the sea does not belong to us, but we have the right to use it to stay alive." These histories of negotiation over rights to use the sea have led to a situation in which the current "owners" of seaweed farming plots are mostly those who claimed parts of the sea early in the establishment of the seaweed farming industry when seaweed prices were much lower. These sea use rights become much more valuable when seaweed prices are high.

The general features presented in this section indicate that both villages have long (>2 decades) histories of reliance on commercial aquaculture but different patterns of engagement in different industries. Laikang has a longer history of marine seaweed farming and more large, specialized farmers, while Pitu Sunggu has lower participation in land-based crops, instead being more specialized in pond farming of fish and shrimp. The next section examines differences in seaweed production methods in these two villages.

12.3.2 Production Methods

12.3.2.1 Longline Systems

All farmers in both villages use the longline production method, which is suitable for deeper waters (Fig. 12.5).

Fig. 12.5 A longline seaweed farm in Pitu Sunggu with insert: individual seaweed longlines

This method involves tying seaweed cuttings to ropes (typically 25 m in length in Pitu Sunggu and 40 m in length in Laikang). These ropes are hung in the ocean at approximately 1 m spacing and are floated to the surface most often using plastic bottles. The ropes are attached to permanent anchors on the ocean floor.

A seaweed farming cycle goes through planting, growth, and harvest stages. During the planting stage, small seaweed cuttings are attached to the ropes using smaller ties. Plastic bottles are also attached to the ropes at regular intervals, and these ropes are then placed in the ocean and attached to permanent anchors on the ocean floor, which define the permanent location of individual farming plots. During the growth cycle, seaweed is left in the ocean for a period of time that varies by location, season, farmer, and species. We received reports of growing times as short as 20 days (for spinosum) and up to 60 days (for cottoni and sacol). Seaweed is typically harvested younger if it is to be used for seeding a new crop, while it may be left longer if it is to be dried or if required by the ocean conditions.

During the growth period, the seaweed might be cleaned of epiphytic algae if it is formed by ocean conditions, and farmers may add additional floats to the seaweed lines to maintain buoyancy as seaweed grows and becomes heavier. After this time, the seaweed is harvested by bringing in the longlines with the seaweed and bottles still attached. The seaweed is removed from the longlines by running the line between two pieces of wood to break off the seaweed. Depending on the quality of the seaweed and the purchase of the seaweed, a portion of the seaweed may be recovered and used to seed the next farming cycle. The remaining portion is sun-dried and sold to local traders.

12.3.2.2 Seasonality

In the two case study villages, farmers reported similar seasonal patterns, although with some notable differences. Laikang farmers reported lower participation in seaweed production in the middle of the year (Fig. 12.6), which may reflect a more distinct seasonality of many of the Laikang plots and/or socioeconomic factors, such as the availability of marine space and of other income-generation activities. Farmers also work different lengths of time during different seasons. In Pitu Sunggu, farmers report working an average of 7.4 h per day and 6.2 days per week in the high season, compared to 4.3 h per day 4 days per week in the low season. In Laikang, farmers report working on average 7.9 h per day and 5.5 days per week in the high season, compared to 3.8 h per day and 2.6 days per week in the low season.

It is notable that farmers do not all agree on seasonal patterns. Even within a single village, farmers cultivate seaweed in defined areas of the ocean that exhibit a wide range of growing conditions. Although February was widely considered the best month for seaweed farming, some farmers report it as the worst month, and others do not farm at all. The difference depends on the species cultivated (sacol or cottonii) and the conditions in the specific plots farmed by different farmers, as even plots in close proximity to each

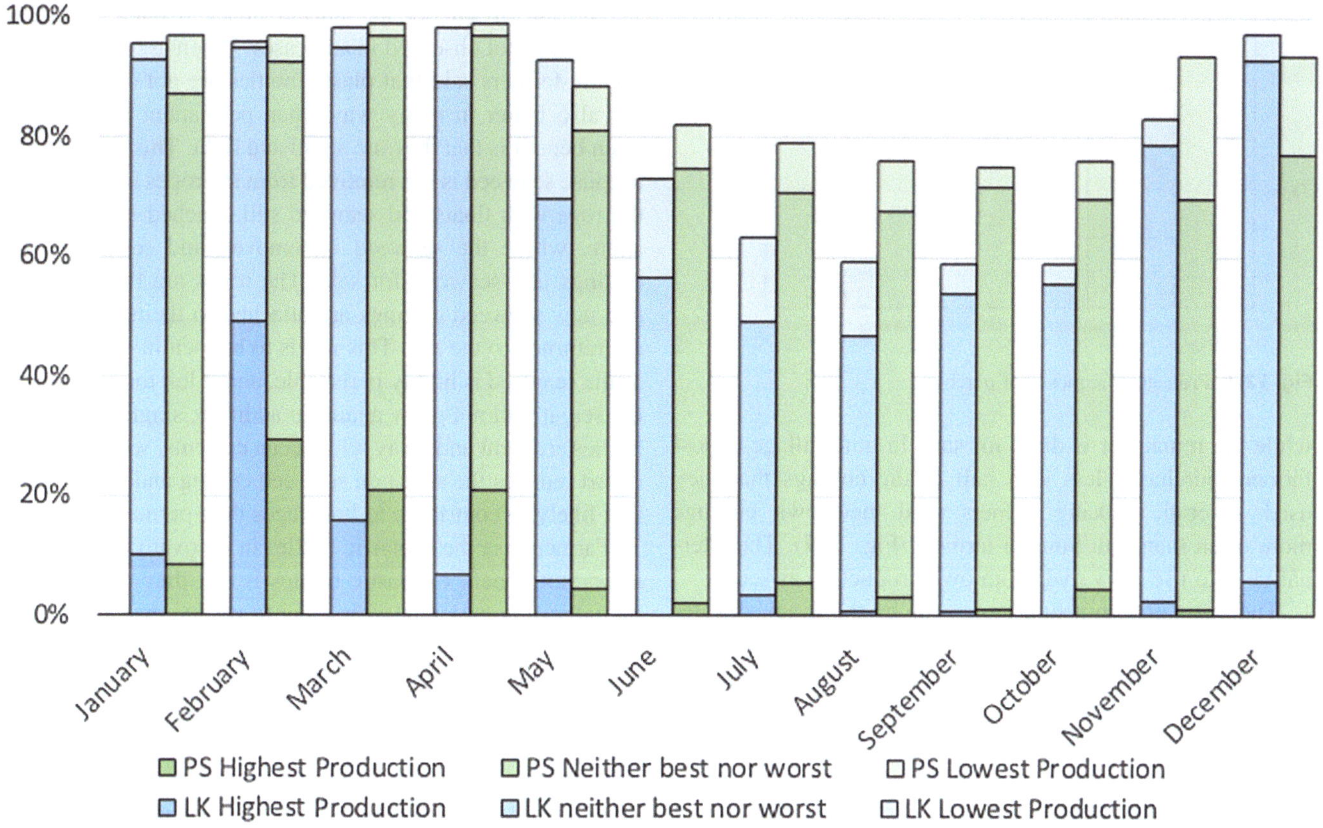

Fig. 12.6 Seasonality of production in Pitu Sunggu (PS) and Laikang (LK)

other can exhibit conditions suitable for the growth of different types of seaweed at different times in the year.

A large range of factors contribute to these differences in seaweed growth throughout the year, including rainfall and river water runoff (which produce small-scale differentials in ocean salinity and turbidity), light, temperature, nutrient levels, and pH (Azanza and Ask 2017). Human activities also affect oceanic conditions. For example, in Laikang, lobsters are fed fish in semicontained ocean nets, releasing organic matter into the surrounding area and contributing to high levels of epiphytic algae growth on nearby seaweed. As seaweed farms are in fixed locations, farmers have little control over these environmental and human activities affecting seaweed growth in their plots. Thus, the seasonality of seaweed farming varies by location, both between and within villages. This has a direct effect on seaweed characteristics such as growth rate and color, and the effect can be pronounced even over small areas. Seaweed is highly morphologically plastic, and seaweed grown in different areas with different oceanic conditions can have vastly different appearances (Simatupang et al. 2021), including different colors, stem thicknesses, and degrees of branching. Variation in seasonal conditions can be managed by cultivating different seaweed species at different times of the year. It is common for farmers in these locations to primarily cultivate cottonii in the wet season and transition to sacol in the dry season to allow year-round cultivation.

12.3.2.3 Propagation

The semiannual transition between seaweed species creates a need for the distribution of new seaweed propagules (often colloquially referred to as "seeds" and known as "bibit" in Indonesia) at the start of each planting season. These seeds are often sourced from different districts where ocean conditions are different. After the first planting cycle, propagules are more likely to be supplied by other farmers. These are typically grown in the ocean for less time than the regular planting cycle before they are harvested, divided, and used to seed new longlines. After farmers have produced sufficient seeds to fill their seaweed farming plots, they will grow seaweed for sale. After harvest, seaweed suitable for propagation will be selected to seed the next planting cycle,

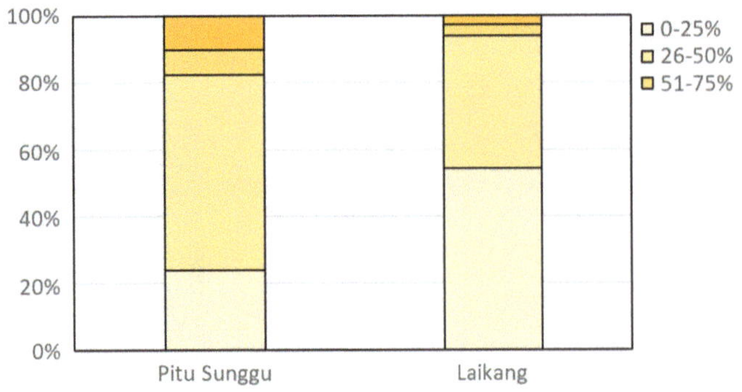

Fig. 12.7 Propagule frequency of purchase

while the remainder is dried for sale. In both villages, most farmers purchased less than half of the cuttings that they used. Overall, Laikang farmers used their own cuttings more often than Pitu Sunggu farmers (Fig. 12.7). The alternative is to use carry over from own production as seeds.

There is a government program run by the Brackishwater Aquaculture Development Centre in Takalar, which aims to produce improved seaweed propagules using tissue culture. The program is operational and distributes seeds to farmer groups around South Sulawesi; however, farmers in our study sites did not report receiving seeds from this channel. Tissue-cultured propagules are expensive to produce using current laboratory-scale methods but are reported to grow more quickly than mature propagules sourced from farmers. Lower-cost seed production could be achieved using spores; however, this method requires further research and development to become operational in South Sulawesi. Further research into intravillage propagule distribution patterns, growth rates, and selection practices could support more sustainable propagule value chains.

12.3.2.4 Floats

All farmers use disposable plastic water bottles to keep their seaweed lines afloat. Occasional use of Styrofoam was also common across both locations, used by 33% of farmers, as was the use of Jerry Cans, used by 31% of farmers. These larger floats are typically used only at the edges of farms to hold the ropes connected to the anchors, while plastic bottles are used at regular intervals along the ropes to keep the lines afloat. Buoys were used by just 7% of farmers, and those that used them indicated that they were used on less than 25% of their ropes. These floats are typically used only on the edges of the plots where longlines are attached to the anchors. Farmers were asked to indicate why they chose not to use permanent floats. The most common response given by 53% of farmers was that they are too expensive. Other common responses were that no one sells them, they easily break away and are lost, are heavy and difficult to use, and are likely to be stolen.

The use of smaller, permanent floats in place of plastic bottles was not observed in any cases. When asked about this option, farmers said that plastic bottles are not only cheaper but also better in many ways than permanent floats. Their main benefit is that they are small and light. This is important because seaweed is not removed from the ropes at sea; rather, the rope with floats and seaweed still attached is brought to shore, where the seaweed is removed and separated into cuttings and seaweed for sale. The ropes are then cleaned, and new seaweed cuttings are attached to them before they are returned to the sea. This needs to happen in a single day, as the seaweed is highly perishable, and if left too long out of the sea, it will not grow again. In addition, single-use plastic bottles are light and sway with ocean currents, which farmers report reduces the stress on seaweed cutting and makes them less likely to contribute to breakages than permanent floats.

Farmers use these plastic bottles in innovative ways. One important benefit of plastic bottles is that they can be filled partially or completely with water to reduce their buoyancy, allowing them to sink their seaweed lines if they wish to. This is an important feature after heavy rains when rivers flow into the sea. As fresh water is lighter than salt water, high rainfall leads to a layer of fresh water on the ocean surface that can destroy seaweed. Some farmers therefore travel to their seaweed farms and sink their seaweed lines after periods of heavy rain in an attempt to minimize the impact of freshwater on their crops.

However, the use of single-use plastic bottles poses risks to marine ecosystems. Approximately 83% of farmers reported that they last two seaweed farming cycles or less (approximately 6–13 weeks) before they need to be replaced due to breakage. These plastic bottles degrade in sensitive coastal marine ecosystems, which also serve as habitats for edible market fish. There may be opportunities for the development of alternatives; however, any technology aiming to replace single-use plastic bottles would need to be designed with careful consideration of the way that these bottles are used. Their prevalence reflects not only a lack of alternatives but also their suitability as a result of being light, low-cost, easily transportable, and containing a buoyancy adjustment mechanism.

12.3.2.5 Labor

Households reported similar patterns of involvement in different activities across both villages (Fig. 12.8). With regard to gender roles, in both villages, male household heads tended to perform the physically demanding ocean-based roles of maintaining, installing, and harvesting seaweed lines and to be involved in drying, bagging, and selling the seaweed. Female household heads reported working mainly on tying seaweed to lines, drying, bagging, and selling the seaweed in both villages. A notable difference between the villages was that men in Pitu Sunggu were more likely to be

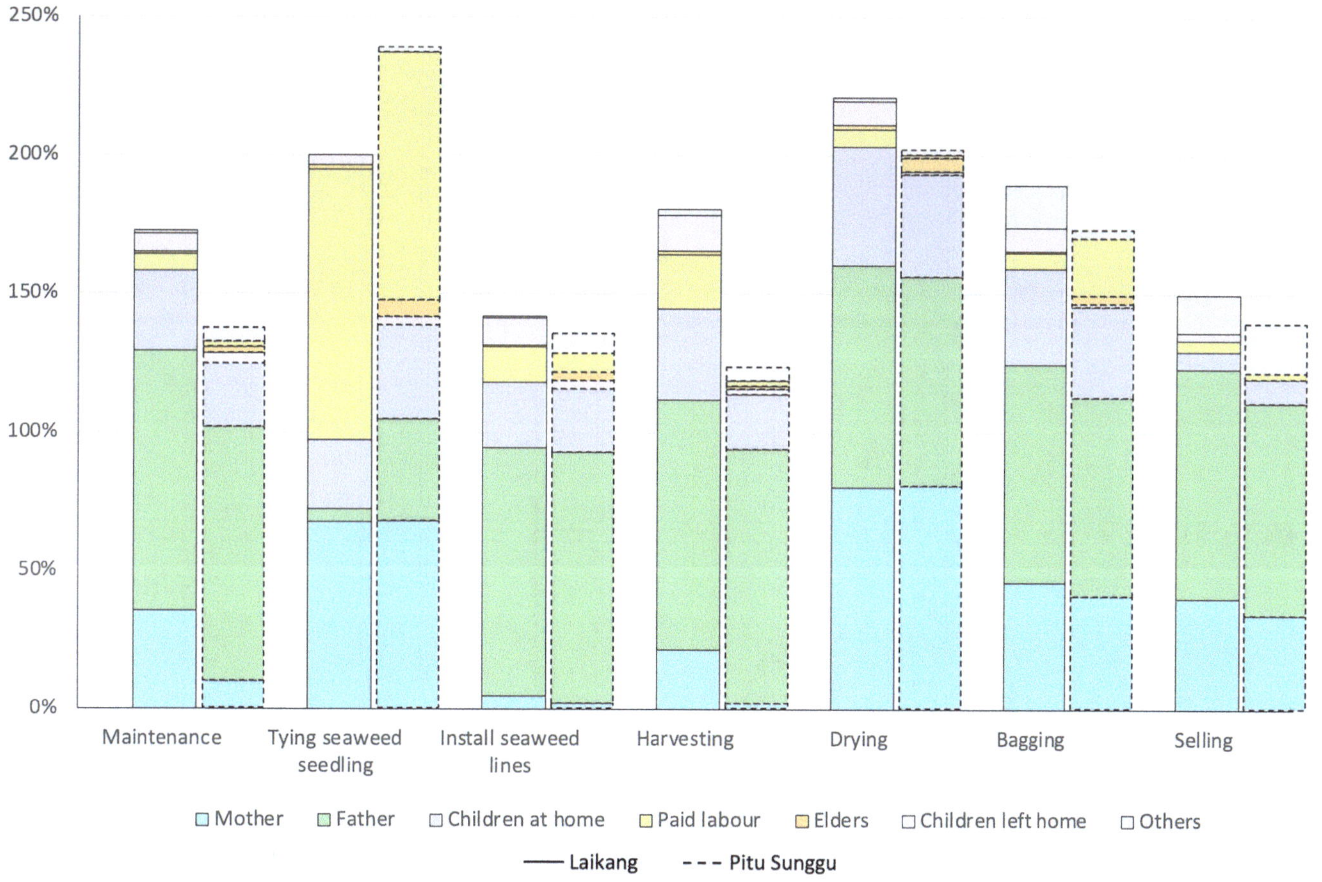

Fig. 12.8 The involvement of different family members in common seaweed jobs. Note that the y-axis shows the proportion of household members in each category reporting participation in these jobs and that these are stacked and therefore exceed 100%

involved in tying seaweed to lines (39% of male household heads did this work, compared to just 5% in Laikang). Women in Laikang were more likely to be involved in seaweed line maintenance and harvesting (36% of female household heads participated in seaweed maintenance and 21% in seaweed harvesting, compared to just 10% and 2%, respectively, in Pitu Sunggu). In both villages, respondents did not appear to consider gender roles to be rigid; rather, there were tendencies toward involvement in different types of work. Binding is a job that can be done while caring for young children, so it was more frequently done by women, while working on the ocean or in the ponds tended to be done by men.

The tying of seaweed cuttings to ropes is a particularly labor-intensive task, and the use of paid labor for this work is common. When an area of seaweed farm is harvested, the rope with the seaweed still attached is brought to shore, where the seaweed is removed from the lines, and some is selected for replanting if suitable. This seaweed is cut into smaller pieces and then tied back onto the longlines using small ropes. This seaweed binding work is very time-consuming—one binder can typically tie seaweed to one longline of approximately 25 m in approximately 1 hour. A group of five people working for 6 h could therefore be expected to complete just 30 longlines. Since farmers need the work completed in a single day so that the seaweed can be returned to the sea before it perishes, they often require the assistance of a team of five or so seaweed binders to complete the work within a day. This means that binders are often hired from outside the household and work in groups, which reportedly also makes the work more enjoyable.

In busy periods, it is common for binders to work long hours, virtually every day of the week, rotating between tying different farmers' crops. At the start of 2022, high rainfall led to crop losses of seaweed, and as a result, many farmers had to harvest and replant their seaweed, which increased the demand for binding labor. During periods of very high demand for labor and high seaweed prices, binders may negotiate higher prices. Some binders report that these prices are maintained even after demand has dropped. However, access to labor is dependent not only on price but also on relationships. Many farmers use their family relationships to

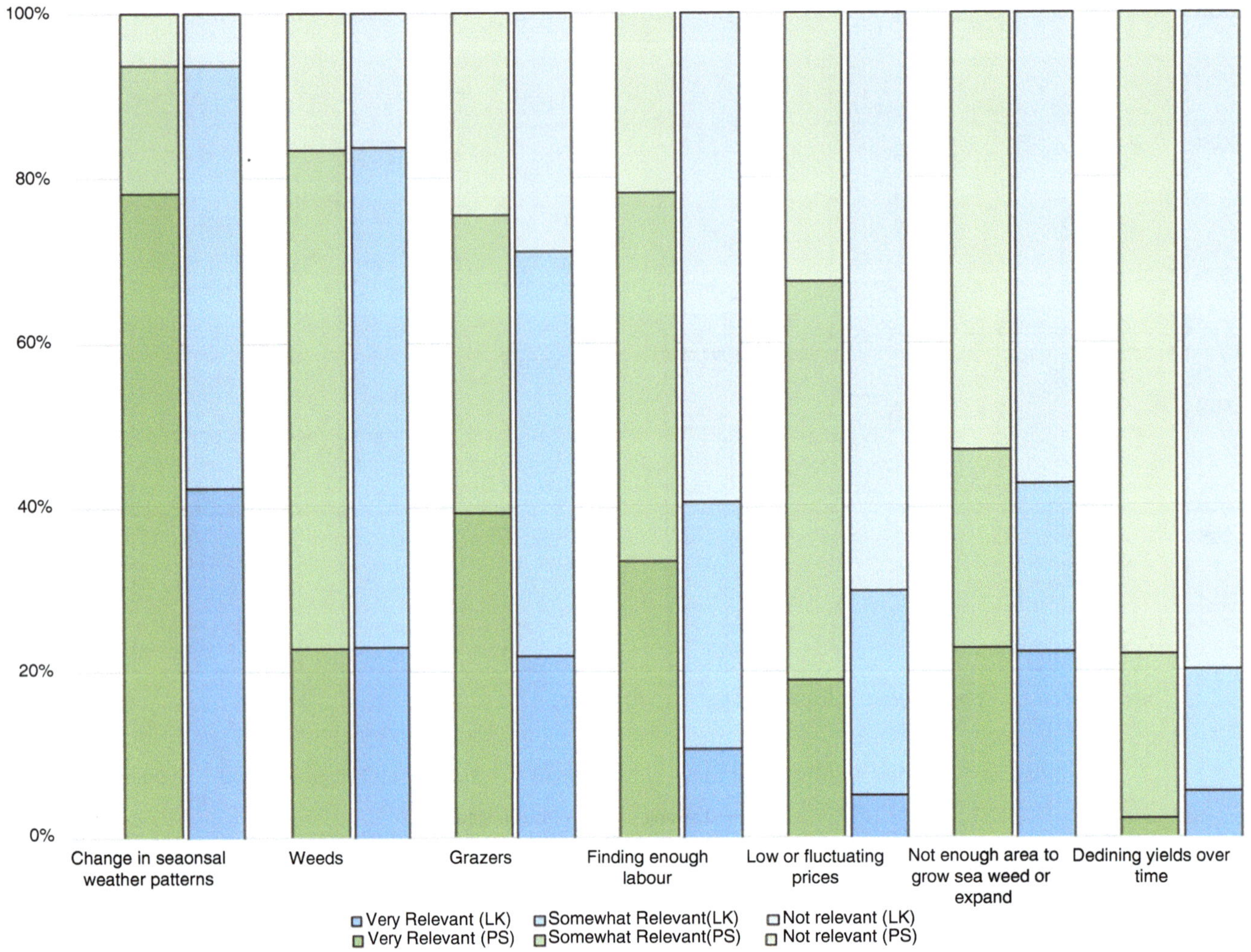

Fig. 12.9 Major production challenges faced by farmers

secure access to binding labor when needed. Some farmers book a group of binders for a few days or weeks during busy periods to ensure their availability.

Seaweed binding is therefore a common and important job in and close to seaweed farming communities, and binders tend to specialize in this work. Many of them do not have other sources of income. As one worker described, "This is my only job, nothing else. If I don't do this, I eat nothing." Binding work provides income to some members of the village who may not otherwise have work, including widows, disabled persons, and people who do not have capital or land. It can also be undertaken by women while they are caring for young children. The work requires no capital or special skill sets. However, it requires binders to sit in the same position for long hours, and binders commonly report mild to severe skin irritation from the work, which some avoid by wearing gloves.

12.3.2.6 Production Risks and Geographic Diversification

Farmers report a range of problems in seaweed production, including highly seasonal and variable weather patterns, weeds, grazers such as fish, binding labor shortages, low and fluctuating prices, a lack of sea space, and declining yields over time (Fig. 12.9). Of these, the most significant production problem reported by farmers is managing around weather patterns, primarily rainfall. During periods of very high rainfall, environmental stress causes seaweed to turn white and die. Known as an "ice-ice" syndrome, it appears to be caused by environmental stressors such as salinity changes, temperature, or sedimentation of the water (Ward et al. 2021).

At the start of 2022, farmers experienced several very high rainfall events that led to extensive seaweed losses. In Pitu Sunggu, after 3 days of rain, one farmer reported that he had visited his seaweed site where he had deployed approximately 600 longlines just a week before to find the stems white and rotting on half of these lines. The farmer's wife estimated the value of the loss at 10 million rupiah ($1000) and noted that they had still not paid the binders who had worked on the ropes. One very large seaweed farmer with 1200 seaweed ropes that were 1 month old reported losing approximately 50 million rupiah ($5000). In Pitu Sunggu, one farmer reported two losses in January. He had been forced to harvest the seaweed after only 10 days because it had been ruined by the fresh water. In the earlier freshwater event, he had lost 200 ropes that he had planned to use as seeds in the following cycle.

Farmers manage this risk in a variety of ways. One is by responding to events to protect the seaweed by bringing the seaweed in quickly after a rain. As one farmer described, "If the seaweed has been hit by a flood, it must be brought in the same day. If left too late, the farmer will have only the rope, without its contents." This means that after a large rainfall event, farmers are very busy going back and forth to the sea, as they need to quickly retrieve hundreds of stretches of seaweed rope. They work from morning to night at this job. Another strategy to reduce losses from high rainfall is to sink the seaweed below the freshwater layer on the top of the ocean by filling the plastic floats with water. This can be a useful approach in some cases, but if submerged for too long, the seaweed will become covered in barnacles and other shellfish that are difficult to remove.

Another common risk reduction strategy used in both villages is for farmers with multiple plots to geographically diversify their plots. By farming seaweed in a range of different locations, which are likely to be differently affected by weather patterns such as high or low rainfall or temperature, farmers manage their exposure to environmental risks.

12.3.2.7 Technology Uptake

Of the 273 farmers surveyed in the two villages, 118 reported implementing at least one new management technology in the last 5 years. The technologies considered most effective were the use of different species in different seasons (18% of all farmers) and the use of the "double rope" technique (reported by 25% of farmers). Notable technologies perceived to be ineffective were the use of bamboo rafts/cages for seaweed cultivation (reported by 4% of farmers). Follow-up interviews indicate that farmers found this technique much more expensive than the longline method, and some farmers indicated that they felt the cages hinder the growth of the seaweed by restricting the flow of ocean currents around it.

Most farmers reported getting information about seaweed production techniques and innovations from traders (88% of Pitu Sunggu farmers and 72% of Laikang farmers) or other farmers (49% of Pitu Sunggu farmers and 73% of Laikang farmers). Just five farmers reported that government extension services were a key channel of information, although it may be the case that a significant amount of information reported as coming from other farmers may have originally derived and spread from extension workshops. Internet groups such as the Facebook group "Rumput Laut Centre" provide a public space for sharing information on prices, propagule purchases, and seaweed sales, as well as to discuss important issues. Although only 3% of farmers report using the Internet to access information, it is likely that Internet-based information has been spread through word of mouth through personal contact with a wider cohort of farmers. One farmer described the importance of this word of mouth to technology uptake: "the farmer learns by eye. Even though they could listen all day to explanations about the potential of ... [a new production technique or technology], if they hadn't seen other farmers earning a lot, they wouldn't have done it. Except those who have money and are willing to bet."

12.3.3 Marketing

12.3.3.1 Postharvest Handing

The first step in the postharvest handling of seaweed is drying. Households dry seaweed after harvest using a range of methods and to varying degrees. This is a critical postharvest activity, as wet seaweed is highly perishable and difficult to handle (Ali et al. 2017). As traders and processors for traditional processes (carrageenan production) are only interested in purchasing dry seaweed, the purchase (by weight) of seaweed with high moisture content imposes costs in yield, handling, and transport. Traders are often required to redry seaweed to lower moisture contents at various stages. As a result, traders attempt to estimate the moisture content of the seaweed—along with other nonseaweed content such as salt, sand, and soil—and adjust offer prices accordingly.

Moisture content is assessed visually by eye using techniques such as squeezing the seaweed, breaking it open to see if the stem is dry the whole way through, or observing the extent of salt crystallization on the dried product. For example, one trader reported extensive salt crystallization in their experience, indicating moisture contents close to 35%, while a glossy surface and an absence of salt crystallization indicated moisture contents of 40–45%. Traders spend the majority of their time working with dried seaweed and thus become quite attuned to subtle differences in appearances that indicate different moisture contents. This is important because there are substantial risks associated with purchasing seaweed with a high moisture content, as this product will shrink and effectively reduce in volume. Managing moisture contents and prices is therefore a core aspect of risk management in the supply chain (Mulyati and Geldermann 2017; Langford et al. 2023a). Traders often manage the risk associated with uncertainty in the quality and consistency of the products they buy by offering discounted prices for high moisture content seaweed; however, offering higher prices for seaweed exceeding moisture standards is not common. Farmers have a price incentive to use adequate postharvest handling methods to avoid receiving a discounted price but not to exceed quality requirements.

Farmers dry their seaweed in a variety of locations. The drying apparatus favored by most farmers is para-para or bamboo platforms. The platforms allow water to drip off through the spaces in the bamboo floor while the seaweed dries in the sun. A total of 66% of farmers reported that they dried their seaweed at least partially on a platform. Other options for seaweed drying are on the pier or roadside (in Pitu Sunggu) or on nets near the beach (in Laikang). Laikang farmers appear to make greater use of drying platforms and benefit from a long sandy coastline over which a large number of drying platforms are built. This coastal space allows farmers to dry seaweed directly after bringing it in from the sea. In contrast, in Pitu Sunggu, the coastline is a mangrove environment, and there are few drying platforms constructed along the shore. As a result, most drying is confined to a small space on the pier or is moved inland between houses and brackish water ponds. In both locations, it is common for farmers to dry their seaweed on drying platforms for only 2 days before transferring it to the ground to finish the drying process, making space for new seaweed on the platform. Drying on the ground can also be quicker due to higher thermal conductivity. It does, however, increase the incidence of sand or soil contamination.

During the wet season, rainfall is frequent, and almost all farmers (99%) protect their seaweed from fresh water by covering it with a tarp. Failure to do so will reduce the seaweed quality and in extreme cases result in loss of the crop. In Laikang, 30% of farmers reported losing a crop in the last 5 years, and this had happened on average three times. In Pitu Sunggu, these rates were almost double, with 62% of farmers reporting a loss, and this happening on average 2.9 times. These rates are relatively low, indicating that the system for drying seaweed works adequately. However, this drying system produces products of varying quality as measured by salt and dirt content. Alternative drying methods producing cleaner, drier seaweed are available and known to farmers (Stone et al. 2023), especially the hang-drying process where lines with seaweed still attached are hung directly for several days until the seaweed is dry. However, farmers did not view this method favorably for three reasons. First, they did not want their seaweed to be too dry, as they felt they would not receive a higher price per kilogram for dryer seaweed. In this case, the incentives are to sell seaweed with a moderate moisture level to maximize weight and overall revenues per harvest. Second, farmers report that hang drying requires longlines to be in use for drying for several days, preventing them from using them for the next cycle and thereby creating redundancies in their equipment. This was a major downfall of the method, as ropes were considered to be a considerable capital expenditure. Third, farmers report that hang drying is time-consuming since the dried seaweed is reportedly more difficult to remove from the rope than wet seaweed, and plastic bottles often must also be tied and untied. It is possible that these difficulties are a result of suboptimal use of the hang-drying method; however, they are nonetheless key reasons for the persistence of other methods of drying. Farmers are also aware of other drying technologies, such as various types of solar and electric ovens, but consider these prohibitively expensive and time-consuming. As a result, they had little interest in alternative drying technologies and considered the combination of bamboo drying platforms followed by ground drying to be ideal.

12.3.3.2 Product Attributes

When asked about perceptions on how product attributes such as moisture and sand or dirt content affected the prices, farmers in Pitu Sunggu were more likely to respond that these factors did not affect their price at all (34% for dirt and 27% for moisture). Less than 7% considered the impact to be significant (Fig. 12.10). In Laikang, however, 23% thought the impact was significant, and just 4% thought that moisture content would not affect their price. These perceptions of price incentives translate into different postharvest handling methods, both between villages and individuals. As a more established and much larger producing region, there are more traders in Laikang that buy more seaweed. There are also larger and more capitalized farmers that can be expected to have better postharvest infrastructure and practices and more incentive to take up price-grade differentials if offered by traders. In Pitu Sunggu, there is a similar diversity in drying practices; however, there are a larger number of farmers who have limited access to drying space. This may mean that Laikang traders have greater scope to offer price-grade differentials.

Other product attributes are also seen to affect the price of seaweed paid by local traders. Of the farmers in Pitu Sunggu and Laikang, 43% and 45%, respectively, responded that quality indicators such as color and age (indicated by thallus thickness and degree of branching) affected the price of seaweed. Of these, 10% of farmers in Laikang and 2% in Pitu Sunggu indicated that these quality attributes would affect the price significantly. An important determinant of the "quality" of seaweed (defined as carrageenan content) is the length of the growing cycle. A longer cycle of 45 days or more would be expected to have higher carrageenan content than a 30-day cycle. However, traders may not be able to accurately measure this content or reward the sellers through prices. Other physical attributes of the seaweed are somewhat out of the control of the farmers, as they are formed by prevailing oceanic conditions. In practice, traders conduct only a general assessment of seaweed quality, do not or

Fig. 12.10 Farmer perceptions of the effect of product attributes on prices

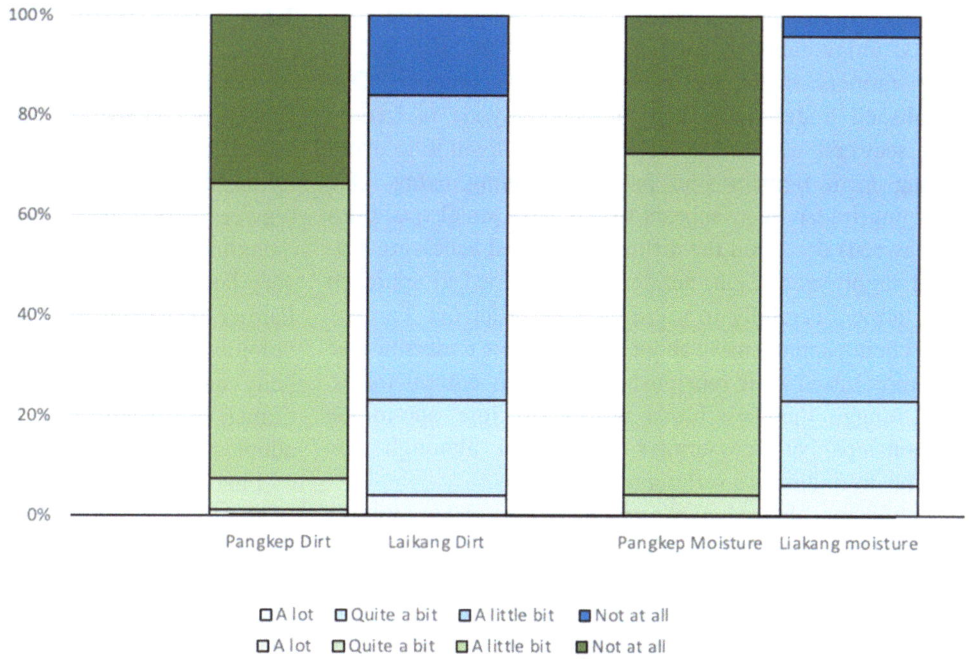

cannot assess other characteristics (such as carrageenan content, viscosity, and gel strength) through a highly differentiated grading system, and with high demand for seaweed, as is currently the case, are only likely to reject very unsuitable seaweed.

These findings support the conclusion of a price analysis (Langford et al. 2022), which found that while the Indonesian seaweed market shows a high level of spatial integration, it shows low levels of transmission of price-grade differentials between farmers and traders. Given the intrinsic difficulties of developing a marketing system to enable accurate pricing, the main form of "quality control" used by traders is to source seaweed from regions known for having good oceanic conditions. Outside of these areas, price incentives for improved postharvest handling appear to be limited.

12.3.3.3 Farmer–Trader Relations

After farmers produce and prepare seaweed, they sell it to a hierarchy of traders, from local to regional levels. The vast majority of farmers sell to a local trader located within their village or subvillage. Most farmers sell to just one trader—74% of Pitu Sunggu farmers and 51% of Laikang farmers. In Pitu Sunggu, the farmers that sell to more than one trader sell to 2–4 different traders, while Laikang farmers reported selling to up to 15 different traders. This may partially reflect differences in the operation of seaweed trading markets in these two villages, with a greater number of traders operating in Laikang and a much larger catchment area for traders. There are a number of small traders in Laikang who then sell to larger traders within the same village. Farmers select the traders they sell to on several criteria. Trust was an important feature in both villages (43% of Pitu Sunggu farmers and 50% of Laikang farmers) and was linked to the fact that the trader lived locally and was known to them. However, farmers reported that the most important factor in their choice of trader was existing credit relations, which drove trader choice for 88% of Pitu Sunggu farmers and 50% of Laikang farmers.

Seaweed farmers require access to credit for a number of reasons. Most commonly, they require credit to buy seeds at the start of the planting season. They may also need to replace ropes or repair drying platforms or boats. The vast majority of seaweed farmers report taking out loans (88% of Pitu Sunggu farmers and 72% of Laikang farmers). In both locations, a loan from a trader obliges a farmer to sell to that trader while they have outstanding debt. However, lending norms vary between villages. While neither location formally charges interest or has fixed maturity dates, in Laikang, farmers with outstanding debt are paid prices reduced by 1000–3000 Rp/kg (approximately 3–9%), whereas in Pitu Sunggu, farmers with debt are paid the same prices as those without debt. This price reduction does not reduce the loan principle but rather is a fee taken in lieu of interest. It has led to substantially lower satisfaction with traders in Laikang than in Pitu Sunggu. In Pitu Sunggu, traders therefore appear to be more interested in capturing supply, whereas in Laikang, traders appear to be more likely to seek to earn additional profits from the provision of financial services.

There are three established traders operating in Pitu Sunggu, and these traders report a sense of loyalty to each other, in that they refuse to buy seaweed from farmers who they know have an outstanding debt to another farmer—as one trader put it, they "don't interfere with each other." During the 2021-22 period of price increases, three new traders commenced operation in Pitu Sunggu, and was viewed unfavorably by existing traders, as the new traders offer farmers higher prices and capture product. One established trader noted that although anyone is free to trade seaweed, new traders have a moral obligation to avoid buying from farmers who have outstanding debts. One emerging trader (who appears to buy very small quantities of seaweed) described the difficulty he faced accessing seaweed supply as a result, noting that he needed to approach new farmers carefully to avoid creating problems. Farmers report being apprehensive about selling to new traders in case it compromised their relationship with their current trader. Pitu Sunggu therefore has a history of close one-to-one relationships between farmers and traders, although new entrants to trading are seeking to disrupt this.

In Laikang, almost half of seaweed farmers sell to multiple different traders, and many sell to a large number of traders. The system of trader credit provision is different and more formalized in Laikang, where prices paid to indebted farmers are typically reduced by a substantial amount (~1000 Rp/kg). Some Laikang farmers seek to avoid the more formalized and costly nature of the arrangement by accessing bank loans (bank loans had been used by 34% of Laikang farmers and 21% of Pitu Sunggu farmers). One Laikang farmer described how they began by borrowing money from a trader for inputs but became frustrated quickly with the system and now seek to save money from previous cycles to use as capital to start a new cycle—as they put it, "better to use my own money rather than rent from others, [then we are] free to sell our seaweed to anyone with the better price." Another farmer reported that rather than borrowing from traders, he would access credit from a pawnshop, using gold jewelry as capital. He preferred this system, as he did not pay interest on the loan and was free to sell to any trader and thereby access better prices.

These different farmer-trader relations—along with the larger size of Laikang village—lead to very different market configurations. In Laikang, 66% of farmers said that price was a key factor affecting their selling decision. The proportion was only 11% in Pitu Sunggu, possibly reflecting obligations to traders. In Laikang, a few large farmers report that they are able to negotiate prices with traders because they have a strong reputation for producing high-quality seaweed. In addition, farmers may be able to negotiate higher prices if they do not have debt and/or are able to sell a large quantity of seaweed in one transaction, and some farmers try to accumulate large quantities before selling to improve their bargaining power. However, for the majority of (smaller) farmers, prices are dictated by traders, and their seaweed trader is preselected by existing credit relations. They maintain autonomy over the timing of sales and are able to store seaweed and wait for higher prices; however, they are typically not in a position to negotiate price.

12.4 Discussion and Conclusion

As this description of seaweed livelihoods shows, seaweed farmers are diverse. Their activities differ between locations, where ocean conditions, capital constraints, and the availability of other on- and off-farm work affect the way they farm seaweed. They are also extremely diverse within villages, including very small farmers, moderately sized farmers, and very large farmers, with different extents of specialization. Farmer extension work, policy recommendations, and technological interventions need to be cognizant of this heterogeneity and consider innovative ways to support seaweed farmer livelihoods that are locally appropriate and easily adopted. Four key findings emerge from this study.

First, seaweed farming villages vary in the diversification of their livelihood activities and, as a result, have different exposures to the risk of changes in the industry. The two villages described here have similar production methods but differ in important ways. In Laikang, seaweed farming has been established for more than two decades, and during this time, some farmers have developed very large, well-organized farming operations. In contrast, in Pitu Sunggu, seaweed farming has mostly commenced within the last 15 years, and although there are some larger farmers, they are not as large as those in Laikang. In Pitu Sunggu, there is a very strong reliance on incomes earned from commercial aquaculture of seaweed and of shrimp and fish in land-based ponds, with little crop production. Laikang, on the other hand, has significant production of food crops such as rice and corn and has a more diversified marine aquaculture sector that also includes lobster grow-out and *Caulerpa* production. In Laikang, seaweed is therefore part of a more diverse array of livelihood activities, while in Pitu Sunggu, farmers are more specialized. Despite the differences, most seaweed farmers in both locations reported that seaweed accounted for the majority of their household income. The diversity of livelihoods therefore differs between the two villages, although in both cases, seaweed farming households tend to be relatively specialized. This is important because it means that different villages have different risk exposures to changes in the seaweed industry, which need to be recognized when considering policy responses to industry changes such as price volatility.

Second, seaweed farmers are highly exposed to risks associated with weather patterns and other changes to oceanic conditions. In both villages, farmers report that they face considerable challenges managing weather-related risk and variability. Ninety-four percent of farmers in both villages reported that managing weather is a somewhat or very relevant challenge, with 78% of Pitu Sunggu farmers and 42% of Laikang farmers reporting it as "very relevant." This makes weather patterns the most frequently reported issue reported by farmers, and as described above, farmers typically go to great lengths to manage the risks of rainfall or changes in ocean conditions. They have developed innovative ways to manage risk, such as by geographically diversifying their seaweed farming plots. They also respond to severe weather events quickly, either by bringing their seaweed in from the ocean and replanting it or by adjusting the buoyancy of their floats to sink the seaweed below the freshwater layer on the surface. Despite this, managing often unpredictable and shifting ocean conditions remains a significant challenge to farmers. Ocean conditions such as turbidity and organic matter concentration can vary greatly even across small areas. Farmers are highly exposed to changes in oceanic conditions that are often out of their control, and as a result, many of the quality attributes of the seaweed they produce are predetermined by the location of their farming plots. This makes it difficult to consistently supply higher value seaweed into value chains where specific quality attributes are demanded.

The third key point relates to the distribution of economic benefits of seaweed farming within a village. Since areas of marine space suitable for seaweed farming in both locations are widely considered to be "full," seaweed farmers with existing rights in the sea are able to benefit from rising seaweed prices, while those without are not. Other users of the sea who have not transitioned into seaweed farming, such as fishermen and crab fishermen, are negatively affected by seaweed farming since the individualization of rights to sea space effectively reduces the remaining public area in which they are allowed to fish and catch crabs. It is worth noting, however, that these effects are not clear-cut, since many fishermen are, or have at one time been, seaweed farmers. In addition, most of the current seaweed farmers were previously fishermen, and many still are. There is therefore substantial overlap in these livelihoods, and different individuals have been differently affected by the changes brought by the development of seaweed farming in coastal villages. They have been able to take advantage of the opportunities brought by seaweed farming to different extents. The extent to which seaweed farming brings economic benefits to nonseaweed farmers is not well studied (but see Langford 2024) – seaweed farmers are only a moderately sized group within a coastal village and are a small minority within a regency, and as such, although seaweed brings high incomes to these farmers, these incomes are not widely distributed at a regional scale. There is some distribution of seaweed industry income within seaweed farming villages and, in some cases, further inland, as seaweed farmers employ teams of laborers to tie seaweed propagules to ropes. The distribution of seaweed income outside of seaweed farming households was not specifically addressed in this study but is an important area for future research. The seaweed industry is widely promoted for its poverty-reducing potential; understanding the way that incomes are distributed is important for understanding the extent and limits of these benefits.

Finally, this study reveals important insights into seaweed marketing chains. Seaweed farmers sell their product into global value chains through layers of local and regional traders. We have elsewhere reported that seaweed prices do not appear to be correlated with product attributes such as moisture and dirt content (Langford et al. 2022). This survey confirms this macrolevel analysis in these two villages, where a majority of farmers reported that they did not think that the moisture and dirt content of their seaweed would significantly affect the price. This was much more common in Pitu Sunggu, where 27% of farmers did not think that a high moisture content would affect the price they received at all. Relatedly, farmers report that price was not a major factor in their choice of buyer, which is determined more by existing debt relations with particular traders. This effect was much greater in Pitu Sunggu, where 89% of farmers reported that price is not a key factor affecting their decision of where to sell their seaweed, compared to 34% in Laikang. Access to credit is therefore a major factor affecting the way that seaweed is marketed in these villages. As financial technology platforms and other microloan facilities expand in emerging markets globally, it will be interesting to see whether access to this type of credit changes the way that these seaweed marketing chains operate, and there may be a role for government extension programs or policy to support the entrance of such platforms into these areas—noting of course that the terms offered by these loans also vary in their favorability to farmers.

These four key observations highlight the need for efforts to upgrade seaweed value chains to take into account farmer diversity both within and between villages. Seaweed farmers are a diverse group, including large, innovating, and entrepreneurial farmers with access to capital who are able to expand their growing region, invest in capital improvements such as drying platforms, and negotiate good prices for their seaweed. It also includes a large number of small farmers who face considerable risks and are incorporated into seaweed value chains through relations of indebtedness that reduce the price received for their product. The seaweed value chain also has the input of large amounts of labor from nonseaweed farming households, which are essential to the harvesting and replanting process. This diverse range of farmers operating in locations with different oceanic conditions produce seaweed of vastly different product attributes and quality indicators, and these products are incorporated into value-adding processes in different ways. Efforts to improve seaweed production and quality and the livelihoods of coastal communities need to be aware of this diversity to design policy and extension programs that benefit seaweed farming communities as a whole.

References

Ali KMA, Fudholi A, Sulaiman J, Muthuvalu MS, Ruslan MH, Yasir SM, Hurtado AQ (2017) PostHarvest handling of Eucheumatoid seaweeds. In: Hurtado AQ, Critchley AT, Neish IC (eds) Tropical seaweed farming trends, problems and opportunities: focus on *Kappaphycus* and *Eucheuma* of commerce. Springer, Cham

Aslan LOM, Iba W, Bolu LOR, Ingram BA, Gooley GJ, de Silva SS (2015) Mariculture in SE Sulawesi, Indonesia: culture practices and the socio economic aspects of the major commodities. Ocean Coastal Manage 116:44–57

Aslan LOM, Supendy R, Taridala SAA, Hafid H, Sifatu WO, Sailan Z, Niampe L (2018) Income of seaweed farming households: a case study from Lemo of Indonesia. IOP conference series earth and environmental. Science 75(012221). https://doi.org/10.1088/1755-1315/175/1/012221

Azanza RV, Ask E (2017) Reproductive biology and eco-physiology of farmed *Kappaphycus* and *Eucheuma*. In: Hurtado AQ, Critchley AT, Neish IC (eds) Tropical seaweed farming trends, problems and opportunities: focus on *Kappaphycus* and *Eucheuma* of commerce. Springer, Cham

BPS (2022a) Hasil Survei Komoditas Perikanan Potensi Rumput Laut 2021 Seri - 2. Badan Pusat Statistic. Available at https://www.bps.go.id/publication/2022/08/29/269de33babc6e3d52bbae5b6/hasil-survei-komoditas-perikanan-potensi-rumput-laut-2021-seri-2.html

BPS (2022b) Statistik Indonesia 2022. Available at www.bps.go.id/publication/2022/02/25/0a2afea4fab72a5d052cb315/statistik-indonesia-2022.html

de La Torre-Castro M, Fröcklin S, Börjesson S, Okupnik J, Jiddawi NS (2017) Gender analysis for better coastal management – increasing our understanding of social-ecological seascapes. Mar Policy 83:62–74. https://doi.org/10.1016/j.marpol.2017.05.015

FAO (Food and Agriculture Organisation of the United Nations) (2023) FishStatJ fao.org/fishery/en/statistics/software/fishstatj

Komarek A, Cahyadi ER, Zhang J, Fariyanti A, Julianto B, Arsyi R, Lapong I, Langford A, Waldron S, Grist M (2023) Increasing incomes in carrageenan seaweed value chains in Takalar, South Sulawesi. Australia-Indonesia Centre. https://pair.australiaindonesiacentre.org/research/increasing-incomes-in-carrageenan-seaweed-value-chains-in-takalar-south-sulawesi/

Langford A, Waldron S, Sulfahri SH (2021) Monitoring the COVID-19-affected Indonesian seaweed industry using remote sensing data. Mar Policy 127(104431):1–10. https://doi.org/10.1016/j.marpol.2021.104431

Langford A, Zhang J, Waldron S, Julianto B, Siradjuddin I, Neish I, Nuryartono N (2022) Price analysis of the Indonesian carrageenan seaweed industry. Aquaculture 550(737828):1–13. https://doi.org/10.1016/j.aquaculture.2021.737828

Langford A, Turupadang W, Waldron S (2023a) Interventionist industry policy to support local value adding: evidence from the eastern Indonesia seaweed industry. Mar Policy 151(105561). https://doi.org/10.1016/j.marpol.2023.105561

Langford A, Waldron S, Nuryartono N, Pasaribu S, Julianto B, Siradjuddin I, Ruhon R, Walyandra Z, Lapong I, Arsyi R (2023b) Sustainable upgrading of the South Sulawesi seaweed industry. Australia Indonesia Centre. https://pair.australiaindonesiacentre.org/research/sustainable-upgrading-of-the-south-sulawesi-seaweed-industry-2/

Langford Z (2024) Globalisation and livelihood transformations in the Indonesian seaweed industry. Routledge, New York

Larson S, Stoeckl N, Fachry ME, Mustafa MD, Lapong I, Purnomo AH, Rimmer MA, Paul NA (2020) Women's well-being and household benefits from seaweed farming in Indonesia. Aquaculture 530(735711)

Mariño M, Breckwoldt A, Teichberg M, Kase A, Reuter H (2019) Livelihood aspects of seaweed farming in Rote Island, Indonesia. Mar Policy 107(103600). https://doi.org/10.1016/j.marpol.2019.103600

Ministry of Marine Affairs and Fisheries Republic of Indonesia (2019) Marine and fisheries figure 2018, Data Information Centre

Mulyati H, Geldermann J (2017) Managing risks in the Indonesian seaweed supply chain. Clean Technol Environ Policy 19:175–189. https://doi.org/10.1007/s10098-016-1219-7

Neish IC (2013) Social and economic dimensions of carrageenan seaweed farming in Indonesia. In: Valderrama D, Cai J, Hishamunda N, Ridler N (eds) Social and economic dimensions of carrageenan seaweed farming, fisheries and aquaculture technical paper no. 580, 204. FAO, Rome, pp 61–90

Neish IC (2015) A diagnostic analysis of seaweed value chains in Sumanep regency, Madura Indonesia. In: report submitted for UNIDO project no 140140 'preparatory assistance to develop a seaweed value chain development project for Madura

Nuryartono N, Waldron S, Tarman K, Siregar U, Pasaribu SH, Langford A, Farid M, Sulfahri (2021) An analysis of the South Sulawesi seaweed industry. The Australia-Indonesia Centre. https://pair.australiaindonesiacentre.org/research/commodities/pilot-project-report-an-analysis-of-the-south-sulawesi-seaweed-industry/

Permani R, Muflikh Y, Nuryartono N, Waldron S, Langford A, Pasaribu S, Sjahruddin F (2023) The policy landscape and supply cahin governance of the Indonesian seaweed industry: a focus on South Sulawesi. Australia-Indonesia Centre. https://pair.australiaindonesiacentre.org/research/the-policy-landscape-and-supply-chain-governance-of-the-indonesian-seaweed-industry-a-focus-on-south-sulawesi/

Rimmer MA, Larson S, Lapong I, Purnomo AH, Pong-Masak PR, Swanepoel L, Paul NA (2021) Seaweed aquaculture in Indonesia contributes to social and economic aspects of livelihoods and community wellbeing. Sustainability 13(10946). https://doi.org/10.3390/su131910946

Salayo ND, Perez ML, Garces LR, Pido MD (2012) Mariculture development and livelihood diversification in The Philippines. Mar Policy 36(4):867–881. https://doi.org/10.1016/j.marpol.2011.12.003

Simatupang NF, Pong-Masak PR, Ratnawati P, Agusman PNA, Rimmer MA (2021) Growth and product quality of the seaweed *Kappaphycus alvarezii* from different farming locations in Indonesia. Aquacult Rep 20(100685):100685. https://doi.org/10.1016/j.aqrep.2021.100685

Stone S, Langford A, Arsyi R, Lapong I, Zach Z, Ruhon R, Julianto B, Siradjudding I, Wong A, Waldron S (2023) Technology adoption by smallholder farmers: the case of drying technology in the Indonesian seaweed industry. J Agribusiness Developing Emerg Economies. https://doi.org/10.1108/JADEE-01-2023-0011

Suyo JGB, Masson VL, Shaxson L, Luhan MRJ, Hurtado AQ (2020) A social network analysis of the Philippine seaweed farming industry: unravelling the web. Mar Policy 118(104007):104007. https://doi.org/10.1016/j.marpol.2020.104007

Ward GM, Kambey CSB, Faisan JP, Tan PL, Daumich CC, Matoju I, Stentiford GD, Bass D, Lim PE, Brodie J, Poong SW (2021) Ice-ice disease: an environmentally and microbiologically driven syndrome in tropical seaweed aquaculture. Rev Aquac. https://doi.org/10.1111/raq.12606

Zhang J, Waldron S, Langford A, Julianto B, Komarek A (2023) China's growing influence in the global carrageenan industry and implications for Indonesia. J Appl Phycology. https://doi.org/10.1007/s10811-023-03004-0

Commercial Farming of *Kappaphycus alvarezii* in Sri Lanka: Current Developments and Opportunity for Becoming a Major Carrageenophyte Producer

13

Shanmugam Munisamy, B. Nirooparaj, and J. M. Asoka

Abstract

The current annual world production of *Kappaphycus alvarezii* is approximately 200,000 metric tonnes, and its value-added product carrageenan is approximately 60,000 t. *Kappaphycus alvarezii* has been farmed successfully since 2012 in Sri Lanka. Over 250 fisher folks from 3 districts, viz., Mannar, Kilinochi, and Jaffna, are currently involved in farming for their livelihood. Floating bamboo rafts and monolines, the two popular cultivation methods, were adapted, and the percentage average daily growth rate (ADGR%) of *K. alvarezii* over 45 days with 150 g of seed propagule across different locations with respect to seasons was 3.50 ± 0.50. The average monthly production of dry seaweed of a cultivator was 450–550 kg with a monthly income of 150–225 USD. The quality of dry seaweed, such as moisture (35 ± 3.0%), clean anhydrous weed (CAW 38 ± 5.0%), soluble salt (28 ± 3.0%), and impurities (< 1.0%), was comparable to the quality of commercial material. The yield of SRC was 30 ± 3.0%, whereas that of refined carrageen was 25 ± 2.5%. The average potassium and water gel strengths of the SRC were 600–950 g cm^{-2} and 250–450 g cm^{-2}, respectively, and the milk gel strength was 90–150 g cm^{-2}. *Kappaphycus alvarezii* farming has been introduced to over 25 countries over the last 35 years. However, only five to seven countries produce commercial volumes (over 1000 t/year) for the carrageenan industry today. Sri Lanka has recently been added to the list of countries farming *K. alvarezii*, and it is estimated that it has the potential to produce over 25,000 t dry weed per year; therefore, there is a huge opportunity to develop carrageenan industries in Sri Lanka. The current status of farming of *K. alvarezii*, scope for expansion, and opportunity to develop carrageenan industries in Sri Lanka are discussed.

Keywords

Seaweed · Carrageenophyte · *Kappaphycus alvarezii* · Growth rate · Raw seaweed quality · Carrageenan · Sri Lanka · Bioeconomy

13.1 Introduction

Seaweed farming is a supplementary activity to existing fishing and its ancillary activities. It could translate into an additional income of approximately Rs. 30,000 (450 USD) per month based on circumstances (personal communication from Dr. Eswaran). It can provide a regular and sustainable income source by leveraging local entrepreneurial talent. One of the major socioeconomic benefits of aqua-agriculture apart from the continued generation of employment opportunities is that it provides employment opportunities for women. Much of the work is currently done by women, thereby providing them with a source of income. Seaweed farming is an environmentally friendly activity that does not involve the use of chemicals, pesticides or any inorganic inputs. In fact, raft cultivation does not hamper access to fishing lanes but may contribute to a more structured fishing industry in coastal communities. In India, the cultivation of *K. alvarezii* started in 2000, and the progress made is summarized by Abhiram and Shanmugam (2016), Vaibhav et al. (2017) and elsewhere in this book.

Kappaphycus is one of the major sources of kappa carrageenan. Due to an increased demand for carrageenan, farming of *Kappaphycus* was first started in the Philippines and later expanded to Indonesia and to more than two dozen other tropical, countries, but at present, only a few countries are

S. Munisamy (✉)
Department of Biomarine Resource Valorisation, Division of Food Production and Society, Norwegian Institute of Bioeconomy Research (NIBIO) Torggarden, Bodo, Norway

Research and Development Division, AquAgri Processing Private Limited (DSIR Lab), Manamadurai, Tamil Nadu, India
e-mail: shanmugam.munisamy@nibio.no

B. Nirooparaj · J. M. Asoka
National Aquatic Resources Research and Development Agency, Colombo, Sri Lanka

A. T. Critchley et al. (eds.), *Tropical Phyconomy Coalition Development*, Developments in Applied Phycology 11,
https://doi.org/10.1007/978-3-031-47806-2_13

growing seaweed successfully. In 1969, world seaweed production was 2.2 million metric tonnes fresh weight, which was derived equally from wild collection and cultivation (Cai 2021). By 2019, the total production volume had increased to 35.8 million metric tonnes, with cultivation accounting for 97% of the production, whereas wild production remained at 1.1 million metric tonnes. The world's geographical area for the cultivation of *Kappaphycus* lies within ±10° latitude range, particularly from the Southeast Asian countries extending to East Africa and Brazil. However, the Southeast Asian region, primarily the Brunei Darussalam–Indonesia–Malaysia–Philippines East Association of Southeast Asian Nations Growth Area, has by far the greatest potential for expanded tropical seaweed cultivation, comprising 60% of the world's production sites. In particular, Indonesia, Malaysia, and the Philippines provide sheltered areas that are favorable for cultivation. Indonesia and the Philippines are the major producers of *Kappaphycus*, and they contribute more than 90% of the total annual production (Neish 2008).

The germplasm of *K. alvarezii* was imported to Sri Lanka from Malaysia in 2011, and commercial farming was successfully established in selected locations (Shanmugam et al. 2017). There was a slowdown in *Kappaphycus* farming for 2 years (2018 and 2019); however, it is now recognized as a major livelihood program and supported by the government of Sri Lanka in terms of investment and regulatory structure. Its Fisheries Department, the National Aquatic Resources Research and Development Agency (NAQDA), provided funding of Rs 75 lakhs LKR (=23,312 UD$) as a subsidy to 60 seaweed growers within the Jaffna, Kilinochi, and Mannar districts and issued necessary permission/licenses to carry out the farming of *K. alvarezii*. Large-scale cultivation was started in June 2012 at Valaipadu and Nainativu, and commercial farming was subsequently established during 2014 and 2015 (Shanmugam et al. 2017). The present study describes the results obtained over the last decade (2012–2022) and the potential of Sri Lanka for expanding seaweed farming and the production of raw *K. alvarezii* for processing.

13.2 Materials and Methods

13.2.1 Cultivation Methods and Farming Operations

Off-bottom monoline and floating bamboo raft methods are being employed based on sea conditions. The size of the monoline plot is 30 m × 20 m, i.e., 600 m of culture line, equivalent to ten rafts. These are constructed using *Casuarina* and bamboo poles, 1.5 m in length with a diameter of 5–7.5 cm; polypropylene rope (3 mm) is securely tied to the stakes at a distance of 0.5 m from the bottom. Approximately 200 g of seedling material is inserted into a loop with 20-cm spacing along the rope. The entire plot was fenced with fishing nets to protect it from grazing (Shanmugam et al. 2017).

In the floating bamboo raft method, a 3 m × 3 m raft is made with four bamboo poles of 3.6 m each, tied together with four diagonal poles in a square shape, and fishing net fixed underneath to avoid fish grazing. Each raft had 20 monolines of polypropylene rope (3 mm), and each line consisted of 15 seedlings weighing approximately 200 g spaced every 15 cm, such that each raft was planted with 60 kg of seedling material. Once the raft was seeded, it was installed at the farming site and anchored with stones. A cluster of 5–10 rafts was held together by one anchor depending on the seabed structure and wave motion characteristics. The entire biomass obtained in a raft, after 45 days, was taken as the final weight, and the average daily growth rate percentage (ADGR%) was calculated.

The farming operation included selection of good seedlings, preparation of seedlings, farm maintenance such as fixing up loose knots, removal of ocean debris from the farm, including that of other seaweeds, relocation of the farm if plants experienced poor growth or grazing, and replantation. Processing included harvesting, drying, and packing of dry seaweed. Procurement of produce is done every week, and payment is made through bank accounts of cultivators.

13.2.2 Average Daily Growth Rate

The average daily growth rate percentage (ADGR%) of *K. alvarezii* was calculated using the following formula of Dawes et al. (1994):

$$\%\text{daily growth} = \frac{\ln\ (\text{Mf}) - \ln\ (\text{Mi}) \times 100}{\text{No.of days}}$$

where Mf = final weight and Mi = initial weight.

The ADGR% was calculated based on the weight obtained at 45 days in a 3 m × 3 m raft (60 m lines) assuming 60 kg as the initial seedling weight.

13.2.3 Analysis of Dry Seaweed

To assess seaweed quality, moisture content, soluble salts (SS), clean anhydrous seaweed (CAW), and impurities were estimated as described by Shanmugam et al. (2017). The moisture content (%) was determined using 50 g of dry seaweed cut to a 25–50 mm length that was oven-dried at 85 ± 2 °C for 16 h. The CAW was estimated by evenly mixing 50 g of sample with 1 L of water in a beaker for 30 min. The washed material was oven-dried at 85 ± 2 °C for

Table 13.1 Status of production *Kappaphycus alvarezii* during the last one decades (2012–2022) in Sri Lanka

Year	Total farming area (raft no.)	No. of beneficiaries	Seaweed production (fresh ton)
2012	Trial	15	55
2013	Trial	15	80
2014	600	18	120
2015	3300	35	1180
2016	3450	40	1470
2017	2770	65	510
2018	500	15	Little[a]
2019	550	15	Little[a]
2020	1400	65	560
2021	3650	120	1600
2022	2700	180	2350

[a]Commercial activity was reduced during 2018 and 2019, and it was revived in 2020

16 h. Impurities were measured by filtering out the washed liquid left post-washing through a pre-weighed muslin cloth (150 mesh). The soluble salt (SS) content was determined by subtracting the sum of CAW, moisture, and impurities from the total sample taken for analysis, and the C/s ratio was calculated as the ratio between CAW and SS.

13.2.4 Manufacturing of Semi-refined Carrageenan

A preparation of a commercial batch of semirefined carrageenan (SRC) was carried out at the facility of AquAgri Processing Private Limited, India. Dry seaweeds produced using both the floating bamboo raft and monoline culture methods from different locations were exported from Sri Lanka to India for processing as described by Shanmugam et al. (2017). The batch size of SRC was 1000 kg, and it was prewashed with plain water processed with 8% KOH at 80 ± 2 °C for 3 h. The KOH-treated material was then washed with water at a 1:6 ratio (w/v) three times to remove the excess KOH, chopped, and sun-dried. The dried SRC chips were ground and sieved with 40, 60, 80, 100, 150, and 200 mesh (180 ASTM) for further quality testing.

13.2.5 Analysis of SRC

The SRC powder (60–80 mesh) was incinerated for 5 h at 525 °C, and the total ash content was estimated (AOAC, 1995). Ester sulfate and insoluble and ash content were determined by the methods as described by Shanmugam et al. (2017). Viscosity was measured at 1.5% in water at 75 °C in 30 rpm, and potassium gel strength was determined with 1.5% SRC in 0.2% KCl solution using a Brookfield Texture Analyzer (Model CT3 4500), and milk gel strengths were measured at 1.5 and 0.2% SRC in water and water plus 10% milk powder, respectively. The 3,6-anhydrogalactose content was estimated by using the phenol-resorcinol method with fructose as the standard (Moses et al. 2015). The microbial profile of SRC samples was carried out according to the procedures of Cruchaga et al. (2001). The FT-IR spectra of SRC samples were analyzed in KBr pellets using FT-IR spectrophotometry (Perkin-Elmer Spectrophotometer GX).

13.2.6 Statistical Analysis

Data were analyzed using analysis of variance (ANOVA; SYSTAT version 7). Correlation and regression were used to analyze the ADGR% data and environmental parameters. Data were considered statistically significant when $p < 0.05$.

13.3 Results

Table 13.1 shows the number of rafts operated, total seaweed farmers involved in this activity, and dry seaweed produced over the last 10 years (2012–2022). During the first two years, a number of trials were carried out to optimize the culture method to generate attractive income for growers. In 2015 and 2016, over 3500 rafts were operated that produced 118 and 147 t dry weight of seaweed, respectively. There were 180 seaweed farmers engaged in this activity, and this number is expected to grow (Table 13.1). Current farming of *K. alvarezii* is being carried out in 27 villages belonging to three districts, i.e., Jaffna, Mannar, and Kilinochi in the northern provinces (Fig. 13.1; Table 13.2). The farming operation, including seedling preparation, farm maintenance, harvest, and drying, is shown in Fig. 13.2. The ADGR% obtained ranged between 3.31 - 4.08%, 3.46 - 4.16%, and 3.38 - 4.18% in Jaffna, Mannar, and Kilinochi districts, respectively. *Kappaphycus alvarezii* is generally brown in color, but in the peak summer season, it became yellowish-brown, and it was also noted that it developed green and red colors in the monsoon season (Fig. 13.3). The plant was also

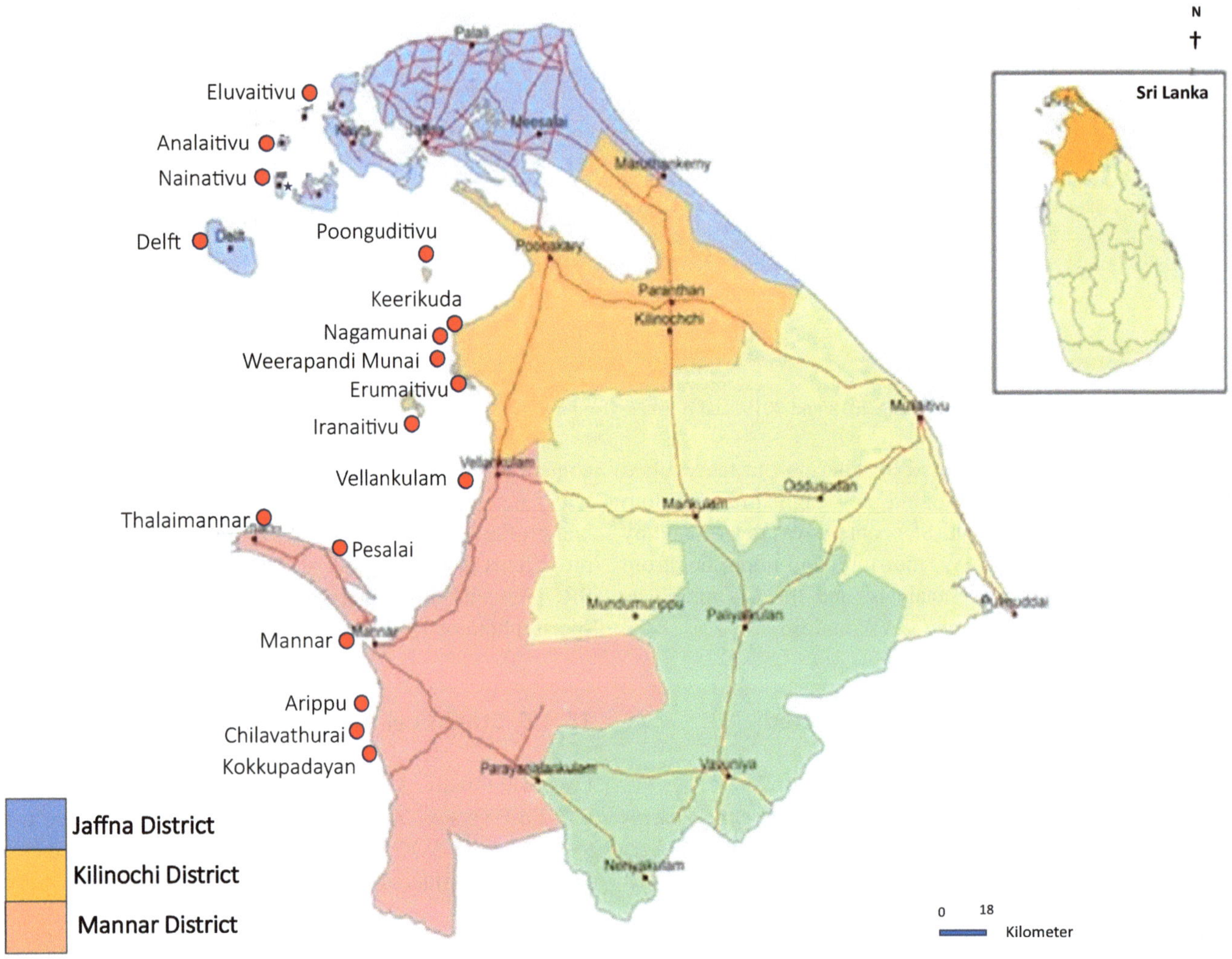

Fig. 13.1 Map showing the locations of *K. alvarezii* farming in northern province of Sri Lanka

affected by ice-ice disease and silting during the peak rainy season (Fig. 13.4). Considerable development was made in the farming of *K. alvarezii* until 2017, and after a break of 2 years, i.e., In 2018 and 2019, further progress was made in farming in terms of the number of seaweed farmers and the volume of seaweed production (Table 13.1). As reported earlier, the ADGR% was higher during the monsoon season but lower in the summer months due to temperature stress (Shanmugam et al. 2017). This was negatively correlated with seawater temperature ($p = -550$, $r = 0.05$), which reached 33 °C during summer, and positively correlated with water motion ($p = 850$, $r = 0.051$). These observations are in agreement with literature reports (Neish, 2008). The average monthly production of dry seaweed per person during 2015 and 2016 was 500–700 kg, and it increased to 600–800 kg during 2021 and 2022, resulting in an income that ranged between 185 and 240 USD per month.

The quality parameters we used to assess raw dry seaweed (RDS) were moisture content, CAW, soluble salt, and impurities. The moisture content of the RDS ranged between 27.47 ± 3.19 and 38.08 ± 3.88 in 2015 and was 30.42 ± 1.71 in 2015. The CAW content was higher in the RDS produced during the summer season of 2014 (Apr 40.48 ± 3.74% and May 42.44 ± 3.66%) compared to the monsoon period (Nov 36.90 ± 4.17% and Dec 34.56 ± 4.07%). Similarly, the CAW content in the summer of 2015 (Apr 42.50 ± 3.06% and May 39.10 ± 4.53%) was higher than in the monsoon period (Nov 36.82 ± 3.10% and Dec 35.84 ± 2.91%). The soluble salt content of RDS produced in both 2014 and 2015 was similar, but a higher C/s ratio was found in the RDS produced during summer months in both 2014 and 2015, and a similar trend was observed in the quality of dry weed produced until 2022 (Table 13.3).

The yield of SRC (60 mesh) manufactured at the commercial scale ranged between 32 - 33%. The water gel strength of SRC ranged from 360 - 485 gm cm^{-2} with a mean of 397 gm cm^{-2} (SD 46.66); the KCl gel strength of SRC ranged from 775 - 880 gm cm^{-2} with a mean of 802 ± 69 gm cm^{-2},

Table 13.2 Average daily growth rate and potential for expansion in different villages in northern province in Sri Lanka

Village	Total rafts operated (2017)	Total fresh produced in 2017 (t)	ADGR %	Total raft potential (no.)	Dry weed production capacity (t)	Potential beneficiary nos.
Jaffna District						
Allapatty	60	8.7	3.67	5000	725	111
Mandativu	55	7.7	3.89	8000	1120	178
Chatty Beach	45	9	3.62	4500	900	100
Velanai	50	9.85	3.57	1000	197	22
Karainagar	30	5.7	4.08	5500	1045	122
Bumbuhar	30	5.88	4.00	5000	980	111
Eluvaitivu	90	18.54	3.46	12,000	2472	267
Nainativu	135	29.03	4.00	15,000	3225.00	333
Analitivu	80	17.6	4.01	1000	220	22
Melingimunai	45	7.965	3.31	2000	354	44
Mannar District						
Iluppakadvai	60	9.48	3.46	3500	553	78
Viduthalaitivu	45	7.47	3.65	3500	581	78
Pesalai	30	5.04	4.04	10,000	2000	222
Nachikuda	30	4.65	3.79	5000	775	111
Thalaimannar	90	17.82	4.16	5000	990	111
Nadukuda	20	0.02		15,000	1665	111
Kilinochi District						
Kiranchi	60	11.82	3.79	4000	788	89
Valaipadu	1200	229.2	3.57	20,000	3820	444
Iranamatha Nagar	60	9.48	3.90	5000	790	111
Iranaithivu	135	25.38	3.48	12,000	2256.00	267
Pallikuda	90	15.75	3.69	3500	612.5	78
Erumaithivu	160	26.72	4.15	4500	751.50	100
Weerapandi Munai	50	7.97	3.38	3000	531.00	67
Nagaha Munai	30	0.11	3.38	3000	11.00	67
Keerikuda	30	5.64	3.38	3000	564.00	67
Ayyanar Kovil	30	6.24	3.49	5000	1040.00	111
Pakkattu Puli	30	6.51	4.18	5000	1085.00	111

whereas the milk gel strength ranged from 125 - 65 gm cm^{-2} with a mean of 139 ± 12.66 gm cm^{-2}. The viscosity of SRC ranged from 72 - 117 cP with a mean of 108 ± 17.8 cP. The water and potassium gel strengths of refined carrageenan (RC) produced from SRC were 25 – 40% more than their respective values for SRC (Table 13.4). There was acid insoluble material in SRC in the range of 0.11 – 0.80%, but it was not present in the RC. The sulfate content, 3,6-anhydrogalactose content, and pH of SRC and RC are reported in Table 13.4.

FT-IR of the refined κ-carrageenan extracted from the five samples was recorded and compared to the k-carrageenan manufactured from *K. alvarezii* cultivated on the southeast coast of India and Sri Lanka, and the FT-IR spectra were identical, which confirmed that the carrageenan extracted under the optimized conditions was consistent with κ-carrageenan (bands at 933 and 847 cm^{-1}). The sulfate bands in the C-4 position (galactopyranose) are shown at 845 – 850 cm^{-1}. Furthermore, a higher intensity of absorbance at 1250 cm^{-1} was observed for carrageenan extracted from *K. alvarezii* in this present study in comparison to commercial carrageenan, which is in agreement with the data of the sulfate content analysis (Table 13.5). All the carrageenan gel samples prepared in 1% KCl showed the

Fig. 13.2 Operation of *K. alvarezii* farming in Delft, Sri Lanka. (**a**) Seedling preparation by fisherwomen; (**b**) *K. alvarezii* farm with off-bottom monoline method; (**c**) harvested fresh material is shifted in a boat; (**d**) drying on the platform; (**e**) dry weed shifted to warehouse, (**f**) baled seaweed ready for export

Fig. 13.3 Different strains of *K. alvarezii* farmed in Delft during 2020. (**a**) Yellowish-brown strain; (**b**) green strains; (**c**) red strain; (**d**) brown strain

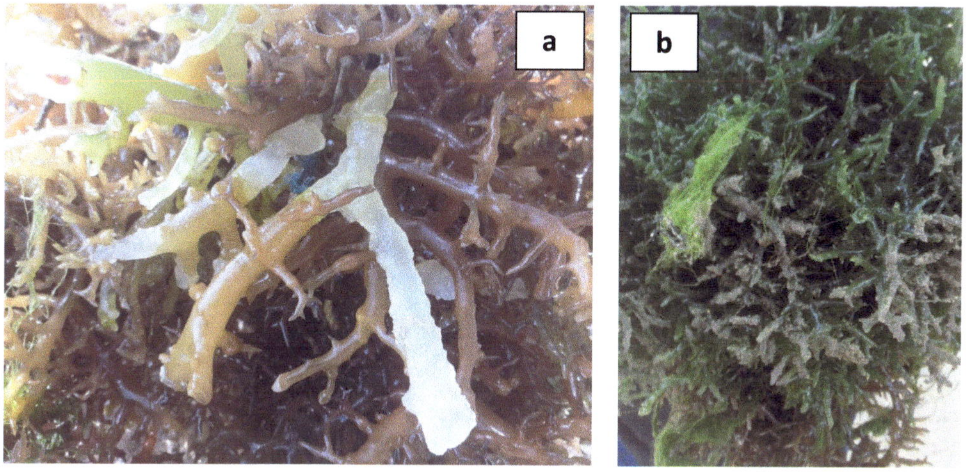

Fig. 13.4 (**a**) Ice-ice disease and (**b**) epiphyte attack and silting on the farmed *K. alvarezii* in Delft, Sri Lanka

Table 13.3 Quality of dry weed *Kappaphycus alvarezii* produced from Valaipadu village during 2016 and 2017

	Moisture %)	CAW (%)	Soluble salt (%)	Impurities (%)	C/S ratio
2016					
Jan	35.54 ± 3.20	37.58 ± 3.80	26.30 ± 1.46	0.58 ± 0.53	1.43 ± 0.33
Feb	36.07 ± 3.23	37.94 ± 3.51	22.87 ± 3.10	2.12 ± 0.71	1.36 ± 0.29
Mar	27.47 ± 3.19	38.26 ± 3.88	34.09 ± 2.78	0.18 ± 0.29	1.12 ± 0.26
Apr	33.34 ± 3.60	40.48 ± 3.74	25.58 ± 2.74	0.6 ± 0.40	1.58 ± 0.30
May	29.78 ± 2.66	42.44 ± 3.66	27.3 ± 1.71	0.48 ± 0.48	1.55 ± 0.42
Jun	31.2 ± 3.06	37.98 ± 3.60	30.44 ± 1.29	0.38 ± 0.41	1.25 ± 0.40
Jul	33.27 ± 3.48	38.98 ± 3.54	25.31 ± 1.65	0.44 ± 0.53	1.54 ± 0.34
Aug	34.52 ± 3.85	39.6 ± 3.49	25.68 ± 3.12	0.2 ± 0.60	1.54 ± 0.45
Sep	33.35 ± 3.91	40.42 ± 3.81	25.97 ± 3.47	0.26 ± 0.44	1.56 ± 0.27
Oct	35.06 ± 4.13	38.08 ± 3.94	24.36 ± 2.09	0.5 ± 0.35	1.44 ± 0.38
Nov	38.08 ± 3.88	36.90 ± 4.17	24.28 ± 2.70	0.74 ± 0.42	1.52 ± 0.41
Dec	35.81 ± 3.40	34.56 ± 4.07	29.33 ± 2.54	0.3 ± 0.40	1.18 ± 0.30
2017					
Jan	35.10 ± 1.50	36.42 ± 3.38	26.68 ± 2.25	1.80 ± 0.55	1.37 ± 0.39
Feb	35.11 ± 1.52	38.42 ± 3.80	24.53 ± 2.30	1.94 ± 0.49	1.57 ± 0.29
Mar	30.42 ± 1.71	37.84 ± 3.48	26.38 ± 1.75	0.70 ± 0.42	1.61 ± 0.45
Apr	34.92 ± 1.44	42.50 ± 3.06	25.26 ± 1.29	1.98 ± 0.70	1.50 ± 0.34
May	33.64 ± 2.15	39.10 ± 4.53	25.74 ± 1.71	1.52 ± 0.29	1.52 ± 0.27
Jun	34.30 ± 2.48	38.12 ± 3.07	25.20 ± 1.47	2.38 ± 0.41	1.51 ± 0.33
Jul	34.90 ± 1,66	38.50 ± 4.33	24.40 ± 2.29	2.20 ± 0.67	1.58 ± 0.40
Aug	34.48 ± 1.65	38.56 ± 3.59	25.78 ± 1.58	1.18 ± 0.75	1.50 ± 0.26
Sep	36.14 ± 1.99	38.22 ± 3.02	24.90 ± 2.06	0.74 ± 0.64	1.53 ± 0.29
Oct	32.59 ± 1.49	36.16 ± 4.18	30.31 ± 1.92	0.94 ± 0.65	1.19 ± 0.36
Nov	33.32 ± 2.14	36.82 ± 3.10	29.38 ± 1.88	0.48 ± 1.22	1.25 ± 0.31
Dec	33.76 ± 2.02	35.84 ± 2.91	26.36 ± 2.33	2.04 ± 1.05	1.44 ± 0.41

Table 13.4 Yield and quality of SRC and RC manufactured on commercial scale from *Kappaphycus. alvarezii*

	Yield (%)	Moisture content (%)	Water gel strength (g cm^{-2})	KCl gel strength (g cm^{-2})	Milk gel strength (g cm^{-2})	Viscosity (cP)	Acid insoluble matter (%)	Acid insoluble ash (%)	Sulfate (%)	3,6-AG (%)	pH (%)
Semirefined carrageenan											
Batch 1	32.45	7.15	375	880	125	72	14.2	0.42	19.7	26.1	10.6
Batch 2	33.55	6.50	360	810	140	117	15.5	0.60	20.2	27.5	10.0
Batch 3	32.11	5.95	375	660	165	115	14.8	0.66	20.5	26.1	10.2
Batch 5	33.40	6.02	350	870	135	94	13.1	0.71	19.4	27.9	10.5
Batch 6	34.42	5.25	485	840	145	108	14.6	0.58	21.2	28.5	9.8
Batch 7	33.00	7.30	385	780	140	126	14.0	0.80	19.8	28.6	9.8
Batch 8	32.72	6.65	450	775	125	125	13.6	0.60	18.8	27.2	9.9
Mean	33.09	6.40	397	802	139	108	14.2	0.62	19.9	27.4	10.1
STD	0.72	0.67	46.66	69.27	12.66	17.87	0.7	0.11	0.7	0.95	0.30
Refined carrageenan											
Batch 1	23.88	5.52	400	1085	130	33	21.8	0	20.5	30.0	9.4
Batch 2	24.75	4.3	460	1080	145	24	22.2	0	21.5	29.3	9.4
Batch 3	24.10	4.5	380	950	125	45	18.4	0	18.5	29.2	9.9
Batch 5	24.87	3.28	425	1010	170	18	21.7	0	19.8	30.3	8.8
Batch 6	24.00	3.45	430	995	155	25	22.5	0	19.9	28.8	8.6
Batch 7	25.38	4.12	410	1190	135	37	21.0	0	20.0	30.0	9.4
Batch 8	23.48	4.10	440	1135	145	35	19.3	0	20.6	29.5	9.5
Mean	24.35	4.18	421	1064	144	31	21.0	0	20.1	29.6	9.3
STD	0.62	0.68	24.56	77.72	14.32	8.50	1.44	0.00	0.85	0.49	0.41

Table 13.5 Physiochemical characterization of refined carrageenan extracted from semirefined carrageenan produced on a commercial scale

Input SRC	Carrageenan yield % ± SD	Gel strength (g/cm^2) ± SD	Zero shear viscosity (mPaS)	Sulfate content (%)
Batch 1	69.94 ± 1.94	634 ± 7	27,900	15.86
Batch 2	62.06 ± 1.10	718 ± 4	56,000	14.85
Batch 3	62.12 ± 0.96	492 ± 5	17,300	13.41
Batch 5	63.32 ± 1.19	462 ± 2	37,700	12.10
Batch 6	58.97 ± 1.00	503 ± 1	26,500	16.21
Batch 8	64.05 ± 0.50	624 ± 2	47,800	17.21

Table 13.6 Water and potassium gel strength of SRC with different particle size distribution

	Particle size distribution (PSD) of SRC											
	40		60		80		100		150		200	
Batch	Water GS	KCl GS	Water GS	KCl GS	Water GS	KCl GS	Water GS	KCl GS	Water GS	KCl GS	Water GS	KCl GS
Batch-1	350.1	770.7	355.2	675.2	316.0	614.3	284.4	583.5	252.8	552.8	237.0	460.7
Batch-2	330.5	640.5	333.4	662.5	326.3	622.3	295.3	588.1	292.1	563.0	259.4	469.8
Batch-3	360.7	652.0	356.2	680.3	324.0	656.9	290.9	635.9	261.8	596.4	229.4	497.9
Batch-4	360.2	705.5	372.8	815.1	346.0	805.2	326.9	786.7	295.8	746.4	261.2	569.3
Batch-5	345.5	702.2	398.0	788.6	366.2	762.0	357.7	738.4	332.1	707.9	284.5	593.6
Batch-6	440.0	840.2	456.5	905.5	406.3	832.6	370.5	815.0	358.3	773.5	313.6	640.3
Batch-7	465.8	862.0	455.8	915.5	389.3	855.5	363.3	826.4	351.6	792.2	312.6	629.7
Batch-8	395.4	781.4	425.4	888.4	399.5	824.1	377.9	792.7	361.9	751.5	322.0	586.0

typical pseudoplastic nature when allowed to flow under an applied shear rate from 10 - 1000 s^{-1}. However, the zero shear viscosity values of the RC obtained from SRC Batches 2 and 8 were higher than that of the carrageenan extracted from SRC of other batches with high gel strength (Table 13.5).

The gel strength of SRC powder with different particle size distributions (PSDs) and different potassium concentrations is given in Table 13.6. SRC powder with lower PSD, i.e., 40–60 mesh, exhibited a higher gel strength of 550 – 675 g cm^{-1}, whereas SRC with finer mesh particles (200 mesh) gave the lowest gel strength (230 – 360 g cm^{-1}) (Fig. 13.5).

13.4 Discussion

The ADGR% of *K. alvarezii* from commercial cultivation in the Philippines, Indonesia, Vietnam, Tanzania, and India are similar to those observed in Sri Lanka over the last 10 years. An ADGR% of 3.31–4.18% was recorded in Valaipadu village in 2016 in the present study, consistent with commercial ADGR% values reported in other countries. The ADGR % of *K. alvarezii* ranged from 2.3 - 4.2% in the Philippines (Hurtado et al., 2019), 2.5 - 3.5% in Indonesia (Adnan and Porse 1987), 3 - 4% in Madagascar (Mollin and Braud 1993), 3.5 - 3.7% in Fiji (Luxton et al. 1987), 1.7 - 6.8% in Zanzibar (Msuya 2013), 3.33 - 4.51% in India (Abhiram and Shanmugam 2016), 3.4 ± 0.3 in Malaysia (Yong et al. 2014), and 3.95 - 6.26% in Vietnam (Ohno et al. 1996). The quality of dry seaweed in terms of its moisture content, CAW, SS, and impurities was good, with a C/s ratio of 1.3, was comparable to that of commercial RDS from other regions. The moisture content of the RDS of *K. alvarezii* in the present study ranged from 27 - 38%, comparable to the moisture content of *K. alvarezii* produced in the Philippines (36 – 42%) or Indonesia (35 – 45%), the major global producers of *K. alvarezii*. The CAW contents reported in the literature are more than 35 (Darmawan et al. 2013), similar to our finding of 34 – 42% in the present work. Similarly, the C/s ratio of *K. alvarezii* produced in Sri Lanka was 1.3, consistent with that for *K. alvarezii* produced in the Philippines, which varies between 1.1 1.3, and Indonesia, which ranges from 0.9 - 1.2.

The yield of SRC manufactured on a commercial scale varied between 32 - 34%; this narrow range could be due to the composite mixture of the RDS used for processing, and this observation is similar to the reported literature (33.5%; Darmawan et al. 2013). The water and KCl gel strengths of

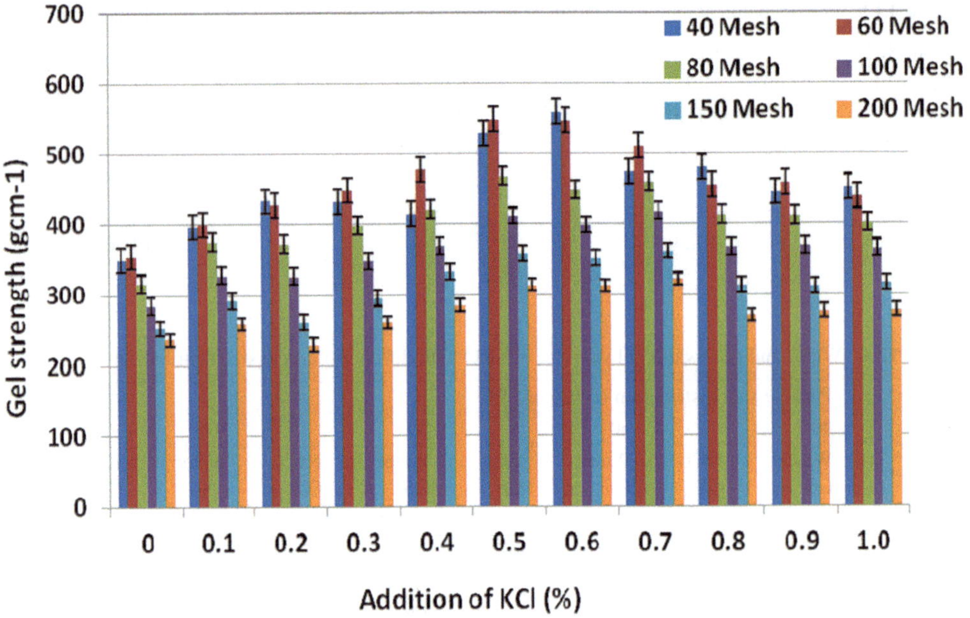

Fig. 13.5 Influence of potassium concentration and particle size (mesh) of semirefined carrageenan on the gel strength

SRC were 397 ± 46 gm cm^{-2} and 802 ± 69 gm cm^{-2}, respectively, and the values are comparable to those of SRC from the same seaweed farmed in the Philippines, Indonesia, Malaysia, and Indonesia. The gel strength of carrageenan is generally influenced by its particle size distribution and the concentration of the cations present in it (Wullandari et al. 2021). In the present study, SRC powder with a larger particle size distribution (40 – 60 mesh) formed a stronger gel strength (550 – 675 g cm^{-1}) than SRC with a finer PSD (200 mesh; 230 – 360 g cm^{-1}). Similarly, the optimum potassium concentration required to form a strong gel strength was 0.5 – 0.6%. The FT-IR spectrum of carrageenan (bands at 933 and 847 cm^{-1}); the sulfate bands in the C-4 position of galactopyranose are shown at 845 – 850 cm^{-1}. Furthermore, a higher intensity of absorbance at 1250 cm^{-1} was observed, suggesting that it was a kappa carrageenan variant (Moses et al. 2015). The total plate count, yeast, and molds of the SRC samples were within specification, and no pathogenic microbes such as *Shigella*, *Salmonella*, and *E. coli* were found in the SRC or carrageenan sample tested.

The total costs of the raft and off-bottom monoline methods are given in Tables 13.7 and 13.8. The total cost per raft was 4536 LKR (29.5 USD), and at 45 rafts per person, it is 204,186 LKR (1325.8 USD) for 12 persons; the total cost amounts to 2,449,710 LKR (15 847.5 USD). The total investment was approximately 0.49 USD/m of line. The Philippine and off-bottom Mexican systems were also comparable (approximately 1.00 USD/m). The most economical investment corresponded to the off-bottom system in Tanzania (0.15 USD/m), and the most expensive systems were found in Mexico (floating) and the Solomon Islands, at approximately 1.40 USD/m (Valderamma et al., 2015); the cost of the off-bottom method was approximately half that of the raft approach (Table 13.8).

13.4.1 Return on Investment (ROI) Economics of Farming

The total infrastructure cost for a group of 12 members was $16,206.98 USD. The net dry seaweed production in a month by the group of 12 members is 6450 kg; thus, the gross t income per month at 0.49 USD kg^{-1} was $3160.5. Therefore, a farmer could earn $263 USD per month (83,108 SLR) per month. The cost of operation and maintenance could be 10–15% of the total income. The ROI was 601.72% with 91.46% of annualized ROI in three years (Table 13.9). The economics of *K. alvarezii* farming are very attractive to farmers in Sri Lanka, consistent with findings for *K. alvarezii* farming in India (Vaibhav 2022).

The average monthly income of a seaweed farmer in the above scenario was US$ 280, and considering eight months of operation in a year, the total annual income per person was US$ 2240. In India, the monthly income of a cultivator ranged between US$ 180 and 200 during 2010 – 2012. The income of seaweed farmers has been reviewed by Valderrama et al. (2015), and those are comparable to the economics of farming *K. alvarezii* in the present study. The tropical seaweed genera *Kappaphycus* and *Eucheuma*, the major sources of carrageenan, are being promoted for expanded cultivation in topical waters (Hayashi et al. (2017). The developments made thus far in Sri Lanka, the

Table 13.7 Capital operation cost for establishment of *Kappaphycus alvarezii* farming to self-help groups (SHG) of 12 members through a floating bamboo raft method

Items	Unit	Qty Req / raft	Unit price (LKR)	Total price (LKR)	Total cost for 540 rafts for 12 members of self-help group @45 rafts /person (LKR)	Total cost for 540 rafts for 12 members of self-help group @45rafts /person (USD). 1USD = 154.58 LKR
(a) Capital cost						
Farming system for 60 m of culture line						
HDPE 1.5" fishing net	Kg	1	1500.00	1500.00	810,000.00	5240.01
1.5 mm HDPE rope	Roll	1	140.00	140.00	75,600.00	489.07
3 mm PP rope	Roll	1.5	255.00	382.50	206,550.00	1336.20
5 mm PP rope	Roll	0.5	650.00	325.00	175,500.00	1135.33
6 mm PP rope	Roll	0.1	750.00	75.00	40,500.00	262.00
10 mm PP rope	Roll	0.02	1800.00	36.00	19,440.00	125.76
12 ft. bamboos	No	6	200.00	1200.00	648,000.00	4192.00
Anchor	No	1/5 raft	1250.00	250.00	135,000.00	873.33
Initial seed cost at five rafts per person	Kg	66	8.00	528.00	31,680.00	204.94
Infrastructure transportation cost	Unit	1	4500.00	100.00	54,000.00	349.33
Total cost/raft of 60 m culture line	Unit			4536.50	2,196,270.00	14,207.98
Total cost for 45 rafts per person					204,142.50	1320.63
Total cost for a 12 members of a SHG					2,449,710.00	15,847.52
(b) Operational cost						
Storage facility @ 10 m × 5 m (4 nos.)					100000.00	646.91
Drying bed (4 nos @ 300sqm size)					120000.00	776.29
Cost of boat for material shifting					400000.00	2587.65
Maintenance of farm and infrastructure (10% of total infrastructure cost)					244971.00	158.47
Total operational cost					864971.00	5595.62
Grand total of capital and operational costs (a + b)					3314681.00	21443.14

potential to expand the farming volume, and availability of seaweed farmers (Table 13.2) are evidence that this island country has huge opportunities to become one of the major producers of seaweed *Kappaphycus alvarezii* in the world.

13.5 Conclusion

The growth trend, extent of farmable area, interest of the coastal communities in seaweed farming, their relative level of economics, as well as support of the government to promote the seaweed domain, and quality of raw seaweed and carrageenan observed in Sri Lanka, for the last decade, demonstrated that Sri Lanka has huge potential to become one of the major producers of high quality *Kappaphycus alvarezii* raw seaweed, for processing industries and provide a significant contribution to the global blue bioeconomy.

Acknowledgments The authors are grateful to Mr. Abhiram Seth and Mr. Tanmaye Seth, AquAgri Processing Private Limited, and the management of IFFCO, India, for the facility. The authors also express their gratitude to the National Aquatic Resource Development Agency (NAQDA) Sri Lanka for providing the required data. The authors are

Table 13.8 Capital and operation cost for establishment of *Kappaphycus(.) alvarezii* farming to self-help groups (SHG) of 12 members through the off-bottom monoline method

Items	Unit	Qty required for 10 rafts equivalent plot	Unit cost (LKR)	Total cost (SLR)	Total cost for 60 plots for 12 members of SHG @5 plots /member (LKR)	Total cost for 60 plots for 12 members of SHG @5 plots / member (USD) 1USD = 154.58 LKR
(a) Capital cost						
1.5 mm HDPE rope	Roll	9.6	140.00	1344.00	80,640.00	521.67
2 mm PP rope	Roll	1.5	160.00	240.00	14,400.00	93.16
3 mm PP rope	Roll	9	255.00	2295.00	137,700.00	890.80
10 mm PP rope	Roll	1	1800.00	1800.00	108,000.00	698.67
Floats	No	174	10.00	1740.00	104,400.00	675.38
HDPE fishnet 0.50" × 22 mm × 300 mm	Kg	5	1500.00	7500.00	450,000.00	2911.11
Anchors (50 kg each)	No	4	130.00	520.00	31,200.00	201.84
Anchors (25 kg each)	No	40	75.00	3000.00	180,000.00	1164.45
Initial seed cost at five rafts/ person	Kg	66	8.00	528.00	31,680.00	204.94
Infrastructure transportation cost	Unit	1	4500.00	100.00	54,000.00	349.33
Total cost/60 m culture line	Unit			3037.60		
Total cost for 45 monolines					136,692.00	884.28
Cost for a SHG of 12 members					1,640,304.00	10,611.36
Cost per meter of culture line					50.63	0.33
(b) Operational cost						
Storage shed (4 nos @ 10 × 5 m)					100000.00	646.91
Drying bed (4 nos @ 300sqm)					120000.00	776.29
Cost of boat for material shifting					400000.00	2587.65
Maintenance of farm and infrastructure (10% of total infrastructure cost)					244971.00	158.47
Total operational cost					864971.00	5595.62
Grand total of capital and operational costs (a + b)					2505275.00	16206.98

Table 13.9 Farming operation and return on investment for self-help group (SHG) of 12 members through farming of *Kappaphycus alvarezii*

Descriptions	
Total rafts or monoline (60 m) operated by 12 members of SGH @ 455 rafts/person (m)	540
Harvest cycle (days)	45
Total raft or monoline handled by 12 persons based on 45 days harvest cycle	12
Total yield of fresh seaweed from 12 raft/day @ 275 kg/raft (kg)	3300
Net produce from 12 rafts/day after replantation @60 kg/raft	2580
Dry produce from 2580 kg of fresh weed @ 10:1 ratio (kg)	258
Dry produce available in a month (258 × 25 days operation) (kg)	6450
Sale price of one kg dry weed (US$)	0.49
Gross monthly income for a self-help group (SHG) of 12 members (US$)	3160.5
Less 10% maintenance and operation cost (US$)	316.05
Net monthly income by a self-help group (SHG) of 12 members (US$)	2844.5
Net monthly income of a person operating 45 rafts (US$)	237.0
Production from total investment assuming the life of infrastructure is 3 years; months 6450*36 (t)	232.2
Total investment ($ USD)	16206.98
Net profit (232.2 t * $USD 490)	113728.00
Investment gain	97,521
ROI	601.72%
Annualized ROI	91.46%
Investment length	3.0 years
Production cost per kg ($USD)	0.07

thankful to Dr. Kamalesh Prasad CSIR-CSMCRI, India, for analyzing the viscosity and FT-IR of carrageenan samples.

Declarations

Funding

No funding.

Competing interests

The authors have no competing interests.

Availability of data

The datasets generated and/or analyzed during the current study are available from the corresponding author on reasonable request.

Author contributions

Conceptualization: [Shanmugam Munisamy]; Methodology: [Shanmugam Munisamy]; Formal analysis and investigation: [Shanmugam Munisamy, Nirooparaj B]; Writing – original draft preparation: [Shanmugam Munisamy, Nirooparaj B, and Asoka JM]; Writing – review and editing: [Shanmugam Munisamy]; Funding acquisition: [Nirooparaj B and Asoka JM]; Resources: [Shanmugam Munisamy, Nirooparaj B, and Asoka JM]; Supervision: [Shanmugam Munisamy and Nirooparaj B.].

References

Abhiram S, Shanmugam M (2016) Seaweeds as agricultural crops in India: new vistas. In: Dagar JC, Sharma PC, Sharma DK, Singh AK (eds) Innovative saline agriculture. Springer India, New Delhi, pp 441–473

Adnan H, Porse H (1987) Culture of *Eucheuma cottonii* and *Eucheuma spinosum* in Indonesia. Hydrobiologia 151(152):355–358

AOAC (1995) Official methods of analysis. Association of Official Analytical Chemists, Washington, DC

Cai J (2021). https://www.competecaribbean.org/wp-content/uploads/2021/05/Global-status-of-seaweed-production-trade-and-utilization-Junning-Cai-FAO.pdf

Cruchaga S, Echeita SA, Aladuena A, Garciapena J, Frias N, Usera MA (2001) Antimicrobial resistance in salmonella from human, food and animals in Spain in 1998. Antimicrob Chemother 47:315–321

Darmawan M, Utomo BSB, Raekal Amral Yuda Mulia RMY (2013) The quality of alkali treated Cottonii (ATC) made from *Eucheuma cottonii* collected from different regions in Indonesia. Squalen Bulletin of Marine & Fisheries Postharvest & Biotechnology 8(3): 117–127

Dawes C, Lluisma AO, Trono GC (1994) Laboratory and field growth studies of commercial strains of *Eucheuma denticulatum* and *Kappaphycus alvarezii* in The Philippines. J Appl Phycol 6:21–24

Hayashi L, Reis RP, Alves dos Santos AA, Castelar B, Robledo D, deVega GB, Msuya FM, Eswaran K, Yasir SM, Ali MKM, Hurtado AQ (2017) The cultivation of *Kappaphycus* and *Eucheuma* in tropical and subtropical waters. In: Hurtado AQ, Critchley AT, Neish IC (eds) Tropical seaweed farming trends, problems and opportunities: focus on spinosum and cottonii of commerce. Springer, Dordrecht, pp 55–90

Hurtado AQ, Neish IC, Critchley AT (2019) Phyconomy: the extensive cultivation of seaweeds, their sustainability, and economic value, with particular reference to important lessons to be learned and transferred from the practice of eucheumatoid farming. Phycologia 58(5):472–483. https://doi.org/10.1080/00318884.2019.1625632

Luxton DM, Robertson M, Kindley MJ (1987) Farming of *Eucheuma* in the south Pacific islands of Fiji. Hydrobiologia 151/152:359–362

Mollin J, Braud JPA (1993) *Eucheuma* (Solieriaceae, Rhodophyta) cultivation test on the south west coast of Madagascar. Hydrobiologia 260(261):373–378

Moses J, Ananthakumar R, Shanmugam M (2015) Effect of alkaline treatment on the sulfate content and quality of semi-refined carrageenan prepared from seaweed *Kappaphycus alvarezii* Doty (Doty) farmed in Indian waters. Afr J Biotech 14:1584–1589

Neish IC (2008) Good agronomy practices for *Kappaphycus* and *Eucheuma* including an over view of basic biology. SEAPlant.net Monograph no. HB2F 1008 V3 GAP, 72pp

Ohno M, Nang HQ, Hirase S (1996) Cultivation and carrageenan yield and quality of *Kappaphycus alvarezii* in the waters of Vietnam. J Appl Phycol 6:431–437

Shanmugam M, Kulendran S, Rajeev S, Pahalawattarachchi V, Chandraratne PN, Asoka JM, Abhiram S (2017) Successful establishment of commercial farming of seaweed *Kappaphycus alvarezii* (Doty) Doty in Sri Lanka: economics of farming and quality of dry-seaweed. J Appl Phycol 29(6):3015–3027

Vaibhav AM, Eswaran K, Shanmugam M, Ganesan M, Veeragurunathan V, Thiruppathi S, Reddy CRK, Seth A (2017) An appraisal on commercial farming of *Kappaphycus alvarezii* in India: success in diversification of livelihood and prospects. J Appl Phycol 29:335–357. https://doi.org/10.1007/s10811-016-0948-7

Vaibhav AM, Ramalingam D, Anshul Y, Eswaran K, Shanmugam M, Tejal KG (2022) Projections for profitability assessment parameters under short-term, medium-term and long-term evaluation for three farming techniques of *Kappaphycus alvarezii* along eastern coast of India. Aquaculture 551:737912

Valderrama D, Cai J, Hishamunda N, Ridler N, Neish IC, Hurtado AQ, Msuya FE, Krishnan M, Narayanakumar R, Kronen M, Robledo D, Gasca-Leyva E, Fraga J (2015) The economics of *Kappaphycus* seaweed cultivation in developing countries: a comparative analysis of farming system. Aquacult Econ Management 19:251–277

Wullandari P, Sedayu BB, Novianto TD, Prasetyo AW (2021) Characteristic of semi refined and refined carrageenan flours used in the making of biofilm (bioplastic) International Conference on Green Agro-industry and Bioeconomy, IOP Conf. Series: Earth and Environmental Science 733 (2021) 012112 IOP publishing https://doi.org/10.1088/1755-1315/733/1/012112

Yong WTL, Chin JYY, Thien VY, Yasir S (2014) Evaluation of growth rate and semirefined carrageenan properties of tissue-cultured *Kappaphycus alvarezii* (Rhodophyta, Gigartinales). Phycol Res 62(4):316–321. https://doi.org/10.1111/pre.12067

14 Eucheumatoid Farming in India: Current Status and Way Forward for Sustainable Production

Shanmugam Munisamy, Thilaga Sethuraman, Doss Ganesh, and C. R. K. Reddy

Abstract

Among the economically important red seaweeds, three tropical seaweeds, *Eucheuma*, *Kappaphycus* and *Gracilaria*, collectively account for more than 43% of the total global seaweed production, which was estimated at 35.76 million (tonnes) t (wet) in 2019 with a value of 16 B USD. The Philippines started the cultivation of *Eucheuma* in early 1970, while Indonesia started in the mid-1980s. The current annual production of *Eucheuma* and *Kappaphycus* together is close to 12 M (T fresh weight). In India, *Kappaphycus alvarezii* cultivation began in 2001, and annual production peaked during 2013 ca. 2000 MT DW (ADGR % 3.3–4.5%). However, after 2013, the seaweed farms were struck with a severe disease, after which the production abruptly decreased to approximately 200 MT dry weight. The market demand for carrageenan in India is approximately 2000 MT yr^{-1}. The *Kappaphycus* extract was found to be an excellent phyco-biostimulant and is marketed successfully in India. Recently, its inclusion in cattle and aquafeed as a feed additive has also shown improved growth and immunity of the animal, indicating a potential new market opportunity. To meet the growing demand for value-added products from *Kappaphycus*, efforts are in progress to improve the vigour of the germplasm through biotechnological interventions such as polyploidy, cellular biotechnology and somatic hybridization while developing innovative farming methods for large-scale production in offshore waters. This paper reviews the problems and possible interventions for sustainable farming of eucheumatoid seaweeds in India.

Keywords

Kappaphycus alvarezii · Eucheumatoid · Carrageenophyte · Carrageenan · Cultivation

S. Munisamy
Department of Biomarine Resource Valorisation, Division of Food Production and Society, Norwegian Institute of Bioeconomy Research (NIBIO), Bodo, Norway

Research and Development Division (DSIR-Lab), AquAgri Processing Private Limited, Manamadurai, Tamil Nadu, India

T. Sethuraman · D. Ganesh (✉)
Department of Plant Biotechnology, School of Biotechnology, Madurai Kamaraj University, Madurai, Tamil Nadu, India
e-mail: dossganesh@mkuuniversity.ac.in

C. R. K. Reddy
DBT-ICT Centre for Energy Biosciences, Institute of Chemical Technology, Mumbai, India

14.1 Introduction

Kappaphycus alvarezii Doty is an important seaweed for the production of carrageenan, a polysaccharide used in a variety of products as an emulsifier, binder, thickener and stabilizer in a wide range of industries ranging from food and beverage processing, dairying and pharmaceuticals to beauty and personal care (Bixler and Porse 2011). The extensive use of *Kappaphycus* in the agriculture and biomedical industries formed the basis for developing innovations in the aquaculture sector for sustainable production (Meinita et al. 2012; Mody et al. 2012; Mondal et al. 2013). Considering the wide range of its applications and ever-growing market demands, the cultivation of *Kappaphycus* has been introduced into more than 30 countries that are geographically located in coastal regions (Neish 2008). It is estimated that approximately 88% of the world's production of *Kappaphycus* is processed for carrageenan production (Bindu and Levine 2011). The major *Kappaphycus*-producing countries are the Philippines, Indonesia and Tanzania (McHugh 2003; Valderrama et al. 2015; Hayashi et al. 2017). *Kappaphycus* farming is predominantly restricted to the most suitable world geographical areas lying within ±10° latitude, primarily from the Southeast Asian countries extending to East Africa and Brazil (Hayashi et al. 2010). The world production of *Kappaphycus* is primarily contributed by the Southeast Asian region, primarily the Brunei–Indonesia–Malaysia–

A. T. Critchley et al. (eds.), *Tropical Phyconomy Coalition Development*, Developments in Applied Phycology 11,
https://doi.org/10.1007/978-3-031-47806-2_14

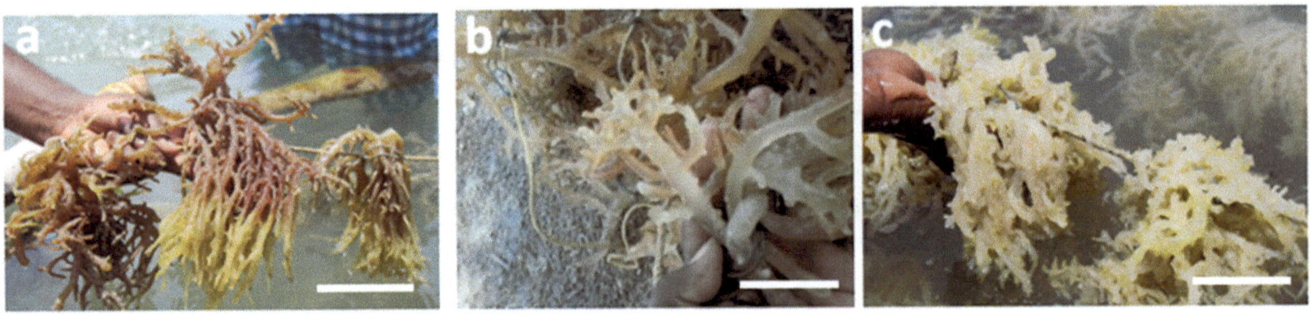

Fig. 14.1 Development of ice-ice disease in *Kappaphycus alvarezii*: (**a-c**) Sequential phases of discoloration and decay of total biomass of during unfavourable growth conditions occurred in 2013. (Scale bar–10 cm)

Philippines (East Association of Southeast Asian Nations (ASEAN) and the Growth Area—BIMP-EAGA—integrated countries). These regions have the highest potential for expanded tropical seaweed cultivation, consisting of approximately 60% of the sites in the world for intensive cultivation. The main reason for the higher production of *Kappaphycus* in Indonesia, Malaysia and the Philippines is primarily due to the most favourable conditions prevailing in these countries, where tropical seaweeds grow abundantly and luxuriantly, and, most importantly, in places where typhoons are seldom experienced. According to Hayashi et al. (2010), the major *Kappaphycus* world production is contributed by the BIMP-EAGA region with 96.5% of the total production, of which 55% is from the Philippines, followed by Indonesia with 38% and Malaysia being the fourth major producer with 2.5%. The rest of the producing countries contribute relatively very little. Subsequently, Indonesia surpassed the Philippines as the dominant producer of carrageenan-yielding eucheumatoid seaweeds (Neish et al. 2017a), and in 2015, Indonesia produced 65% of the global total of *Kappaphycus* and 56% of the global total of *Eucheuma* (Porse and Rudolph 2017).

Since the introduction *of Kappaphycus* in the Philippines by Prof. Maxwell Doty of the University of Hawaii during the mid-1960s, rigorous selection coupled with the development of meticulous culture techniques paved the way for the successful farming of *Kappaphycus* and other commercial eucheumoids in 1972 (Doty and Alvarez 1981). The introduction of this seaweed to over 31 countries, including India during 1989 ((Mandal et al. 2010) for research and commercial farming (Ask 2001), led to the widespread use of this seaweed in several coastal regions of the world. However, only Indonesia, the Philippines, Malaysia and the United Republic of Tanzania are major contributors (Cai et al. 2013). The single genetic stock acquired by the Central Salt and Marine Chemical Research Institute in 1984 (Mandal et al. 2010) became the source for commercial farming of *Kappaphycus* in India. The introduction of *Kappaphycus* as an outdoor culture in India by the Central Salt and Marine Chemical Research Institute and subsequent field experiments carried out from 1989 to 1996 at Port Okha in Gujarat led to commercial farming in many coastal regions (Mairh et al. 1995). Extensive field trials coupled with commercial farming of *Kappaphycus* at Mandapam from 1995 to 1997 were laid out until 2002 (Eswaran et al. 2002). An integrated and systematic approach between governmental and non-governmental organizations such as M/s. AquAgri Processing Pvt. and PepsiCo India Holdings Ltd. laid a remarkable foundation for the establishment of commercial seaweed farms of *Kappaphycus* in many regions in the southern coastal parts of India. Although the annual biomass production of *Kappaphycus* reached 20,000 MT fwt in 2013, there was mass mortality of the crop in late 2013 due to unfavourable climatic conditions or loss in vigour of the seaweed (Fig. 14.1). Therefore, the dry biomass production of *Kappaphycus* in India was only approximately 200 MT during 2021, which is much below the 100,000 MT fresh demand. The local market demand is increasing from 8 - 10% per year. Global fresh biomass production is increasing primarily due to the increased utilization of carrageenan by food and nonfood industries (Cai et al. 2013). In this chapter, several aspects of cultivation, production, diverse application, problems and prospects for sustainability of the *Kappaphycus* seaweed industry with a primary focus on India are emphasized.

14.2 Recap of Eucheumatoid Farming in India

14.2.1 Commercial Farming: The Genesis

The initial experimental outdoor culture studies followed by systematic outplanting experiments from 1989 to 1996 at Port Okha (Mairh et al. 1995) led to extensive pilot field trials at Mandapam from 1995 to 1997 (Eswaran et al. 2002). The encouraging results from initial pilot trials conducted until November 2000 offered a scope for large-scale cultivation of *Kappaphycus* by M/s. Pepsi Foods Ltd. (a subsidiary of PepsiCo India Holdings Pvt. Ltd.). A collaborative approach between PepsiCo India Holdings Pvt. Ltd.

and CSIR-CSMCRI in 2000 led to the farming of *Hypnea musciformis* in the Gulf of Mannar. Since the carrageenan from *H. musciformis* was found to be inferior, its cultivation was abandoned. Alternative farming with a good quality of carrageenan from *K. alvarezii* was explored. A systematic evaluation of cultivation in the shrimp culture pond and open sea cultivation led to the selection of open sea cultivation since it yielded a 3.9% growth rate over 1% in the shrimp culture pond. With this effort, PepsiCo India Holdings Pvt. Ltd. took up large-scale farming on the Palk Bay side of the Mandapam coast in 2001. Transfer of farming business rights to M/s. AquAgri Processing Pvt. by PepsiCo Holdings India Ltd. in 2008 led to significant development in *Kappaphycus* farming by partnering self-help groups (SHGs) with a commitment to improving the socio-economic status of several thousand people associated with coastal regions by providing on-farm training coupled with acceleration of biomass production. India was successful in expanding the cultivation *of Kappaphycus* in several areas of coastal regions with an additional focus on agricultural by-products, i.e., a plant growth stimulant (sap) from fresh seaweed biomass (Agarwal et al. 2016; Singh et al. 2016a, b). The farming system of *Kappaphycus* has evolved simultaneously with innovations in biomass processing. To date, farming has been successfully implemented in over 30 locations along the Indian coast. With improved cultivation practices, *Kappaphycus* farming was highly successful and improved the livelihoods of economically poor communities on both the east and west coasts of India until 2013 (Mantri et al. 2017). However, a sudden decrease in biomass production due to poor growth has become a major challenge for seaweed industries, and efforts are being made to improve biomass production.

14.2.2 *Kappaphycus*: Production Potential

With the advancement of cultivation techniques coupled with the rapid expansion of the area under cultivation of *Kappaphycus*, there was a significant enhancement of production from 944,000 T fwt in 2000 to 5.6 M fwt in 2010, accounting for over a 63% increase in total red seaweed cultivation (Cai et al. 2013). Impressive growth in production was witnessed from 20 T in 2001 to approximately 1046 T dwt in 2011, corresponding to 489% growth. Commercial cultivation of *Kappaphycus* in India realized its yield potential of 7187 dry t biomass worth Rs 140 million farm gate values between 2005 and 2015. *Kappaphycus* biomass expanded from 20 tonnes dry weight in 2001 to 700 tonnes dry weight in 2013. Correspondingly, the number of seaweed growers in the southern part of India significantly increased from 6 000 in 2001 to 950 000 t in 2013. The progressive trend of export of biomass has been subsequently reversed gradually since a major percentage of biomass has been utilized domestically for the production of value-added products such as biostimulants (Karthikeyan and Shanmugam 2017) and animal feed (Rossita et al. 2015). The highest annual production of 1490 T of dry weight was realized in 2013, but a sharp decline in production in the subsequent years owing to poor vegetative growth became a serious concern in the seaweed industry. In 2016, the production level declined heavily from 1490 to 181 (T) t dry weight. In addition, efforts to expand the cultivation of *Kappaphycus* were also scuttled due to the declaration of the Gulf of Mannar as a Marine Protected Area and named a Marine National Park. Presently, the production of *Kappaphycus* biomass is seriously impaired and is attributed to climate changes and to the loss of genetic vigour in the seed stock (Reddy et al. 2010).

14.2.3 Status of *Kappaphycus* Cultivation: The Past and Present

In India, commercial cultivation of *Kappaphycus alvarezii* was initiated in 2001 along the southeast coast of India. This activity was first initiated by PepsiCo India Holdings (P) Ltd. India after licensing the cultivation technology from CSIR-CSMCRI. The contract-farming model with a buy-back agreement with self-help groups (SHGs) led to higher production of biomass as well as employment for several thousand people in the coastal regions. The contract farming model with a distribution of a sizeable number of rafts to SHGs facilitated the harvest of fresh biomass every 45-day cycle. The production of biomass, with an average of 240 kg of fresh seaweed raft^{-1} from an input of 40 kg seaweed raft^{-1}, was highly profitable until 2013. More than 1500 households along the southern part of Indian coastal regions were engaged in seaweed farming as an alternative livelihood, each earning approximately US $300 per month. Seaweed cultivation has emerged as a viable option for improving the economic status of low-income coastal communities in India (Krishnan and Narayanakumar 2010; Periyasami et al. 2014, 2015; Mantri et al. 2017). An increase in production from 21 T dwt in 2001 to 1490 dry tonnes in 2013 was realized with a total of 7187 T dwt of biomass, worth US $2.5 million during the span of 13 years of cultivation (Mantri et al. 2017). Unfortunately, commercial farming of *Kappaphycus alvarezii* in India has faced a significant decline since 2013 (Fig. 14.2). A systematic survey coupled with research on this aspect revealed that a sudden rise in seawater temperature to 36°C and subsequent onset of bacterial disease are the causes of decline in the production of fresh biomass. Commercial production of *Kappaphycus* has expanded to new coastal districts in the southern state of India and is expected to improve commercial seaweed farming activities. The status of cultivation during the past and present is presented in Table 14.1.

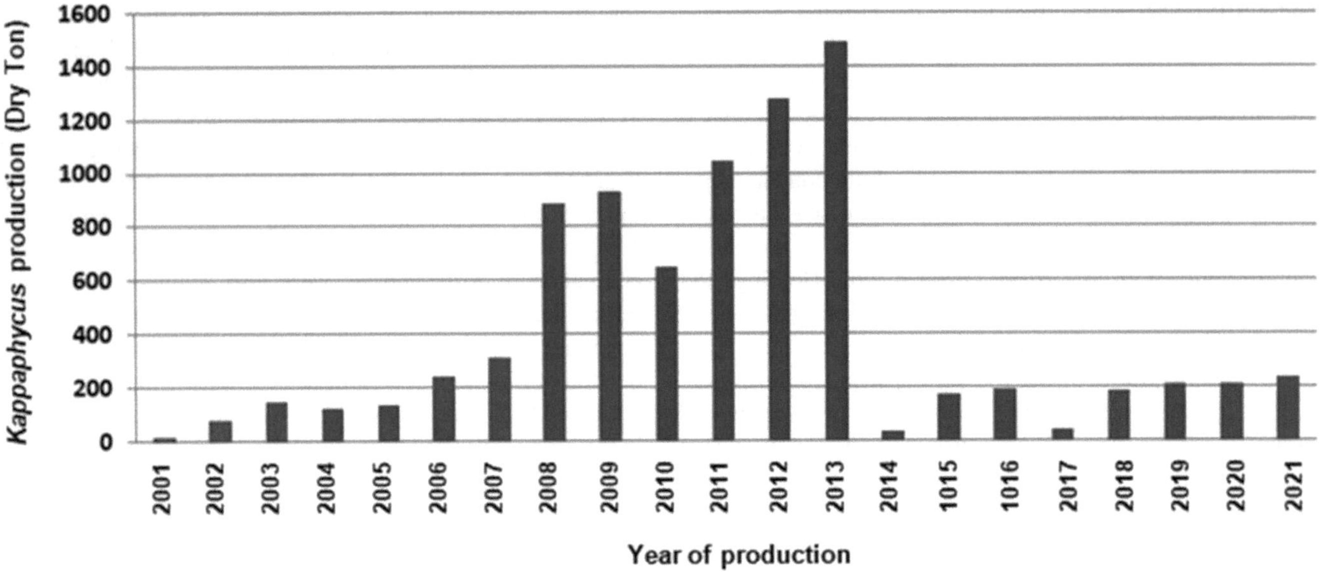

Fig. 14.2 Two decades of biomass production of *Kappaphycus alvarezii* in India. Source: AquAgri Processing Pvt. Ltd., Mana Madurai, Tamil Nadu, India (2022)

14.2.4 Learning from Research and Commercial Farming of *Kappaphycus*

The lack of diverse genetic resources of *Kappaphycus* in India has been recognized as one of the major problems in recent times. The development of newer genetic resources of *Kappaphycus* by an extensive survey is expected to identify naturally available and potential genetic stocks in India. A global network programme with multidisciplinary research aimed at investigating genetic improvement for improved production of biomass, high-quality and disease-resistant seed production and refinement of downstream processing and exploring the possibility of diversifying the value-added products and molecules in healthcare needs to be developed with mutual benefits among *Kappaphycus*-producing countries. Relying on limited genetic resources is highly risky in commodity crops, and thus, searching for *Kappaphycus* germplasm coupled with the import of exotic genotypes with desirable agronomic values would farm a stronger baseline for research to achieve a tangible outcome.

14.3 Issues of *Kappaphycus* Farming and their Management

14.3.1 Epiphytism and Disease

Kappaphycus farming is a labour-intensive activity, and production is often dependent on cultivation conditions and the prevalence of disease during the cultivation season. Two significant diseases that are often encountered in high-density commercial farming are ice-ice disease and epiphyte infection. The emergence of 'ice-ice' disease was first reported in 1974 in the Philippines by Trono (1974). The infestation and development of this disease on this seaweed were described in detail by Uyenco et al. (1981) and Largo et al. (1995a, b). Their findings led to the understanding of how stress induced by culture conditions could trigger pathogenesis and lead to loss of biomass production. The causative microbes were identified as the *Vibrio-Aeromonas* complex and *Cytophaga-Flavobacterium* complex, and their mode of action was based on the lytic enzymatic activities of bacterial exudates (Largo et al. 1995a).

Epiphytic infection of red filamentous microalgae are known to exist naturally along with *Kappaphycus*. An epiphytic outbreak in commercial farming of *Kappaphycus* was reported as early as 1975 by Doty and Alvarez (1975). However, not much interest has been focused on research to address this problem until now (Critchley et al. 2004; Hurtado et al. 2006; Vairappan 2006). Investigation of this problem indicated that *Neosiphonia (savateri) savatieri* is the causative filamentous red alga and revealed some insights into its symptoms and seasonality during *Kappaphycus* farming in Sabah, Malaysia (Vairappan 2006). Hurtado et al. (2006) described *Polysiphonia* sp. as the causative organism in the Philippine farms, with different types of symptoms. The occurrence of epiphyte outbreaks in Malaysia and the Philippines resulted in a significant reduction in biomass production and declines in carrageenan quality and yield (Vairappan et al. 2007). Due to the continuous incidence and economic loss caused by epiphytes, research collaboration was initiated by leading *Kappaphycus*-producing

Table 14.1 Status of production of *Kappaphycus alvarezii* during the last two decades (2001–2022) in India

Year	Total farming area (raft no.)	No. of beneficiaries	Seaweed production (dry tonne)	Percentage increase over previous year	Production per meter (dry kg)
2001	Trial	6	21	NA	1.25
2002	5275	6	82	290.47	1.33
2003	5529	12	147	79.26	1.35
2004	3469	40	126	−14.28	1.12
2005	3450	150	135	7.14	1.39
2006	8100	280	244	80.74	1.35
2007	9500	320	315	29.09	1.36
2008	12,000	525	888	181.9	1.25
2009	14,000	650	930	4.72	1.32
2010	16,000	700	650	−30.1	1.36
2011	17,000	900	1046	60.92	1.25
2012	19,950	950	1273	21.7	1.22
2013	21,000	950	1490	17.04	1.22
2014	450	200	40	−97.31	0.33
2015	9000	250	176	340	0.35
2016	7500	250	195	10.79	0.30
2017	3500	110	41	−78.97	0.18
2018	4000	150	185	351.21	0.22
2019	5500	250	210	18.31	0.25
2020	5200	230	210	0	0.28
2021	5800	250	235	19	0.36
2022	5500	250	240	2.12	1.37

countries. In India, loss of biomass production in *Kappaphycus* due to epiphytism and ice-ice disease is one of the serious problems that has emerged, and these problems need to be addressed systematically.

14.3.2 Poor "Seed" Quality

Seed quality is one of the primary drivers of higher biomass production in *Kappaphycus* and comes with many challenges. High-quality seed production in *Kappaphycus* is faced with a serious problem due to the incidence of 'ice-ice' malaise and macro-epiphytism, which are predominant during seed production. Epiphytic filamentous algae (EFA) infestations adversely affect the quality and quantity of seaweed during seed production. EFA infestation by *Melanothamnus* sp. (formerly *Neosiphonia* a filamentous red alga) and a concomitant 'ice-ice' disease are the most disastrous problems in the eucheumatoid industry and need to be addressed on priority. Recent reports showed that dipping the source of seeds in *Ascophyllum* marine plant extract powder (AMPEP), a commercial extract of the brown seaweed *Ascophyllum nodosum,* resulted in the development of natural defence by producing more phenols and free radicals to resist against a variety of epiphytes (Loureiro et al. 2011; Ali et al. 2020). However, several areas of research, such as the isolation of resistant genotypes through intensive research on genetic diversity and the development of newer genotypes through genetic manipulation using biotechnological approaches, are needed to address these issues.

14.3.3 Uncertainty and Risk Factors

Several factors affect the yield and quality of biomass production in *Kappaphycus.* In India, the uncertainty of favourable seasons for the development of seedlings and commercial farming is one of the major problems faced by the farming community. This factor either directly or indirectly affects the vegetative vigour and growth of *Kappaphycus* seedlings, leading to slow growth and reduction in biomass production, and very often, the healthy growing biomass loses its pigments and decays during unfavourable conditions. Changes in environmental conditions often influence the normal biochemical constituents in addition to the unusual physiological functioning of seaweed in tropical regions. A similar problem has been encountered in *Gracilaria cornea* (Orduna et al. 2002). Pascoe et al. (2002) and Hean and Cacho (2002) reported that delays in suitable climatic conditions normally delay the harvesting time or force farmers to prematurely harvest, which eventually decreases the quality of biomass. Unfavourable or unpredictable events in the farming process could make the farmer be cautious and adopt fewer harvesting times than the proposed schedule of harvest. Cyclones in the Indian Ocean are very common, and this unexpected and sudden event damages the rafts and vanishes the entire biomass, leading to heavy loss to the farming community. As many farming communities are economically challenged, this problem is a serious issue in India. Seasonal dependency on Indian seaweed cultivation poses a crucial challenge to overcome. The southwest monsoon (June to September) and northeast monsoon (October to December) in India are often associated with the occurrence of cyclones and typhoons and create high tidal fluctuations, causing an unfavourable situation for cultivation.

14.3.4 Persistent Crop Failures

Low yield of biomass continuously over the years due to grazing and nibbling by herbivores such as siganids, acanthurids, sea urchins and starfish on the growing apices of the young thallus is often considered a major problem faced by the farming community in India. During the winter season, farmers experience intensive grazing, which affects the yield by up to 50–80%. Epiphytism also accounts for the persistent failure of biomass production since most undesirable seaweeds dominate thallus growth during vegetative growth at the onset of monsoons. In India, *Kappaphycus* has been mostly cultivated using 3 × 3 m bamboo rafts in shallow coastal areas under the lowest tide (0.1–0.2 m), which often encourages grazing primarily by sea urchins. Generally, lower water temperature, wind and water movement enhance the growth of epiphytes, which unfortunately coincides with the ideal season for thallus growth and multiplication. Epiphytic algae compete with *Kappaphycus* for light and nutrients. Due to the firm attachment of epiphytic filamentous algae over the thallus of *Kappaphycus*, the growing thallus always suffers limitations of space, nutrients and sunlight. A decline in salinity and an increase in seawater temperature and light intensity are the reasons for the aggravation of diseases. The thallus of *Kappaphycus* often turns white to pinkish under stress conditions. In addition, natural calamities such as heavy storms and cyclones cause complete damage to farming.

14.3.5 Climate Change Impact

Elevation of sea surface temperature (SST) owing to global warming is the most likely reason for the increase in strength and frequency of occurrences of tropical cyclones (Chan and Liu 2004; Stowasser et al. 2007). Fixed off-bottom mono-line cultivation is the most popular cultivation method applied in shallow water of *Kappaphycus*. An average of 34°C prevails in the shallow areas of sea water where cultivation using the bamboo raft method is most common. This method is followed in India in some areas where grazing is very limited. However, in other areas, such as Rameswaram, Tuticorin and Pudukkottai in the state of Tamil Nadu in India, the most popular bamboo raft method of cultivation is followed. However, both methods of cultivation are most vulnerable even to ordinary cyclones, and entire farms are lost during the cyclone seasons. Similar experiences were reported in major *Kappaphycus*-producing countries. This situation often forces farmers to harvest seaweeds either at the premature stage or before they can reach their maximum biomass.

14.4 Strategies for Revival and Expansion of Farming

14.4.1 Role of Government

India has a very long coastline with diverse geoclimatic conditions, and several thousand farmers have become dependent on seaweed farming, particularly in the coastal regions of Tamil Nadu, which includes three districts, namely, Ramnad, Pudukkottai and Tuticorin. *Kappaphycus* is the major alga cultivated for diverse applications in India (Ganesan et al. 2019). The potential of seaweed cultivation and its associated industries for producing several commodity products has increased in recent years (Mantri et al. 2017). By realizing the potential of seaweeds, the Government of India launched a mega scheme called 'Seaweed Park' in the coastal region of the southern part of India. This unique project broadens the platform of the seaweed industry in India by expanding employment opportunities through science and technology development. The National Fisheries and Development Board, Government of India, offers a wide range of financial support as well as a policy framework for further promoting *Kappaphycus* farming in the country with a partnership with regional state departments. More conscious efforts by the government to implement the project through the SHG model have yielded better coordination and productivity of biomass in *Kappaphycus* cultivation. This was initiated through the Kutumbam (family) model of cultivation (KMC). In the KMC, members of SHGs normally belong to the same family but may also include members from the same community (Krishnan and Narayanakumar 2010). National financial institutions such as the State Bank of India (SBI) and the National Bank for Agriculture and Rural Development (NABARD) support commercial farming by providing loans to farmers. The Department of Biotechnology (DBT), Ministry of Science and Technology, was very keen on aquaculture-related research and development activities by providing funds to national institutions and academic institutes.

14.4.2 Invigoration of Seeds Through Biotechnological Tools

India experienced steady growth in the farming of *Kappaphycus* with self-sufficient production of biomass to meet domestic demands until 2013. However, the decline in biomass production during the past few years forced seaweed industries to import *Kappaphycus* raw materials from major producing countries. AquAgri Processing Private Limited, one of the prominent seaweed industries in the southern part of India, imports approximately 100 MT (dry weight) annually to compensate for the raw material demand. It has been commonly interpreted that the presently used genetic stock of *Kappaphycus* lost its genetic vigour owing to repeated multiplication from the very narrow genetic stock (Robinson et al. 2013) in addition to climatic change being one of the reasons (Terada et al. 2015). Import of improved genetic stock possessing good vegetative vigour with relative tolerance to biotic and abiotic stresses from neighbouring countries such as the Philippines, Indonesia, etc., is a possible action at the governmental level to resolve these issues. In addition, a systematic short- and long-term network research programme among the government, including universities and non-governmental organizations, is expected to address several issues related to the genetic improvement of *Kappaphycus* with available expertise and infrastructure facilities available in India. Currently, *Kappaphycus* cultivation has been improved due to favourable conditions for cultivation in addition to careful planning and management and treatment with the seaweed extract from *Ascophyllum* (AMPEP) (Figs. 14.3 and 14.4).

A systematic survey, selection and several possible means of genetic improvement of *Kappaphycus* through conventional and biotechnological approaches are needed to address the various issues. Red seaweeds are cultivated primarily for hydrocolloid production and provide income for many developing nations, including India. A strategy for the maintenance of diverse genetic resources of *Kappaphycus* is needed to explore the possibility of developing genetic improvement in the near future. Natural selection processes under diverse climatic conditions coupled with systematic screening of elite genotypes are expected to offer several sporulating and nonsporulating types that allow sexual as well as nonsporulating methods of propagation on the basis of preference. This approach offers the possibility of widening the genetic resources for overcoming biotic and abiotic stresses in addition to cloning elite genotypes with high carrageenan contents. In India, research needs to also be focused on replacing the vegetative system for reproduction with sexual propagation of *Kappaphycus* to broaden the genetic base to enhance the natural breeding behaviour followed by continuous selection of wild types either for direct exploitation or breeding involving conventional and biotechnological approaches.

14.4.3 Offshore Cultivation and Mechanization of Farming

It is worth exploring the potential of offshore waters for the production of seaweed biomass for food and feedstock for biofuel. Offshore waters are unpolluted and largely remain underutilized for seaweed farming. The southern part of the coast in India has been continuously used for commercial cultivation for the last two decades, and the intertidal areas are intensely farmed; thus, there is a need to expand the area under cultivation to offshore regions. The offshore waters are relatively stable and are more productive due to nutrient upwelling. The feasibility of offshore seaweed farming (at approximately 8 m depth) was optimized by CSMCRI–Marine Algal Research Station during 2010–2012 over 170 floating bamboo rafts of 2 × 2 m along the Palk Bay coast of Mandapam, Tamil Nadu, India, with financial support from the Ministry of New and Renewable Energy. The growth differed substantially during the experiments, yielding 127.5 kg fwt from 22.5 kg planting material, achieving a DGR of 3.85% in a 45-day cultivation cycle in nearshore waters, whereas the rafts farmed in offshore regions yielded 219 kg fwt with a DGR of 5.10%. Nevertheless, structures for offshore sea farming require novel designs that are simple, expandable and manoeuvrable, manipulated or modified need no specialized skills for assembly and could be managed with ease by the individual cultivator.

The technical challenge is to develop a robust aquaculture system with a permanent strong anchor without compromising the flexibility to withstand turbulent sea conditions. A 'mooring farm' method for offshore farming is under evaluation. The farm would be 500 m long and 25 m wide. The farm is held intact within the water column with the support of anchors and floats of an appropriate size adequate to give required stability, buoyancy and intactness to the entire structure. A pair of primary anchors (800 kg) is lowered and kept apart at a 500-m distance with a similar pair in parallel at 25-m width to set up a rectangular shape farm. A small-scale trial is being conducted in India. However, commercial utility of this could be possible only after systematic field trial and evaluation.

To enhance stability, extra anchors of the same capacity are provided between two primary anchors on either side. These anchors are fastened with a chain to a vertically floating buoy. The buoys running through a 500-m length are further connected with another steel chain and held in approximately 30 cm below the surface. To give adequate buoyancy to this line, several floats are also tied at equal

Fig. 14.3 Treatment of seedlings of *Kappaphycus alvarezii* with biostimulant AMPEP (**a** & **b**) *K. alvarezii* is treated with AMPEP; (**c**) aerial view of *K. alvarezii* farming; (**d**) AMPEP-treated plant looking healthy with multiple branches. (Shanmugam et al. 2023)

distances. The seeded tube nets of 25-m long with small floats at an equal distance are tied between these two chains, which are held 25 m apart and tied with anchors at the bottom. This forms a solid supporting frame for fastening several 25-m-long tube nets in parallel at 1-m intervals. The farm is oriented in a direction such that current impacts are the bare minimum. The above-proposed farm can hold 500 tube nets of 25-m long in parallel at 1-m intervals. Each tube is seeded with approximately 20 kg of biomass and is expected to yield 100–125 kg in a 40–45-day growth cycle. A few

Fig. 14.4 Restoration of cultivation of *Kappaphycus alvarezii* in India (Site: Mullakadu Village, Tuticorin, 2021–2022)

prototypes set at Simar, Gujarat coastal region, are under evaluation for their stability, biomass production and suitability for harvest.

14.4.4 Status of the Extension of Cultivation in Gulf Areas

A long-pending issue of cultivation of *Kappaphycus* in the region of the Gulf of Mannar has yet to be resolved. Although the Gulf of Mannar is declared a reserved zone for the conservation of marine genetic resources, deliberation is in progress at the governmental level to encourage a few areas in the Gulf of Mannar for the cultivation of *Kappaphycus*. There is a dialogue in progress between environmentalists and *Kappaphycus* cultivating industries on this issue. From the farming point of view, *Kappaphycus* can be safely cultivated in any place in the sea, as this seaweed is never reported to be dominant over any life forms under the sea, as vegetative growth is limited during the unfavourable season for farming for more than 6 months in a year. There is no scientific evidence that *Kappaphycus* sporulates in the Indian Sea under natural conditions, thus limiting the multiplication of *Kappaphycus* only to vegetative means. Expansion of the area under cultivation of *Kappaphycus* in this region, at a safe level, would boost the seaweed industry in India. It is our considered opinion that there are enough scientific reasons that *Kappaphycus* can be safely cultivated in the Gulf of Mannar regions.

More recently, it was confirmed that the carpospores and tetraspores of *K. alvarezii* typically have low viability under various environmental conditions, and they rarely develop into young thalli (Mairh et al. 1995; Azanza and Aliaza 1999). A study by a team of investigators from the USA and Australia suggests that coral reefs are more resistant to seaweeds. Their study is the first global-scale analysis of thousands of individual reef surveys (Bruno et al. 2009). In another study, Castelar et al. (2009) evaluated the invasive potential of *K. alvarezii* in tropical regions, such as Rio de Janeiro, Brazil, and concluded that cultivating *K. alvarezii* was environmentally safe and secure in tropical regions. Based on a conceptual model proposed by Colautti and MacIsaac (2004), it was concluded that the determinants, viz. propagule pressure, physiochemical requirements of the species and community interactions, act on exotic species to make them invasive. Therefore, it appears that the establishment of *K. alvarezii* at Kurusadai Island is restricted to stage three (i.e., localized and numerically rare), as all the determinants, viz. (i) seawater temperature, (ii) turbidity or seawater transparency and (iii) propagule pressure (due to high grazing pressure), are acting negatively. The observed occurrence of *K. alvarezii* on corals at Kurusadai Island could merely be accidental, and its confinement over a relatively small area might be due to a combination of factors, particularly those mentioned above. Therefore, there is an expectation from the cultivars and seaweed industries that the government initiate *Kappaphycus* in the Gulf of Mannar regions on a trial basis and possibly that it can be extended further.

14.4.5 Multiplication of Quality Seed Material and Seed Bank Maintenance

Quality seeds contribute to the better production of biomass in *Kappaphycus* and ultimately improve the quality after the downstream process of various primary and by-products. In India, seed production is practised by random sampling of propagules based on visual observation of thallus. However, systematic strategies need to be drawn for the maintenance of quality seed stock by a careful selection process from commercial farms followed by multiplication of those source materials without any diseases or epiphytic algal growth. An ideal location needs to be identified for every zone of commercial farming for the maintenance of seedlings. Propagules need to be treated systematically with effective controlling agents for bacterial diseases and to control epiphytic algal growth to ensure that the farming community receives quality seeds with proven genetic vigour and disease and pest-free propagules. The establishment of seed propagation facilities with the input of scientific advancement through governmental and non-governmental agencies is needed to boost seaweed cultivation in India.

14.4.6 Promotion of Domestic Seaweeds

The southeast Indian coast is considered to be the most suitable area for commercial seaweed cultivation due to its shallow, wide and flat coastal shorelines with moderate wave action and richness in nutrients, including phosphorus and nitrogen (Ganesan et al. 2019). Seawater salinities and temperatures are optimal for seaweed production, making the region suitable for commercial seaweed cultivation year round, except during the monsoon season. Essentially, cultivation input requirements include bamboo, rope and anchors that are readily available at reasonable prices close to the cultivation sites. A large number of cultivators between the age groups of 14 and 35 in the coastal region makes farming activities viable (Ganesan et al. 2017). However, the promotion of native seaweeds needs systematic research and extension activities. As of today, there is not much progress on scientific advancement in *Kappaphycus* in the area of genetic improvement. Sound knowledge is needed to understand the genetic diversity and how it may be associated with qualitative and quantitative traits for higher biomass production.

The lack of germplasm through indigenous and exotic collection processes needs to be intensified. Systematic seed production procedures need to be followed to maintain genetic vigour and purity to ensure the quality of various products at the end of downstream processes. Presently, biomass production can be sustained for commercial production by enhancing the growth rate using AMPEP.

14.4.7 Import of Other Tropical Seaweeds and Germplasm

The *Kappaphycus* seaweed industry in India has faced a serious setback for a decade. The possible reasons for this crisis are attributed to climate change coupled with the loss of genetic vigour in *Kappaphycus*. This issue needs to be addressed systematically to restore the production of biomass. Currently, AquAgri Processing Private Limited, one of the leading *Kappaphycus* processing Industries in India, imports dry biomass from Indonesia, as the current production of biomass in the southern part of India is limited to 210 tons annually, which is not sufficient to meet domestic consumer market demands. Similar problems are faced by other seaweed industries in India. To overcome this problem, there is an urgent need to import exotic genotypes of *Kappaphycus* from leading *Kappaphycus*-cultivating countries such as the Philippines and Indonesia. This could be possible with a series of dialogues between India and other *Kappaphycus*-producing countries for mutual sharing of potential germplasm under bilateral agreements that could be useful both the countries either for direct commercial exploitation and to utilize them in genetic improvement programs. Some of the potential germplasm could be directly exploited for commercial cultivation.

14.4.8 Role of Research Institutes in Seaweed Genetic Improvement

Several governmental and non-governmental organizations have been involved in *Kappaphycus* cultivation for more than three decades. However, the majority of the activities are primarily centred on the optimization of cultivation methods, downstream processes for carrageenan production and value-added products (Anon 2003). The problem of depletion of genetic vigour and subsequent yield loss has never been anticipated in India. The importance of genetic improvement in the context of generating improved genetic stock as future material for commercial cultivation has never been seriously considered, and as a result, seaweed industries are facing a serious problem. Reddy et al. (2003, 2008a & b, 2010, 2017) performed extensive studies on improving the vigour of *Kappaphycus* seeds through cellular biotechnology. However, this technology could be used for multiplication if India has such high vigour genetic stocks. More recently, AquAgri Processing Private Limited seriously explored the possibility of developing new lines of polyploids of *Kappaphycus* with a primary interest in improving the genetic and vegetative vigour for enhanced production of biomass with the School of Biotechnology, Madurai Kamaraj University, in the southern part of India. This effort is expected to offer some polyploids with desirable agronomic value (unpublished data). Joint network activities among governmental and non-governmental organizations to explore infrastructure facilities and the field of specialization among scientists and entrepreneurs and industrialists are expected to create a stronger platform to address the various issues related to *Kappaphycus* genetic improvement. Some of the promising research, and development centres actively involved in *Kappaphycus* research, are summarized in the table. More recently, research funding by the National Fisheries Development Board, Govt. India, to the Central Salt and Marine Chemical Research Institute at Algal Research Station at Mandapam, Tamil Nadu, resulted in the development of micropropagation technology for the commercial multiplication of *Kappaphycus* seeds.

14.4.9 International Research Collaboration for Technology Development

Kappaphycus is a global, tropical seaweed, and more than 30 countries are involved in the cultivation and development of various products (Neish 2008). Therefore, advancement in research and development in any applied aspects of *Kappaphycus* is expected to have a greater impact on global seaweed production and technology in the food, biomedical and agriculture sectors. Therefore, exploring the possibility of international cooperation among *Kappaphycus*-producing and *Kappaphycus*-consuming countries would pave the way for advancement in *Kappaphycus* research that leads to need-based science and technology development. Such an international platform would be much stronger in achieving the targeted research and technology because intellectuals and infrastructure resources are shared to speed up and target the envisaged objectives and issues.

In India, several governments and non-governmental organizations are pursuing research on *Kappaphycus* and other seaweeds. In addition to research activities, these institutes extend their contribution to assisting farmers by offering training, financing and collaborative development (Krishnan and Narayanakumar 2010). The contribution of research and development in *Kappaphycus* leads to sustainable farming with a wide range of product development to earn revenues from various sectors (Baghel et al. 2015; Trivedi et al. 2016). However, these activities need to be extended globally with various institutions that are actively pursuing research on *Kappaphycus*. Some of the prominent global organizations rendering services to seaweed industries through research and development are highlighted. In the Philippines, academic experiments on *Kappaphycus* have been spearheaded by the University of the Philippines as a leading research institution, the Southeast Asian Fisheries Development Center-Aquaculture Department and Cargill Texturizing Solution SAS France, a multinational carrageenan company based in the country (Hurtado 2013). In Mexico, the pilot project was an academic experiment, and the Center for Advanced Studies and Research (CINVESTAV) has emerged as the leading research centre in Mexico (Robledo et al. 2013). In India, the case study describes the activities being carried out by the Seaweed Research Association, the Aquaculture Foundation and other NGOs. Not only have they undertaken research, but they have also assisted farmers with training, financing and collaborative development (Krishnan and Narayanakumar 2010).

14.4.10 Stewardship for Organized Farming

It has been estimated that India has the potential to produce one million tonnes of dried seaweed (Krishnan and Narayanakumar 2010). Cultivation of fresh seaweed is usually performed by several small-scale, independent seaweed growers. India's seaweed farming primarily depends on economically poor farming communities in coastal regions with poor literacy. Therefore, the development of educated personnel with in-depth knowledge on cultivation with the ability to transfer the current knowledge on cultivation, technology, market avenues and future scope needs to be transferred to the farming community. Cultivation of *Kappaphycus* through individual farmers and SHGs has been successful in seaweed farming over the decades. However, a more organized structure with the intervention of science and technology is needed for the improvement of farming. Dissemination of knowledge on the advancement of *Kappaphycus* through electronic media was already in practice in leading *Kappaphycus*-producing countries. For example, global carrageenan seaweed farming has benefited from technical research and dissemination undertaken by research communities through electronic media (Msuya 2013). The University of the South Pacific has tested quarantine protocols for introducing seaweed and developed training manuals for prospective farmers to develop skills. Manuals and monographs on *Kappaphycus* have been published by SEAPlant.Net based in Indonesia, and these are available on the Internet. Such activities in India transform the farming community into more educative to improve the quality of cultivation and processing methods. Similar activities need to be intensified in India.

14.4.11 Innovations and Scope for Newer Products in Downstream Processing

The development of efficient industrial machinery for lowering the consumption of fuels and electricity, speeding up downstream processes and innovative and improved protocols for improving the purity and quality of final produce on par with international standards are the key areas to be focused on. Conventional phycocolloid seaweed processors, which do not perform biorefinery process technologies, use only approximately 15–30% of the total dry seaweed biomass, and the remaining 70–85% is degraded during phycocolloid extraction and discarded. Recent studies have reported that seaweeds are rich in valuable metabolites, such as natural pigments, proteins, lipids, minerals and cellulose, which can be extracted and utilized through biorefinery processing methodologies (Ingle et al. 2011; Baghel et al. 2015; Bikker et al. 2016; Trivedi et al. 2016' Gajaria et al. 2017). Seaweeds are reported to contain low lipids, but their composition is rich in C18 (linoleic and alpha-linolenic) fatty acids and low in C20 PUFAs (polyunsaturated fatty acids), indicating the possibility of their application in the prevention of cardiovascular diseases, osteoarthritis and diabetes. Proteins synthesized by macroalgae are known to contain all the essential amino acids (EAA) with a fluctuation of their total contents owing to seasonal effects. Most macroalgae are found to contain high protein content comparable to traditional high protein plant sources for their direct utility in human or animal nutrition (Fleurence 1999; Bleakley and Hayes 2017). Cellulose and other polysaccharide fractions extracted from macroalgae can be converted to ethanol, levulinic acid and its derivatives, including ethyl levulinate, which can be used as a fuel additive for both diesel and biodiesel (Jeong et al. 2015; Francavilla et al. 2016; Trivedi et al. 2016). Presently, India is striving to improve the biomass coupled with the high content of carrageenan by a systematic method of cultivation, such as treating the planting material with the biostimulant AMPEP before outplanting to the sea (Tables 14.2 & 14.3) to obtain good-quality raw material for the production of carrageenan as given in Table 14.4 (Shanmugam et al. 2023).

14.4.12 Market Penetration in International Markets

Indian seaweed industries have developed many products for diverse uses in the food industry, animal feedstocks, biostimulants and several other value-added products (Bixler

Table 14.2 Name of the research centres participating in *Kappaphycus* research in India

Name of the centre	Research institutions/universities
Applied Phycology and Biotechnology Division	CSIR—Central Salt and Marine Chemicals Research Institute, Bhavnagar, Gujarat, India
Marine Algal Research Station	Mandapam Camp, CSIR-CSMCRI, Ramnad District, Tamil Nadu, India
Marine Biotechnology,	ICAR – Central Marine Fisheries Research Institute, Kochi, Kerala India
Regional Research Centre	Regional Centre of Central Marine, Fisheries Research Institute, Mandapam, Tamil Nadu, India
Biotechnology	ICAR—Central Institute of Fisheries Education, Mumbai, Maharashtra, India
Marine Biotechnology	National Institute of Ocean Technology, Chennai , Tamil Nadu, India
School of Marine Sciences	Cochin University of Science and Technology, Cochin, Kerala, India
School of Biotechnology	Madurai Kamaraj University, Palkalai Nagar, Madurai, Tamil Nadu (India)
Centre for Advanced Studies	Faculty of Marine Sciences, Annamalai University in Marine Biology, Tamil Nadu, India
Department of Biotechnology	University of Madras, Guindy Campus, Chennai, Tamil Nadu, India

CSIR Central Scientific & Industrial Research; *ICAR* Indian Council of Agricultural Research

Table 14.3 Average daily growth of *K. alvarezii* treated with different concentrations of AMPEP during different seasons of 2021–2022 (Shanmugam et al. 2023)

Seasonality	Control	Concentration of AMPEP (%)					
		0.025	0.05	0.10	0.15	0.20	0.25
Summer	1.02 ± 0.11	1.33 ± 0.15	1.79 ± 0.11	1.98 ± 0.17	2.07 ± 0.17	2.12 ± 0.15	1.90 ± 0.18
	0.55 ± 0.10	1.35 ± 0.21	1.26 ± 0.20	2.30 ± 0.20	2.62 ± 0.25	2.38 ± 0.10	2.30 ± 0.20
	1.11 ± 0.35	1.93 ± 0.11	2.05 ± 0.13	2.67 ± 0.24	2.74 ± 0.25	2.56 ± 0.27	2.82 ± 0.25
Pre-monsoon	2.14 ± 0.18	2.29 ± 0.17	2.01 ± 0.20	2.95 ± 0.22	2.75 ± 0.21	2.66 ± 0.25	2.91 ± 0.27
	2.26 ± 0.25	2.63 ± 0.08	3.11 ± 0.22	3.22 ± 0.30	3.02 ± 0.27	2.89 ± 0.27	3.11 ± 0.30
	2.91 ± 0.20	1.95 ± 0.10	2.48 ± 0.26	3.57 ± 0.32	3.28 ± 0.30	3.23 ± 0.33	3.42 ± 0.31
Rainy	3.22 ± 0.40	2.90 ± 0.20	3.22 ± 0.29	3.58 ± 0.33	3.54 ± 0.36	3.68 ± 0.31	3.39 ± 0.37
	3.40 ± 0.29	3.50 ± 0.31	3.67 ± 0.39	3.60 ± 0.32	3.77 ± 0.35	3.57 ± 0.38	3.47 ± 0.40
	3.35 ± 0.33	3.15 ± 0.25	3.67 ± 0.35	3.26 ± 0.19	3.42 ± 0.29	3.57 ± 0.37	3.81 ± 0.38
Winter	3.00 ± 0.28	3.15 ± 0.30	3.40 ± 0.34	3.22 ± 0.28	3.40 ± 0.28	3.30 ± 0.25	3.31 ± 0.30
	2.13 ± 0.24	2.95 ± 0.28	2.72 ± 0.12	3.02 ± 0.22	2.96 ± 0.25	2.94 ± 0.27	3.02 ± 0.31
	2.26 ± 0.21	3.07 ± 0.33	2.29 ± 0.10	3.29 ± 0.20	3.00 ± 0.20	2.68 ± 0.23	3.12 ± 0.19

Table 14.4 Quality of raw material and semirefined carrageenan of *K. alvarezii* treated with AMPEP (composite mixture) during different seasons of 2021–2022 (Shanmugam et al. 2023)

AMPEP treatment (%)	Moisture (%)	Impurities (%)	Soluble salt (%)	CAW (%)	SRC yield (%)	Water gel strength (g cm^{-2})	KCl gel strength (g cm^{-2})	Viscosity (cP)
Control	32.16	1.45	28.73	37.66	31.3	255	590	85
0.025	34.22	1.89	24.4	39.49	30.65	275	610	105
0.05	31.85	2.1	30.05	36	30.8	280	585	130
0.1	35.05	1.85	24.18	38.92	33.25	335	625	125
0.15	31.12	2	27.27	39.61	31.25	340	575	95
0.2	32.15	1.7	25.18	40.97	33.15	330	700	155
0.25	34.76	0.8	23.16	41.28	33.33	285	705	85
Mean	33.04	1.68	26.14	39.13	31.96	300.00	627.14	111.43
STD	1.47	0.41	2.39	1.71	1.13	31.62	50.06	24.31

and Porse 2011; Mody et al. 2012; Mondal et al. 2013). These products offered benefits to improve the overall status of several seaweed industries and their allied sectors. The quality of these products can be fine-tuned to compete in international markets for revenue generation and employment opportunities. More recently, AquAgri Processing Private Limited in India had a marketing tie-up with IFFCO, India, in promoting their products to agriculture and food industries. Such efforts undoubtedly provide more awareness of the products and make clear visibility of the industry at the domestic and global levels.

14.5 Market Opportunity of Seaweed Products in India

14.5.1 Biostimulant: Domestic Market and Export Opportunity

Kappaphycus and *Eucheuma* have been the main sources of carrageenan, but recently, they have been proven to be sources of potent agri-biostimulants (Shanmugam and Abhiram 2018). Technology for the production of plant biostimulants (sap) and carrageenan-rich raw material has evolved to reduce the cost of transportation. The sap contains macronutrients, micronutrients and plant growth hormones such as indole 3-acetic acid, gibberellin GA3, kinetin and zeatin (Mantri et al. 2017). Their estimated concentrations reported by AquAgri Processing Private Limited, India, are 56.04 ± 8.68, 77.20 ± 17.38 and 128.42 ± 9.98 ppm for total auxins, cytokinins and gibberellins, respectively. Studies have also confirmed the presence of choline (60.71 mg L^{-1}) and glycine betaine (78.47 mg L^{-1}) in pristine sap, which is associated with tolerance against abiotic stresses in crops (Mondal et al. 2015). Seaweed sap is an ideal organic fertilizer for improving grain size and biomass yield in a wide range of agricultural crops under varied geographical and climatic conditions. This has been confirmed by systematic field trials by pan-India trials over 20 states with the support of 43 agriculture centres. Sap production of M/s. AquAgri Processing Pvt. Ltd. has escalated from 5.25 KL in 2008 to 540 KL in 2020. The US Agency for International Development (USAID) has supported its promotion in Africa under the brand name 'BParas'. The Union Minister of Science and Technology, Government of India, released yet another plant biostimulant based on CSIR-CSMCRI technology under the brand name 'Sagarika' on Technology Day (May 11, 2016) for marketing. Currently marketed products of *Kappaphycus* are shown (Fig. 14.5).

14.5.2 Animal Feed Additives

Aquatic plants such as seaweeds are the classic example of feed resources, which are available in coastal areas throughout the year (Christaki et al. 2010). Seaweed extracts contain major and trace nutrients, amino acids, vitamins and biologically active compounds. Various seaweeds contain growth-promoting substances that stimulate the growth and yield of plants. The consumption of seaweeds by animals and humans has proven to be health promoting, and its benefits are well documented (El-Gamal 2014; Evans and Critchley 2014). Seaweeds are a rich source of natural antioxidants (Matanjun et al. 2008). Sulphated polysaccharides present in *K. alvarezii* inhibit the activity of many bacterial species (Leonard et al. 2010). In view of these findings, *Kappaphycus* forms an important source of animal feedstock and has potential applications in poultry and dairy farming in India.

14.5.3 Food Additives

M/s. AquAgri Processing Pvt. Ltd. located at Manamadurai in Southern Tamil Nadu produces 50 T semirefined carrageenan annually for the domestic beverage industry as a clarifying agent (Mantri et al. 2017). A nutritious and affordable health drink is under process and development from fresh *K. alvarezii.* It is rich in potassium (2500–5000 mg L^{-1}) with adequate quantities of iodine (1000–2000 mg L^{-1}) and other minerals, namely, sodium (150–200 mg L^{-1}), magnesium (750–1250 mg L^{-1}), calcium (250–350 mg L^{-1}), zinc (3–6 mg L^{-1}), phosphorous (25–35 mg L^{-1}), iron (50–210 mg $^{L-1}$) and manganese (1.5–2.5 mg L^{-1}). It also contains 0.2–0.3 ppm thiamine and 0.1–0.2 ppm riboflavin with only a trace amount of fat with a total energy content of 70–100 kcal L^{-1}. Naturally, free-flowing salt contains 30–40% w/w potassium chloride, 20–30% w/w sodium chloride, 0.02–0.5% w/w calcium, 0.02–1.2% w/w magnesium and 0.0005–0.0007% w/w iodine, and the remaining salt is consumed as an ideal health drink against hypertension (Ghosh et al. 2005). The salt is marketed under the brand name 'Saloni K' by M/s. N M S Pharma, Bhavnagar, India. An innovative stand-alone method to obtain 5-hydroxymethyl furfural (HMF) (fuel intermediate), potassium sulphate (K_2SO_4) (agricultural nutrient) and pure water in an integrated manner has been developed from fresh biomass (Mondal et al. 2013) as new products for future markets.

14.5.4 Biomedicals

A few recent studies have demonstrated the beneficial properties of *Kappaphycus* in preventing diabetes (Cyriac and Eswaran 2016) and the ability of its dietary fibres to bind mutagenic amines (Raman and Dobale 2014). Thus, biomass could soon be utilized for developing low-cost pharmaceuticals. Functional foods such as red seaweeds could serve as potential therapeutic options for metabolic syndrome (Brown et al. 2015), Type 2 diabetes and cardiovascular disease. There is some evidence for the antioxidant and blood lipid-lowering properties of *Kappaphycus* (Suresh Kumar et al. 2008; Matanjun et al. 2010). Furthermore, an *in vitro* study using the extracted sulphated fractions of κ-carrageenan from *Kappaphycus* demonstrated the potential of

Fig. 14.5 Commercial products from *Kappaphycus alvarezii.* (**a**) Refined carrageenan powder; (**b**) packed carrageenan; (**c**) animal feed; (**d**) biostimulant Sagarika liquid for foliar application; (**e**) Sagarika granule for soil application (Source: AquAgri Processing Private Limited, India)

this seaweed as an intervention for colon cancer (Raman and Doble 2015). Isolated carrageenan has been used to induce paw oedema to test anti-inflammatory compounds (Posadas et al. 2004; Cong et al. 2015; Morris 2003). As carrageenan is a soluble fibre, it has the properties of slow and steady digestion, which helps improve metabolism (MacArtain et al. 2007). Sulphated polysaccharides, including carrageenans, alginates and porphyrans derived from red seaweeds and fucoidans derived from brown seaweeds, have shown anti-obesity, blood glucose-lowering, and blood lipid-lowering effects in different experimental models (Kim et al. 2012, 2014). This observation reveals enormous scope and possibilities for utilizing *Kappaphycus* as one of the sources in the biomedical industry.

14.5.5 Biodegradable Thin Films

Biodegradable hydrophobic ropes for numerous domestic applications have been prepared (Chaudhary et al. 2015). The grafting of carbohydrate polymers with different substrates exhibits improved properties and functions. An eco-friendly biodegradable thin film from semirefined carrageenan was developed by M/s AquAgri Processing Pvt. Ltd., India. These films are 50–2000 μm in thickness with a tensile strength of 100–1000 kg cm^{-2} and 50–350% elongation ability. The water vapour permeability is 0.01–0.15 kg m^{-2} h^{-1} bar^{-1} (Ghosh et al. 2006). Pouches prepared by heat sealing the film were found to be useful in the storage of edible vegetable oils. The functional modification of carrageenan is yet another area for developing newer biodegradable films and packaging materials depending upon the nature of needs and applications. The cross-linked blend of agar and kappa-carrageenan with natural cross-linker genipin in an aqueous medium was found to exhibit higher viscosity, thermal stability, swelling ability and low weight loss ratio. These results provide yet another opportunity for developing innovative products as super absorbents in food industries, sustained release drugs/formulations in biomedical applications and in several other pH-specific applications (Meena et al. 2007, 2009). Product diversification in line with emerging markets through integrated research and development is one of the areas to focus on in India.

14.6 Conclusion and Future Prospects

The advancement in R&D and innovations are undoubtedly the reason for the successful implementation of commercial cultivation of *K. alvarezii* in India. However, continuous efforts aimed at developing innovative farming methods for onshore, nearshore and offshore production of seaweeds are needed and assume significant importance in the backdrop of burgeoning demand for raw material in India. There is also a need for value chain development for the sustainability of the seaweed industry in the country. Such a holistic approach not only caters to the growing market demands and offers young entrepreneurs and new investments but also contributes to the inclusive economic growth of the country. The geographical expansion of commercial farming is inevitable to explore the underutilized coastal regions for higher biomass production despite some technical, environmental and social challenges. The active participation of the government helps to formulate an exclusive policy framework together with infrastructure development under special schemes, incentives, seaweed parks and seaweed clusters, giving a jumpstart to the seaweed sector. Intensive seaweed cultivation provides higher and more stable income to fishermen, contributing to improved living standards of their families, children's education, health and accessibility to better amenities. Pan-India multi-institutional trials have recorded a 15–40% increase in the yield of several crops in response to the application of biostimulants from seaweeds. The selective accumulation of potash (KCl 30–35% on a dry weight basis) by *Kappaphycus* has opened up a new avenue for the production of liquid phyco-biostimulants for agricultural industries. Population expansion leading to unemployment persuades the migration of the younger generation in search of better livelihood prospects. The establishment of seaweed farming in rural areas would help reduce emigration trends and enhance the quality of life through seaweed cultivation and processing.

Acknowledgements The authors are very grateful to Mr. Abiram Seth, Managing Director, Mr. Tanmaye Seth, Director of AquAgri Processing Private Limited, and Management of IFFCO (Indian Farmers Fertilizer Cooperative Limited) for their constant encouragement and support during the preparation of this chapter. Prof. Doss Ganesh acknowledges the Rashtriya Uchchatar Shiksha Abhiyan (RUSA) programme of Madurai Kamaraj University, Tamil Nadu, India for support during the preparation of this chapter.

References

Agarwal P, Patel K, Das AK, Ghosh A, Agarwal PK (2016) Insights into the role of seaweed *Kappaphycus alvarezii* sap towards phytohormone signalling and regulating defence responsive genes in *Lycopersicon esculentum*. J Appl Phycol 28:2529–2537

Ali MKM, Critchley AT, Hurtado AQ (2020) The impacts of AMPEP K^+ (*Ascophyllum* marine plant extract, enhanced with potassium) on the growth rate, carrageenan quality, and percentage incidence of the damaging epiphyte *Neosiphonia apiculata* on four strains of the commercially important carrageenophyte *Kappaphycus*, as developed by micropropagation techniques. J Appl Phycol 32:1907–1916

Anon (2003) Seaweed cultivation and utilization. National Academy of Agricultural Sciences, vol 22. Policy Paper, India, p 5

Ask EI (2001) Creating sustainable commercial *Eucheuma* cultivation industry: the importance and necessity of human factor. In: Chapman ARO, Anderson RJ, Vreeland VJ, Davison IR (eds) Proceedings 17th International Seaweed Symposium, Cape Town South Africa, pp 13–18

Azanza RV, Aliaza TT (1999) *In vitro* carpospores release and germination in *Kappaphycus alvarezii* (Doty) Doty from Tawi-Tawi, Philippines. Bot Mar 42:281–284

Baghel RS, Trivedi N, Gupta V, Neori A, Reddy CRK, Lali A, Jha B (2015) Biorefining of marine macroalgal biomass for production of biofuel and commodity chemicals. Green Chem 17:2436–2443

Bikker P, Krimpen MM, Wikselaar P, Houweling-Tan B, Scaccia N, Hal JW (2016) Biorefinery of the green seaweed *Ulva lactuca* to produce animal feed, chemicals and biofuels. J Appl Phycol 28:3511–3525

Bindu MS, Levine IA (2011) The commercial red seaweed *Kappaphycus alvarezii*—an overview on farming and environment. J Appl Phycol 23:789–796

Bixler HJ, Porse H (2011) A decade of change in the seaweed hydrocolloids industry. J Appl Phycol 23:321–335

Bleakley S, Hayes M (2017) Algal proteins: extraction, application, and challenges concerning production. Foods 6:33–35

Brown L, Poudyal H, Panchal SK (2015) Functional foods as potential therapeutic options for metabolic syndrome. Obes Rev 16:914–941

Bruno JF, Sweatman H, Precht WF, Selig ER, Schutte VGW (2009) Assessing evidence of phase shifts from coral to macro algal dominance on coral reefs. Ecol 90:1478–1484

Cai J, Hishamunda N, Ridler N (2013) Social and economic dimensions of carrageenan seaweed farming: a global synthesis. In: Valderrama D, Cai J, Hishamunda N, Ridler N (eds) Social and economic dimensions of carrageenan seaweed farming. FAO, Rome, pp 5–59. Fisheries and Aquaculture Technical Paper No. 580

Castelar B, Reis RP, Moura AL, Kirk R (2009) Invasive potential of *Kappaphycus alvarezii* off the south coast of Rio de Janeiro state, Brazil: a contribution to environmentally secure cultivation in the tropics. Bot Mar 52:283–289

Chan JCL, Liu KS (2004) Global warming and Western North Pacific typhoon activity from an observational perspective. J Clim 17:4590–4602

Chaudhary JP, Chejara DR, Eswaran K, Meena R, Ghosh PK (2015) Seaweed-derived polymeric materials for multi applications including marine algal cultivation. RSC Adv 5:19426–19431

Christaki E, Karatzia M, Florou-Paneri P (2010) The use of algae in animal nutrition. J Hellenic Vet Med Soc 61:267–276

Colautti RI, MacIsaac HJ (2004) A neutral terminology to define 'invasive' species. Divers Distrib 10:135–141

Cong HH, Khaziakhmetova VN, Zigashina LE (2015) Rat paw oedema modelling and NSAIDs: timing of effects. Int J Risk Saf Med 27(1): 76–77

Critchley AT, Largo DB, Wee W, Bleicher LG, Hurtado AQ (2004) A preliminary summary on *Kappaphycus* farming and the impact of epiphytes. Jpn J Phycol 52:231–232

Cyriac B, Eswaran K (2016) Anti- hyperglycemic effect of aqueous extract of *Kappaphycus alvarezii* (Doty) Doty ex. P. Silva in alloxaninduced diabetic rats. J Appl Phycol 28:2507–2513

Doty MS, Alvarez VB (1975) Status, problems, advances and economics of *Eucheuma* farms. Mar Technol Soc J 9:30–35

Doty MS, Alvarez VB (1981) *Eucheuma* farm productivity. Int Seaweed Symp 8:688–691

El-Gamal AA (2014) Biological importance of marine algae. In: Se-Kwon K (ed) Handbook of marine macroalgae: biotechnology and applied phycology. John Wiley and Sons, p 567

Evans FD, Critchley AT (2014) Seaweeds for animal production use. J Appl Phycol 26:891–899

Eswaran K, Ghosh PK, Mairh OP (2002) Experimental field cultivation of *Kappaphycus alvarezii* (Doty) Doty ex Silva at Mandapam region. Seaweed Res Utilin 24:67–72

Fleurence J (1999) Seaweed proteins: biochemical, nutritional aspects and potential uses. Trends Food Sci Technol 10:25–28

Francavilla M, Intini S, Luchetti L, Luque R (2016) Tunable microwave-assisted aqueous conversion of seaweed-derived agarose for the selective production of 5-hydroxymethyl furfural/levulinic acid. Green Chem 18:5971–5977

Gajaria TK, Suthar P, Baghel RS, Balar NB, Sharnagat P, Mantri VA, Reddy CRK (2017) Integration of protein extraction with a stream of byproducts from marine macroalgae: a model forms the basis for marine bioeconomy. Bioresour Technol 243:867–873

Ganesan M, Eswaran K, Reddy CRK (2017) Farming of agarophytes in India–a long time 2248

Ganesan M, Trivedi N, Gupta S, Madhav V, Reddy CRK (2019) Levine LA (2019) seaweed resources in India–current status of diversity and cultivation: prospects and challenges. Bot Mar 62(5):463–482

Ghosh PK, Mody KH, Reddy MP, Patolia JS, Eswaran K, Shah R, Barot B, Gandhi MR, Mehta AS, Bhatt AM, Reddy AVR (2005) Low sodium salt of botanic origin. US patent, US20050220975A1

Ghosh PK, Siddhanta AK, Prasad K, Meena R, Bhattacharya A (2006) Process of preparation of biodegradable films from semi refined kappa carrageenan. US patent US7067568B1

Hayashi L, Reis RP, dos Santos AA, Castelar B, Robledo D, de Vega GB, Msuya FE, Eswaran K, Yasir SM, Ali MKM, Hurtado AQ (2017) The cultivation of *Kappaphycus* and *Eucheuma* in tropical and subtropical waters. In: Hurtado A, Critchley A, Neish I (eds) Tropical seaweed farming trends, problems and opportunities. Developments in applied phycology, vol 9. Springer, Cham, pp 55–90

Hayashi L, Hurtado AQ, Msuya FE, Bleicher-Lhonneur G, Critchley AT (2010) A review of *Kappaphycus* farming: prospects and constraints. In: Seckbach J (ed) Seaweeds and their role in globally changing environments. Springer, Dordrecht, pp 251–283

Hean RL, Cacho O (2002) Mariculture of giant clams, *Tridacna crocea* and *T. derasa*: management of maximum profit by small holders in Solomon Islands. Aquac Econ Manag 6:373–395

Hurtado AQ (2013) Social and economic dimensions of carrageenan seaweed farming in The Philippines. In: Valderrama D, Cai J, Hishamunda N, Ridler N (eds) Social and economic dimensions of carrageenan seaweed farming, Fish Aqua Tech Paper, vol 580. FAO, Rome, pp 91–113

Hurtado AQ, Critchley AT, Trespoey A, Lhonneur GB (2006) Occurrence of *Polysiphonia* epiphytes in *Kappaphycus* farms at Calaguas is., Camarines Norte Phillippines. J Appl Phycol 18:301–306

Ingle K, Vitkin E, Robin A, Yakhini Z, Mishori D, Golberg A (2011) Macroalgae biorefinery from *Kappaphycus alvarezii*: conversion modelling and performance prediction for India and Philippines as examples. Bioenergy Res 11:22–28

Jeong GT, Ra CH, Hong KY, Kim JK, Kong IS, Kim SK, Park DH (2015) Conversion of red-algae *Gracilaria verrucosa* to sugars, levulinic acid and 5-hydroxymethylfurfural. Bioprocess Biosyst Eng 38:207–217

Karthikeyan K, Shanmugam M (2017) The effect of potassium-rich biostimulant from seaweed *Kappaphycus alvarezii* on yield and quality of cane and cane juice of sugarcane var. co 86032 under plantation and ratoon crops. J Appl Phycol 29:3245–3252

Kim KJ, Yoon KY, Lee BY (2012) Fucoidan regulate blood glucose homeostasis in C57BL/KSJ m+/+db and C57BL/KSJ db/db mice. Fitoterapia 83:1105–1109

Kim MJ, Jeon J, Lee JS (2014) Fucoidan prevents high-fat diet-induced obesity in animals by suppression of fat accumulation. Phytother Res 28:137–143

Krishnan M, Narayanakumar R (2010) Socioeconomics of seaweed farming in India. Central Marine Fisheries Research Institute, Kochi, India, Special Bulletin No 104:103

Largo DB, Fukami F, Nishijima T, Olino M (1995a) Laboratory induced development of the ice-ice disease of the farmed red algae *Kappaphycus alvarezii* and *Eucheuma denticulatum* (Solieriaceae, Gigartinales, Rhodophyta). J Appl Phycol 7:539–543

Largo DB, Fukami F, Nishijima T (1995b) Occasional bacteria promoting ice-ice disease of the farmed red algae *Kappaphycus alvarezii* and *Eucheuma denticulatum* (Solieriaceae, Gigartinales, Rhodophyta). J Appl Phycol 7:545–554

Leonard SG, Sweeney T, Pierce KM, Bahar B, Lynch BP, O'Doherty JV (2010) The effects of supplementing the diet of the sow with seaweed extracts and fish oil on aspects of gastrointestinal health and performance of the weaned piglet. Livest Sci 134:135–138

Loureiro RR, Reis RP, Berrogain FD, Critchley AT (2011) Extract powder from the brown alga *Ascophyllum nodosum* (Linnaeus) Le Jolis (AMPEP): a "vaccine-like" effect on *Kappaphycus alvarezii* (Doty) Doty ex P.C. Silva. J Appl Phycol. https://doi.org/10.1007/s10811-011-9735-7

MacArtain P, Gill CIR, Brooks M, Campbell R, Rowland IR (2007) Nutritional value of edible seaweeds. Nutr Rev 65:535–543

Mairh OP, Zodape ST, Tewari A, Rajyaguru MR (1995) Culture of marine red algae *Kappaphycus striatum* (Schmitz) Doty on the Saurashtra region, west coast of India. Indian J Mar Sci 24:24–31

Mandal SK, Mantri VA, Haldar S, Eswran K, Ganesan M (2010) Invasive potential of *Kappaphycus alvarezii* on corals at Kurusadai Island, gulf of Mannar, India. Algae 25:205–216

Mantri VA, Eswaran K, Shanmugam M, Ganesan M, Veeragurunathan V, Thiruppathi S, Reddy CRK, Seth A (2017) An appraisal on commercial farming of *Kappaphycus alvarezii* in India: success in the diversification of livelihood and prospects. J Appl Phycol 29:335–357

Matanjun P, Mohamed S, Muhammad K, Mustapha NM (2010) Comparison of cardiovascular protective effects of tropical seaweeds, *Kappaphycus alvarezii*, *Caulerpa lentillifera*, and *Sargassum polycystum*, on high cholesterol/high-fat diet in rats. J Med Food 13:792–800

Matanjun P, Mohamed S, Mustapha NM, Muhammad K, Ming CH (2008) Antioxidant activities and phenolics content of eight species of seaweeds from North Borneo. J Appl Phycol 20:367–373

McHugh DJ (2003) A guide to the seaweed industry. FAO Fisheries Technical Paper. Food and Agriculture Organization of the United Nations, Rome

Meena R, Prasad K, Siddhanta AK (2007) Effect of genipin, on the properties of *kappa*-carrageenan. Int J Biol Macromol 41:94–101

Meena R, Prasad K, Siddhanta AK (2009) Development of a stable hydrogel network based on agar–*kappa*-carrageenan blend cross linked with genipin. Food Hydrocolloid 23:497–509

Meinita MDN, Kang JY, Jeong GT, Koo HM, Park SM, Hong YK (2012) Bioethanol production from the acid hydrolysate of the carrageenophyte *Kappaphycus alvarezii* (cottonii). J Appl Phycol 24:857–862

Mody KH, Ghosh PK, Barindra S, Gnanasekaran G, Shukla AD, Eswaran K (2012) A process for integrated production of ethanol and seaweed sap from *Kappaphycus alvarezii*. European Patent EP2475776 A1

Mondal D, Sharma M, Maiti P, Prasad K, Meena M, Siddhanta AK, Bhatt P, Ijardar S, Mohandas VP, Ghosh A, Eswaran K, Shah BG, Ghosh PK (2013) Fuel intermediates, agricultural nutrients and pure water from *Kappaphycus alvarezii* seaweed. RSC Adv 3:17989–17997

Mondal D, Ghosh A, Prasad K, Singh S, Bhatt N, Zodape ST, Chaudhary JP, Chaudhari J, Chatterjee PB, Seth A, Ghosh PK (2015) Elimination of gibberellin from *Kappaphycus alvarezii* seaweed sap foliar spray enhances corn Stover production without compromising the grain yield advantage. Plant Growth Regul 75: 657–666

Morris CJ (2003) Carrageenan-induced paw edema in the rat and mouse. In: Winyard PG, Willoughby DA (eds) Inflammation protocols. Humana Press, Totowa, NJ, USA, pp 115–121

Msuya FE (2013) Social and economic dimensions of carrageenan seaweed farming in the United Republic of Tanzania. In: Valderrama D, Cai J, Hishamunda N, Ridler N (eds) Social and economic dimensions of carrageenan seaweed farming, fish aqua tech paper 580. FAO, Rome, pp 115–146

Neish IC (2008) Good agronomy practices for *Kappaphycus* and *Euchema*: including an overview of basic biology. SEAPlant.net monograph no. HB2F 1008 V3 GAP. Makassar, Indonesia. 72 pp.

Neish IC, Sepulveda M, Hurtado AQ, Critchley AT (2017a) Reflection of commercial development of eucheumatoid seaweed farming. In: Hurtado AQ, Critchley AT, Neish IC (eds) Tropical seaweed farming trends, problems and opportunities: focus on *Kappaphycus* and *Eucheuma*. Springer, Cham, pp 1–27

Orduna J, Robledo D, Dawes C (2002) Studies on the tropical–agarophyte *Gracilaria cornea* J. Agardh (Rhodophyta, Gracilariales) from Yucatan, Mexico: I. Seasonal, physiological and biochemical responses. Bot Mar 45:453–458

Pascoe S, Wattage P, Naik D (2002) Optimal harvesting strategies: practice versus theory. Aquac Econ Manag 6:295–230

Periyasami C, Anantharaman P, Balasubramanian T (2014) Social upliftment of coastal fisher women through seaweed (*Kappaphycus alvarezii* (Doty) Doty) farming in Tamil Nadu, India. J Appl Phycol 26:775–781

Periyasami C, Anantharaman P, Subba Rao PV (2015) Experimental farming of *Kappaphycus alvarezii* (Doty) Doty with income estimates at different sites in the Mandapam region, Palk Bay, southeast coast of India. J Appl Phycol 27:935–944

Porse H, Rudolph B (2017) The seaweed hydrocolloid industry: 2016 updates, requirements, and outlook. J Appl Phycol 29:2187–2200

Posadas I, Bucci M, Roviezzo F, Rossi A, Parente L, Sautebin L, Cirino G (2004) Carrageenan-induced mouse paw oedema is biphasic, age-weight dependent and displays differential nitric oxide cyclooxygenase-2 expression. Br J Pharmacol 142:331–338

Raman M, Dobale M (2014) Physicochemical and structural characterization of marine algae *Kappaphycus alvarezii* and the ability of its dietary fibres to bind mutagenic amines. J Appl Phycol 26:2183–2191

Raman M, Doble M (2015) κ-Carrageenan from marine red algae, *Kappaphycus alvarezii*—a functional food to prevent colon carcinogenesis. J Funct Food 15:354–364

Reddy CRK, Kumar GRK, Siddhanta AK, Tewari A, Eswaran K (2003) *In vitro* somatic embryogenesis and regeneration of somatic embryos from pigmented callus of *Kappaphycus alvarezii* (Doty) Doty (Rhodophyta, Gigartinales). J Phycol 39(3):610–616

Reddy CRK, Jha B, Fujita Y, Ohno M (2008a) Seaweed micropropagation techniques and their potentials: an overview. J Appl Phycol 20(5):609–617

Reddy CRK, Gupta MK, Mantri VA, Jha B (2008b) Seaweed protoplasts: status, biotechnological perspectives and needs. J Appl Phycol 20(5):619–632

Reddy CRK, Gupta V, Jha B (2010) Developments in biotechnology of red algae. In: Chapman DJ, Seckbach J (eds) Red algae in genomic age: cellular origin, life in extreme habitats and astrobiology. Springer, New York, pp 307–341

Robinson N, Winberg P, Kirkendale L (2013) Genetic improvement of macroalgae: status to date and needs for the future. J Appl Phycol 25: 703–716

Robledo D, Gasca-Leyva E, Fraga J (2013) Social and economic dimensions of carrageenan seaweed farming in Mexico. In: Valderrama D, Cai J, Hishamunda N, Ridler N (eds) Social and economic dimensions of carrageenan seaweed farming, Fish aqua tech paper, vol 580. FAO, Rome, pp 185–204

Rossita S, Safiin NSZ, Senoo S (2015) Improving dietary reds seaweed *Kappaphycus alvarezii* (Doty)Doty ex. P. Silva meal utilization in Asian seabass *Lates calcarifer*. J Appl Phycol 27:1681–1688

Shanmugam M, Abhiram S (2018) Recovery ratio and quality of an agricultural biostimulant and semiquined carrageenan coproduced from the fresh biomass of *Kappaphycus alvarezii* with respect to seasonality. Algal Res 32:262–271

Shanmugam M, Senthil Rajan T, Eswaran K, Abhiram S, Hurtado AQ (2023) Application of the brown seaweed-derived biostimulant AMPEP to the commercial farming of the carrageenophyte *Kappaphycus alvarezii* in India: growth enhancement and production of quality raw material. Algal Res 71:103041

Singh S, Singh MK, Pal SK, Trivedi K, Yesuraj D, Singh CS, Anand KGV, Chandramohan M, Patidar R, Kubavat D, Zodape ST, Ghosh A (2016a) Sustainable enhancement in yield and quality of rain-fed maize through *Gracilaria edulis* and *Kappaphycus alvarezii* seaweed sap. J Appl Phycol 28:2099–2112

Singh S, Singh MK, Pal SK, Trivedi K, Yesuraj D, Singh CS, Anand KGV, Chandramohan M, Patidar R, Kubavat D, Zodape ST, Ghosh A (2016b) Erratum to: sustainable enhancement in yield and quality of rain-fed maize through *Gracilaria edulis* and *Kappaphycus alvarezii* seaweed sap. J Appl Phycol 28:2113

Stowasser M, Wang Y, Hamilton K (2007) Tropical cyclone changes in the Western North Pacific in a global warming scenario. J Clim 20: 2378–2396

Suresh Kumar K, Ganesan K, Subba Rao PV (2008) Antioxidant potential of solvent extracts of *Kappaphycus alvarezii* (Doty) Doty- an edible seaweed. Food Chem 107:289–295

Terada R, Duy Vo T, Nishihara GN, Shioya K, Shimada S, Kawaguchi S (2015) The effect of irradiance and temperature on the photosynthesis and growth of a cultivated red alga *Kappaphycus alvarezii* (Solieriaceae) from Vietnam, based on *in situ* and *in vitro* measurements. J Appl Phycol 28:457–467

Trivedi N, Baghel RS, Bothwell J, Reddy CRK, Lali A, Jha B (2016) An integrated process for the extraction of fuel and chemicals from marine macro algal biomass. Sci Rep 6:1–8

Trono GC Jr (1974) *Eucheuma* farming in The Philippines. U. P. Natural Science Research Center, Quezon City

Uyenco FR, Saniel LS, Jacinto GS (1981) The "ice-ice" problem in seaweed farming. Proc Int Seaweed Symp 10:625–630

Vairappan CS (2006) Seasonal occurrences of epiphytic algae on the commercially cultivated red alga *Kappaphycus alvarezii* (Solieriaceae, Gigartinales, Rhodophyta). J Appl Phycol 18:611–661

Vairappan CS, Chong SC, Lhonneur GB, Critchley AT (2007) Epiphyte outbreak in *Kappaphycus alvarezii* (Gigartinales, Rhodophyta) farms in South East Asia: impact on the yield and quality of carrageenan. Int Seaweed Symp 19:80

Valderrama D, Cai J, Hishamunda N, Ridler N, Neish IC, Hurtado AQ, Msuya FE, Krishnan M, Narayanakumar R, Kronen M, Robledo D, Gasca-Leyva E, Fraga J (2015) The economics of *Kappaphycus* seaweed cultivation in developing countries: a comparative analysis of farming systems. Aquac Econ Manag 19:251–277

Boutique-Type Cultivation of *Kappaphycus alvarezii* (Doty) L.M. Liao in the Subtropical Waters of Tosa Bay, Shikoku, Japan

15

Masao Ohno, Danilo B. Largo, and Christine A. Orosco

Abstract

The red seaweed, *Kappaphycus alvarezii*, has been introduced to several countries (e.g., Brazil and Tunisia) outside of its origin in the Philippines, mainly for the extraction of the polysaccharide carrageenan. Unlike other countries, Japan started to cultivate the seaweed in 1991 in the subtropical waters of Tosa Bay, in Kochi Prefecture, to produce fresh seaweed to meet the demand from specialized gourmet restaurants offering seaweeds on their menus for phycogastronomy. Harvested biomass is also being used as a source of biostimulant for agricultural/horticultural applications. Initial research on *K. alvarezii* in Japan also discovered the seaweed as a source of lectin. Although relatively small in terms of production scale compared to other countries where *K. alvarezii* has been introduced, the innovation made on the culture of *K. alvarezii* developed by the local seaweed farmers in Tosa Bay involving over-wintering process in a land-based facility was able to sustain a niche market for this seaweed in Japan. The innovative method consists of placing small, palm-size fragments of the cultivars in plastic containers inside nursery tanks maintained at 23° C under natural light. This practically allowed the survival of the seaweed in Japan during cold winter months until the next growing season in the field. Given the name "nettai kirinsai", or "tropical kirinsai", to differentiate it from other kirinsai species in Japan, *K. alvarezii* is now cultivated in Uranouchi Inlet in Tosa Bay, from late spring in May, when surface seawater temperature (SST) begin to rise to 23 °C, doubling growth upon reaching 28–29°C in August. However, it suffers from ice-ice disease when the SST reaches an extreme of 32°C in the peak of summer and when the SST falls down again to below 23°C towards winter. The seaweed is planted by suspending the cultivars on horizontal lines which sink slightly (0.5–1.0 m) below the surface from a floating platform in which fish is co-cultured in net cages. The cultured seaweed not only supplies the gourmet market and the agricultural/horticultural sectors, but it also serves to bioremediate nitrogenous wastes released from fish culture in Tosa Bay.

Keywords

Boutique-type seaweed cultivation · *Kappaphycus alvarezii* · Phycogastronomy · Over-wintering · Tosa Bay · Japan

M. Ohno
Usa Marine Biological Institute, Kochi University, Tosa, Japan

D. B. Largo (✉)
Marine Biology Section, Department of Biology, School of Arts and Sciences, University of San Carlos, Cebu City, Philippines
e-mail: dblargo@usc.edu.ph

C. A. Orosco
South China Sea Repository & Reference Center, Institute of Oceanography and Environment (INOS), Universiti Malaysia Terengganu (UMT), Kuala Nerus, Terengganu, Malaysia

15.1 Introduction

Kappaphycus alvarezii is a tropical red seaweed that was introduced to Japan in the early 1990s from the Philippines and from which other countries received clones for test planting that eventually became part of their boutique-type farming to obtain high-value products. With the success of this type of farming in Tosa Bay of Kochi Prefecture on Shikoku Island, Japan is now a *K. alvarezii*-producing subtropical country in the same group as Brazil and Tunisia. However, Japan is slightly different, as it is smaller in scale and is not farming for the carrageenan market but for niche markets, mainly in phycogastronomy. Tunisia, on the other hand, has private companies (e.g. Selt Marine Group) that grow its own seaweeds, including eucheumatoids, and has its own modern factories to process them into niche products such as cosmetics, delicatessens, water fruits and confectioneries. (https://seltmarinegroup.com/a-propos/). Brazil succeeded in farming *K. alvarezii* from a vegetative clone introduced to

A. T. Critchley et al. (eds.), *Tropical Phyconomy Coalition Development*, Developments in Applied Phycology 11,
https://doi.org/10.1007/978-3-031-47806-2_15

Japan from the Philippines in the 1990s into its tropical northeastern region. Here, farming can be performed throughout the year, whereas its south and southeast waters, which are subtropical, seedstocks of seaweeds, spend the winter in land-based facilities before the next farming season (Pedra et al. 2017; Simioni et al. 2019). Concerns were expressed about the possible negative consequences of farming-introduced eucheumatoids. However, there have only been two cases in Hawaii and India, where carrageenophytes were considered invasive (Glenn and Doty 1990; Chandrasekaran et al. 2014).

In southern Japan, where small-scale cultivation has been undertaken, harvested biomass serves to provide a steady supply of raw materials for special purposes. For instance, *K. alvarezii* has been used for the extraction of lectin, which has applications in biotechnology and the gourmet food market. A large-scale joint investment project between the Ministry of Agriculture, Forestry and Fisheries and the private sector was established in 1991. A decade of research to extract and commercialize has begun. Because lectin is a protein, the study required a supply of *K. alvarezii* raw samples throughout the year. However, the seaweed grows only for a short period of time in a year, when the water temperature reaches a level that is optimal for the seaweed, i.e. between 23 and 30°C. The cultivation of *K. alvarezii* in Tosa Bay, Shikoku Island, Japan, has continued to the present.

15.2 The Cultivation Site (Tosa Bay, Shikoku, Japan)

Tosa Bay is an open curved bay along the entire Pacific coast of Kochi Prefecture on Shikoku Island, southwestern Japan, in the western North Pacific. The shoreline is mostly sandy and intermittent with rocky shores. It is strongly influenced by the warm Kuroshio Current from the Equator flowing southwest to the northeast at the opening of Tosa Bay, making the bay a boundary zone between temperate and subtropical areas. Warm-temperate areas and subtropical districts are homes to distinct flora and faunal communities. Hermatypic corals are abundant (Iwase 2004). The brown seaweed *Sargassum* covers the shallow rocky areas throughout the year, with peak growth in summer.

Uranouchi Inlet, which is Tosa Bay's innermost labyrinth (Fig. 15.1) that extends 12 km inwards of protected waters with a maximum depth of 20 m, is a highly suitable site for *Kappaphycus* farming. Several aquaculture activities are carried out at this site, including the Kochi Prefecture Fisheries Research and Experimental Station. The seaweed flora here belongs to Region II with a temperate northern area and a subtropical southern area (Tokuda and Takeda 1989). Various seaweed species grow from spring to autumn, when the water temperature is warmer than that in most parts of the year.

In the mid-1980s to early 1990s, the sea surface temperature (SST) in this area ranged from 13.8°C in January to 29.4°C in August. However, reports of changes in seaweed community composition, loss of temperate brown species and increasing dominance of tropical species attributed to increased SST are increasing (Nagai et al. 2011; Tanaka et al. 2012; Hiraoka et al. 2020).

15.3 Eucheumatoid Cultivation Trials in Tosa Bay and Vicinity

Studies on the cultivation of eucheumatoids in Tosa Bay started in 1983 using *Kappaphycus striatus (*as *Eucheuma striatum)* obtained from Iloilo, Philippines (Mairh et al. 1986). Subsequently, other species, namely, *Kappaphycus alvarezii* and *Eucheuma denticulatum*, were also introduced for cultivation trials in Tosa Bay.

In 1991, a large-scale joint investment project between the Ministry of Agriculture, Forestry and Fisheries and the private sector was established with extraction of lectin from freshly harvested eucheumatoids as the main activity (Ohno and Largo 2020). Two kilograms of *K. alvarezii* (as *K. muricatum* in Ohno and Largo 2020) consisting of green and brown morphotypes were brought in from a seaweed farm in Danajon Reef in Bohol, Philippines, in 1992, which served as a starting material for biomass production at Uranouchi Inlet in Tosa Bay, Shikoku, Japan. From May 1992, this species was grown in the sea every year from late spring in May to early autumn or October to November for 12 years (Ohno et al. 1994) and used as experimental samples. Initially, *K. alvarezii* did not have a Japanese name, but to differentiate it from the other kirinsai species in Japan, it was given the name "nettai kirinsai", meaning "tropical kirinsai" (Yano et al. 2010).

15.4 Commercial Cultivation of *Kappaphycus alvarezii*

Yearly cultivation trials in Tosa Bay of *Kappaphycus alvarezii* have been performed during the warmer months in Japan from late spring to early autumn. Starting in late autumn, seed stocks were originally maintained in an incubator (Aquatron™, Kochi University) at a slightly lower temperature of 23°C than at its optimal temperature of 25°C due to the difficulty of raising the Aquatron temperature to a higher level. There were problems with growth and health, especially towards winter, due to low temperatures. By 2015, the group decided that the farming method it had practised for

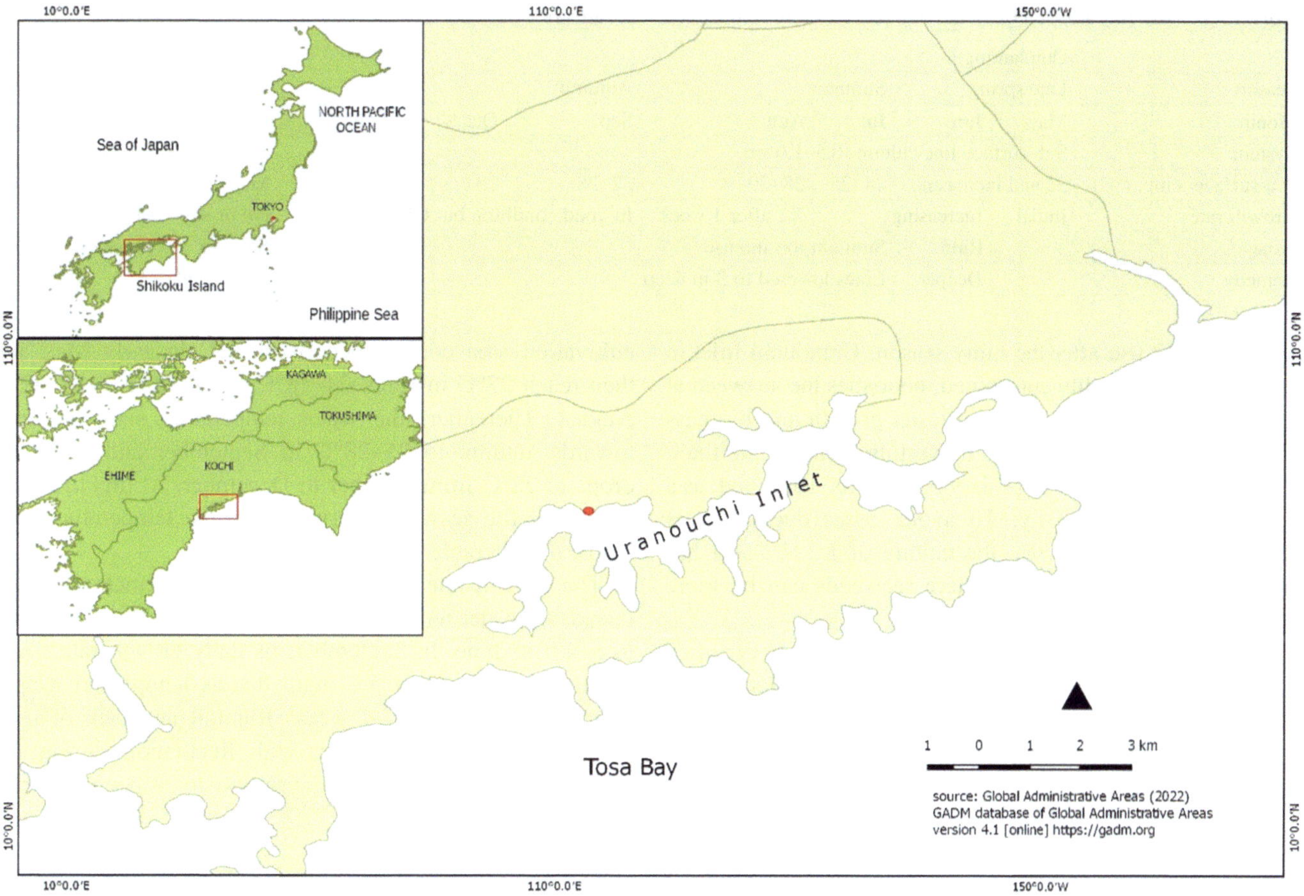

Fig. 15.1 Map of southwestern portion of Japan showing Uranouchi Inlet, Tosa Bay, in Shikoku Is. where small-scale cultivation of *Kappaphycus alvarezii* has been going on until today

several years was now ripe for expansion to commercial scale. This time, the main purpose of cultivating the eucheumatoid in Tosa Bay shifted its use to direct human food, or as functional ingredients (Ohno and Largo 2020). For five consecutive years since then, harvesting of *K. alvarezii* in Tosa Bay cultivation trials has not failed, as long as over-wintering of enough materials, in controlled conditions, is carried out to make seedlings available for the next planting season. Therefore, a two-step farming protocol was followed, consisting mainly of outplanting in the field during the warmer months from spring to autumn and over-wintering in indoor nursery tanks during the winter months when seawater temperature is too low for *K. alvarezii* to survive (Table 15.1).

15.5 Outplanting

Prior to field cultivation, the *Kappaphycus* thalli were prepared by tying palm-sized seedlings on long lines (Fig. 15.2a). The lines are then taken out to sea in spring when seawater starts to warm up. When the SST reaches 23° C, the seaweeds are kept at the same temperature during their over-wintering period in on-land tanks. Growth increases as SST increases as the summer season progresses. The seaweeds grow faster in June when SST is 24–25°C, resulting in a mean daily growth rate of 3–4% day^{-1}, reaching a peak (i.e., doubling their fresh weight in a week; approx. 5% day^{-1}, or higher) in August when SST reaches 28–29°C. However, growth decreases when SST in summer reaches 30°C and peaks at 32°C, which is the species' upper limit for optimal growth (Ohno et al. 1994).

To counter the high light intensity from the intense summer sun, which often results in the occurrence of ice-ice disease (Largo et al. 1995), the lines with seaweeds are usually lowered deeper to 1 m below the surface (Fig. 15.2b). However, in June, when the salinity decreases to 19–24 psu, the seaweed lines are sunk further to more than 1 m. *Kappaphycus* prefers depths below the mixed layer resulting from turbulence by wind or waves. Growth improves and continues even as the SST decreases towards autumn from 28 to 23°C in September. Salinity increases

Table 15.1 Tosa Bay *Kappaphycus* farming protocol followed by seaweed farmers

	Outplanting							Over-wintering				
Season	Late spring		Summer		Autumn			Winter to early spring				
Month	May	Jun	Jul	Aug	Sep	Oct	Nov	Dec	Jan	Feb	Mar	Apr
System	Sub-surface line culture (0.5–1.0 m)							Tank culture				
Sea-surface temp. (°C)	23 and increasing		24–25	28–29	32–28	23		Maintain at 23				
Growth rate	Initial	Increasing		X2 after 1 week	In good condition but GR decreasing			Nil or slow but healthy				
Issues		Rain	Sunlight too intense					Sunlight too strong				
Remedy		Deeper	Lines lowered to 3 m deep					Cover tanks				

back to 29–32 psu after the rainy season. Uranouchi Inlet in Tosa Bay, being highly eutrophied, nourishes the seaweed at optimal temperature, resulting in faster growth and biomass accumulation that necessitates harvesting some of the thalli between June and August, as they become too large and, as a consequence, break easily. To avoid losses due to water turbulence in the open sea, the culture of *K. alvarezii* has moved to fish farming rafts where seaweeds can be more protected (Ohno and Largo 2020).

Cultivation with fish culture. Co-cultivation of *K. alvarezii* with fish culture was attempted in the Uranouchi Inlet in 10 m × 10 m floating net cages. Here, the seaweed cultivation ropes are sunk at a depth of approximately 50 cm with sinkers attached to both ends. The water temperature in the cultivation area could reach over 20°C in early May and then reach 23°C in June, 24–28°C in July and 29–32°C in August. Thereafter, the water temperature starts to drop towards autumn to 25–28°C in September and gradually drops to 22°C from October to December. At this time, the culture of the seaweed is over when the temperature falls below the tolerable level of *K. alvarezii*.

The growth rate of *K. alvarezii* cultivars increases with increasing water temperature, with a daily growth rate of up to 3–4% in June. In September, its daily growth rate could increase to more than 5%, with thallus length and weight more than doubling in 1 week. Rainfall and lack of light could slow the growth rate, with fluctuation during the rainy season in June causing the salinity to drop significantly

Fig. 15.2 The culture of *Kappaphycus alvarezii* in Tosa Bay. (**a**) Green *K. alvarezii* as the only morphotype being cultivated in Japan, (**b**) rope with cultivars of *K. alvarezii* sunk 0.5 m below sea surface, (**c**) over-wintering of *K. alvarezii* in land-based tanks where palm-sized seedlings are placed in plastic containers, (**d**) land-based tanks with cover of plastic sheets to reduce light intensity. (Photocredit: M. Ohno)

to 19–24 psu. Salinity increases back to 29–32 psu in summer, which is quite favourable to *K. alvarezii*. However, when the water temperature exceeds 30°C in summer, the growth rate slows down. The period of highest growth rate is from September to November, when the water temperature is 23–28°C.

During the harvesting period, half of the harvested seaweeds are cut into palm-sized cultivars and are attached back to the ropes after cleaning. In the past, 1 kg of seedlings from over-wintered materials could have a yield of as much as 70–100 kg at the end of the culture period in December. During the culture period, the decrease in salinity during the rainy season and the high temperature in summer greatly affected the production.

Before the SST drops to 23°C in September, harvesting is performed with the premium *Kappaphycus* thalli selected and used for over-wintering at 23°C in semi-outdoor nursery tanks under a roofed structure. This conserved biomass becomes the seedstock for the following outplanting season. The remainder of the biomass is taken as harvest. The harvested biomass could stay fresh for up to a week as long as it is kept moist with seawater (Ohno and Largo 2020).

Net cultivation method. The current method of culturing eucheumatoid species in Tosa Bay commonly uses the floating rope method. This method is carried out in the open sea, but a large number of fronds fall off from the rope during rough seas. Therefore, an alternative method was tested by placing the cultivars in a tube net. In this method, new branches are produced and can elongate outside the net. This, however, depends on the kind of material the net is composed of. If the net is made of thick, synthetic cotton thread, it will easily become fouled and very dirty. Nets used for nori farming have been tested for *K. alvarezii* cultivation and proved to be a far better material with less fouling and easier insertion of larger seedlings. In the future, the development of better cultivation methods that allow the insertion of seedings at minimal loss in open sea cultivation will be a main priority.

15.6 Over-wintering

When the water temperature in Tosa Bay drops to below 20° C, the growth of *K. alvarezii* stops, and the cultivars become softer and start to lose colour. They then become yellowish green. Every year when the water temperature in the cultivation site approaches 20°C, seaweed farmers start preparing materials for over-wintering to ensure the availability of seedlings for the next growing season. Approximately 20 kg of harvested seaweed is maintained in a one-tonne tank installed in a nursery facility with a roof. Seawater, warmed using a heater, is added into the tank little by little until the temperature reaches 23°C. The culture seawater in the tank was then continuously aerated. During this period in winter, the seaweed thalli are maintained in the outdoor tank in a state in which the thalli grow only slightly. A new method of over-wintering in a culture tank is cutting the seaweed into palm-sized seedlings, which are kept in individual plastic containers (Fig. 15.2c). The bright sun could be a problem, and this is solved by covering the tanks with light-intensity-controlling plastic netting to lessen the light intensity (Fig. 15.2d). The prospect of cultivating tropical *K. alvarezii* in Japan is promising. Its growth rate could double in a week when the water temperature reaches 23–28°C in Tosa Bay, at which time the daily growth rate could often exceed 4%. The daily growth rate of *K. alvarezii* cultivated in the Philippines is between 4 - 5% (Trono and Ohno 1989); therefore, the growth rate of *K. alvarezii* cultured in Tosa Bay is not very far from its source of origin. It was observed by the farmers of Tosa Bay that *K. alvarezii*'s growth rate slows down when the water temperature drops below 23°C in November, causing the green colour of the alga to become pale. However, even at a water temperature of 18°C in winter, although growth is retarded, it can still grow at a normal rate. Therefore, if seedlings are secured by tank culture, it will be possible to cultivate *K. alvarezii* in many sea areas in Japan where the temperature is over 20°C from May to November. Although growth is better if salinity is high, it was also found that even at a low salinity of 15 psu, they can still survive and continue to grow. Sandy sea areas where algae-eating fish do not exist are suitable for farming (Yano and Ohno 2007).

15.7 Harvesting and Post-Harvesting

The harvested seaweeds are transported, cleaned and prepared for marketing or for the production of certain products. A cycle in a good year that started with 20 kg of seedlings for over-wintering could yield a total of 2 metric tonnes of fresh seaweed harvest. The decreased salinity during the rainy season and the high temperature in summer could have a major impact on the harvest, one that exhibits soft thalli and/or pale to whitened branches similar to ice-ice conditions.

15.8 Uses of Japanese Eucheumatoid Products

Kappaphycus products can be used either as food or in nonfood applications. These products come in four basic forms:

1. Live/fresh *kirinsai* that can be kept fresh for a week and used as ingredients to prepare salads

Fig. 15.3 *Kappaphycus alvarezii* in Japanese phycogastronomy. (**a**) Slightly blanched *K. alvarezii*, (**b**) *K. alvarezii* as seaweed salad, (**c**) Italian-style cuisine with *K. alvarezii*, (**d**) kirinsai spaghetti. (Photocredit: M. Ohno)

2. Frozen *kirinsai* that can be kept for a long time
3. *Kirinsai*-enzyme mix with unique herbs growing in Kochi
4. Powder form from freeze-dried *kirinsai* for extraction of novel products (e.g. lectin)

As ingredients in phycogastronomy. The reason why *K. alvarezii* cultivation thrives in Tosa Bay is due to its demand as an ingredient in Italian- and French-style cuisines (Fig. 15.3). Approximately 30% of the harvest goes into the gourmet market for phycogastronomy. Adding a small amount of the dry powder to kamaboko (fish paste) and jelly is popular for its viscosity and texture-enhancing properties, so it is likely that more applications in this field will be developed in the future.

The current world trend started with high-level chefs working with scientists towards the use of local seaweeds in regional cuisines around the world based on seaweeds as a healthy local food material produced in a sustainable manner (Mouritsen et al. 2018). This gave rise to a new brand of small companies that gather, collect, pack and market local seaweeds or specific groups of seaweeds for use in unique recipes or dishes. These companies sell their products online or in small live markets.

As liquid fertilizer in agriculture and horticulture. Some of the biomass in the harvesting of *Kappaphycus* sometimes does not meet the high-quality standard required for phycogastronomy and other food uses. This below-standard quality material is usually put aside and processed into liquid fertilizer or biostimulant. Following the method of extracting liquid fertilizer from kelp species such as *Ascophyllum nodosum* and *Ecklonia maxima*, which are growing in Northern Europe and South Africa, respectively, liquid fertilizer from *K. alvarezii* can be produced by squeezing the thallus of fresh seaweed to extract its natural juice. The solid remains of the squeezed fresh material are removed by filtration to produce a clear and pale yellow-green extract (Fig. 15.4a), while the juice from frozen fronds is pink (Fig. 15.4b). This pink pigment is sensitive to light and, when stored at room

Fig. 15.4 Application of *K. alvarezii* juice to various plants. (**a**) Filtered *K. alvarezii* juice from fresh thallus showing clear colour, (**b**) filtered juice of frozen *K. alvarezii* fronds showing pinkish colour, (**c**) flowers of daisies fertilized with seaweed extract characterized by thicker stems and longer leaves (left) compared to daisies with no application of the seaweed extract (right) where stems are thinner and leave are shorter, (**d**) tomatoes applied with *K. alvarezii* juice has darker fruit colour and tastes sweeter, (**e**) greenhouse farming of melon injecting liquid fertilizer of *K. alvarezii* at the base of the tube, (**f**) melon with larger size and sweeter taste, (**g**), sweet potato, bigger and sweeter. (Photocredit: M. Ohno)

temperature, could gradually fade to a pale, transparent yellow-green colour similar to that of fresh *K. alvarezii*. Interestingly, there is no difference in the fertilizing effect towards the test samples. The liquid from fresh *K. alvarezii* prepared in this way is transparent with a slight odour and can be stored at room temperature for a long period of time.

Table 15.2 Composition of seaweed liquid fertilizer based on 1000 g sample indicated in some commercial organic fertilizers

Content	*Kappaphycus alvarezii*	*Saccharina* (formerly *Laminaria*)	*Sargassum horneri*	*Ascophyllum nodosum*
Moisture	88.8 g	95 g	*	*
Protein	1.2 g	1.1 g	1.62 g	0.53 g
Lipid	0.1 g	0.1 g	0.15 g	0.24 g
Carbohydrate	5.3 g	6.4 g	5.11 g	7.33 g
Ash	4.6 g	2.0 g	3.0 g	1.9 g
Sodium	280 mg	260 mg	51.9 mg	*
Potassium	1600 mg	610 mg	117 mg	*
Calcium	13 mg	78 mg	18 mg	*
Magnesium	32 mg	53 mg	19 mg	*
Phosphorus	16 mg	18 mg	*	*
Iron	0.7 mg	0.3 mg	*	*

*no information

The effect of *Kappaphycus* extract has been demonstrated in several terrestrial plants:

Daisies. *Kappaphycus alvarezii* liquid fertilizer applied once a week to daisies at 500-fold dilution could produce daisies with thicker stems and slightly longer leaves compared to daisies of similar size but receiving no liquid fertilizer (Fig. 15.4c). Tomato. *K. alvarezii* juice, at 500-fold dilution, and added to the conventionally used liquid fertilizer and applied to the roots yields sweeter and richer tomatoes (Fig. 15.4d).

Watermelon. *Kappaphycus* juice has also been applied by farmers to melon cultivation at 500-fold dilution from the time of planting under greenhouse conditions (Fig. 15.4e). Melon cultivation usually uses a plastic tube that is run along the base of the melon, and a moderately small hole is opened in the tube. *Kappaphycus alvarezii* liquid fertilizer was added to the conventional liquid fertilizer once a week. For the melon fruit, in the area where the juice is added, the treated fruits are found to be larger than the control (Fig. 15.4f). The pericarp is thin, sweet and sufficiently soft.

Sweet potato: Since the spring of 2020, an agricultural corporation in Okayama has been experimenting with sweet potato cultivation in sandy fields by applying *K. alvarezii* liquid fertilizer diluted 500 times. The growth of sweet potatoes appeared to be better than that of the control plants (Fig. 15.4g). Additionally, the growth of the vine, including its roots and rhizomes, is remarkably good. Sweetness and umami flavours were judged after the potatoes were dried and fully ripe (i.e., 4–6 months after lifting) and were also found to be better than those of the control.

Kappaphycus alvarezii liquid fertilizer was found to contain higher amounts of potassium than other seaweed extracts, as shown in Table 15.2. It features a number of constituents. It is presumed that the high potassium content of *K. alvarezii*, which is higher than that of kelp and *Sargassum* species, promotes the growth of roots, stems and leaves and increases the sweetness component. Iron is a component of chloroplasts and activates photosynthesis. Carrageenan from *K. alvarezii* (also alginic acid from brown seaweeds such as *Sargassum*) is a viscous polysaccharide that is very absorbent to water and forms a highly viscous aqueous solution. This property works effectively when soil improvement is carried out. When the content of viscous polysaccharides in the soil increases, the water-holding capacity increases, the soil moisture stabilizes, the propagation of useful microorganisms is promoted, and the treated soil becomes a better environment for plant growth. In ordinary fertilizers, amino acids and enzymes are synthesized in the plant body from nitrogen components absorbed from the roots, and then phosphoric acid is combined to synthesize proteins and the like. After the amino acids of seaweed are absorbed by the treated leaves and roots, they are immediately involved in the synthesis of proteins. Seaweed liquid fertilizer contains amino acids that promote the growth of crops more efficiently than inorganic fertilizers. Plant hormones play a very important role in the growth of plants in general, such as increasing flowering, promoting the branching and elongation of roots and stems, improving cold resistance, promoting fruit maturation and preventing aging. Plant hormones are also necessary ingredients to produce amino acids. Liquid fertilizer from the kelp *Ecklonia maxima* collected in Cape Town is described as containing auxin (11 mg/L) and cytokinin (0.03 mg/L) (Moncada et al. 2022). *Kappaphycus alvarezii* grows at almost the same growth rate as this species, so it is speculated that *K. alvarezii* may also contain a large amount of beneficial growth hormones.

Lectin from Kappaphycus alvarezii. Kappaphycus alvarezii is one eucheumatoid where lectin has been found. Lectins are carbohydrate-binding proteins that have important applications in biotechnology. *K. alvarezii* (as *Eucheuma cottonii*) was found by Kawakubo et al. (1997, 1999) to have a high content of lectin at 1 g per 100 g of dried tissue using the aqueous ethanol extraction method. Two other species, *Eucheuma amakusaensis* and *E. serra,* were also found to have equally high contents of lectin, making these three species a potential source of this protein in Japan. Since it requires a large volume of materials to extract viable amounts of lectin, the logical source will have to be from culture, and thus far, only *K. alvarezii* cultivated in Tosa Bay satisfies this requirement.

15.9 Future Prospects of Eucheumatoid Utilization in Japan

Despite the limited growing period for *K. alvarezii* in Japan, this seaweed has become an additional source of products that have the potential to contribute to at least three sectoral industries in Japan: phycogastronomy, agriculture and biomedical/biotechnology. With warming seas brought by global warming and climate change, this trend has already facilitated the spread of eucheumatoids to non-traditional grounds in sub- to temperate regions such as Japan from their more tropical origin (Hayashi et al. 2010). Therefore, it will not be a surprise that *K. alvarezii* will hold an important place in the boutique-type production of high-value products in Japan in years to come.

Acknowledgement The authors wish to thank Dr. Alan Critchley for his patience. Ms. Annie Diola is well appreciated in providing the map in this paper.

References

Chandrasekaran S, Nagendran NA, Pandiaraja D, Krishnankutty N, Kamalakannan B (2014) Bioinvasion of *Kappaphycus alvarezii* on corals in the Gulf of Mannar, India. Curr Sci 94(9):1167–1172

Glenn EP, Doty MS (1990) Growth of the seaweeds *Kappaphycus alvarezii, K. striatum* and *Eucheuma denticulatum* as affected by environment in Hawaii. Aquaculture 84:245–255

Hayashi L, Hurtado AQ, Msuya FE, Bleicher-Lhonneur G, Critchley AT (2010) A review of *Kappaphycus* farming: prospects and constraints. In: Israel A, Einav R, Seckbach J (eds) Seaweeds and their role in globally changing environments. Springer, Doldrecht, pp 225–283

Hiraoka M, Tanaka K, Yamasaki T, Miura O (2020) Replacement of *Ulva ohnoi* in the type locality under rapid ocean warming in southwestern Japan. J Appl Phycol 32:2489–2494

Iwase F (2004) East of Muroto. Coral reefs of Japan:260–269

Kawakubo A, Makino H, Ohnishi J, Hirohara H, Hori K (1997) The marine red alga *Eucheuma serra* J. Agardh, a high yielding source of two isolectins. J Appl Phycol 9:331–338

Kawakubo A, Makino H, Ohnishi J, Hirohara H, Hori K (1999) Occurrence of highly yielded lectins homologous within the genus *Eucheuma*. J Appl Phycol 11:149–156

Largo DB, Fukami K, Nishijima T, Ohno M (1995) Notes on the thalli whitening called *ice ice* in red algae *Eucheuma/Kappaphycus* and *Gracilaria*. Bull Mar Sci Fish Kochi Univ 15:39–42

Mairh OP, Soe-Htun U, Ohno M (1986) Culture of *Eucheuma striatum* (Rhodophyta, Solieriaceae) in sub-tropical waters of Shikoku, Japan. Bot Mar 29:185–291

Moncada A, Vetrano F, Esposito A, Miceli A (2022) Effects of NAA and *Ecklonia maxima* extracts on lettuce and tomato transplant production. Agronomy 12:329

Mouritsen OG, Rhatigan P, Pérez-Lloréns JL (2018) The rise of seaweed gastronomy: phycogastronomy. Bot Mar 62(3):195–209

Nagai S, Yoshida G, Tarutani K (2011) Change in species composition and distribution of algae in the coastal waters of western Japan. Global warming impacts-case studies on the economy, human health, and on urban and natural environments, 209–237

Ohno M, Largo DB, Ikumoto T (1994) Growth rate, carrageenan yield and gel properties of cultured kappa-carrageenan producing red alga *Kappaphycus alvarezii* (Doty) Doty in the subtropical waters of Shikoku, Japan. J Appl Phycol 6:1–5

Ohno M, Largo DB (2020) Footprints of tropical seaweed *Kappaphycus alvarezii* cultivation in Tosa Bay. Algal Resour 13:127–132. (in Japanese)

Pedra AG, Ramlov F, Maraschin M, Hayashi L (2017) Cultivation of the red seaweed *Kappaphycus alvarezii* with effluents from shrimp cultivation and brown seaweed extract: effects on growth and secondary metabolism. Aquaculture 479:297–303

Simioni C, Hayashi L, Oliveira MC (2019) Seaweed resources of Brazil: what has changed in 20 years? Bot Mar 62(5):433–441

Tanaka K, Taino S, Haraguchi H, Prendergast G, Hiraoka M (2012) Warming off southwestern Japan linked to distributional shifts of subtidal canopy-forming seaweeds. Ecol Evol 2(11):2854–2865

Tokuda M, Takeda A (1989) Monitoring and prediction of wind, wave and anomalous sea states. J Remote Sensing Soc Jap 9(4):466–471. (In Japanese)

Trono GC, Ohno M (1989) Seasonality in the biomass production of the *Eucheuma* strains in northern Bohol Philippines. In: Umezaki I (ed) Scientific survey of the marine algae and their resources in the Philippine Islands. Japanese Ministry of Education, Science and Culture, pp 71–81

Yano M, Ohno M (2007) The growth rates of cultivated green and brown type of *Kappaphycus alvarezii*. Bull Mar Sci Fish Kochi Univ 23:59–62. (in Japanese)

Yano M, Ohno M, Hiraoka M (2010) A new technology of the cultivation and outdoor culture on *Kappaphycus alvarezii* (Japanese name: Nettaikirinsai) in the warm waters of Japan. Bull Mar Sci Fish Kochi Univ 26:23–28

Status and Trends of Eucheumatoid and Carrageenan Production in China

16

Jing Wang, Yumeng Wu, Quanbin Zhang, and Delin Duan

Abstract

Eucheumatoids are the main resources for carrageenan production in China and Southeast Asian countries, and many efforts have gone towards eucheumatoid seaweed cultivation and carrageenan production. In this chapter, we reviewed the status of eucheumatoid cultivation in China and pointed out problems that exist. We also summarized carrageenan extraction methods, quality and applications in China. Since the 1980s, eucheumatoid seaweed cultivation, including the genera *Eucheuma* and *Kappaphycus,* was started in China, and cultivation techniques have been sufficient for seaweed farmers to achieve some crop production, but annual production has declined since 2000. Nevertheless, with its huge market needs, carrageenan production has increased, and China is already ranked as the largest carrageenan producer in the world. Nevertheless, there are improvements that can be made in China's carrageenan products and qualities. To accomplish this, multiple disciplines must be integrated for the analyses of the genetic, physiological and chemical properties of carrageenophyte red seaweeds. This will guarantee the sustainable development of eucheumatoid seaweed for the carrageenan industry of China.

Keywords

Eucheumatoids · Sea farming · Carrageenan production · China

J. Wang · Q. Zhang · D. Duan (✉)
Key Lab of Breeding Biotechnology & Sustainable Aquaculture, Shandong Province Key Lab of Experimental Marine Biology, Institute of Oceanology, Chinese Academy of Sciences, Qingdao, People's Republic of China

Laboratory for Marine Biology and Biotechnology, Qingdao National Laboratory for Marine Science and Technology, Qingdao, People's Republic of China
e-mail: dlduan@qdio.ac.cn

Y. Wu
Key Lab of Breeding Biotechnology & Sustainable Aquaculture, Shandong Province Key Lab of Experimental Marine Biology, Institute of Oceanology, Chinese Academy of Sciences, Qingdao, People's Republic of China

Laboratory for Marine Biology and Biotechnology, Qingdao National Laboratory for Marine Science and Technology, Qingdao, People's Republic of China

University of Chinese Academy of Sciences, Beijing, People's Republic of China

16.1 Introduction

Eucheumatoids are the main resources for carrageenan production in Indonesia, the Philippines and Malaysia, and presently, these countries contribute significantly to world seaweed production. To date, eight eucheumatoid seaweed species have been identified in southern China (Table 16.1), and *Eucheuma gelatinae* and *Kappaphycus alvarezii* are the two main cultivated species in China. Since the 1980s, the cultivation of *Eucheuma* and *Kappaphycus* started in Hainan, and annual production became significant, but it declined rapidly in recent years.

Carrageenan is a type of natural linear polysaccharide extracted from *Eucheuma, Kappaphycus, Chondrus and Gigartina*. Chemically, it is a sulphated polysaccharide composed of galactose, 3,6-anhydro-galactose and sulphate groups, with a molecular weight of 100–500 kDa, and is classified into κ, λ, τ and μ types (Fig. 16.1). Different generations of carrageenophytes have different types of polysaccharides and exhibit different properties; usually, gametophytic plants contain κ/λ-type carrageenan, while sporophytes contain λ-type carrageenan (Chen et al. 1973; Pickmere et al. 1973; Waaland 1973). Various types of carrageenans exhibit different gelling and viscosity properties in aqueous solution (Velde 2008), and today, they are used in the food, printing, textile and drug fields (Liu et al. 2021; Pangestuti et al. 2021).

In this chapter, we address cultivation techniques, status and problems relevant to eucheumatoids in China. Moreover, we reviewed the development of carrageenan extraction

A. T. Critchley et al. (eds.), *Tropical Phyconomy Coalition Development*, Developments in Applied Phycology 11,
https://doi.org/10.1007/978-3-031-47806-2_16

Table 16.1 Natural eucheumatoid species distributions in China

Species	Distribution	References
E. crustaefore[a]	Taiwan	Kintarou (1936)
E. muricatum	Taiwan	Kintarou (1936)
E. gelatinae	Hainan island	Tseng (1935)
E. serra	Taiwan	Yamada (1936)
E. crassum	Taiwan	Yamada (1936)
E. okamurai	Taiwan	Yamada (1936)
E. cupressoideum	Taiwan	Shen and Fan (1950)
K. cottonii	South China Sea	Xia (1965)
K. striatum	South China Sea	Xia (1965)

[a] Revised as *E. okamurai* by Yamada 1936

methods, quality and applications and reviewed progress towards future prospects and improvements.

16.2 Taxonomic and Biogeographic Study

Kappaphycus spp. belong to the Rhodophyta, Florideae, Gigartinales and Solieriaceae (Cai 1985), and members of the genus are distributed among tropical and subtropical environments. Species of eucheumatoids are found mainly in the Indo-western Pacific Ocean and the Atlantic East Pacific Ocean. Ecologically, there are many eucheumatoids found in *Acropora* communities where seawater is below 1 ~ 2 m, and the optimal seawater temperature is between 20 and 35°C. Growth is inhibited when the seawater temperatures are lower than 20°C (Cai 1985). To date, approximately eight eucheumatoid species (Xia 1965), including *Kappaphycus* spp. and *Eucheuma denticulatum*, were verified in southern China, such as Hainan, the South China Sea and Taiwan (Table 16.1).

16.3 Eucheumatoid Cultivation in China

Historically, in mainland China, commercial cultivation of eucheumatoids mainly occurred in Wenchang, Qionghai, Lingshui, Sanya, Changjiang, Danzhou and Chengmai in Hainan (Fig. 16.2). *Eucheuma gelatinae* and *K. alvarezii* were the main cultivated species in China (Fig. 16.3). Early in 1960, with increased market needs for eucheumatoids, the local sea farmers in Hainan tried to cultivate *E. gelatinae*. *Eucheuma striatum* (now renamed *K. alvarezii*) was introduced from the Philippines in 1985 (Wu et al. 1988), and cultivation trials were conducted in Qionghai in Hainan. Although fast growth of vegetative branches was observed, most of the crop was fragile in a way that caused branches to break and eventually be lost. Moreover, high winds and large waves caused farming failures in some places.

During early trials in China, seedlings were tied to various substrata. The cement frame mesh and longline floating raft cultivation methods were developed later and became the most popular cultivation techniques.

During sea farming in Hainan, coral reefs were used as substrata for *E. gelatinae* cultivation. Plants were transported to cultivation sites during the lowest tides, with the planting density being approximately 9–10 thalli per m^2. After deploying the seedlings for 15 days, they firmly attached to the coral reef. Then the crop could be harvested after 15–20 months of growth (Zhuang 1965).

Due to the lack of available coral reefs locally, many sea farmers adopted sea bottom cultivation methods for later trials. This could yield 100–200 kg (dry weight) per square metre with less labour and input costs. Generally, there are two types of substrata used in cultivation: cement columns and dead coral branches (Li and Bai 2016). On the cement column substratum, areas of 10 cm × 10 cm × 100 cm with 16 seedling spacings of 15 cm were fixed on each cement column with rubber bands. The crop could be harvested after 5~7 months, and the yield of *E. gelatinae* increased approximately six fold (wet weight) beyond the initial planted biomass. On coral substrata, branches were approximately 7~8 cm in diameter and weighed approximately 20~30 g. The seedlings were fixed to coral branches by rubber bands and were inserted into staghorn coral, *Acropora* coral or *Cyathophyllum*. When the tide flow and waves became stable, the seawater depth was 0~2 m during the high dry tide, the planted density was approximately 9–10 seedlings/m^2, and the crop could be harvested after 15~20 months. In Hainan, production was conducted from late May to late September.

When Zhuang (1965) used longline floating raft methods for *E. gelatinae* (original name at the time of publication) cultivation, they yielded fast crop growth. Within 1 m of the sea surface, this method could resist strong wind and high temperature influences (Fig. 16.4). Generally, the cost was lower, and the system was suitable for *E. striatum* (*Kappaphycus*) cultivation. The growth rate reached 10% under 26°C and 140 E/m^2s^{-1} conditions (Wu et al. 1988). During the period of 2006–2007, the cement frame mesh method was applied for *E. gelatinae* cultivation in Changjiang and Tanyang, Hainan. It was used for supporting and fixing the mesh, while the mesh with a polyethylene net

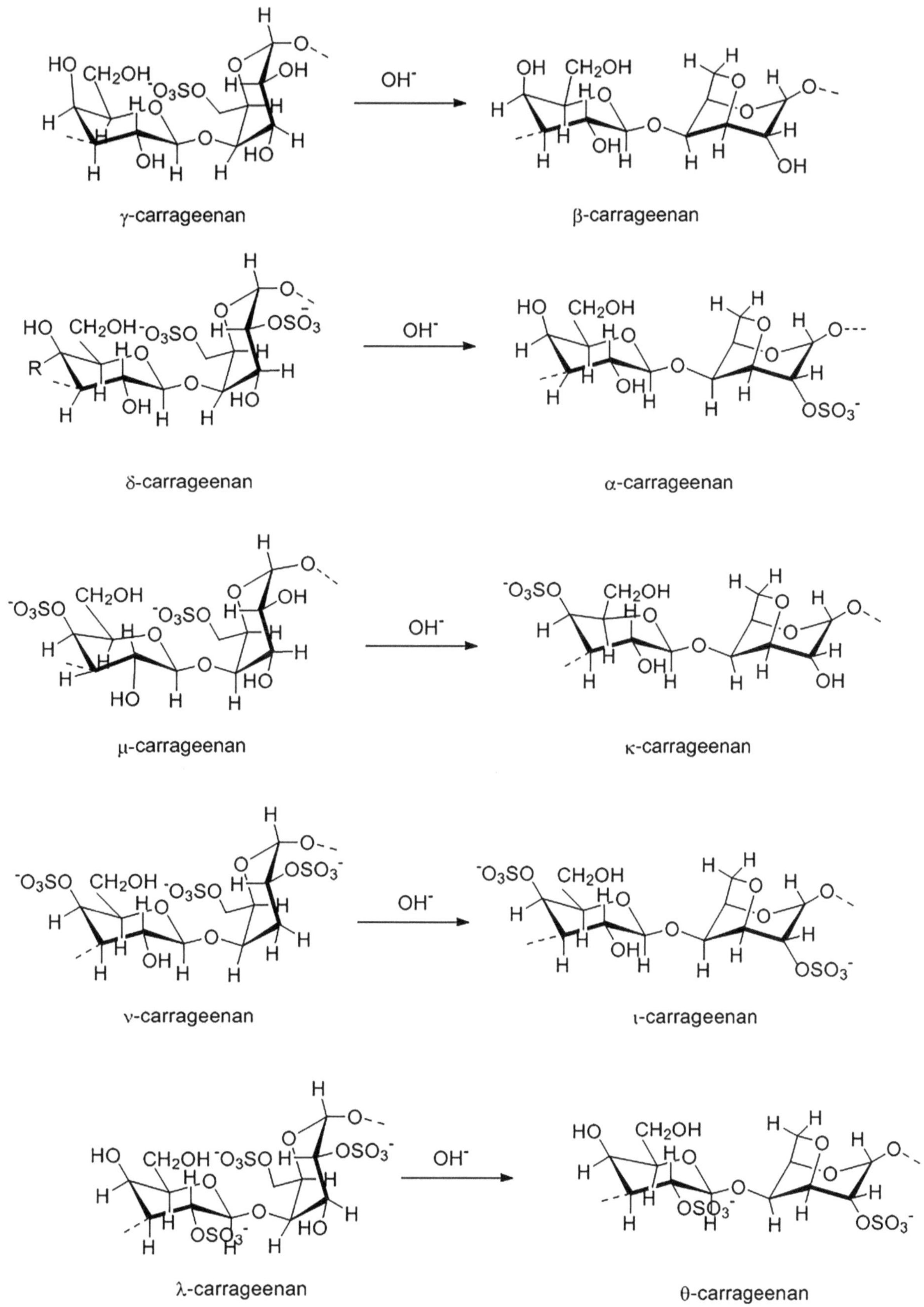

Fig. 16.1 Structures of carrageenans

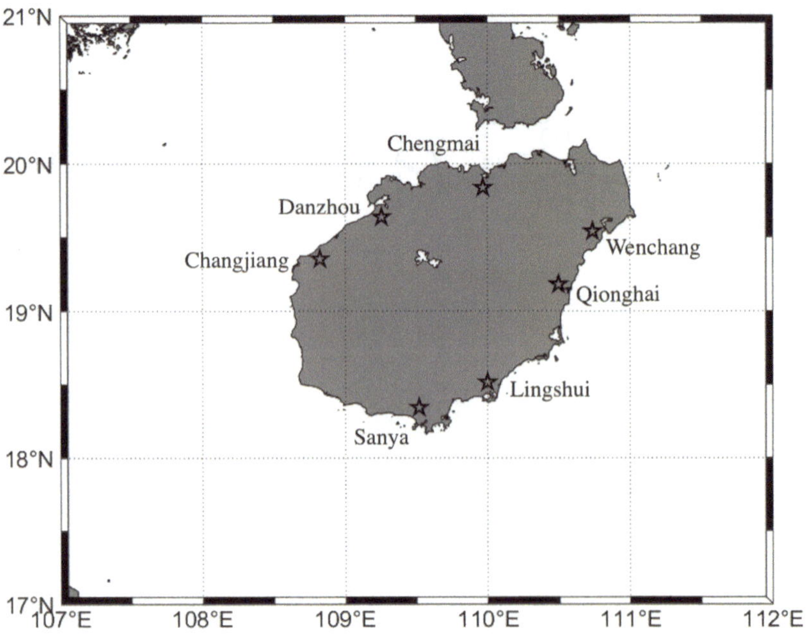

Fig. 16.2 Biogeographic locations of eucheumatoids in China

Fig. 16.3 Eucheumatoid species in China

of 1.5~2.0 cm was placed on top of the cement frame with two layers, and the seedlings were placed in the middle. The mesh played the role of fixing and protecting the seedlings (Fig. 16.5) (Zhao 2008).

Comparisons of different cultivation methods are listed in Table 16.2. Generally, the seedling insertion and seedling casting methods were the most efficient and lowest cost, but these methods needed reef sea bottom areas, and it was

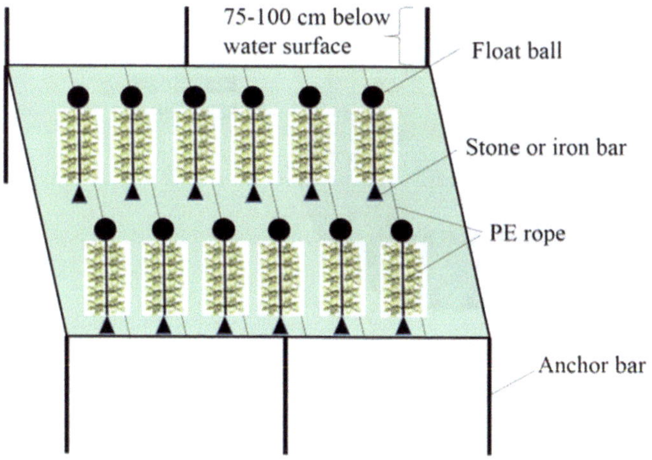

Fig. 16.4 Structure of floating raft cultivation system

difficult to enlarge the farming areas. Longline floating raft and cement frame mesh methods are widely used in China. These two methods gave high yields and profits, which are suitable to promote farming widely. The cost analysis of the three farming modes of *Eucheuma* is shown in Fig. 16.6. The cost of seeding accounted for 40 and 55%, respectively, compared with the cement frame mesh and tied seedling sowing methods. The depreciation cost was 10% in the tied seedling sowing method and was lower when compared with the rest of the methods. Therefore, we should consider the marine environment, the costs and the profits for each method for suitable cultivation implementation in China.

The sea warming area and annual production of eucheumatoids in 2016–2020 are shown in Fig. 16.7. During 2016–2020, the eucheumatoid farmed areas and production decreased continuously. Farmed areas decreased from more than 400 to 100 ha, and the annual production decreased from 5000 to 500 tonnes (fw). The culture of eucheumatoids mainly occurred in Hainan in 2016–2018, while in 2019–2020, there was no eucheumatoid farming in Guangdong.

Several problems exist during the cultivation of eucheumatoids in China, including the following: (1) Long duration vegetative propagation would cause the eucheumatoids to degrade, and there was frequent disease occurrence that caused yields to decline. Means were not available to maintain seed stock essential to support eucheumatoid production. (2) Unfavourable cultivation conditions, such as nutritional supplementation, irradiance, seawater exchange, typhoons and fish attack, would cause

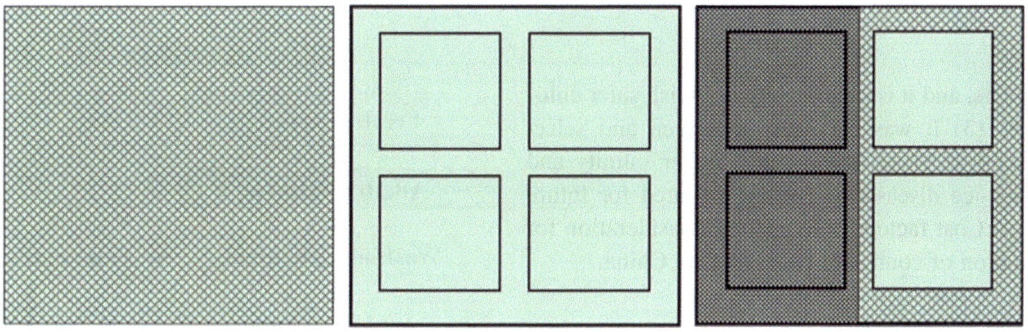

Fig. 16.5 Net mesh structure of cement screen frame clamping seeding cultivation systems

Table 16.2 Cultivation methods for eucheumatoids in China

Cultivation methods	Advantages	Disadvantages
Seeding insertion method	Practical and efficient, low cost	Dive must, high labour intensity, limited to the reef bottom of the sea area
Seeding casting method	Practical and efficient, low cost	Dive must, high labour intensity, limited to the reef bottom of the sea area
Seeding tied sowing method	The operation is simple, less input	Higher labour cost
Longline floating raft method	Short harvest period, high yield, low cost and good economic benefits	Strict requirements for the culture area
Cement frame mesh method	High yield, fast growth, safe, low investment and high profit	High cost

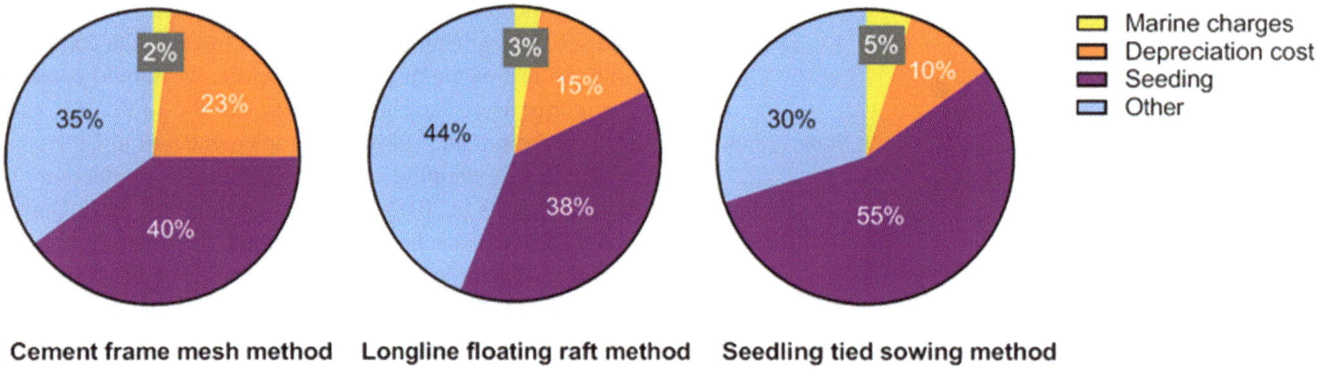

Fig. 16.6 Cost comparisons of three cultivation systems of *Eucheuma*

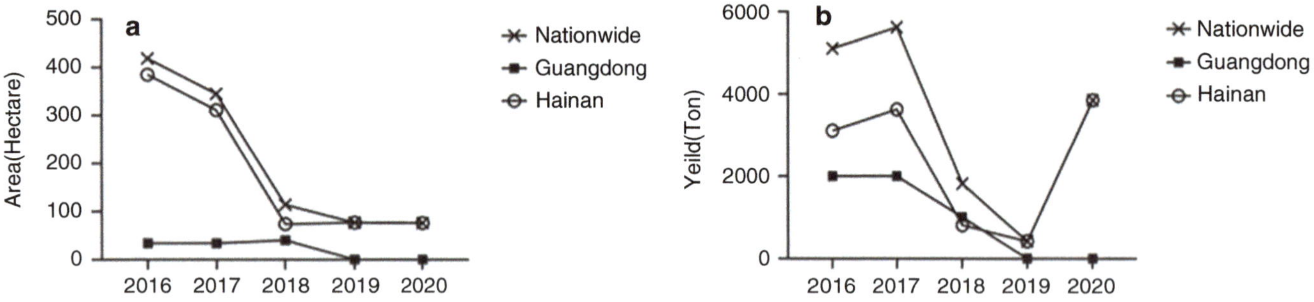

Fig. 16.7 Eucheumatoid farming area (**a**) and output (**b**) (from the Chinese Fishery Statistical Yearbook, 2016–2021)

deleterious effects, and it is necessary to establish safer cultivation systems. (3) It was necessary to screen and select cultivars resistant to lower temperature, higher salinity and resistance to ice-ice diseases and to be prepared for future applications. (4) Cost factors were another consideration for the implementation of continued cultivation in China.

16.4 Extraction Methods of Carrageenan

Carrageenan is a linear polysaccharide composed of galactose, 3,6-anhydro-galactose and sulphate groups. Initially, it was mainly obtained from the cold and cool temperate genera *Chondrus* and *Gigartina*. Later, it was extracted from the tropical genus then collectively known as *Eucheuma*. The yield, chemical composition and physical properties of carrageenan varied with alkali extraction methods and the eucheumatoid species.

A flow chart for the carrageenan extraction process is shown in Fig. 16.8. It was concluded that the alkali treatment, acidification bleaching, carrageenan extraction, dewatering and drying steps most influenced the carrageenan properties and extraction rate.

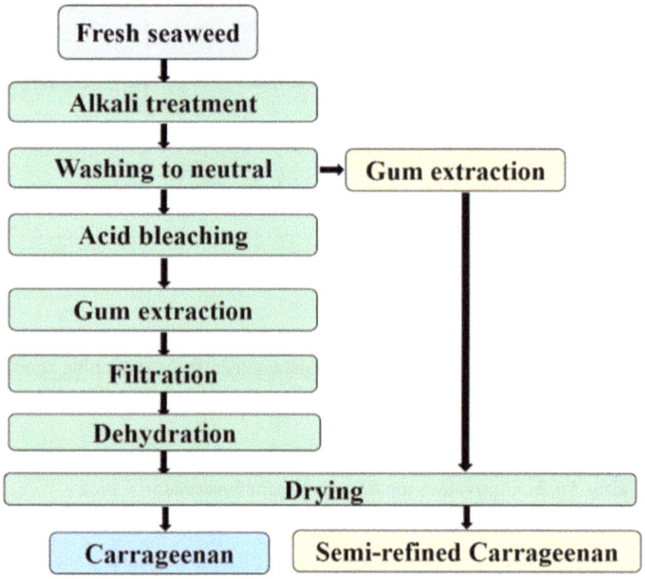

Fig. 16.8 Flow chart for carrageenan and semirefined carrageenan extraction

Shi et al. (1986) first extracted carrageenan from eucheumatoid species in Hainan and compared their yield and chemical properties. The direct extraction methods were adopted with different concentrations of NaOH or CaO (Tables 16.3 and 16.4). Seaweed biomass was pre-treated with dilute or concentrated alkali solution and washed with

Table 16.3 Production rate and properties of carrageenan extracted from *Eucheuma gelatinae* through water or alkaline extraction (Shi et al. 1986)

Species of algae and sampling locale	Extracting methods	Production rate (%)	Strength of gel (g/cm^2)	Gelation temperature (°C)	Transparency	SO_4^{2-} (%)	3,6-anhydro -galactose (%)	Galactose (%)
Eucheuma gelatinae (Qinglan, Wenchang)	Water	74.6	57	39.3	1.5	14.7	22.4	46.9
	NaOH, 0.1%	66.9	72	38.6	1.5	14.5	22.5	47.1
	NaOH, 0.5%	65.7	114	40.1	1.2	14.0	27.6	42.6
	NaOH, 1.0%	62.8	187	44.7	0.7	11.0	28.6	41.7
	NaOH, 2.0%	58.9	244	47.4	0.3	10.2	29.9	40.8
	CaO, 0.1%	68.5	75	38.4	1.5	14.9	22.7	46.7
	CaO, 0.5%	58.3	80	39.6	1.2	14.0	22.8	48.9
	CaO, 0.1%	59.8	87	39.4	1.0	14.2	24.0	44.6
	CaO, 2.0%	60.9	78	40.0	0.1	14.4	23.4	45.3
Eucheuma gelatinae(Wengtian, Wenchang)	Water	59.1	45	40.0	1.5	17.9	22.5	46.2
	NaOH, 0.1%	52.5	58	39.9	1.5	18.1	22.4	45.0
	NaOH, 0.5%	47.0	107	42.3	0.7	14.5	25.4	43.1
	NaOH, 1.0%	44.9	152	45.6	0.5	14.4	27.4	41.4
	NaOH, 2.0%	44.8	145	46.5	0.1	14.4	27.3	41.9
	CaO, 0.1%	55.6	67	39.2	1.5	17.5	25.0	43.9
	CaO, 0.5%	52.9	70	39.9	1.5	18.0	23.0	44.8
	CaO, 0.1%	52.4	30	38.9	0.7	18.9	22.9	44.5
	CaO, 2.0%	49.5	34	39.8	0.4	18.7	22.6	45.0

Table 16.4 Production rate and properties of carrageenan extracted from *Eucheuma cottonii* and *Eucheuma okamurai* through water or alkaline liquid (Shi et al. 1986; taxonomic names at the time)

Species of algae and sampling locale	Extracting methods	Production rate (%)	Strength of gel (g/cm^2)	Gelation temperature (°C)	Transparency	SO_4^{2-} (%)	3,6-anhydro -galactose (%)	Galactose (%)
Eucheuma cottonii (Shalao, Qionghai)	Water	49.4	22	34.1	1.5	22.8	21.6	44.7
	NaOH, 0.1%	39.7	21	32.5	1.5	23.0	22.3	44.8
	NaOH, 0.5%	35.1	44	32.0	1.5	19.6	22.7	43.8
	NaOH,1.0%	34.8	129	32.8	1.5	18.4	27.8	39.6
	NaOH, 2.0%	35.2	145	32.6	1.5	17.7	27.6	39.4
	CaO, 0.1%	39.0	33	34.0	1.5	19.8	22.6	44.9
	CaO, 0.5%	37.8	48	33.9	1.5	19.1	22.4	45.1
	CaO, 0.1%	37.8	33	33.2	1.2	20.4	22.1	45.0
	CaO, 2.0%	37.5	44	33.7	0.4	19.1	22.2	44.9
Eucheuma okamurai (Shalao, Qionghai)	Water	48.7	30	35.4	1.5	19.4	22.8	42.5
	NaOH, 0.1%	39.7	21	33.8	1.5	20.3	22.3	44.2
	NaOH, 0.5%	37.0	54	32.9	1.5	19.3	25.5	39.9
	NaOH, 1.0%	36.4	148	32.7	1.5	16.6	27.4	36.4
	NaOH, 2.0%	37.3	173	31.9	1.5	16.4	29.8	37.3
	CaO, 0.1%	39.1	62	34.4	1.2	19.3	22.2	39.1
	CaO, 0.5%	37.5	87	33.8	1.2	19.3	25.1	37.5
	CaO, 0.1%	36.9	41	32.3	1.0	24.3	22.9	36.9
	CaO, 2.0%	33.4	39	33.5	0.8	24.2	23.0	33.4

Fig. 16.9 Carrageenan production equipment (Greenfresh Food Ltd. Co. China)

water to remove alkali. The gel strength was improved after alkali treatment, and the carrageenan strength from *K. cottonii* and *E. okamurae* (names applied at the time) reached 500 g/cm^2. Alkali treatment made the 1,4-linked D-galactose-6-sulfate into 1,4-linked 3,6-anhydro-D-galactose (Freile-Pelegrín and Robledo 1997) and enabled the conversion of μ-carrageenan into κ-carrageenan or ν-carrageenan into ι-carrageenan; moreover, it could increase the gel strength, yield and chemical properties and variations (Rees 1969). The molecular weight of carrageenan was lowest when treated with alkali solutions. While high-temperature and dilute alkali treatment could avoid pollution from the excessive alkali solution and short extraction influences and caused a 5% extraction ratio loss. Alkali accelerated the dissolution of small molecules such as cellulose, pigment and proteins and improved the quality of extracted carrageenan.

Acid bleaching is another important step during carrageenan extraction because it influences carrageenan colour, strength and yield (Sadi et al. 2019). There was balance between the structural degradation and carrageenan strength during the extraction. Under weak alkali treatment and subsequent acidification, higher carrageenan contents and high gel strength could be obtained. This indicated that $H_2C_2O_4$ was positive for carrageenan extraction. In addition, peracetic acid (PAA) is a strong oxidizing agent for bleaching, and it could yield more fine and better carrageenan than using a sodium hypochlorite (NaClO) bleaching method (Sadi et al. 2019) (Fig. 16.9).

The carrageenan was cooled through filtration, and many dehydration methods were tried, including pressing dehydration, drum dehydration, freezing and melting dehydration and freeze drying (Tari and Pekcan 2008). Usually, freeze-thaw methods are used for dehydration. These required −20 to −15°C for 30~48 h and thawing carrageenan solution in water for quick removal of impurities. To obtain coagulated carrageenan, potassium chloride solution was added into the filtered gel solution after cooling. This enabled the separated

Table 16.5 New development of carrageenan extraction methods

Extraction methods	Species	Characteristic	References
Deep eutectic solvents (DESs)	*Kappaphycus alvarezii*	The highest yield of carrageenan was 60.25%, rheological properties of carrageenan was superior in quality	Das et al. (2016)
Microwave-assisted method	*Hypnea musciformis*	Although the yield of this method was low, the method could reduce consumption of solvents, energy and extraction time; it is a green method	Vazquez-Delfin et al. (2014)
Ohmic based	*Eucheuma spinosum*	The yield could reach 29.6–62.4% (dry weight) with gel strength between 259–290 mPa.s and 42.54–70.41 g/cm^2, respectively. It is one faster extraction process due to the fast and uniform heating and more energy efficient	Hasizah et al. (2018)
Ultrasound-assisted methods	*Kappaphycus alvarezii, Euchema denticulatum*	The carrageenans with yield of 50–55%. In a short time (15–30 min), NMR analysis indicated that the non-degradate carrageenan are complete	Youssouf et al. (2017)
Subcritical water extraction (SWE)	*Kappaphycus alvarezii*	The highest yield was 78.75%, gel strength and viscosity were minimal, but emulsification index was relatively high	Gereniu et al. (2018)
Enzymatic extraction	*Eucheuma cottonii*	Higher yield was obtained through cellulase and recombinant kappa-carrageenase extraction (38%)	Duan et al. (2016)

Fig. 16.10 High water retention carrageenan

carrageenan to dehydrate more easily. In addition, an alcohol dehydration method was introduced as an alternative choice. After squeezing for several hours, the carrageenan gel sheets were chopped, dried in an air dryer and milled to an appropriate particle size. Carrageenan extraction methods are summarized in Table 16.5.

16.5 Carrageenan Qualities and Properties

The market needs for carrageenan are huge, and viscosity is one important parameter of carrageenan applications. Although the required viscosity in China is not less than 5 m Pa.s, most products were over 100 m Pa.s, and the average viscosity of carrageenan ranged from 60 - 120 m Pa.s. Presently, high-strength gels with low heavy metals are required by the markets (Dong et al. 2020). A common product is κ-carrageenan (Gul et al. 2016), and approximately 100 types of carrageenan products exist on the market (Krempel et al. 2019; Blanco-Pascual et al. 2014; Pereira 2013; Villanueva et al. 2009). Generally, high water retention and high acid resistance carrageenan are most appreciated.

With higher water retention features (Fig. 16.10) (Aga et al. 2021), binding to proteins enhances carrageenan in stabilizing freeze-thaw cycles and reducing water loss (Sen and Erboz 2010; Zhou et al. 2021). After the addition of maleic anhydride, the water retention properties of carrageenan can be changed (Sumathra et al. 2020; Zhou et al. 2021). Higher acid-resistant (>pH 3.2) carrageenan was used for protection of fruit flavours and other acidic products (Fig. 16.11).

16.6 Applications of Carrageenan

16.6.1 Food Industry

Different types of carrageenans were used as additives in many types of foods (Tables 16.6 and 16.7). Generally, they function as carriers, emulsifiers, thickeners and gelling or glazing agents. In ice cream and cold drink products, it could prevent milk component separation (Seo and Yoo 2021). For wheat flour dough and bread products, it was used for texture, water retention and gel formation (Linares et al. 2022). During brewing processes, it can function as a clarifying and anti-clouding agent (Li et al. 2021; Ratnayake et al. 2019). With good gelation properties, low melting temperature and transparency, it can also be used in jelly products (Bartlová et al. 2021; Valado et al. 2020).

16.6.2 Feed and Fertilizer Applications

Carrageenan can be used as a feed additive for cows, pigs and other livestock. Generally, it is suggested that one 10 g/kg diet of carrageenan could enhance immunity and the overall

Fig. 16.11 High acid-resistant carrageenan (Fujian Gold Swallow Ocean Biotech Co., China)

Table 16.6 Characteristics and applications of carrageenan products

Product categories of carrageenan products	Benefits and characteristics	Applications
Refined kappa-carrageenan	– As a gelling agent	Food medicinal (including jelly, fudge, ice cream, baked goods and meat products) Non-food medicinal products (including cosmetics and personal care products)
	– It has moderate viscosity, stability and high transparency	
	– Gelling agent and thickening agent in food medicine	
	– Maintain the best shelf life, viscosity and taste stability	
	– Improve food texture	
Refined iota-carrageenan	– As water retention agent and soft gel forming	Food medicinal (including dairy, meat products, confectionery, jellies and puddings) Non-food medicinal (including dietary supplements, pharmaceuticals, cosmetics and healthcare products)
	– As a binding agent.	
	– Gel, thickening, film formation and stable dispersion	
	– The gelation of meat products forms and retains moisture, thereby enhancing texture by reducing firmness and increasing juiciness	

Table 16.7 Further characteristics and applications of carrageenan products

Product categories of carrageenan products	Benefits and characteristics	Applications
Semi-refined kappa-carrageenan	– Promotes gelation and retention of moisture in meat products, thereby enhancing texture by reducing firmness and increasing juiciness	Food medicinal (including dairy, meat products, confectionery, jellies and puddings) Non-food medicinal (including dietary supplements, pharmaceuticals, cosmetics and healthcare products)
	– As a binding agent	
	– As a beer clarifying agent	
Semi-refined iota-carrageenan	– To form a soft gel – Form a thin film – As a thickening agent – As a water retention agent – As a beer clarifying agent	Food medicinal (including dairy products, meat products, confectionery, jellies, puddings, ice cream and beer) Non-food medicinal (including dietary supplements, healthcare products and cosmetics)

health status in *Labeo rohita* (Kumar et al. 2014). Diets containing carrageenan showed growth-promoting and immune-stimulating effects against salinity stress in black tiger shrimp, post larvae (Jumah et al. 2020). Diets supplemented with carrageenan increased the resistance of Pacific white shrimp to white spot syndrome virus (WSSV) without changing their growth parameters (Mariot et al. 2021). Bampidis et al. (2022) believed that carrageenan could be efficacious as a gelling agent, thickener and contributor to the stabilization of canned pet feed. Carrageenan can be used as a fertilizer carrier for agriculture because of its slow release, excellent water retention, nontoxicity and good

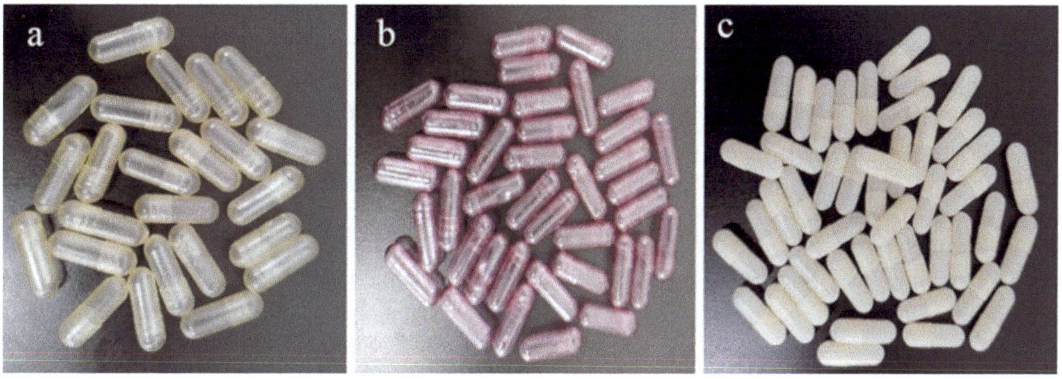

Fig. 16.12 Carrageenan-starch hollow capsules. (**a**) Hollow capsules, (**b**) hollow capsules with added pigments, (**c**) hollow capsules with masking agents (Courtesy by Dr. Qu CF)

Table 16.8 The compaction properties of different dietary fibres (Courtesy by Prof. Yang XQ)

Index product	Expansive force (mL/g)	Retention ability (%)	Moisture(%)
Carrageenan dietary fibre	15	1000	≤5
Apple dietary fibre	8	600	≤5
Oat dietary fibre	6	500	≤5
Konjac dietary fibre	8	700	≤5
Wheat dietary fibre	4	400	≤5

biodegradability properties (Wang et al. 2012). Rozo et al. (2019) proved that carrageenan has the ability to control the release of nitrogen and phosphorous. Vanegas et al. (2019) found that fertilizers with gels had slow release behaviour of N, P and K and adequate provision of nutrients for land plants.

16.6.3 Biomedical Applications

With nontoxicity, moisture retention, flexibility, permeability, biocompatibility, biodegradability and excellent film-forming properties, carrageenan has been widely used in biomedical applications (Jaiswal et al. 2020; Polat et al. 2020), such as gelatin capsules (Fig. 16.12). The properties of different dietary fibres are listed in Table 16.8. Supplementation with carrageenan could reduce body weight, adipocyte size, blood glucose and lipids in model mice (Wang et al. 2021), and carrageenan is reported to attenuate the symptoms of diet-induced metabolic syndrome in model rats (Preez et al. 2020). Chen et al. (2020) proved that carrageenan was effective in retarding the photo- and thermal degradation of encapsulated curcumin and piperine and delayed the release of nutraceuticals in *in vitro* gastrointestinal microenvironments. Carrageenan/chitosan complexes exhibit high swelling properties and cause the release of methotrexate from hydrogels (Mandavinia et al. 2017). We confirmed that carrageenans with different ratios of glycerol could feasibly promote wound healing (Fig. 16.13).

Fig. 16.13 Carrageenan film developed for wound healing

16.6.4 Cosmetics Applications

With moisture-resistant and drying properties, carrageenan has been applied in cosmetic fields to maintain skin moisture. It can form hydrogels with different intensities and is added to toothpaste, shampoos, sun blockers, shaving creams, deodorant sticks, sprays, foams and lotions (Morais et al. 2021). In addition, carrageenan is a green, eco-friendly and native compound, and its bioactive substance was proven to be useful in skincare and cosmetic fields. It preserves antioxidant, anti-inflammatory and anti-photoaging activities and has been used in cosmetic-pharmaceutical hybrids with skin health benefits (Pangestuti et al. 2021). Vanayagam et al. (2014) elucidated the protective roles against ultraviolet B (UVB)-induced cell killing and reactive oxygen species (ROS) release.

16.6.5 Carrageenan Industry in China

During 2013 and 2018 in China, carrageenan sales increased significantly from 17,100 to 31,000 tonnes, with an annual growth rate of 12.7% (Fig. 16.14). It was expected that this growth could be maintained with a compound annual growth rate (CAGR) of 14.6% during 2019 and 2023 (estimated 61,600 tonnes). In 2018, carrageenan sales accounted for 22.0% of the world's total hydrocolloid volume, and it was estimated that this level would reach 28.7% in 2023 (Fig. 16.14).

According to the Frost and Sullivan Report 2019, the value of carrageenan sales in China from 2013 - 2018 increased from 853.0 to 1658.4 million RMB, which accounted for a CAGR of 14.2%. It was estimated that the values would reach approximately 3467.5 million RMB in 2023 with a growth rate of 12.4% (Fig. 16.15).

China became the largest global exporter of carrageenan, accounting for approximately 35.7% of the total by 2018 (Fig. 16.16). The United States and European countries were the major import countries for vegetable-based derived gums. In 2018, the United States was the largest importer with 11.7% of the total, followed by Germany and Spain with 6.6% and 5.5%, respectively.

From 2013 - 2018 in China, carrageenan production reached an estimated 11,400–16,800 tonnes, which was approximately 8.1% of the carrageenan market (Fig. 16.17). It was expected to grow at a CAGR of 12.2% from 2019 - 2023 and reach approximately 29,900 tonnes in 2023. Spain was the main country for China's carrageenan exports and accounted for 29.3% of exports in 2018, while the Philippines and Russia accounted for approximately 9.8% and 9.7%, respectively.

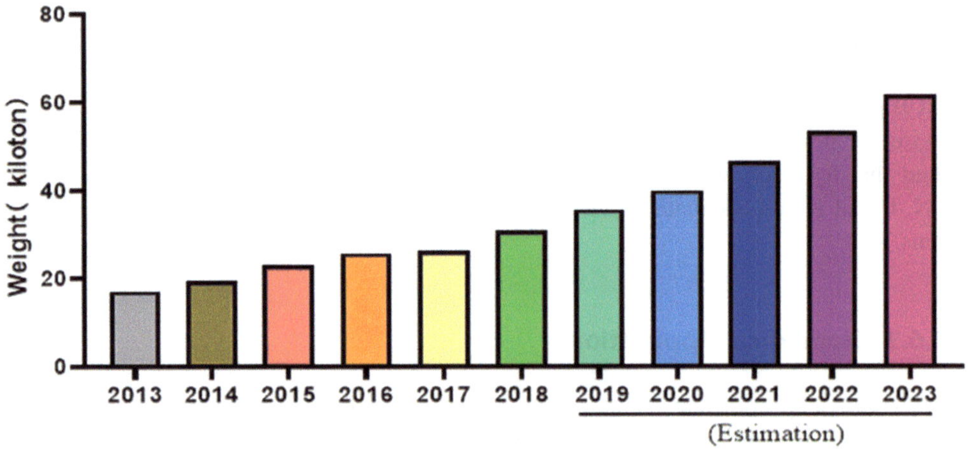

Fig. 16.14 Sales of carrageenan in China (**a**) and the breakdown by carrageenan type in 2018 (**b**) (From Frost and Sullivan Report 2019)

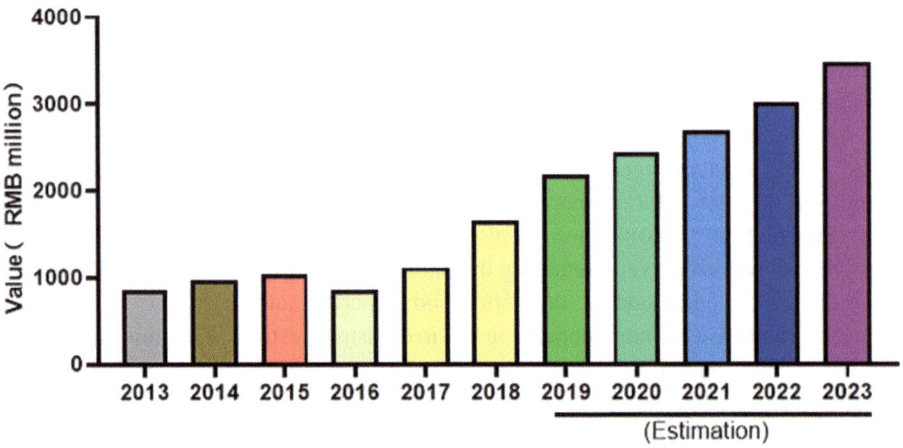

Fig. 16.15 The sales value of carrageenan (China) (From Frost and Sullivan Report 2019)

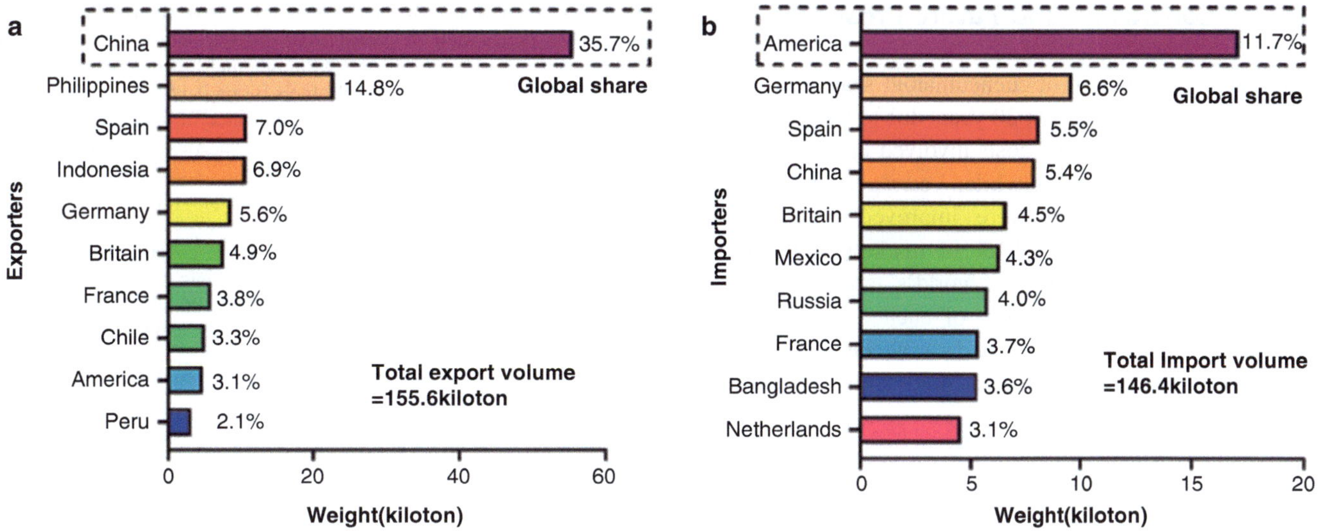

Fig. 16.16 The top ten exporters of gum liquids and thickeners in terms of trade volume in 2018 (**a**) and the top ten importers of gum liquids and thickeners in terms of trade volume in 2018 (**b**) (From Frost and Sullivan Report 2019)

Fig. 16.17 China carrageenan export volume and value 2013–2023 (estimation) (**a**) and a breakdown of exports by destination 2018 (**b**) (From Frost and Sullivan Report 2019)

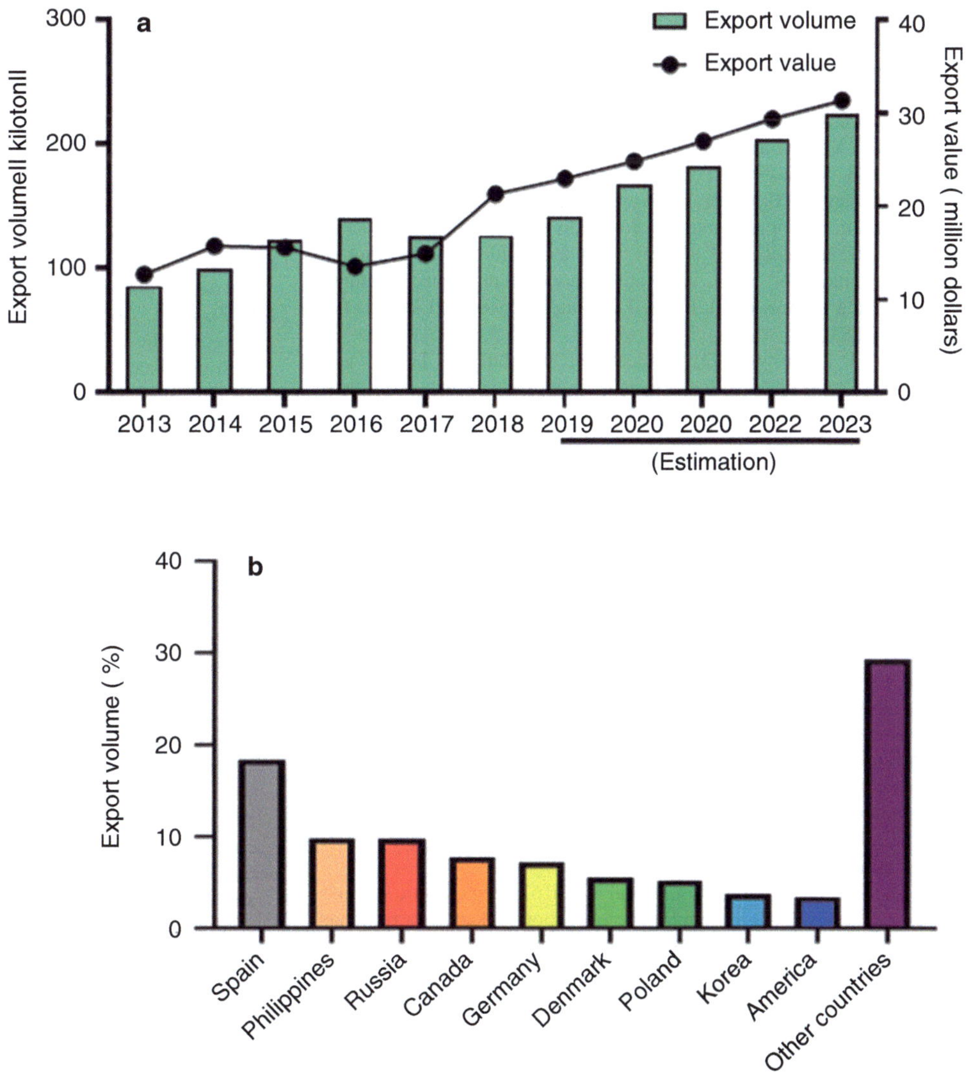

16.7 Conclusion and Future Prospects

The annual production of eucheumatoid seaweeds in China has decreased over the past 20 years due to the unfavourable cultivation conditions and costs involved in tropical red seaweed production. Meanwhile, carrageenan extraction methods and properties were improved greatly, and it occupied approximately 72.1% of markets at the time of writing, when even with the world's largest carrageenan production, China's utilization of eucheumatoid seaweeds requires increased efficiency, and carrageenan production is still in need of new and special innovations for sustaining global market needs in the future.

Acknowledgements This research was supported by the Science and Technology project of Fujian Province (No. 2022 T3024) and Asia Collaboration Project on Development of Ecological Marine Ranching.

References

Aga MB, Dar AH, Nayik GA, Panesar PS, Allai F, Khan SA, Shams R, Kennedy JF, Altaf A (2021) Recent insights into carrageenan-based bionanocomposite polymers in food applications: a review. Int J Biol Macromol 192:197–209

Bampidis V, Azimonti G, Bastos MD, Christensen H, Dusemund B, Durjava MF, Kouba M, López-Alonso M, Puente SL, Marcon F, Mayo B, Pechová A, Petkova M, Ramos F, Sanz Y, Villa RE, Woutersen R, Brantom P, Svensson K, Brozzi R, Galobart J, Gregoretti L, López-Gálvez G, Sofianidis K, Vettori MV, Innocenti ML (2022) Safety and efficacy of a feed additive consisting of carrageenan for pets and other nonfood-producing animals (Marinalg international). EFSA J 20. https://doi.org/10.2903/j.efsa.2022.7285

Bartlová M, Ziółkowska D, Pospiech M, Shyichuk A, Tremlova B (2021) Determination of carrageenan in jellies with new methylene blue dye using spectrophotometry, smartphone-based colorimetry and spectrophotometric titration. Food Sci Technol 41:81–90

Blanco-Pascual N, Alemán A, Gomez-Guillén MC, Montero MP (2014) Enzyme-assisted extraction of κ/ι-hybrid carrageenan from *Mastocarpus stellatus* for obtaining bioactive ingredients and their application for edible active film development. Food Funct 5:319–329

Cai ZL (1985) Underwater crop-eucheumatoids. Fisheries Sci Technol Inf 4:20–21

Chen LM, McLachlan J, Neish AC, Shacklock PF (1973) Ratio of kappa-carrageenan to lambda-carrageenan in nuclear phases of Rhodophycean algae, *Chondrus crispus* and *Gigartina stellata*. J Mar Biol Assoc UK 53(1):11–16

Chen S, Li Q, McClements DJ, Han YH, Dai L, Mao L, Gao YX (2020) Codelivery of curcumin and piperine in zein-carrageenan coreshell nanoparticles: formation, structure, stability and in vitro gastrointestinal digestion. Food Hydrocoll 99:105334. https://doi.org/10.1016/j.foodhyd.2019.105334

Das AK, Sharma M, Mondal D, Prasad K (2016) Deep eutectic solvents as efficient solvent system for the extraction of kappa-carrageenan from *Kappaphycus alvarezii*. Carbohydr Polym 136:930–935

Dong M, Zhang K, Wang LL, Han J, Wang YX, Xue ZX, Xia YZ (2020) High-strength carrageenan fibres with compactly packed chain structure induced by combination of Ba^{2+} and ethanol. Carbohydr Polym 236:116057. https://doi.org/10.1016/j.carbpol.2020.116057

Duan FY, Yu Y, Liu ZM, Tian L, Mou HJ (2016) An effective method for the preparation of carrageenan oligosaccharides directly from *Eucheuma cottonii* using cellulase and recombinant kappa-carrageenase. Algal Res-BioBiofuels Bioprod 15:93–99

Freile-Pelegrín Y, Robledo D (1997) Influence of alkali treatment on agar from *Gracilaria cornea* from Yucatán, México. J Appl Phycol 9(6):533–539

Frost & Sulliva Report (2019) Surveys and analysis reports on the global and Chinese hydrophilic gels market

Gereniu CRN, Saravana PS, Chun BS (2018) Recovery of carrageenan from Solomon Islands red seaweed using ionic liquid-assisted subcritical water extraction. Sep Purif Technol 196:309–317

Gul K, Singh AK, Jabeen R (2016) Nutraceuticals and functional foods: the foods for the future world. Crit Rev Food Sci Nutr 56(16):2617–2627

Hasizah A, Mahendradatta M, Laga A, Netusalach M, Supratomo WA, Salengke S (2018) A novel omic-based technology for seaweed processing. Int Food Res J 25(4):1341–1348

Jaiswal LL, Shiv S, Rhim JW, Hahm DH (2020) Lignin-mediated green synthesis of AgNPs in carrageenan matrix for wound dressing applications. Int J Biol Macromol 159:859–869

Jumah YU, Tumbokon BLM, Serrano AE (2020) Effects of dietary kappa-carrageenan on growth and resistance to acute salinity stress in the black tiger shrimp *Penaeus monodon* post larvae. Israeli J Aquacult Bamidgeh 72. https://doi.org/10.46989/001c.21464

Kintarou W (1936) Marine algal Flora of Japan. Uchida Old Tsurugi, Tokyo

Krempel M, Griffin K, Khouryieh H (2019) 13-hydrocolloids as emulsifiers and stabilizers in beverage preservation. Preservatives and preservation approaches in beverages. Academic Press 15:427–465

Kumar V, Kumar S, Pandey PK, Raman RP, Prasad KP, Roy S, Kumar A, Kumar K (2014) Growth and hemato-immunological response to dietary i-carrageenan in *Labeo rohita* (Hamilton, 1822) juveniles. Israeli J Aquacult Bamidgeh 66:917–927

Li R, Bai ZZ (2016) Biological characteristics and culture technique of *Eucheuma muricatum*. Mod Agric Sci Technol 18:244–245

Li Y, Zhang Z, Abbaspourrad A (2021) Improved thermal stability of phycocyanin under acidic conditions by forming soluble complexes with polysaccharides. Food Hydrocoll 119:106852. https://doi.org/10.1016/j.foodhyd.2021.106852

Linares MB, Peñaranda I, Iniesta CM, Egea M, Garrido MD (2022) Development of edible gels and films as potential strategy to revalorize entire male pork. Food Hydrocoll 123:107182. https://doi.org/10.1016/j.foodhyd.2021.107182

Liu F, Hou PF, Zhang H, Tang QJ, Xue CH, Li RW (2021) Food-grade carrageenans and their implications in health and disease. Compr Rev Food Sci Food Saf 20(4):3918–3936

Mandavinia GR, Mosallanezhad A, Soleymani M, Sabzi M (2017) Magnetic-and pH-responsive kappa-carrageenan/chitosan complexes for controlled release of methotrexate anticancer drug. Int J Biol Macromol 97:209–217

Mariot LV, Bolivar N, Coelho JD, Goncalves P, Colombo SM, Nascimento FV, Schleder DD, Hayashi L (2021) Diets supplemented with carrageenan increase the resistance of the Pacific white shrimp to WSSV without changing its growth performance parameters. Aquaculture 545:737172. https://doi.org/10.1016/j.aquaculture.2021.737172

Morais T, Cotas J, Pacheco D, Pereira L (2021) Seaweeds compounds: an ecosustainable source of cosmetic ingredients? Cosmetics 8. https://doi.org/10.3390/cosmetics8010008

Pangestuti R, Shin KH, Kim SK (2021) Anti-photoaging and potential skin health benefits of seaweeds. Mar Drugs 19. https://doi.org/10.3390/md19030172

Pereira L (2013) Population studies and carrageenan properties in eight Gigartinales (Rhodophyta) from western coast of Portugal. Sci World J 2013:1. https://doi.org/10.1155/2013/939830

Pickmere SE, Parsons MJ, Bailey RW (1973) Composition of *Gigartina* carrageenan in relation to sporophyte and gametophyte stages of life-cycle. Phytochemistry 12(10):2441–2444

Polat TG, Duman O, Tunc S (2020) Agar/kappa-carrageenan/montmorillonite nanocomposite hydrogels for wound dressing applications. Int J Biol Macromol 164:4591–4602

Preez R, Paul N, Mouatt P, Majzoub ME, Thomas T, Panchal SK, Brown L (2020) Carrageenans from the red seaweed *Sarconema filiforme* attenuate symptoms of diet-induced metabolic syndrome in rats. Mar Drugs 18. https://doi.org/10.3390/md18020097

Ratnayake S, Stockdale V, Grafton S, Munro P, Robinson AL, Pearson W, McRae JM, Bacic A (2019) Carrageenans as heat stabilizers of white wine. Aust J Grape Wine Res 25(4):439–450

Rees DA (1969) Structure, conformation, and mechanism in the formation of polysaccharide gels and networks. In Adv Carbohydr Chem Biochem 24:267–332

Rozo G, Bohorques L, Santamaria J (2019) Controlled release fertilizer encapsulated by a kappa-carrageenan hydrogel. Polimeros-Ciencia Tecnologia 29. https://doi.org/10.1590/0104-1428.02719

Sadi U, Prasetyaningsih E, Santoso J, Kamiya N, Oshima T. (2019) Preparation and characterization of semirefined carrageenan from *Kappaphycus alvarezii* seaweed bleached by peracetic acid. IOP Conference Series: Earth Environ Sci. https://doi.org/10.1088/1755-1315/278/1/012077

Sen M, Erboz EN (2010) Determination of critical gelation conditions of kappa-carrageenan by viscosimetric and FT-IR analyses. Food Res Int 43(5):1361–1364

Seo CW, Yoo BS (2021) Effect of kappa-carrageenan/milk protein interaction on rheology and microstructure in dairy emulsion systems with different milk protein types and kappa-carrageenan concentrations. J Food Process Preserv 45. https://doi.org/10.1111/jfpp.15038

Shen YF, Fan KC (1950) Taiwania 1(2/4):317–345

Shi SY, Li WQ, Li ZE, Zhang YX (1986) Preparation and properties of carrageenan of the eucheumatoids from Hainan Island, China. J Fisheries China 10(1):1–9

Sumathra M, Rajan M, Praphakar RA, Marraiki N, Elgorban AM (2020) In vivo assessment of a hydroxyapatite/kappa-carrageenan-maleic anhydride-casein/doxorubicin composite-coated titanium bone implant. ACS Biomater Sci Eng 6(3):1650–1662

Tari O, Pekcan Ö (2008) Study of drying of κ-carrageenan gel at various temperatures using a fluorescence technique. Dry Technol 26:101–107

Tseng CK (1935) Economic seaweeds of Kwangtung province. South China Lingn Sci J 14(1):93–104

Valado A, Pereira M, Caseiro A, Figueiredo JP, Loureiro H, Almeida C, Cotas J, Pereira L (2020) Effect of carrageenans on vegetable jelly in humans with hypercholesterolemia. Mar Drugs 18. https://doi.org/10.3390/md18010019

Vanayagam H, Mohamed SM, Chu WL (2014) Assessment of UVB-photoprotective and antioxidative activities of carrageenan in keratinocytes. J Appl Phycol 26(4):1813–1821

Vanegas JS, Torres GR, Campos BB (2019) Characterization of a κ-carrageenan hydrogel and its evaluation as a coating material for fertilizers. J Polym Environ 27(4):774–783

Vazquez-Delfin E, Robledo D, Freile-Pelegrin Y (2014) Microwave-assisted extraction of the carrageenan from *Hypnea musciformis* (Cystocloniaceae, Rhodophyta). J Appl Phycol 26(2):901–907

Velde F (2008) Structure and function of hybrid carrageenans. Food Hydrocoll 22(5):727–734

Villanueva RD, Montaño MNE, Romero JB (2009) Iota-carrageenan from a newly farmed rare variety of Eucheumoid seaweed-"endong". J Appl Phycol 21(1):27–30

Waaland JR (1973) Experimental studies on propagation of *Iridaea* and *Gigartina*. J Phycol 19:12–12

Wang Q, Zhang L, He YL, Zeng LR, He JC, Yang Y, Zhang TC (2021) Effect of kappa-carrageenan on glucolipid metabolism and gut microbiota in high-fat diet-fed mice. J Funct Foods 86:104707. https://doi.org/10.1016/j.jff.2021.104707

Wang YF, Liu MZ, Ni BL, Xie LH (2012) Kappa-carrageenan-sodium alginate beads and superabsorbent coated nitrogen fertilizer with slow-release, water-retention, and anti compaction properties. Ind Eng Chem Res 51(3):1413–1422

Wu CY, Xia EZ, Peng ZS, Tan SZ, LiJ WZC, Huang XH, Cai ZL, Chen GJ (1988) On the transplantation and cultivation of *Eucheuma striatum*. Oceanologia et Limnologia Sinica 5:410–418

Xia EZ (1965) A preliminary phytogeographical studies on Chinese species of *Eucheuma*. Oceanologia et Limnologia Sinica 5(1):52–55

Yamada Y (1936) The species of *Eucheuma* from Ryuku and Formosa. Sci Paper Inst Alg Res, Fac Sci Hokkaido Imp Univ 1(2):119–134

Youssouf L, Lallemand L, Giraud P, Soule F, Bhaw-Luximon A, Meilhac O, D'Hellencourt CL, Jhurry D, Couprie J (2017) Ultrasound-assisted extraction and structural characterization by NMR of alginates and carrageenans from seaweeds. Carbohydr Polym 166:55–63

Zhao YF (2008) High-efficiency technology for Eucheumatoids cultivation in cement frame mesh. Sci Fish Farming 8:27–28

Zhou Y, Chen FQ, Chen S, Xiao Q, Weng HF, Yang QM, Xiao AF (2021) Preparation and characterization of kappa-carrageenan modified with maleic anhydride and its application in films. Mar Drugs 19. https://doi.org/10.3390/md19090486

Zhuang P (1965) A preliminary experiment in the culture of *Eucheuma gelatinae* by the drooping method. J Fisheries China 2(4):72–74

Opportunities for Strengthening the Indonesian Seaweed Penta-Helix Through Collaboration 17

Jamaluddin Jompa, Nadiarti Nurdin Kadir, Amanda Priscella Putri, and Abigail Mary Moore

Abstract

Seaweeds are increasingly seen as commodities that can support sustainable development, including many of the Sustainable Development Goals (SDGs). Already the second-largest seaweed producer globally, Indonesia has great opportunities for expanding and enhancing the value of seaweed production in terms of quantity, quality and diversification. Furthermore, the potential for added value through downstream product development is immense, for example, in the food, cosmetics, pharmaceutical and health care, animal feed, fertiliser, energy and other industries. Recognising the potential of seaweeds to contribute to national and global goals, the Indonesian government has set an ambitious target to become the top global producer of processed seaweed products, with a Road Map set out in Presidential Decree 33/2019. The government also realised the need for a collaborative approach to develop and realise this potential. The innovative Penta-Helix model involves government, academia/research, industry, community and media as the five key stakeholder groups who need to work together to achieve this national goal. Hasanuddin University (UNHAS) is committed to embracing the penta-helix approach through collaboration. Innovations from the Centre of Excellence (CoE) for Development and Utilization of Seaweed at Hasanuddin University (CEDUS-UNHAS), a multidisciplinary CoE with many partners across the Penta-Helix, and a wide range of other research and development initiatives under other UNHAS units and partnerships, are leading to advances in seaweed-related fields.

Keywords

Algae · Sustainable development · Innovation · Multi-stakeholder collaboration · Diversification · Added value

J. Jompa (✉)
Faculty of Marine Science and Fisheries, Hasanuddin University, Makassar, South Sulawesi, Indonesia

Indonesian Academy of Sciences, Jakarta, Indonesia

Center of Excellence for Development and Utilization of Seaweed, Hasanuddin University, Makassar, Indonesia

Center of Excellence for Marine Resilience and Sustainable Development, Hasanuddin University, Makassar, Indonesia
e-mail: j.jompa@unhas.ac.id

N. N. Kadir
Faculty of Marine Science and Fisheries, Hasanuddin University, Makassar, South Sulawesi, Indonesia

Center of Excellence for Marine Resilience and Sustainable Development, Hasanuddin University, Makassar, Indonesia

A. P. Putri
Center of Excellence for Marine Resilience and Sustainable Development, Hasanuddin University, Makassar, Indonesia

Center of Excellence for Interdisciplinary and Sustainability Science, Hasanuddin University, Makassar, South Sulawesi, Indonesia

A. M. Moore
Center of Excellence for Marine Resilience and Sustainable Development, Hasanuddin University, Makassar, Indonesia

Graduate School, Hasanuddin University, Makassar, South Sulawesi, Indonesia

17.1 Introduction

Seaweeds are increasingly seen as commodities that can support sustainable development from local to global scales (FAO 2018; Hoegh-Guldberg et al. 2019; Zuccarello and Paul 2019; Neish 2021), as well as key components of the biosphere in this era of global change (Seckbach et al. 2010). In particular, well-planned and implemented seaweed-based development could contribute to at least nine of the Sustainable Development Goals (SDGs): SDG 1 No Poverty; SDG 2 Zero Hunger; SDG 3 Good Health and Wellbeing; SDG 4 Gender Equality; SDG 7 Affordable and Clean Energy; SDG 8 Decent Work and Economic Growth; SDG 10 Reduced Inequalities; SDG 13 Climate Action; and SDG 14 Life Below water (Duarte et al. 2017; Hoegh-Guldberg

A. T. Critchley et al. (eds.), *Tropical Phyconomy Coalition Development*, Developments in Applied Phycology 11,
https://doi.org/10.1007/978-3-031-47806-2_17

et al. 2019; Froehlich et al. 2019; Larson et al. 2021; Rimmer et al. 2021), although care is needed to avoid potential negative impacts, particularly on SDG 3 and SDG 14 (Hoegh-Guldberg et al. 2019), for example, through integrated aquascape approaches (Neish 2021).

Already the second-largest seaweed producer globally (after China), but with a much lower ranking in terms of seaweed export value (FAO 2018), Indonesia still has great opportunities for further advances (Neish 2021). In terms of seaweed production volume, there is room to expand the spatial extent and efficiency of seaweed cultivation, as well as the diversity of the species or strains cultivated in Indonesia (Hurtado et al. 2019; Neish 2021; Rimmer et al. 2021; Sangeetha and Thangadurai 2022; Tresnati et al. 2023; Tuwo et al. 2023; Kasmiati et al. 2023). Furthermore, Indonesia exports mainly raw materials (dried seaweed), while importing considerable quantities of processed seaweed products (President RI 2019). Meanwhile, in-country seaweed value chains and socio-economic impacts are often far from optimal, and the sector has been impacted by the COVID-19 pandemic (Nuryartono et al. 2021; Rimmer et al. 2021).

There is considerable opportunity to increase the economic, social and environmental value of Indonesian seaweeds by addressing issues with low or variable seaweed quality through the development and implementation of better farming and post-harvest practices, including biosecurity (Hurtado et al. 2019; Kambey et al. 2020; Nuryartono et al. 2021; Neish 2021; Rimmer et al. 2021). Furthermore, the potential for added value through downstream seaweed product development is immense, for example, in food, cosmetics, pharmaceutical and health care, animal feed, fertiliser, energy and other industries, as well as fields such as bioremediation and climate change mitigation/adaptation (Seckbach et al. 2010; FAO 2018; Hoegh-Guldberg et al. 2019; Froehlich et al. 2019; Neish 2021; Tresnati et al. 2023; Tuwo et al. 2023).

17.2 Indonesian Government Policy: Presidential Decree 33/2019

Recognising the potential of seaweeds to contribute to national and global development goals, the Indonesian government has set an ambitious target to become a market leader in the field of processed seaweed products and the top global producer of semi-refined carrageenan (SRC) and alkali-treated seaweed (ATS), with a roadmap set out in Presidential Decree 33/2019 (President RI 2019). In particular, this Decree addresses (1) enhancing the economic value of seaweed through cultivation and post-harvest development; (2) promoting the development of seaweed-based processing industries, especially at strategic sites close to producers; (3) market development for seaweed as a raw material and seaweed products; and (4) research and development (R&D) for the cultivation of new seaweed species/strains, innovation in products and processing technology, and national and global seaweed marketing. Out of over 1.1 million hectares of coastal waters considered suitable for farming one or both of the main commodities, eucheumatoid seaweeds and Gracilaria sp. (Fig. 17.1), approximately 20% are currently utilised for this purpose.

17.3 The Penta-Helix Model

The implementation matrix appended to Presidential Decree 33/2019 includes research, development and innovation through technology transfer and diversification of semi-refined and finished seaweed products, leading to technology packages and marketable seaweed products. Collaboration is recognised as key to achieving seaweed development targets. The triple helix (academia-government-industry) is a widely used institutional collaboration model (Etzkowitz and Leydesdorff 2000) that has been influential in the development of some programmes and policies in Indonesia (Martini et al. 2012; Moeliodihardjo et al. 2012; Sunitiyoso et al. 2012). However, in some situations, there is a recognition that more "helices" are needed (Fitriati et al. 2012).

Models proposed for "smart cities" include the Triple-Helix (Public-Private-Academia), Quadruple-Helix (Public-Private-Academia-Civic Society) and Penta-Helix (Public-Private-Academia-Civic Society-Social Entrepreneurs or/and Activists) models (Calzada 2020). The model advocated for seaweed development (Fig. 17.2) is also a penta-helix model recognising five core stakeholder groups or "helices". However, this model differs from the Penta-Helix model advocated for smart cities by Calzada (2020) in that the social entrepreneurs or activists (also referred to as bricoleurs, brokers and assemblers) forming the fifth "helix" are considered part of the community or civil society (fourth "helix"). In the seaweed penta-helix model, the media in a broad sense is the fifth "helix", as proposed for Indonesian "smarty city" development (Effendi et al. 2016) and the "flagship" industrial centres to accelerate development (Muhyi et al. 2017). In the Indonesian tourism industry, this Penta-Helix model has proven beneficial in sustaining the industry during the COVID-19 pandemic (Purnomo et al. 2021). The penta-helix can collaborate to promote entrepreneurship and innovation, including through state-of-the-art incubators and counselling/consulting services in technical and operational aspects of seaweed development. Severally and together, the Penta-Helix partners can all contribute towards improving the competitiveness and performance of individual seaweed businesses and the seaweed sector as a whole.

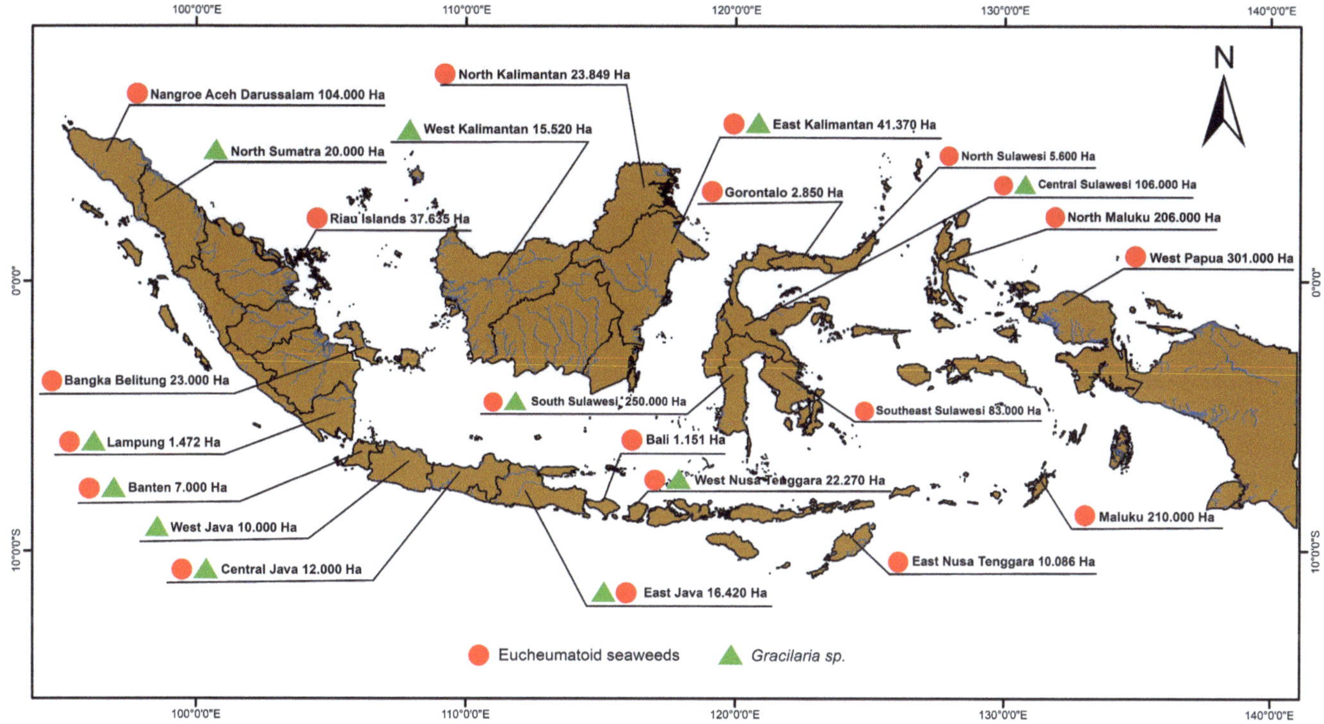

Fig. 17.1 Approximate extent of Indonesian coastal waters considered suitable for seaweed farming by province based on Ministerial Decree Number 1/KEPMEN-KP/2019 (Ministry of Marine Affairs and Fisheries of the Republic of Indonesia 2019)

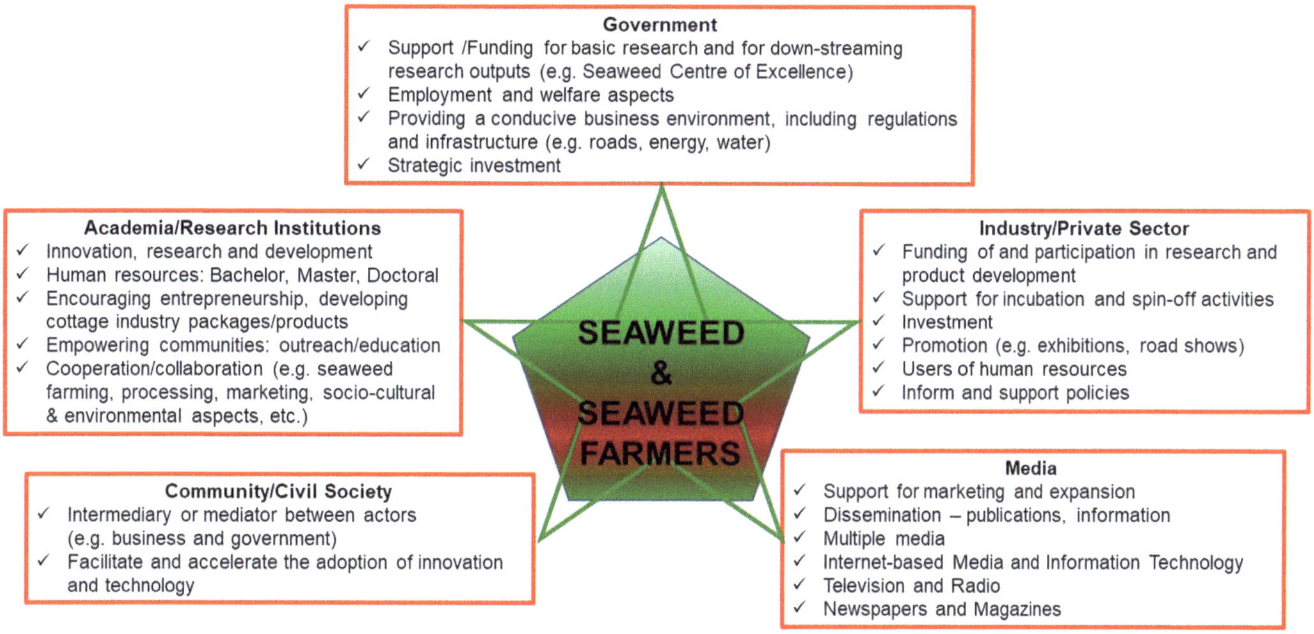

Fig. 17.2 Proposed Indonesian seaweed penta-helix model

The role of each Penta-Helix group or component is vital, as are the synergistic relations between them. Under this model, the main role of academic and research institutions is as leaders in research and innovation. However, under the Indonesian tertiary education "*Tri Dharma*" philosophy, knowledge transfer, outreach and community service are also important roles of universities, such as the UNHAS. Human resources development and knowledge transfer have typically focused on the three levels of formal tertiary education (bachelor, master and doctoral programmes), but there is an expansion in the need and opportunity for providing innovative educational services, including skills directly applicable to practical industry applications.

17.4 Hasanuddin University Seaweed Programmes

Seaweed-related research, development and outreach at Hasanuddin University, including many collaborative activities, have involved researchers from many disciplines (Fig. 17.3) from at least eight faculties (Table 17.1), in addition to activities under the CEDUS-UNHAS (Kasmiati et al. 2023). Collaboration with parties from other Penta-Helix components has been a key element of many of these activities, especially government institutions, industry and community groups (Fig. 17.4; Kasmiati et al. 2023), with comparatively limited media partnerships to date. However, in the context of Industry 4.0 and beyond, the media component will be increasingly important, including collaborations under Australia-Indonesia Centre Partnership for Australia-Indonesia Research (AIC-PAIR) initiatives.

Bearing in mind the key Penta-Helix roles of academia, recent seaweed research conducted at Hasanuddin University has covered a wide range of seaweed-focused and seaweed-related topics and is set to expand even further. Topics include but are not limited to seaweed diversity and ecology, seaweed culture, processing and marketable products, and environmental services (Tuwo et al. 2023; Kasmiati et al. 2023). Basic research on seaweed identification, distribution and characteristics has focused on both phytoplankton (Tambaru et al. 2020; Lestari et al. 2021) and macroalgae (Hartono and Sitepu 2016), with a herbarium being established for the latter. Some aspects covered by ecological

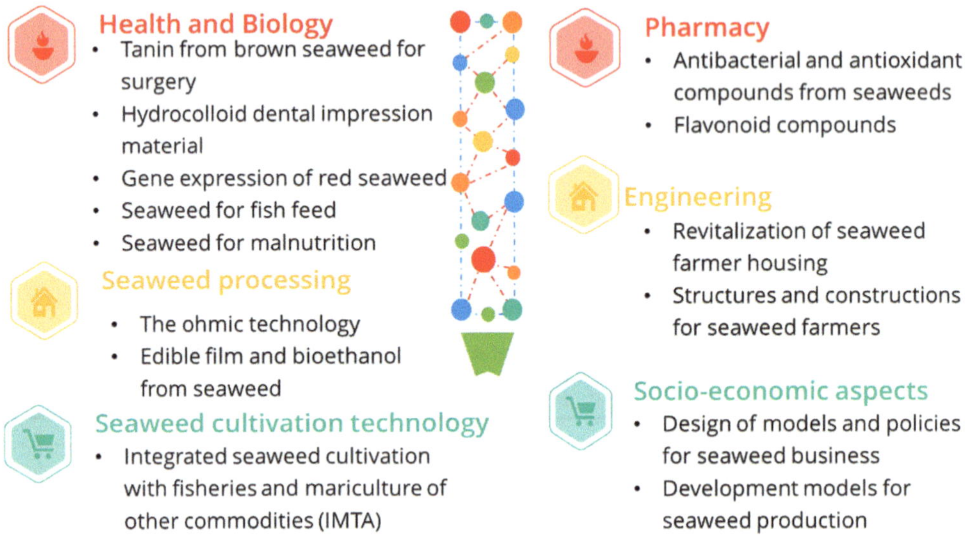

Fig. 17.3 Key areas of seaweed research, development and outreach at UNHAS

Table 17.1 Synopsis of seaweed-related research, development and outreach activities supported by grants and registered at the UNHAS Research and Outreach Unit (LPPM) in 2018–2020

Faculty/Unit	Number of items by year and type (R = research, O = outreach)					
	2018		2019		2020	
	R	O	R	O	R	O
Agriculture	2	2	1	–	1	–
Engineering	2	–	1	1	1	–
Marine Science & Fisheries	11	1	8	1	7	3
Economy and business	1	–	–	–	–	–
Pharmacy	1	–	1	–	–	–
Social & Politics	1	1	1	–	–	–
Public health	1	1	1	–	1	–
Mathematics & Natural Sciences	–	–	1	–	2	–
Total	19	5	14	2	11	3

Fig. 17.4 Some domestic and international UNHAS seaweed-related partners

studies are coral–algal interactions (Hasim et al. 2020) and bioconcentration by seaweeds (Abidin et al. 2016).

17.5 Key Areas of Opportunity

The scope of seaweed-based development is constantly expanding globally and in Indonesia (Tuwo et al. 2023). Key areas of research and utilisation where Hasanuddin University can contribute to strengthening the Penta-Helix range from upstream to downstream products, applications and technologies that could be mainstreamed through Penta-Helix collaboration. Here, we present a few examples at various levels in seaweed production and the value chain.

In seaweed production, including diversification, this will include building on past research to support seaweed culture, including aquaculture methods, technology and species diversification (Supriadi et al. 2016; Syamsuddin et al. 2019b; Syamsuddin 2023), seaweed diseases (Nasmia et al. 2014; Aqmal et al. 2016; Zainuddin et al. 2019; Darma et al. 2021), integrated multitrophic aquaculture (IMTA) and other polyculture systems (Syamsuddin et al. 2016, 2019a; Tuwo et al. 2019; Muchlis et al. 2020; Yasir et al. 2021; Tresnati et al. 2023), as well as supporting equipment (Mustafa et al. 2020). In the wider aquaculture sector, another major opportunity being explored is the potential of seaweeds in aquaculture feed (Idris et al. 2016; Salam et al. 2017; Moore et al. 2019; Lestari et al. 2020; Saade et al. 2021). One aspect of seaweed cultivation sustainability being addressed through an AIC-PAIR collaboration is the fate or "end of life" of plastics used in seaweed aquaculture in South Sulawesi. Other collaborations with AIC-PAIR are examining the potential of Indonesia-Australia partnerships for seaweed industry growth in Indonesia, including the application of multidisciplinary approaches, such as the use of remote sensing, as presented at the TPDC-1 workshop by PAIR researcher Zannie Langford (see elsewhere in this book).

In the health care sector, one growing research sector is the search for and application of bioactive compounds from seaweeds, for example, as anti-viral, anti-bacterial, anti-oxidant, anti-cancer, analgesic and anti-inflammatory agents, and for reducing blood loss in surgery (Wariz et al. 2016; Fauzi et al. 2018; Ruslin et al. 2018; Marinda et al. 2019; Hamrun et al. 2020a, 2022a; Tajrin et al. 2020; Darfiah et al. 2021; Akbar et al. 2021; Bahrun et al. 2021; Tassakka et al. 2023a, b). The Penta-Helix can build on such research for pharmaceutical and nutraceutical products, as well as to inform health professionals and the public about health benefits from seaweeds with specific compounds. Hasanuddin University has also been active in exploring the role of brown seaweeds in dentistry, including the field of dental prosthetics (Asmawati et al. 2016; Hamrun and Rachman 2016; Hamrun et al. 2018a, b, 2020b, 2021, 2022b; Dharmautama et al. 2019; Utama et al. 2020; Trilaksana and Kirana 2020).

Downstream seaweed processing technology and processed products can include a wide range of foods and food industry applications. Seaweed-based food and drink products include fresh and dried salad, seaweed powders, sheets or wafers, soup, noodles, drinks and jellies, snacks and an increasing range of creative culinary products and health foods. In Indonesia, raw seaweeds such as *Caulerpa* and *Ulva* are part of several culinary traditions. Hasanuddin University's contributions to expanding this use include post-harvesting technology to improve product shelf life, while other penta-helix partners can help overcome other limiting factors, such as resource access and rights, logistics and consumer awareness. Seaweeds have also long been used in deserts (van der Heijden et al. 2022). In particular, jellies are made from agar–agar, which, like several of the ever-widening range of seaweed-based foods and snacks, is still often imported. In this field, research and development at Hasanuddin University is leading the way, from creative formulations of traditional foods (e.g., noodles, cakes, crackers and other snacks) to innovative high-protein and other nutritionally enriched seaweed-based foods (Salma et al. 2021) and new commodities for consumers and the food industry, such as biosugar (Tassakka et al. 2021) (Kasmiati et al. 2023) and nata de *Sargassum* (Utami et al. 2021). Packaging and other items associated with food are also an area of active research, including seaweed-based edible films for prolonging product shelf life, as well as biodegradable packaging and utensils.

Other non-food uses of seaweed covered to date include personal care products (Pakki et al. 2018; Kasmiati et al. 2023), biofuels (Sari et al. 2017; Wadi et al. 2019; Sulfahri et al. 2020), bioremediation (Palayukan et al. 2016; Mukhtar et al. 2017), biofouling (Wulandari et al. 2019; Fahruddin et al. 2020), cutting-edge research into applications in arable and livestock farming (Prajogo et al. 2023) and carbon sequestration (Mashoreng et al. 2019). This diversity reflects the view that, in addition to developing the food security and economic potential of seaweeds, it is also vital to strengthen the scientific and technological basis for enhancing and leveraging the current and potential roles of seaweeds to address the multi-dimensional climate and biodiversity crises through penta-helix collaboration.

17.6 Conclusion

Seaweeds are already making a significant contribution to the Indonesian economy and society. Strengthening the Penta-Helix can be the key to unlocking further opportunities and advances in the volume, quality and diversity, as well as the societal and environmental benefits of the Indonesian seaweed sector, and to meeting the ambitious target of Indonesia as a world leader in processed seaweed commodities. Enhancing seaweed production and products can contribute to national and global food security and support healthy diets based on sustainably produced ingredients. Seaweed expansion can also contribute towards mitigating some anthropogenic impacts, especially eutrophication and some other forms of pollution, as well as climate change mitigation. Realising these benefits will require synergising the roles and contributions of government, academia and research institutions, industry/private sector, communities/civil society and the media through the Penta-Helix model. As a member of the academia "Helix", Hasanuddin University is committed to building on and developing capacity and networks to strengthen the Indonesian seaweed Penta-Helix through collaboration.

Declarations Competing interests—The authors have no competing interests to declare.

Availability of data and material—Data sharing is not applicable to this article, as no datasets were generated or analysed during the study.

Code availability—not applicable as no code was generated or analysed.

Authors' contributions:

Jamaluddin Jompa: conceived and wrote the paper, reviewed and edited the final version.

Nadiarti Nurdin Kadir: contributed content, reviewed and edited the final version.

Amanda Pricella Putri: contributed content, reviewed and edited the final version.

Abigail Mary Moore: contributed content, English language proofreading of all versions, reviewed and edited the final version.

The authors approved the manuscript and agree to be accountable for all aspects of the work in ensuring that questions related to the accuracy or integrity of any part of the work are appropriately investigated and resolved.

Acknowledgements The authors thank the Tropical Phyconomy Coalition Development (TPCD) sponsors, GenialG, represented by Philippe Potin, and the TPCD-1 initiators.

References

Abidin F, Werorilangi S, Tambaru R (2016) Bioconcentration of lead (Pb) and copper (cu) in fleshy macroalgae from islands of Bonebatang, Barranglompo, and Lae-Lae Caddi, Makassar City. J Rumput Laut Indones 1:8–16

Akbar A, Soekamto NH, Firdaus B (2021) Antioxidant of n-hexane, ethyl acetate and methanol extracts of *Padina* sp with DPPH method. IOP Conf Ser Earth Environ Sci 800:012019

Aqmal A, Tuwo A, Haryati H (2016) Analisis hubungan antara keberadaan alga filamen kompetitor terhadap pertumbuhan dan kandungan karaginan rumput laut *Kappaphycus* sp. di Provinsi Sulawesi Selatan. J Rumput Laut Indones 1:94–102

Asmawati A, Hasyim R, Lianingsih AI, Ariani DF (2016) The difference of anti-inflammatory effect of brown algae extract *Padina* sp. and *Sargassum* sp. that is derived from Punaga Beach, South Sulawesi. J Dentomaxillofacial Sci 1:116. https://doi.org/10.15562/jdmfs.v1i2.8

Bahrun, Soekamto NH, Firdaus (2021) In vitro and in silico analysis for antibacterial activity of various extracts of *Gracilaria salicornia* (Rhodophyta) from Selayar Islands, Indonesia. Egypt J Chem 64: 7203–7212

Calzada I (2020) Democratising smart cities? Penta-helix multistakeholder social innovation framework. Smart Cities 3: 1145–1172

Darfiah D, Kasmiati K, Latama G (2021) Antibacterial activity and identification of active compounds of seaweed extract *Sargassum* sp., *Halimeda opuntia* and *Halymenia* sp. from Lae-Lae Island of South Sulawesi. Int J Environ Agric Biotechnol 6:187–195. https://doi.org/10.22161/ijeab.66.23

Darma TA, Kasim M, Nur I et al (2021) Bacteria population on *Eucheuma denticulatum* (Rhodophyta) thallus infected by ice-ice disease cultivated on horizontal net cage. AACL Bioflux 14:399–411

Dharmautama M, Ikhriahni, Manggau MA et al (2019) The effectiveness of *Sargassum polycystum* extract against *Streptococcus mutans* and *Candida albicans* as denture cleanser. J Int Dent Med Res 12: 528–532

Duarte CM, Wu J, Xiao X et al (2017) Can seaweed farming play a role in climate change mitigation and adaptation? Front Mar Sci 4:1–8

Effendi D, Syukri F, Subiyanto AF, Utdityasan RN (2016) Smart city Nusantara development through the application of penta helix model (a practical study to develop smart city based on local wisdom). In: 2016 International Conference on ICT For Smart Society (ICISS). IEEE, pp 80–85. https://doi.org/10.1109/ICTSS.2016.7792856

Etzkowitz H, Leydesdorff L (2000) The dynamics of innovation: from national systems and "mode 2" to a triple helix of university–industry–government relations. Res Policy 29:109–123

Fahruddin F, Johannes E, Dwyana Z, Haedar N (2020) Antimacrofouling assays of *Sargassum duplicatum* extract in the settlement inhibition of *Balanus* sp. on wooden plank substrate. Chiang Mai Univ J Nat Sci 19:531–547

FAO (2018) The state of world fisheries and aquaculture 2018–meeting the sustainable development goals. Food and Agriculture Organization of the United Nations, Rome

Fauzi A, Lamma S, Ruslin M (2018) Total tannin levels analysis of brown algae (*Sargassum* sp and *Padina* sp) to prevent blood loss in a surgery. J Dentomaxillofacial Sci 3:37. https://doi.org/10.15562/jdmfs.v3i1.621

Fitriati R, Rahmayanti KP, Salomo RV (2012) Critical review of triple helix to quanto-tuple helix: lesson learned from social security system act implementation. Procedia–Soc Behav Sci 52:217–225

Froehlich HE, Afflerbach JC, Frazier M, Halpern BS (2019) Blue growth potential to mitigate climate change through seaweed offsetting. Curr Biol 29:3087–3093.e3

Hamrun N, Djamaluddin N, Dahri INA (2022a) Antioxidant activity of red algae extract (Rhodophyta) *Eucheuma spinosum* with DPPH (2.2-diphenyl-1- picrylhydrazyl) method. J Dentomaxillofacial Sci 7:14–19

Hamrun N, Hasyim R, Mude AH et al (2021) Compressive and tensile strength measurement of irreversible hydrocolloid impression material made of brown algae *Padina* sp. J Dentomaxillofacial Sci 6:51–53

Hamrun N, Nabilah T, Hasyim R et al (2020a) Toxicity test of bioactive red alga extract *Eucheuma spinosum* on shrimp *Artemia salina* leach. Syst Rev Pharm 11:672–676

Hamrun N, Oktawati S, Asmawati et al (2020b) Effectiveness of fucoidan extract from brown algae to inhibit bacteria causes of oral cavity damage. Syst Rev Pharm 11:686–693

Hamrun N, Rachman SA (2016) Measuring sodium alginate content of brown algae species *Padina* sp. as the basic matter for making dental impression material (irreversible hydrocolloid impression material). J Dentomaxillofacial Sci 1:129

Hamrun N, Talib B, Ruslin M et al (2022b) A promising potential of brown algae *Sargassum polycystum* as irreversible hydrocolloid impression material. Mar Drugs 20:55

Hamrun N, Thalib B, Tahir D et al (2018a) Physical properties of irreversible hydrocolloid dental impression materials obtained from brown algae species *Padina* sp. J Phys Conf Ser 1073:052018

Hamrun N, Thalib B, Tahir D et al (2018b) Physical characteristics test (water content and viscosity) of extraction sodium alginate brown algae (Phaeophyta) species *Padina* sp. as basic material for production dental impression material. J Dentomaxillofacial Sci 3:84

Hartono YK, Sitepu FG (2016) Seaweed species diversity in aquatic littoral of Tamalabba, Punaga Village, district of Mangarabombang, Takalar regency. J Rumput Laut Indones 1:77–81

Hasim H, Zamani NP, Jompa J (2020) The growth of turf alga on hard corals in Barrang Lompo Island, Makassar. IOP Conf Ser Earth Environ Sci 575:012105

Hoegh-Guldberg O, Caldeira K, Chopin T et al (2019) The ocean as a solution to climate change: five opportunities for action. World Resources Institute, Washington DC. http://www.oceanpanel.org/climate

Hurtado AQ, Neish IC, Critchley AT (2019) Phyconomy: the extensive cultivation of seaweeds, their sustainability and economic value, with particular reference to important lessons to be learned and transferred from the practice of eucheumatoid farming. Phycologia 58:472–483

Idris APS, Saade E, Haryati (2016) The use of industrial seaweed disposal as binder on the stability of food, growth, and survival of Nile gift (*Oreochromis niloticus*). J Rumput Laut Indones 2:132–142

Kambey CSB, Campbell I, Sondak CFA et al (2020) An analysis of the current status and future of biosecurity frameworks for the Indonesian seaweed industry. J Appl Phycol 32:2147–2160

Larson S, Stoeckl N, Fachry ME et al (2021) Women's Well-being and household benefits from seaweed farming in Indonesia. Aquaculture 530:735711

Lestari D, Saade E, Achmad M (2020) The biological evaluation of mud crabs (*Scylla olivacea*) in fattening activities using seaweed *Kappaphycus alvarezii* in ponds. AACL Bioflux 13:1090–1097

Lestari HA, Samawi MF, Faizal A et al (2021) Diversity and abundance of phytoplankton in the coastal waters of South Sulawesi. Hayati J Biosci 28:199

Marinda AA, Yusriadi et al (2019) Isolation and identification of bioactive proteins from microsymbiont algae *Eucheuma cottonii* and their potential as an anticancer. J Phys Conf Ser 1341:032024

Martini L, Tjakraatmadja JH, Anggoro Y et al (2012) Triple helix collaboration to develop economic corridors as knowledge hub in Indonesia. Procedia–Soc Behav Sci 52:130–139

Mashoreng S, La Nafie YA, Isyrini R (2019) Cultivated seaweed carbon sequestration capacity. IOP Conf Ser Earth Environ Sci 370:012017

Ministry of marine Affairs and Fisheries of the Republic of Indonesia (2019) Keputusasan Menteri Kelautan dan Perikanan Republik Indonesia Nomor 1/Kepmen-KP/2019 Tentang Pedoman Umum Pembudidayaan Rumput Laut [Decree of the Minister for Marine Affairs and Fisheries Number 1/Kepmen-KP/2019 on General Guidelines for Seaweed Farming]. Available from https://peraturan.bpk.go.id/Home/Download/152052/Kepmen/KKP/Nomor1/Tahun2019.pdf

Moeliodihardjo BY, Soemardi BW, Brodjonegoro SS, Hatakenaka S (2012) University, industry, and government partnership: its present and future challenges in Indonesia. Procedia–Soc Behav Sci 52:307–316

Moore A, Ndobe S, Ambo-Rappe R et al (2019) Dietary preference of key microhabitat *Diadema setosum*: a step towards holistic Banggai cardinalfish conservation. IOP Conf Ser Earth Environ Sci 235: 012054

Muchlis AM, Fachry ME, Adhawati SS (2020) The effect of production factors on the productivity of seaweed Gracilaria sp. in the polyculture cultivation system with milkfish (*Chanos chanos*) in Luwu District. Int J Environ Agric Biotechnol 5:1589–1595. https://doi.org/10.22161/ijeab.56.24

Muhyi HA, Chan A, Sukoco I, Herawaty T (2017) The penta helix collaboration model in developing centers of flagship industry in Bandung City. Rev Integr Bus Econ Res 6:412–417

Mukhtar M, Zainuddin EN, Yaqin K (2017) Adsorpsi bahan pencemar logam Pb menggunakan vermikompos *Sargassum*. J Rumput Laut Indones 2:12–18

Mustafa W, Asri S, Firmansyah MR et al (2020) An improvement for the design of the small fiberglass seaweed farmers boat to minimize the sinkable risk. IOP Conf Ser Mater Sci Eng 875:012077

Nasmia N, Syamsuddin R, Rantetondok A, Zainuddin EN (2014) Characterization and identification of bacteria isolated from seaweed *Gracilaria verrucosa* (Linn., 1758) infected by ice-ice. Int J Aquac 4:2014

Neish IC (2021) Adaptive phyconomy for sustainable management of coastal ecoscapes in Indonesia. IOP Conf Ser Earth Environ Sci 763: 012009

Nuryartono N, Waldron S, Tarman K et al (2021) An analysis of the South Sulawesi seaweed industry. The Australia- Indonesia Centre, Melbourne, Jakarta, Makassar. https://pair.australiaindonesiacentre.org/

Pakki E, Murdifin M, Wijoyo N, Sumarheni S (2018) Study of sunscreen and antioxidant activity of combination extracts from the red algae *Eucheuma cottonii* and *Eucheuma spinosum*. Drug Invent Today 10:1827–1830

Palayukan RA, Bandraeni AHY, Tuwo A (2016) The effectiveness of seaweed *Gracilaria* sp. as bioremediator for the change of N and P in cultivation container of *Litopenaeus vannamei*. J Rumput Laut Indones 1:88–92

Prajogo D, Alam JF, Malina AC et al (2023) Perdagangan rumput laut : Membuka potensi perdagangan industri rumput laut Australia-Indonesia. The Australia-Indonesia Centre, Melbourne, Jakarta, Makassar. https://pair.australiaindonesiacentre.org/

President RI (2019) Peraturan Presiden Republik Indonesia Nomor 33 tahun 2019 tentang Peta panduan (road map) pengembangan industri rumput laut nasional tahun 2018-2021. 1–60

Purnomo EP, Fathani AT, Setiawan D et al (2021) Penta-helix model in sustaining Indonesia's tourism industry. In: Antipova T (ed) Advances in digital science. ICADS 2021. Advances in intelligent systems and computing, vol 1352. Springer, Cham, pp 477–486. https://doi.org/10.1007/978-3-030-71782-7_42

Rimmer MA, Larson S, Lapong I et al (2021) Seaweed aquaculture in Indonesia contributes to social and economic aspects of livelihoods and community wellbeing. Sustainability 13:10946

Ruslin M, Husain AF, As HY, Subehan S (2018) Analysis of total flavonoid levels in brown algae (*Sargassum* sp. and *Padina* sp.) as analgesic drug therapy. Asian J Pharm Clin Res 11:81

Saade E, Usman U, Sulaeman HA, Jannah NM (2021) The effect of corn starch substitution with *Sargassum* sp. starch in diet on grow-out of cultivated rabbitfish, (*Siganus guttatus*) in floating net cages. J Aquac Fish Heal 11:10–20

Salam NI, Aslamyah S, Saade E (2017) Efek berbagai konsentrasi tepung rumput laut *Kappaphycus alvarezii* dalam pakan buatan terhadap kadar kolesterol dan komposisi kimia tubuh udang windu. J Rumput Laut Indones 2:1–11

Salma S, Hadju V, Jompa J et al (2021) The effect of red seaweed (*Kappaphycus alvarezii*) biscuits on hemoglobin levels and body weight among the first trimester pregnant women. Open Access Maced J Med Sci 9:1019–1023

Sangeetha J, Thangadurai D (eds) (2022) Seaweed biotechnology: biodiversity and biotechnology of seaweeds and their applications. Apple Academic Press, Boca Raton

Sari AP, Ahmad A, Usman H, Tuwo A (2017) Produksi bioetanol dari selulosa alga merah dengan sistem fermentasi simultan menggunakan bakteri *Clostridium acetobutylicum*. J Rumput Laut Indones 2:58–62

Seckbach J, Einav R, Israel A (eds) (2010) Seaweeds and their role in globally changing environments. Springer, Netherlands, Dordrecht

Sulfahri MS, Langford A, Tassakka ACMAR (2020) Ozonolysis as an effective pretreatment strategy for bioethanol production from marine algae. Bioenergy Res 13:1269–1279

Sunitiyoso Y, Wicaksono A, Utomo DS et al (2012) Developing strategic initiatives through triple helix interactions: systems modelling for policy development. Procedia–Soc Behav Sci 52:140–149

Supriadi SR, Abustang A, Yasir I (2016) Pertumbuhan dan kandungan karotenoid lawi-lawi *Caulerpa racemosa* yang ditumbuhkan pada tipe substrat berbeda. J Rumput Laut Indones 1:117–122

Syamsuddin R, Abustang R et al (2019a) Weight gain and carrageenan content of *Kappaphycus alvarezii* (Rhodophyta, Solierisceae) polycultured with *Sargassum polycystum* (Paeophyta, Sargassaceae). IOP Conf Ser Earth Environ Sci 253:012005. https://doi.org/10.1088/1755-1315/253/1/012005

Syamsuddin R, Azis HY, Badraeni R (2019b) Comparative study on the growth, carotenoid, fibre and mineral content of the seaweed *Caulerpa lentillifera* cultivated indoors and in the sea. IOP Conf Ser Earth Environ Sci 370:012019

Syamsuddin R, Nasmia RA, Zainuddin EN (2016) *Caulerpa racemosa* (Chlorophyta, Caulerpaceae) extract increases growth and biomass production of *Gracilaria verrucosa* (Rhodophyta, Gracilariaceae). AACL Bioflux 9:1044–1052

Tajrin A, Chandha MH, Mappangara S et al (2020) Analysis of the total level of flavonoids in the brown algae (Phaeophyceae) extract as analgesic and anti-inflammatory drugs. Pesqui Bras Odontopediatria Clin Integr 20:1–6

Tambaru R, Nafie YAL, Junaidi AW (2020) Proportion of HABs in Losari coastal waters of Makassar. IOP Conf Ser Earth Environ Sci 564:012018

Tassakka ACMAR, Iskandar IW, Alam JF et al (2023a) Docking studies and molecular dynamics simulations of potential inhibitors from the brown seaweed *Sargassum polycystum* (Phaeophyceae) against PLpro of SARS-CoV-2. Biotech 12:46. https://doi.org/10.3390/biotech12020046

Tassakka ACMAR, Iskandar IW, Juniyazaki ABA et al (2023b) Green algae *Caulerpa racemosa* compounds as antiviral candidates for SARS-CoV-2: in silico study. Narra J 3.:in press. https://doi.org/10.52225/narra.v3i2.179

Tassakka ACMAR, Sulfahri IDM et al (2021) Biosugar production from marine algae *Kappaphycus alvarezii* by acid hydrolysis. IOP Conf Ser Earth Environ Sci 860:012078

Tresnati J, Kiyokazu U, Yasir I et al (2023) Seaweed cultivation for community welfare in coastal areas and small islands. In: Sangeetha J, Thangadurai D (eds) Seaweed biotechnology : biodiversity and biotechnology of seaweeds and their applications. Innovations in Biotechnology. Apple Academic Press Inc., Palm Bay, pp 15–45

Trilaksana AC, Kirana I, Arisandi (2020) Effectiveness of brown algae extract (*Sargassum* sp) 15% in dissolving root canal smear layer (a SEM study). Med Clínica Práctica 3:100095

Tuwo A, Nakajima M, Tresnati J et al (2023) Seaweeds for food security and environmental mitigation. In: Sangeetha J, Thangadurai D (eds) Seaweed biotechnology : biodiversity and biotechnology of seaweeds and their applications. Innovations in Biotechnology. Apple Academic Press Inc., Palm Bay, pp 335–368

Tuwo A, Yasir I, Tresnati J et al (2019) A microcosm multitrophic aquaculture system. IOP Conf Ser Earth Environ Sci 370:012012

Utama MD, Akbar FH, Kartika A (2020) Compressive strength of acrylic resin plate after immersing in denture cleanser alga chocolate. Syst Rev Pharm 11:37–42

Utami D, Kasmiati K, Metusalach M (2021) Effect of concentration of *Sargassum polycystum* and fermentation time on thickness and yield of nata de *Sargassum*. Int J Environ Agric Biotechnol 6:163–167. https://doi.org/10.22161/ijeab.65.21

van der Heijden PGM, Lansbergen R, Axmann H et al (2022) Seaweed in Indonesia: farming, utilization and research. Report WCDI-22-220. Wageningen University & Research, Wageningen. https://doi.org/10.18174/578007

Wadi A, Ahmad A, Tompo M et al (2019) Production of bioethanol from seaweed, *Gracilaria verrucosa* and *Eucheuma cottonii*, by simultaneous saccharification and fermentation methods. J Phys Conf Ser 1341:032031

Wariz R, Asfa NW, Fauzi A (2016) The toxicity of brown algae (*Sargassum* sp) extract to mice (*Mus muscullus*). J Dentomaxillofacial Sci 1:109

Wulandari V, Latama G, Zainuddin EN (2019) Antibacterial activity of *Sargassum polycystum* and *Ulva reticulata* methanol extract against marine fouling bacteria. Int J Sci Res Publ 9:p91100

Yasir I, Tresnati J, Aprianto R, Tuwo A (2021) Survival rate of brown-marbled grouper *Epinephelus fuscoguttatus* cultured with seaweed *Gracilaria changii* in multitrophic microcosm models. IOP Conf Ser Earth Environ Sci 763:012020

Zainuddin EN, Anshary H, Huyyirnah H et al (2019) Antibacterial activity of *Caulerpa racemosa* against pathogenic bacteria promoting "ice-ice" disease in the red alga *Gracilaria verrucosa*. J Appl Phycol 31:3201–3212

Zuccarello GC, Paul NA (2019) A beginner's guide to molecular identification of seaweed. Squalen Bull Mar Fish Postharvest Biotechnol 14:43

Seaweed Production in Kenya amid Environmental, Market, and COVID-19 Pandemic Challenges

18

Alex Kimathi Gabriel, James Mwaluma, David Mirera, James Kairo, and Joseph Wakibia

Abstract

Despite being a widely accepted, viable, and sustainable livelihood for coastal communities in the Western Indian Ocean region, seaweed farming in Kenya has faced several challenges, including environmental, marketing, limited space for expansion, and, recently, the impact of the COVID-19 pandemic, among others. In the present study, we discuss the results of a survey conducted to assess the status of seaweed farms, seaweed biomass production, and the household economic status of a small-scale coastal seaweed farming village on the southern coast of Kenya upon the outbreak of the COVID-19 pandemic. The study established that the immediate containment measures imposed by the government to curb the spread of the COVID-19 pandemic had extremely negative impacts on the general performance of seaweed production strategies in Kenya. The smooth flow of farm management was interrupted, with 80% of the farms being in a dilapidated state and seed preservation structures in deep water collapsed. The year's total sales results showed a 6% decline in seaweed biomass in 2019 after losing approximately 135 tonnes (dwt) of seaweed, worth approximately KSh. 3.0 million. The loss of this revenue, coupled with the loss of clients for locally made seaweed value-added products, adversely affected the economic livelihood of seaweed farmers in Kibuyuni village. Based on the experience of COVID-19 impacts on seaweed farming in Kenya, a multidisciplinary approach composed of governments, development partners, donor agencies, researchers, and farmers has a substantial role in mitigating these challenges to ensure the sustainability and resilience of this important livelihood in the WIO region.

Keywords

Seaweed farming · Seaweed production · COVID-19 pandemic · Livelihood · Kibuyuni

18.1 Introduction

Two species of red seaweed, *Eucheuma denticulatum* and *Kappaphycus alvarezii*, are widely cultivated to meet the growing demand for hydrocolloids (kappa and iota carrageenans), which have a wide range of applications in the food, cosmetic, and pharmaceutical industries (Sukiman et al. 2014; FAO 2018). Carrageenan is used as a stabilizer, thickener, suspending agent, and gelling agent in food industries, while in non-food industries, it is used to make products such as toothpaste, cosmetics, paints, and textile dyes (Angka and Suhartono 2000; Reine and Trono 2002). Although the giant producers of these seaweeds are from Asia, e.g., China, the Republic of Southern Korea, Indonesia, the Philippines, etc., FAO (2018), these seaweeds have been introduced in recent decades in the Western Indian Ocean (WIO) region as an alternative livelihood (Msuya et al. 2014; Kimathi et al. 2018).

Farmers on the southern coast of Kenya have been widely trained in various seaweed farming techniques, including the fixed off-bottom line, floating raft, tubular net, and modified off-bottom method (Lirasan and Twide 1993; Renata et al. 2014; Kimathi et al. 2018). The same farmers have adopted a model farm system characterized by the establishment of six blocks of farmed seaweeds using a fixed off-bottom line culture technique. Seaweed is cultivated for 42 days, harvested, dried, and sold to local buyers.

A. K. Gabriel (✉) · J. Mwaluma · J. Kairo
Kenya Marine and Fisheries Research Institute, Mombasa, Kenya

D. Mirera
Kenya Marine and Fisheries Research Institute, Mombasa, Kenya

Department of Botany, Jomo Kenyatta University of Agriculture and Technology, Nairobi, Kenya

J. Wakibia
Department of Botany, Jomo Kenyatta University of Agriculture and Technology, Nairobi, Kenya
e-mail: jwakibia@jkuat.ac.ke

A. T. Critchley et al. (eds.), *Tropical Phyconomy Coalition Development*, Developments in Applied Phycology 11,
https://doi.org/10.1007/978-3-031-47806-2_18

However, seaweed farming in Kenya has faced several challenges, including environmental and marketing challenges, limited space for expansion, and, recently, the impact of the COVID-19 pandemic, among others. Environmental changes characterized by an increase in surface seawater temperature (above 30 °C) between December and February have been associated with infection of cultivated seaweeds with "ice-ice" syndrome (Msuya et al. 2014). The term "ice-ice" is described as an unhealthy condition of seaweeds, manifested by the degradation of seaweed thalli, exhibited by rotting and turning from brown to a pale white color similar to ice (Wakibia et al. 2006). During this period, the eucheumatoids experience a massive die-off; hence, farmers incur enormous economic losses. It has also been established that environmental challenges are more lethal to the growth of *Kapppahycus alvarezii* than *Eucheuma denticulatum* (Kimathi et al. 2018). As a result, farmers focus on the cultivation of *E. denticulatum* because of its greater resistance to environmental stress and ease of farming using the relatively cheaper method (the fixed-off bottom) in shallow water lagoons. The market price of seaweed cultivated on the southern coast of Kenya has persistently remained low because it is purchased by one buyer, hence portraying a monopolistic type of business. Therefore, farmers have a limited opportunity to negotiate the price of their products and hence can only accept what is offered by the buyer. On the other hand, irregular price fluctuations (Ksh15–30) and delays in purchasing dried seaweed are common phenomena. The lack of effective marine spatial plans in coastal county governments has left seaweed farming in Kenya being carried out without the guidance of such vital management documents. Hence, the space for seaweed cultivation and expansion has remained limited, as seaweed farmers have frequently found themselves in conflicts with other resource users, such as fishermen, boat operators, and government and private developers. COVID-19 was declared a global pandemic in March 2019. The mitigation measures adopted by different countries to arrest the spread of the COVID-19 pandemic have also created unprecedented negative impacts on government economies worldwide (FAO 2020). The main sectors driving rapid economic growth, such as tourism, transport, manufacturing, and agriculture, were adversely affected (FAO 2020). Small-scale fisheries and aquaculture food systems are not exceptional, with reports indicating a reduction in fishing activities in different parts of Africa, Asia, and Europe (FAO 2021). For example, introductions of physical distancing have been associated with a limited supply of inputs and labor and incapacitated fish farmers to sell their harvest and acquire seed and feed to restock their farms (FAO 2021).

The immediate measures taken by the government through the Ministry of Health to prevent the spread of the COVID-19 pandemic in Kenya included maintaining a social distance between persons, restricting movement from county to county, imposing a twelve (12)-hour lockdown (6 pm to 5 am), wearing face masks, and frequent hand washing and sanitization. Although seaweed farming has been widely accepted as a sustainable livelihood in the WIO region because of its ability to cope with and recover from stressors and shocks (Msuya et al. 2014), there is scarce information on how the outbreak of COVID-19, coupled with other prevailing challenges, affected the integrity of seaweed farming in the WIO region. In the present study, we discuss the results of a survey conducted in April 2019 to assess the status of seaweed biomass production on seaweed farms and the household economic status of small-scale coastal seaweed farming villages on the southern coast of Kenya since the outbreak of the COVID-19 pandemic. The objective of the study was to assess the impact of COVID-19 on the dynamics of seaweed production. Sustainable mitigation measures for challenges that require urgent attention to salvage livelihoods from imminent collapse are also highlighted. The study recommended policy formulation for mitigating similar challenges in the future and for sustainable seaweed utilization and development in the WIO region.

18.2 Materials and Methods

18.2.1 Study Site

The study was conducted in April 2019 at Kibuyuni, a seaweed cultivation site on the southern coast of Kenya (Fig. 18.1). Kibuyuni village has well-established commercial seaweed farms supported by a robust local management system and today has been branded as the model farm site for training new farmers in the coastal region. It has over 300 farmers composed of 240 women and 80 men. Seaweed farmers have a formally registered group (Kibuyuni Seaweed Association) with duly elected leaders. The vigor and zeal demonstrated by farmers at this site for seaweed cultivation have attracted tremendous attention from many seaweed stakeholders, including researchers in the mariculture sector.

18.2.2 Methodology

This study adopted a composite of qualitative and quantitative methods. The qualitative methods included the review of documentation and interviews, while the quantitative methods included observations, semi-structured questionnaires, and scoring. The study was also guided by the research principles set by Bennett et al. (2020) and COBI (2020), which stated that researchers working with small-scale fisheries (SSF) and in coastal communities can use their expertise, resources, and networks to research the

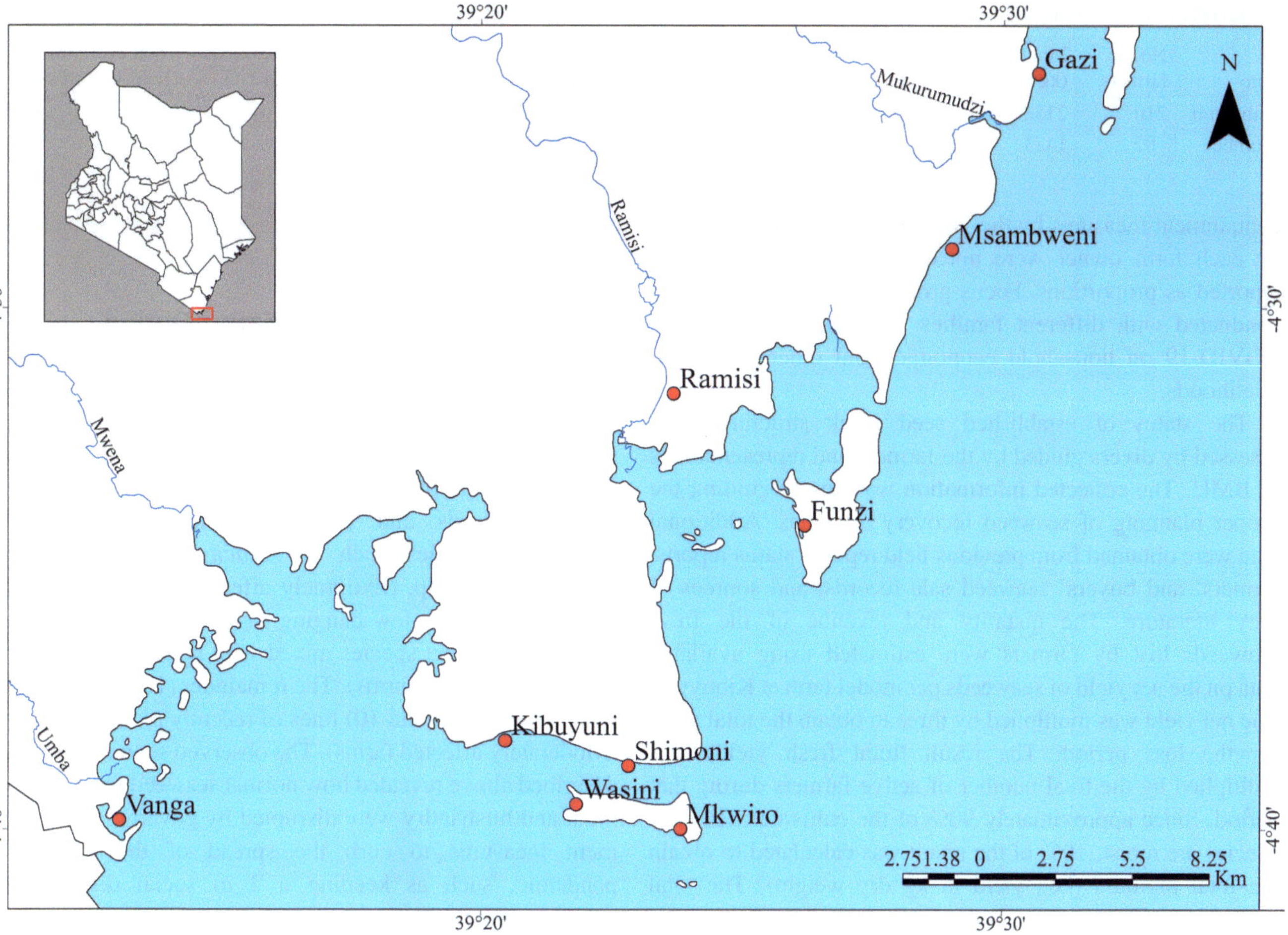

Fig. 18.1 Map of the South Coast of Kenya showing seaweed farming areas

immediate economic, social, and food security impacts of the pandemic while following institutional and community research and social distancing protocols to respect community safety. This principle complements the observation made earlier by Chevalier and Buckles (2019), who noted that the immediate priority of research in the pandemic era should be conducting action-oriented research that meets needs identified by community partners. Therefore, through strict observation of these protocols, the background is provided upon which the underlying issues are discussed, sustainable solutions are identified, and an appropriate policy framework is proposed.

To initiate the study, KMFRI applied for a travel permit from the Ministry of Health to allow movement from Mombasa County to Kibuyuni village in Kwale County. After the approval was granted, a seaweed multidisciplinary recovery team composed of KMFRI researchers, divers, technicians, seaweed farmers, and seaweed stakeholders, including the Kenya Fisheries Service (KFS), County Fisheries Department, and Beach Management Units (BMU), was formed. KMFRI researchers guided the participants in collecting the requisite data in the field. The data on the status of individual farms were collected by the researchers and technicians by conducting a rapid survey of 30 farms and interviewing the farm owners. The individual farm status (IFS) was grouped into three broad categories "Extremely affected" (farms with hanging ropes devoid of target seaweeds but densely colonized by epiphytes), "Seriously affected" (farms with hanging ropes with few remnants of cultivated seaweed and colonized by epiphytes), and "Moderately affected" (farms with hanging ropes and evidence of recently planted target seaweeds). On the farm site, the farmer owner was asked to confirm the observations made by the researcher. Upon confirmation, the researcher read a list of possible reasons, structured in a questionnaire, that could be attributable to the farm status and asked the farm owner to choose one of the reasons. The six reasons were as follows: (1) Strict COVID-19 containment measures taken by the government, (2) Low profit accrued from seaweed farming, (3) Engagement with alternative livelihood activities, (4) Lack of initial capital input, (5) Environmental challenges, (6) Environmental challenges and COVID-19

Table 18.1 The economic loss suffered by seaweed farmers at Kibuyuni between March and June 2019 due to the COVID-19 pandemic

Site	No. of farmers	Monthly net yield (kg wet wt/model farm)	Gross yield loss (kg wet wt) in 3 months	Gross yield loss (kg d wt) in 3 months	Price of seaweeds (Ksh/kg dwt)	Total economic loss (Ksh) in 3 months
Kibuyuni	210	2152	1,355,760	135,576	25	3,389,400
Tumbe	87	1473	384,453	38445.3	25	961,132

containment measures by the government. The reasons cited by each farm owner were noted for further analysis and reported as proportions. Focus group discussions were then conducted with different families to assess the impact of COVID-19 on household economies and other associated livelihoods.

The status of established seed stock structures was assessed by divers guided by the farmers and representatives of BMU. The collected information was key to guiding the proper planning of seaweed recovery activities. Additional data were obtained from previous field reports, status reports, farmers' and buyers' seaweed sale records, and sources of grey literature. The quantity and revenue of the fresh seaweeds lost by farmers were estimated using available data on the net yield of seaweeds per model farm at Kibuyuni. The net yield was multiplied by three to obtain the total yield for the loss period. The result (total fresh yield) was multiplied by the total number of active farmers during that period. Since approximately 90% of the cultivated seaweed species are moist, 10% of the result was calculated to obtain the total biomass (i.e., yield in kg dry weight). The total biomass (dwt) was then multiplied by the prevailing gate price of seaweed for that particular site (Table 18.1) as revenue gained.

Both subjective and objective dimensions related to people's overall quality of life and the factors affecting it were adopted to assess the well-being of Kibuyuni seaweed farmers (Diener et al. 1999). Subjective elements indicate how a condition is perceived by participants and are distinct from an objective and independently observable assessment of conditions (Smith and Clay 2010). Based on the methodology described by Diener and Suh (1997) and Max-Neef (1995), subjective well-being was measured by asking the seaweed farmers to respond to questions about their satisfaction with life overall and with the various factors that are likely to influence it. The approach takes into account individual experiences and helps to understand and communicate the interpretations, priorities, and needs of people.

18.2.3 Data Analysis

Descriptive statistics were generated using a Microsoft Excel spreadsheet. Statistical Packages for Social Sciences (SPSS) software version 26.0 was used to analyze the data obtained from the interviewees and the net yield of seaweed.

18.3 Results

18.3.1 Status of Seaweed Farms During the COVID-19 Pandemic

The status of seaweed farms in Kibuyuni village is displayed in Fig. 18.2. Out of the 30 model farms visited and assessed, more than 80% of the farms had hanging ropes devoid of target seaweeds and densely colonized by epiphytic macroalgal species such as *Gracilaria*, *Padina*, *Ulva*, *Chaetomorpha* spp. (Extremely affected farms), while 10% of the farms had a few hanging ropes with a few and weak remnants of target species mixed with macroalgal epiphytes (Seriously affected farms). The remainder (10%) of the farms had a few (fewer than 10) lines of recently planted seaweeds (Moderately affected farms). The observed status of the farms described above revealed how normal seaweed farm management and husbandry were disrupted by government containment measures to curb the spread of the COVID-19 pandemic, such as keeping a 2 m social distance and restricting movement.

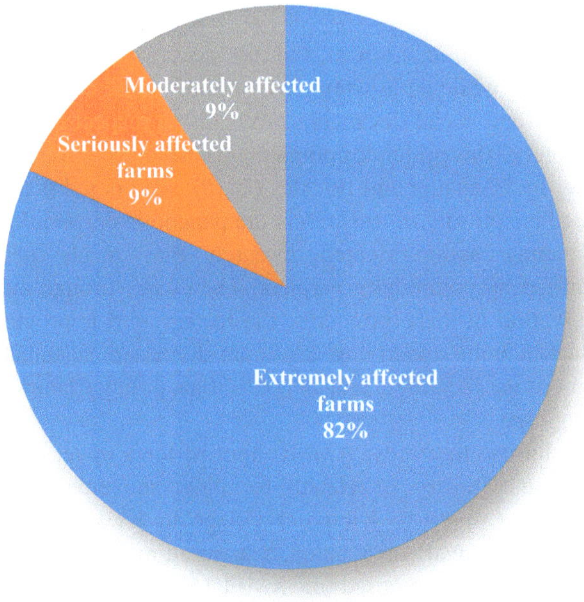

Fig. 18.2 The status of seaweed farms on the southern coast of Kenya during the COVID-19 pandemic

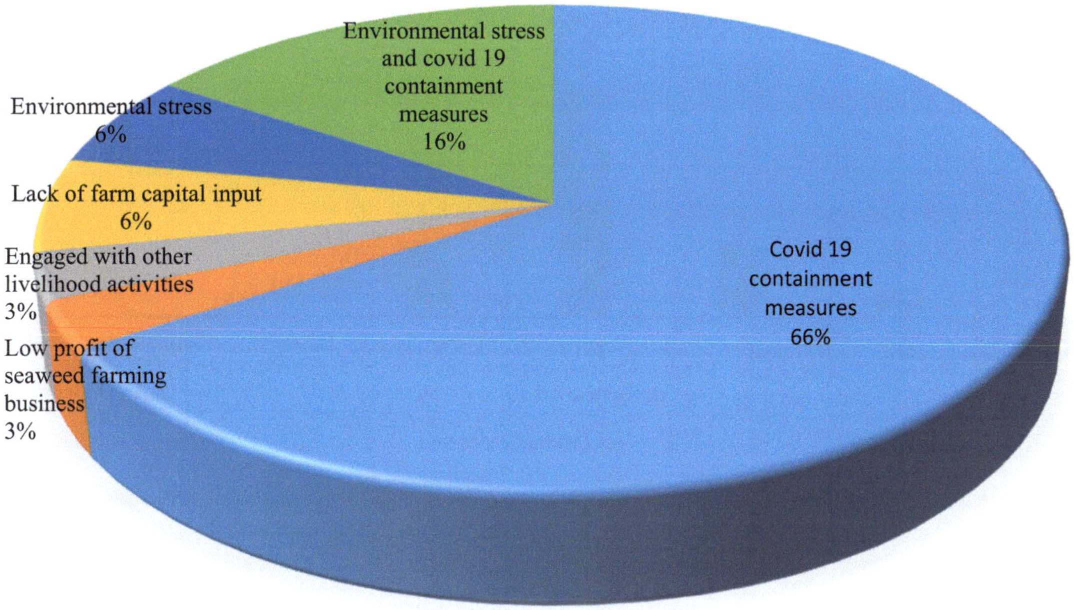

Fig. 18.3 The reasons cited by the farmer for the prevailing status of the seaweed farms at Kibuyuni village

In Fig. 18.3, several reasons attributed to the dilapidated state of the farms were cited by the 30 farm owners. The majority of the farmers (66%) associated the containment measures taken by the government to curb the spread of the COVID-19 pandemic with the dilapidated state of the farms. They stated that farmers were forced to stay at home, hence abandoning their normal farm management practices. Sixteen percent of the farmers blamed the environmental stress and COVID-19 containment measures for the loss of seaweed on their farms. Lack of farm capital inputs (polypropylene ropes and tie-ties) and environmental changes were each blamed by 6% of the respondents. A similar percentage of farmers (6%) was obtained for those who attributed the poor state of farms to the abandonment of seaweed farms due to engagements with other livelihood activities and those who were discouraged from farming due to the low profit accrued from the seaweed farming business.

The results obtained from the established seed banks in deep water showed that the culture facilities (floating rafts) were badly damaged (beyond repair), as evidenced by the presence of bamboo wood loosely floating in pieces and holding empty seaweed ropes. Furthermore, the effort demonstrated by the diving team confirmed the presence of large masses of fresh seaweed lying on the sea bottom at low-tide water depths of between 4 - 6 m at Kibuyuni.

18.3.2 The Declines in the Seaweed Farming Population, Seaweed Biomass Production, and Revenue During the COVID-19 Pandemic

Seaweed farming on the south coast of Kenya started at a low level in Kibuyuni village, with fewer than 70 farmers in 2015, and increased to 620 farmers by 2018 (see Fig. 18.4). The steady growth curve changed to a lag phase from 2018 to 2019. During the time of the study, a low percentage (2%) increase in the number of farmers was observed between 2018 and 2019, compared to the previous annual growth of 9% (see Fig. 18.4). Seaweed biomass production on the south coast has been growing gradually in volume output, from a mere biomass production of 5.2 tonnes in 2012 to just 43.9 tonnes in December 2018, before a decline in 2019 was observed (see Fig. 18.5). An economic analysis of loss suffered by farmers due to the COVID-19 pandemic showed that farmers had lost seaweed biomass on their model farms for three months, valued at approximately Ksh 3,389,400 and Kshs 961,132 at Kibuyuni and Tumbe, respectively (Table 18.1). The loss of the 3-month revenue consequently led to an overall decline in revenue generated from seaweed sales between 2018 and 2019, with sales ranging from Ksh 96,704 in 2018 to Ksh 922,508 in 2019 (see Fig. 18.6). The loss of seaweed biomass in both communities curtailed the

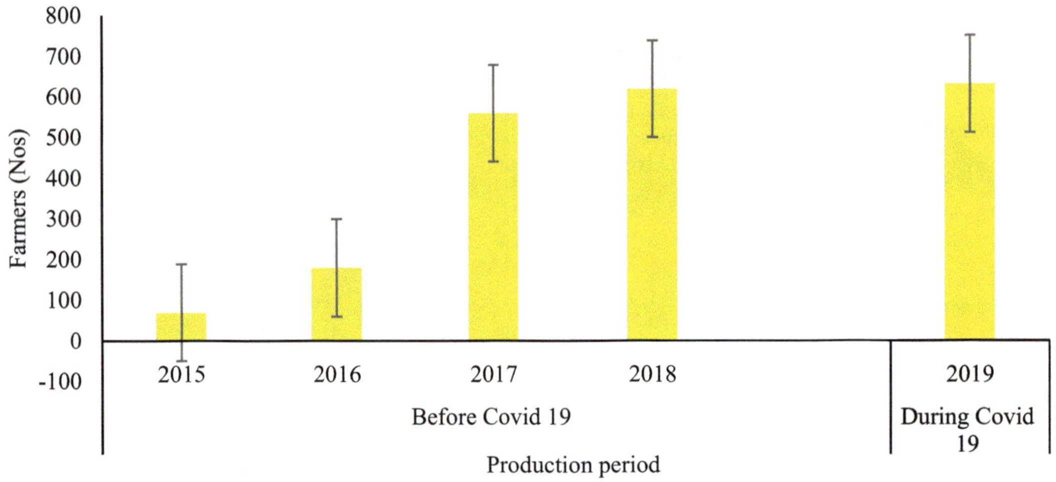

Fig. 18.4 The growth in the number of seaweed farmers before and during the Covid-19 pandemic on the southern coast of Kenya

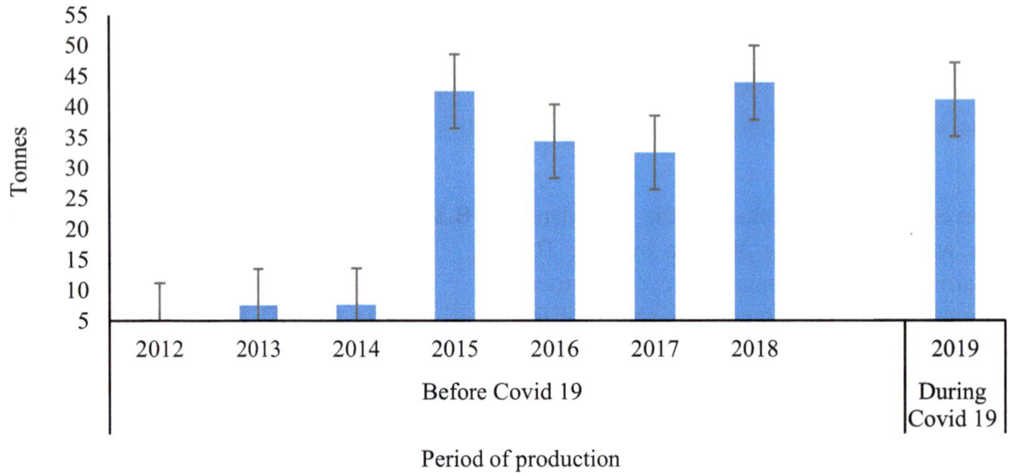

Fig. 18.5 Seaweed production (tonnes dwt) on the southern coast of Kenya before and during the COVID-19 pandemic

Fig. 18.6 The revenue from seaweed sales before and during the COVID-19 pandemic on the southern coast of Kenya

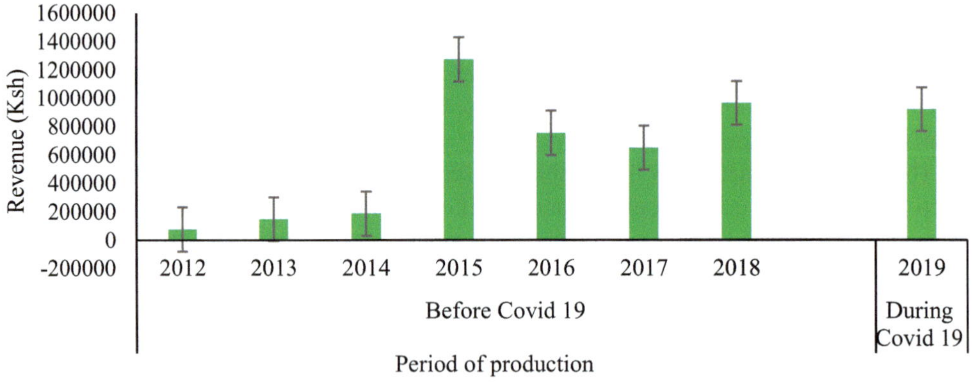

effort to meet the threshold (10,000 kg dwt/month) set for sale by the buyer, hence aggravating the economic suffering of both communities. It was also established that the regular circulation of money within the communities was adversely affected, leading to small local businesses indirectly or directly supported by seaweed farming also closing down.

18.3.3 The Household Economy of Seaweed Farmers

The economic status of the seaweed farmers was assessed by measuring the priority of three basic needs, including food, education, health, and education for their children, during the COVID-19 pandemic. According to the results displayed in Fig. 18.7, 85% of the seaweed farmers stated that ease of access to food was the most important need, while access to health and children's education trailed with 10 and 5%, respectively.

However, less than 50% of the respondents confirmed having guaranteed access to food due to the dwindling household savings and the uncertainty of the recovery period of the seaweed farming business. Due to movement restrictions, the farmers at Kibuyuni also lost clients for seaweed value-added products, such as seaweed soap and seaweed shampoo, which were locally manufactured and sold to visitors and the neighboring villages. Therefore, the economy that had groomed at Kibuyuni village since the inception of seaweed cultivation was suddenly collapsing. On the other hand, the effect of inter-county movement restriction was equally hurting since some of the materials for repairing farms, for instance, the polypropylene ropes, could not be accessed, as they were only available in Mombasa County, a distance of approximately 60 km from Kibuyuni. More than 50% of the respondents stated that the economic challenge had forced them to review their feeding habits, e.g., reducing the ratio served to each member of the family and adults omitting some meals to preserve young children. The majority (90%) of the farmers argued that, although the COVID-19 pandemic outbreak was a dangerous disease, they were confident that they would emerge victoriously, as they strictly adhered to the established protocols and had access to health facilities to treat other diseases near their villages. On the other hand, all respondents stated that access to education was not a priority because all schools were closed by then.

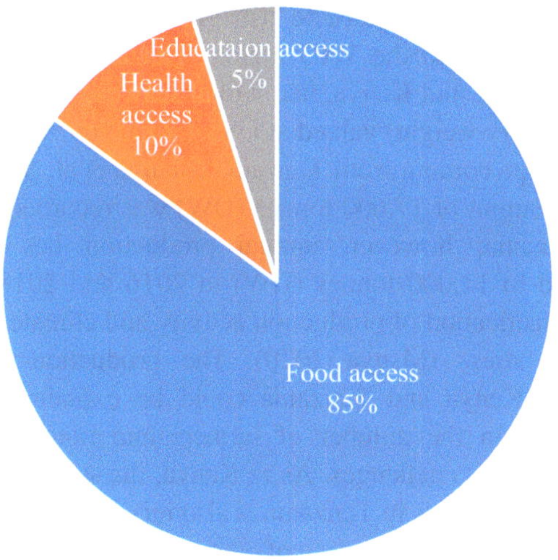

Fig. 18.7 The priority needs of seaweed farmers in Kibuyuni village during the COVID-19 pandemic

18.3.4 The Immediate Measures Adopted to Mitigate the Effects of the COVID-19 Pandemic on Seaweed Farming in Kenya

18.3.4.1 Government Intervention Initiatives

In an attempt to cushion vulnerable communities such as those involved in fishing and fish processing, the government of Kenya provided relief food, direct financial assistance, including cash stipends via mobile funds transfer, and tax relief in some food commodities and salaries for civil servants. However, 80% of the people interviewed in Kibuyuni village had very different experiences in receiving aid and support. Apart from a few (10%) members of the community who confessed to having received a one-off food aid package that included maize flour, beans, sugar, and rice, the seaweed farmers never benefited from these donations, as they were perceived to be more economically stable than the other needy population. Despite more than 50% of the members registering with the relevant agencies to benefit from cash transfers through mobile, no positive response was forthcoming. Therefore, families in the village developed a stronger culture of generosity, manifested by sharing with neighbors any little that was available from any source.

18.3.4.2 Recovery of Lost Seaweeds from the Deep Coralline Environment Using a Multidisciplinary Approach

The seaweed multidisciplinary recovery team composed of KMFRI researchers, divers, technicians, seaweed farmers, and other seaweed stakeholders, including the Kenya Fisheries Service (KFS), County Fisheries Department, and Beach Management Units (BMU), managed to recover a total of 1964 kg (wet weight) of seaweed. The recovered seaweed was eventually distributed to more than 400 farmers in Kibuyuni and Tumbe villages who shared similar farming challenges. Due to the sensitivity of seaweeds to open air, the farmers who were supplied with seaweed seeds were advised to plant them immediately on their respective farms. Close monitoring of the growth performance of the stocked seaweeds by the KMFRI technical team showed that after 42 days of the growth period, the seaweed biomass had increased tenfold and was able to be substantially shared with fellow farmers.

18.4 Discussion

The present study showed that the poor status of both seaweed farms and seed bank facilities established in the deep-water environment was partially attributed to the containment measures taken by the government to curb the spread of the COVID-19 pandemic and environmental stress. The COVID-19 pandemic outbreak occurred when farmers were already suffering an economic loss due to massive seaweed die-offs on their farms due to the ice-ice and epiphyte invasion of seaweeds. The negative effects of ice-ice and epiphyte invasion of seaweeds have been observed in previous reports (Kimathi et al. 2018; Msuya et al. 2014; Wakibia et al. 2011). Based on the results obtained from the farmers interviewed, over 60% of the seaweed farmers were confident that the COVID-19 pandemic was the main challenge to their farms. The low percentage of farmers who associated the prevailing farm status with environmental challenges (6%) and lack of capital input (6%) suggested that the majority of the farmers were familiar with these challenges and probably had knowledge of how to mitigate them to avoid adverse economic losses. The immediate measures taken by the government through the Ministry of Health to prevent the spread of COVID-19 in Kenya included maintaining a social distance of 2 m between persons, restricting movement from County to County, imposing a twelve (12)-hour lockdown (6 pm to 5 am), wearing face masks, and frequent hand washing and sanitization. Although these measures had good intentions for the health of seaweed farmers and the communities at large, they created a more complicated crisis in seaweed farming since farmers were denied access to the farms and established seed banks. Due to the delay in harvesting seaweeds from the established culture structures, the seaweed weight overwhelmed the established structures (floating devices). As a result, the entire structure sank to the sea bottom and drifted to deep areas beyond the reach of an ordinary farmer. Other floating rafts were dislodged from their anchored positions and swept away by strong tidal waves of the southeast monsoon (SEM) winds. It was therefore evident that farmers lost all the seeds that were expected to facilitate the stocking of farms from March when water conditions started improving for seaweed farming. The situation of seaweed farms in Kenya could be compared to a scenario described in aquaculture by the FAO (2020), where the impact of the COVID-19 pandemic resulted in reduced consumer demand for fish products and the closure of post-harvest enterprises, rendering women jobless since they form the bulk of the workforce in the processing sector (FAO 2020).

The seaweed multidisciplinary recovery team, guided by KMFRI researchers, yielded promising results toward the rehabilitation of seaweed farms on the southern coast of Kenya. Similar initiatives have been reported in several countries where seaweed is grown, such as Tanzania, Madagascar, and India. In India, seaweed is supported by private investments, industries, financial institutions, and NGOs, led by the Aquaculture Foundation of India, and today it has become a potential employment opportunity and income-earning activity, which is practiced by more than a thousand members of Self Help Groups in Ramanathapuram District (Narayankumar and Krishnan 2011).

The decline in seaweed production (43.97 tonnes dwt) in 2018 by 6.3% (41.2 tonnes dwt) in 2019 in Kenya was attributed to the loss of cultivated seaweeds on the farms during the first three months of the COVID-19 pandemic outbreak, as indicated by the analysis in Table 18.1. In the WIO region, Msuya and Kyewalyanga (2006) observed that the negative effects of "ice-ice" disease and epiphytes were responsible for a drop in biomass production of *K. alvarezii* from 1000 tonnes (dry wt) to only 13 tonnes (dwt) in Tanzania, while in Kenya, epiphytic algae and "ice-ice" syndrome infestations seriously hampered the growth of *Kappaphycus alvarezii* (Wakibia et al. 2006; Msuya et al. 2014; Kimathi et al. 2018). Based on the results of economic analysis, farmers at Kibuyuni lost more than 135.5 tonnes (dwt) of seaweed. However, it should be noted that if the normal tendering of the farms had taken place, approximately three-quarters of this seaweed would have been on the farms as restocked seaweeds, while a quarter of it would have been dried for sale. Nevertheless, the production of 41.2 tonnes was higher than the 2.0 tonnes previously produced in 2009, valued at USD 320 (Troell et al. 2011). The highest production observed in Kenya by over 600 farmers was lower than the 119 million tonnes in Tanzania produced by over 26,000 farmers (Msuya and Hurtado 2017; FAO 2018). In 2012, the three countries in the WIO region, Tanzania, Madagascar, Mozambique, and Kenya, had a total seaweed production of 15,966 t (dry weight) valued at US$ 4.2 million, with 95% of the tonnage coming from Tanzania (Msuya et al. 2014). The highest output of 17,600 tonnes (DW) was recorded in 2015 in Tanzania; however, annual production has steadily declined to 11,000 tonnes (DW) in 2016 and 2018 due to the intensification of production activity and climate change-induced stress (Msuya 2020). The production disparity between Kenya and Tanzania could be explained by the difference in the number of farmers and not merely the environmental challenges. As in Kenya, the seaweed farmed in subtidal lagoons in Tanzania is also reported to experience die-offs at certain times of the year, related to high temperatures during the spring tides of the hot season and low salinity during the rainy season (Mmochi et al. 2005). The solution to this problem was found to be the use of alternative farming methods such as the floating line system and the initiation of research interventions to establish alternative seaweed species with potential for commercial farming in the WIO region (Msuya et al. 2007, 2014).

Seaweed biomass of approximately 52 tonnes, worth approximately Ksh four million, was estimated to be lost within three months (March to May) in two villages on the southern coast of Kenya. The loss of this revenue from a marginalized community could have created far-reaching household challenges and precipitated a national crisis. At Kibuyuni and Tumbe, seaweed farmers could not even sell the stored seaweeds since the threshold (>30 tonnes dry weight) set by the buyer was not met, seaweed value-added products, such as seaweed soap and seaweed shampoo, manufactured locally could not be sold, and local businesses were closing down, leaving communities in sheer poverty as money in circulation substantially dropped. The economic gains of seaweed cultivation were being threatened at a time when the coastal artisanal fisheries were also suffering the effects of travel restrictions, lockdown, and a curfew. Therefore, families may have suffered a general lack of fish protein in their daily meals, which could result in malnutrition and other health-related challenges. With such scenarios, it is clear that the impact of COVID-19 on food security, especially for communities dependent on small-scale fisheries and aquaculture, was threatened (FAO 2020).

18.5 Conclusion and Recommendations

The emergence of the COVID-19 pandemic had extremely negative impacts on the general performance of production strategies, as evidenced by 90% of seaweed farms being in dilapidated states, the collapse of seed preservation structures in deep water, and the decline in seaweed biomass in 2019. The immediate measures taken by the government to prevent the spread of COVID-19 in Kenya created a more complicated crisis in seaweed farming, since farmers were denied access to established seed banks in time to harvest mature seaweeds and restock their already dilapidated farms. Eventually, farmers lost all the seeds, which rendered women jobless as they formed the bulk of farmers.

The progressive increase in the population of seaweed farmers and seaweed biomass production before the COVID-19 pandemic indicates that the structure and management strategies adopted by seaweed farmers to sustain seaweed production and business operations before the COVID-19 pandemic were effective and should be emphasized. The roles played by various stakeholders in seaweed farming technology before and during the COVID-19 pandemic are also associated with the growth of seaweed farming on the south coast of Kenya. However, the decline in seaweed biomass in 2019 by 6% and the results of revenue loss analysis observed in this study demonstrate that the seaweed industry could be grossly affected by natural phenomena such as the COVID-19 pandemic if an immediate mitigation strategy is not adopted. The research interventions made by KMFRI and the collaboration agreements made between the farmers, donor agencies, GOK, NGOs, and seaweed buyers have been critical in the prosperity of seaweed farming on the southern coast of Kenya.

Seaweed farming technology is an economically viable alternative livelihood for the coastal communities in Kenya, particularly providing employment and adequate income to women. However, seaweed farming is vulnerable to environmental, market, and pandemic outbreaks. The sudden outbreak of COVID-19 had a serious negative economic impact on the livelihood of Kibuyuni seaweed farmers, resulting from the loss of seaweeds, the inability to repair their farms, and an extended period of general poverty in the village as ancillary businesses closed down.

Based on the impact of COVID-19 on seaweed farming in Kenya, a multidisciplinary approach involving governments, development partners, donor agencies, researchers, and farmers has a substantial role in mitigating these challenges to ensure the sustainability and resilience of this important livelihood. Governments should create targeted economic relief packages, provide financial aid, and advocate for the extension of the loan repayment period given to seaweed farmers by banks. Meanwhile, there is an urgent need to mobilize resources to cushion the economic loss incurred by seaweed farmers and to facilitate the restoration of seaweed farms. To realize the economic viability of the seaweed farming business, farmers should focus on achieving economies of scale to enhance market access and negotiate better prices. Furthermore, countries engaged in commercial seaweed farming in the WIO region should also focus on innovations, including the value addition of seaweed and encouraging the local consumption of seaweed as food and products to maximize the profits of seaweed farming. Coastal county governments should prioritize formulating effective and sustainable policies on marine resource utilization and development, e.g., marine spatial plans, to provide a sustainable solution to the space limitation of seaweed expansion for sustainable livelihood development. This document is also important because it will guide private investors in the WIO region on the recommended opportunities for a particular resource and protect the rights of other interested resource users.

References

Angka SL, Suhartono MT (2000) Marine products biotechnology. Bogor Agricultural University, Bogor, Indonesia

Bennett JN, Finkbeiner EM, Ban NC, Belhabib D, Jupiter DS, Kittinger JN, Mangubhai S, Scholtens J, Gill D, Christie P (2020) The COVID-19 pandemic, Small-scale fisheries and coastal fishing communities. Coastal Management. https://doi.org/10.1080/08920753.2020.1766937

Chevalier JM, Buckles DJ (2019) Participatory action research: theory and methods for engaged inquiry. Routledge, New York. https://doi.org/10.4324/9781351033268

COBI (2020) Mexican fishing communities' resilience to COVID-19: economic and social impacts. Comunidady Biodiversidad AC, Mexico. www.cobi.org.mx

Diener E, Suh EM (1997) Measuring quality of life: economic, social, and subjective indicators. Soc Indic Res 40:189–216

Diener E, Suh EM, Lucas RE, Smith HL (1999) Subjective Well-being: three decades of progress. Psychol Bull 125:276–302

FAO (2018) The state of world fisheries and aquaculture 2018–meeting the sustainable development goals. Rome. https://www.fao.org/3/ca5162en/ca5162en.pdf

FAO (2020) Q and A: COVID-19 pandemic–impact on food and agriculture. https://nutritionconnect.org/fao-qa-covid-19-pandemic-impact-food-and-agriculture

FAO (2021) The impact of COVID-19 on fisheries and aquaculture food systems, possible responses: information paper, November 2020. Rome https://doi.org/10.4060/cb2537en

Kimathi AG, Wakibia JG, Gichua MK (2018) Growth rates of *Eucheuma denticulatum* and *Kappaphycus alvarezii* (Rhodophyta; Gigartinales) cultured using modified off-bottom and floating raft techniques in the Kenyan coast. Western Indian Ocean J Mar Sci 17(2):11–24

Lirasan T, Twide P (1993) Farming *Eucheuma* in Zanzibar. Tanzania Hydrobiologia 260(261):353–355

Max-Neef M (1995) Economic growth and quality of life: a threshold hypothesis. Ecol Econ 15:115–118

Mmochi AJ, Shaghude YW, Msuya FE (2005) Comparative study of seaweed farms in Tanga, Tanzania. Report submitted to ACDI/VOCA SEEGAAD for a short-term technical assistance contract No. J 409. 40 pp.

Msuya FE (2020) Seaweed resources of Tanzania: status, potential species, challenges and development potentials. Bot Mar 63:371–380

Msuya FE, Buriyo A, Omar I, Pascal B, Narrain K, Ravina JM, Mrabu E, Wakibia JG (2014) Cultivation and utilization of red seaweeds in the Western Indian Ocean (WIO) region. J Appl Phycol 26(2):69–705

Msuya FE, Hurtado AQ (2017) The role of women in seaweed aquaculture in the Western Indian Ocean and South-East Asia. Eur J Phycol 52(4):482–494

Msuya FE, Shalli MS, Sullivan K, Crawford B, Tobey J, Mmochi AJ (2007) A comparative economic analysis of two seaweed farming methods in Tanzania. The sustainable coastal communities and ecosystems program. Coastal Resources Centre, University of Rhode Island and the western Indian Ocean Marine Science Association, 27 pp

Msuya, F. E., Kyewalyanga, M. S. (2006). Quality and quantity of phycocolloid carrageenan in the seaweeds *Kappaphycus alvarezii* and *Eucheuma denticulatum* as affected by grow-out period, seasonality, and nutrient concentration in Zanzibar, Tanzania. Report submitted to Cargill Texturizing Solutions, 46 pp.

Narayankumar R, Krishnan M (2011) Seaweed mariculture: an economically viable alternate livelihood option (ALO) for fishers. Indian J Fish 58(1):79–84

Reine WFPV, Trono GC (2002) Plant resources of South-East Asia: algae. Prosea Foundation, Bogor. Indonesia

Renata PR, Roberta R, Pereira HG (2014) The efficiency of the tubular netting method of cultivation for *Kappaphycus alvarezii* (Rhodophyta, Gigartinales) on the southeastern Brazilian coast. J Appl Phycol 27:421. https://doi.org/10.1007/s10811-014-0330-6

Smith CL, Clay PM (2010) Measuring subjective and objective Well-being: analyses from five marine commercial fisheries. Hum Organ 69:158–168

Sukiman F, Rohyani IS, Ahyadi I (2014) Growth of seaweed *Eucheuma cottonii* in multi-trophic sea farming systems at Gerupuk Bay, Central Lombok, Indonesia. Bioscience 6(1):82–85

Troell M, Hecht T, Beveridge M, Stead S, Bryceson I, Kautsky N, Mmochi A, Ollivier F (eds.) (2011) Mariculture in the WIO region–challenges and prospects. WIOMSA Book Series No 11(8) 59

Wakibia JG, Bolton JJ, Keats DW, Raitt LM (2006) Factors influencing the growth rates of three commercial eucheumoids at coastal sites in southern Kenya. J Appl Phycol 18:565–573

Wakibia JG, Ochiewo J, Bolton JJ (2011) Economic analysis of Eucheumoid algae farming in Kenya. Western Indian Ocean J Mar Sci 10(1):13–24

Integration of Precision Technology into Adaptive Phyconomy Systems for Extensive Tropical Red Seaweed Farming

19

Nelson Vadassery and Iain C. Neish

Abstract

Efforts are underway among phyconomists to blend precision techniques into the adaptive phyconomy systems that are prevalent in extensive commercial farming of tropical red seaweeds. Phyconomy is the branch of applied phycology that comprises systems of art, science, and technology applied to production systems that yield crops of algae. Adaptive phyconomy is a low-cost, low-control approach that is practiced in situations where the biology of algal crop organisms is poorly understood and/or where confounding variables render precision technologies impractical. Precision phyconomy, on the other hand, is a high-cost, high-control approach to algal crop production that is practiced in situations where the biology of crop organisms is well understood, and culture conditions can be comprehensively controlled. With respect to extensive ocean farming of tropical red seaweeds, adaptive phyconomy methods prevail among coastal seaweed farmers, even as the science and technology sector aspires to apply precision techniques to commercial seaweed farming systems, especially in the cases of farm mechanization, information technology, biosecurity, tissue culture techniques, and molecular taxonomy as applied to cultivar development and propagation. As phyconomy systems evolve, precision techniques will progressively displace adaptive approaches to yield extensive farm systems that are intermediate on the adaptation-to-precision heuristic model curve.

Keywords

Biosecurity · Phyconomy · Adaptive · Precision · Seaweed management

N. Vadassery (✉)
Sea6 Energy Pvt. Ltd., Bengaluru, India
e-mail: nelson@sea6energy.com

I. C. Neish
PT Sea Six Energy Indonesia, Bali, Indonesia

19.1 Phyconomy Defined

As discussed in the Neish phycosecurity chapter in this volume, considering the phyconomy as discussed in Hurtado et al. (2019) led to a notional definition of the phyconomy as "a branch of applied phycology that comprises systems of art, science and technology applied to biosecure production systems that yield crops of algae."

In the context of extensive commercial farming of tropical red seaweeds, it is more useful to define the phyconomy operationally as "a branch of applied phycology that strives to manage seaweed ecosystem services valorized per ecoeconomic principles in phycosecure, multibiomass, socioecological production ecoscapes (SEPE) that are subject to integrated coastal area management (ICAM)."

Marine phyconomy in tropical ecoscapes generally comprises the production of seaweeds (a.k.a. macroalgae) in free-flowing seawater. In contrast, agronomy generally comprises the production of rooted vascular plants in managed, fixed soils. Phyconomy and agronomy have fundamental differences related to the characteristics of the organisms cultivated and the properties of the growth media where crops grow. The (literally) fluid nature of coastal seawater as a growth medium is a compelling factor in support of adaptive approaches to the phyconomy. Some features of the agronomy and marine phyconomy are compared in Fig. 19.1.

19.2 The Continuum from Adaptive to Precision Phyconomy

Adaptive phyconomy is a low-cost, low-control approach to seaweed farming (Fig. 19.2). It is practiced in situations where the biology of algal crops is poorly understood and/or where confounding variables render precision technologies impractical. In contrast, precision phyconomy is a high-cost, high-control approach to algal crop production. It is practiced in situations where the biology of algal

A. T. Critchley et al. (eds.), *Tropical Phyconomy Coalition Development*, Developments in Applied Phycology 11,
https://doi.org/10.1007/978-3-031-47806-2_19

Fig. 19.1. A comparison of some features of agronomy and marine phyconomy. Note that in systems with extreme precision and control, such as hydroponic systems, agronomy, and phyconomy can converge

Agronomy features	Common features	Marine phyconomy features
Generally flowering plants that produce seeds and seedlings.	Vegetative propagules can be used in both cases.	More-or-less complex life cycles produce haploid and diploid spores and sporelings.
Vascular plants rooted in soils.	symbiotic microbes are a feature of both	Crops are macroalgae and/or microalgae.
Crops planted in soils that are fixed in place and managed.	Wetland farming can be a blend of agronomy and phyconomy	Crops grow in unmanaged flowing waters.

crops is well understood, and culture conditions can be comprehensively controlled.

As depicted in Fig. 19.2, there is a continuum of cost/control options that lies between the extremes of adaptive and precision technologies. The salient point for the present document is that the adaptive phyconomy of red seaweeds in tropical marine production ecoscapes is toward the lowest end of the cost/control range. The only approach to seaweed production that is generally lower in cost and control is managed harvesting of seaweed from natural stands.

A second heuristic for the continuum of adaptive and precision phyconomy is shown in Fig. 19.3. This matrix depicts ten-point qualitative scales with the degree of knowledge and control pertaining to crop biology on the ordinate and the degree of knowledge and control pertaining to the culture environment on the abscissa.

The lower left quadrant of Fig. 19.3 matrix includes more or less adaptive phyconomy strategies rated above 0/0 and below 5/5, whereas the upper right quadrant includes more or less precision strategies rated above 5/5 up to 10/10. The upper left quadrant represents cases where crop biology is understood, but the phyconomy is adaptive with respect to environmental management (ratings 1/10 to 5/5). The upper left quadrant represents cases where crop biology is understood, but phyconomy is adaptive with respect to cultural environment management (median rating 2.5/7.5), while the lower right quadrant represents cases where the cultural environment is understood, but the phyconomy is adaptive with respect to crop organism management (median rating 7.5/2.5). Exemplary ratings shown in Fig. 19.3 include 1/1 (for *Halymenia* farming); 9/9 (for typical micropropagation in bioreactors); and 6/1 (for eucheumatoid marine phyconomy).

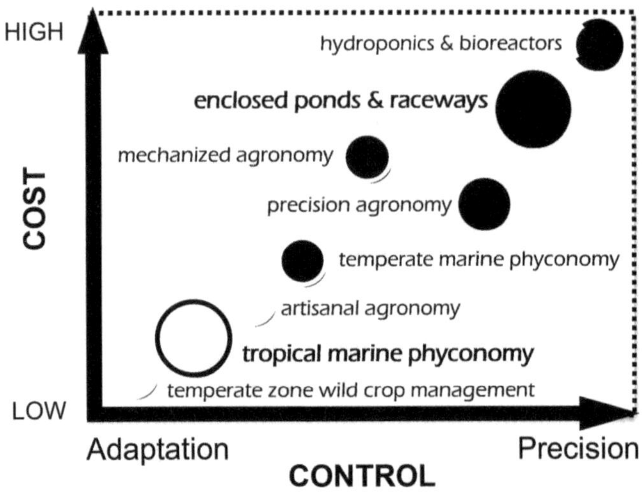

Fig. 19.2 From the adaptive phyconomy to the precision phyconomy, control comes with costs. This figure is a qualitative heuristic where several types of agronomy and phyconomy are plotted with the degree of system control on the abscissa and production cost on the ordinate. The placement of systems on the matrix was based on the authors' experience as managers of commercial phyconomy and agronomy systems

19.3 Origins of Adaptive Phyconomy

The adaptive phyconomy for extensive tropical red seaweed farming is an outstanding example of large-scale crop production developed from simple methods refined by farmers in the sea.

Neish (2020a) has described how initial notions toward adaptive phyconomy stemmed from late 1960s fisheries management modeling at the Institute of Animal Resource Ecology at the University of British Columbia (UBC), even as first steps were made with eucheumatoid seaweed farming at the University of Hawaii and in the Philippines. The UBC projects led to concepts of C.S. Holling (Holling 1978, 1993) and C.J. Walters, which resulted in Walters' seminal book entitled "Adaptable Management of Renewable Resources" (Walters 1986). Meanwhile, eucheumatoid seaweed phyconomy development led by Doty and Alvarez resulted in commercial farm systems in the Philippines by 1974 (Doty and Alvarez 1981). The two schools of thought had converged by the time of the 13th International Seaweed Symposium held at UBC in August 1989. The term "adaptive phyconomy" had not yet been coined in 1989, but many of the discipline's earliest practitioners in tropical red seaweed farming convened at the 13th ISS.

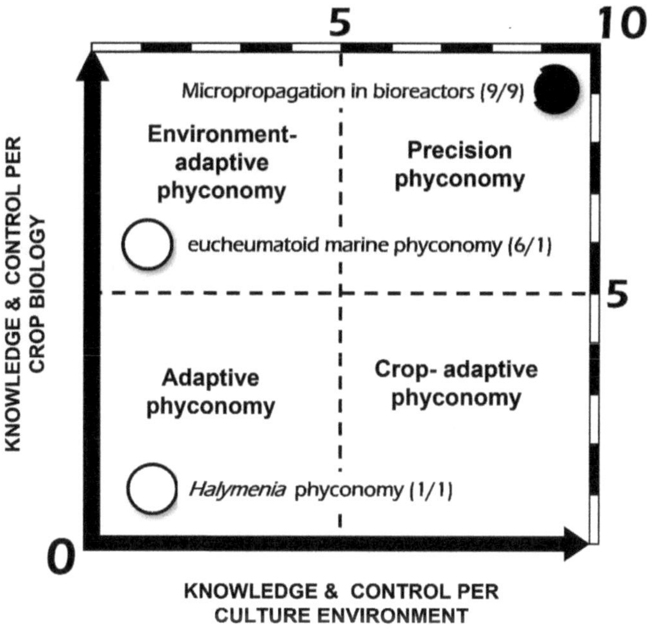

Fig. 19.3 Degrees of adaptation and precision can be approximately rated on scales of 0–10 with respect to knowledge and control of the culture environment and/or crop organism biology. For example, a 5/5 rating indicates an intermediate state; a 1/1 rating indicates a near-fully adaptive state (e.g., *Halymenia* phyconomy); and a 9/9 rating indicates a near-fully precision state (e.g., micropropagation in bioreactors). Ocean farming of eucheumatoid seaweeds tends to rate at approximately 6/1 since crop organism biology is moderately well understood (if not controlled), but there is little or no understanding and control over open sea farm areas

The work of Holling and Walters led to the further development of adaptive management paradigms by Lee (1999), Salafsky et al. (2019), and others to the point where adaptive management principles can be expressed in phyconomy terms. Thus, adaptive phyconomy is "a process of dealing with uncertainty in the management of ecoscape crops and adjoining renewable resources." It is a process of treating human interventions in the phyconomy and ecoscape systems as experimental probes, and it incorporates research into phyconomy initiatives and stewardship actions.

Models of adaptive management were initially developed by ecologists, and they have mainly been applied in Australia and Canada for the management of fisheries, forests, and waterfowl in large watersheds (Morris et al. 2011). There has also been some development of adaptive agronomy based on the principles of adaptive management, especially since climate change has created increasing conditions of risk and uncertainty for farmers (Morris 2008: Morris et al. 2011).

As adaptive agronomy applies to agriculture, adaptive phyconomy applies to aquaculture. The need for an adaptive approach is especially essential because marine seaweed farms are in situations where confounding variables foil predictions (Fig. 19.4).

Regarding the unknowns of adaptive management, Holling (1978) wrote, "Man has always lived in a sea of the unknown and yet has prospered. His customary method of dealing with the unknown has been trial-and-error. Existing information is used to set up a trial. Any errors provide additional information to modify subsequent efforts. Such 'failures' create the experience and information upon which new knowledge is built. Both prehistoric man's exploration of fire and the modem scientist's development of hypotheses and experiments are in this tradition. The success of this time-honored method, however, depends on some minimum conditions. The experiment should not, ideally, destroy the experimenter—or at least someone must be left to learn from it. Nor should the experiment cause irreversible changes in the environment. The experimenter should be able to start again, having been humbled and enlightened by a 'failure.' And, finally, the experimenter must be willing to start again."

Every seaweed crop cycle is a gamble on ocean farms. If the crop grows well, farmers are happy. If the crop grows badly, farmers are unhappy. If the crop dies, farmers are devastated, but they seldom know why events unfolded as they did and often have no idea how to fix problems that arise. This is typical for adaptive phyconomy in tropical ecoscapes.

Regarding the use of models in adaptive management, Walters (1986) wrote: "Few of the models developed in [Adaptive Environmental Assessment] workshops have been used directly for policy analysis. Most have been put where they belong as mechanical instruments for prediction . . . [in the trash can] . . . having served the essential purpose of promoting clearer thinking by and communication among the workshop participants. Some have provided a starting point or broad frame-work of relationships for organizing sequences of more focused workshops and meetings, leading finally to serious policy recommendations."

During the development of adaptive phyconomy, the authors have found that farmer workshops have been extremely valuable. The very act of "joining the dots" in model heuristics has been beneficial in helping diverse phyconomy system stakeholders harmonize their mindsets. Some aspects of adaptive phyconomy, such as yield curves and event probabilities, can be amenable to mathematical treatment, but most aspects can only be effectively communicated using analogy-based heuristic models.

19.4 Adaptive Phyconomy Driven by Farmers

The concept of "antifragility," promulgated by Taleb approximately 2012, meshed well into adaptive phyconomy mindsets that were evolving in seaweed farming initiatives throughout Southeast Asia by that time. Taleb's virtual "Black Swans" represented the unpredictability of

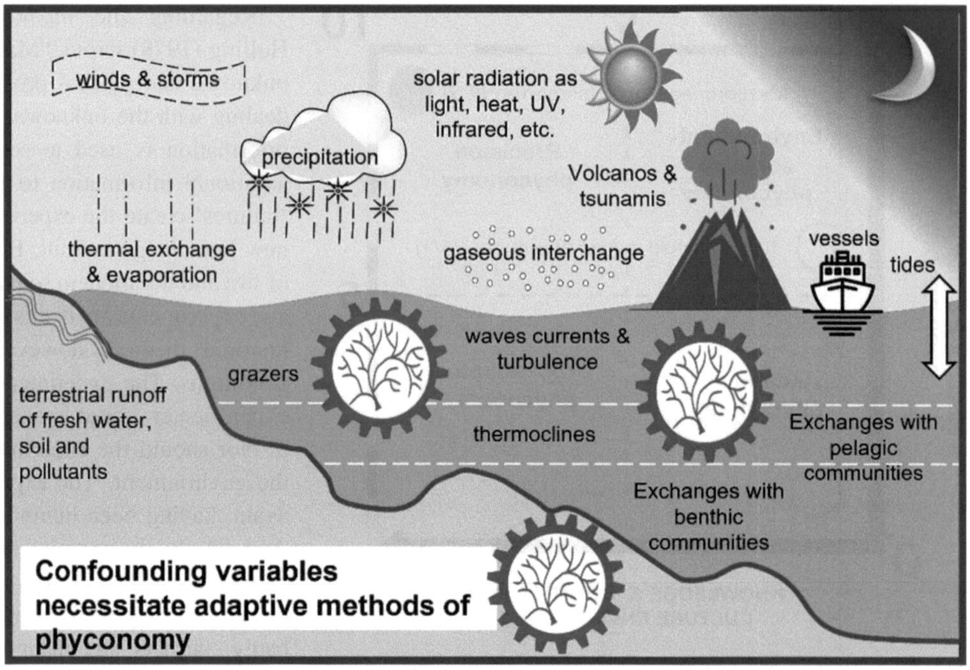

Fig. 19.4 Uncontrollable processes and events are confounding variables for all adaptive phyconomy in marine ecoscapes. Confounding variables can make crop yields unpredictable. Ecocosm factors, such as sun, moon, weather, and oceanography, combine with socioeconomic factors in a complex way that makes extrapolation from knowledge of the past useful but risky. Courses of phyconomy action must be adjusted based on constant vigilance during frequent field presence

environments where phyconomists operate. They were, as Taleb wrote, "large-scale unpredictable and irregular events of massive consequence," whether positive or negative. Embracing antifragility meant avoiding complacency based on an overly optimistic valuation of system strengths. Antifragility involves developing strength through adversity. It was beyond resilience or robustness. The resilient resist shocks and stay the same, while the antifragile resist shocks. The essence of the concept is expressed in the aphorism "what doesn't kill you makes you stronger" (attributed to Friedrich Nietzsche).

The reflexive nature of adaptive phyconomy was demonstrated as simple methods were refined by the experiences of farmers in the sea. The term "reflexive," simply put, means "bottom-up" as opposed to "top-down" (Fig. 19.5). As suggested by Taleb, antifragility was encouraged in systems where phycotechnology was evolved by practicing farmers.

The research of Scoones and his teams found that in agriculture systems, successful development tended to be associated with reflexive action rather than with top-down equilibrium or non-equilibrium approaches (Scoones 1998; Scoones et al. 2007). Since the late 1990s, the Scoones paradigm of reflexivity in development has been integrated into models of adaptive management, as in the case of Voss and Bornemann's 2011 paper entitled "The Politics of

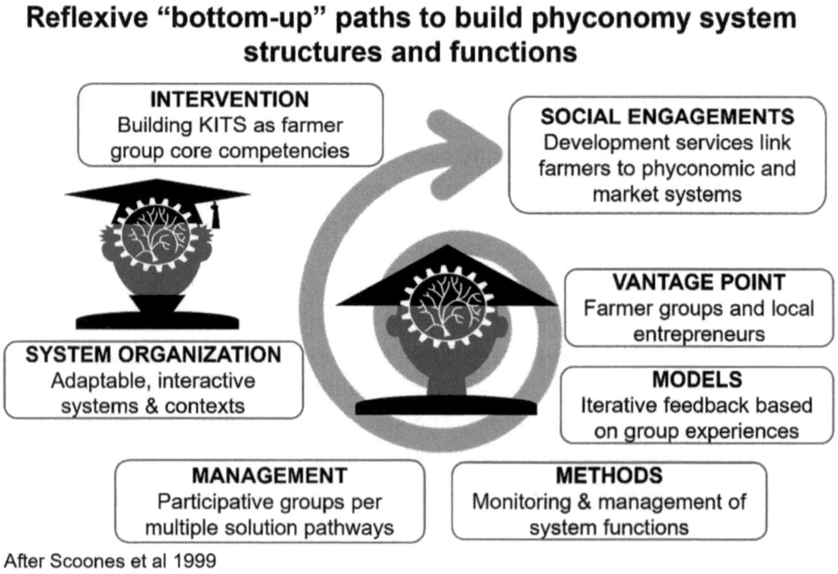

Fig. 19.5 Reflexive ("bottom-up") paths build adaptive phyconomy systems. The term "KITS" is an acronym for Knowledge, Information, Tools, and Solutions. VC is short for "Value Chain"

Elements		Monitoring functions	Control functions
People	The social environment where phyconomy functions happen, including stakeholders such as employees, support providers and governance enforcers.	Awareness of interactions with ecoscape and value chain stakeholders.	Use best management practices for farm enterprises.
Energy	Sunlight, water motion, wind, electricity, fuel and human labor are the major forms of energy managed by the phyconomist.	Measurement, recording and data handling of energy-related meteorological and oceanographic parameters and of purchased energy products including fuel, electricity and human labor.	Pro-active minimization of negative energy impacts and maximization of positive impacts.
Materials	Capital items and consumable materials including crop biomass; physical farm structures; chemical inputs such as nutrients and crop treatments; and equipment and materials used in value chain functions.	Operation of crop logging and crop production information systems; oversight and inspection leading to design, installation, operation and maintenance of farm structures and equipment.	Operation of farm systems through manually applied operational protocols and automated control systems.
Transactions	Mediated by money; by barter; and by social currency such as favors, knowledge, information, and trust.	Maintenance and operation of financial records and records of social transactions including contracts and agreements.	Build balanced profitability per legacy, ecoeconomy and crypto economy models.

Fig. 19.6 Elements of the bricolage engine, monitoring, and control functions per Fig. 19.7

Reflexive Governance: Challenges for Designing Adaptive Management and Transition Management." They discussed how concepts of development governance involve reflexivity during adaptive management, which involves collective experimentation and learning.

As a discipline, tropical marine phyconomy seems to have been adaptive and reflexive from its origins in the 1960s. It has been the authors' experience that successful, extensive, tropical seaweed farm development has been characterized by the same reflexive patterns that had positive impacts on terrestrial systems. This is an aspect of seaweed farm development that merits further study and analysis.

19.5 Bricolage-Driven Virtuous Cycles Evolve Adaptive Phyconomy Systems

Adaptive marine phyconomy of red seaweeds in tropical ecoscapes is substantially based on bricolage (a.k.a. trial and error). A "bricolage engine" drives virtuous cycles of adaptation that are at the core of adaptive phyconomy (Figs. 19.6 and 19.7). The matrix in Fig. 19.6 indicates how the bricolage engine elements of people, energy, materials, and transactions are monitored and controlled during virtuous cycles of innovation and adaptation.

Bricolage is a process undertaken with the use of purpose-built test plots, but at the present state of adaptive phyconomy, it is fair to say that farmers are installing test plots every time they plant crops in the sea. Bricolage occurs as phyconomists learn through iterative manipulation of elements and functions of farm systems. Virtuous cycles of adaptation occur as they manipulate elements and functions, monitor and control what happens, understand what happens, improve systems based on lessons learned, and then try again.

19.6 Phyconomy Knowledge, Information, Tools, and Solutions (KITS)

"KITS" is an acronym for the Knowledge, Information, Tools, and Solutions that are embedded in the mindsets of farmers and phyconomists (Fig. 19.8). KITS are the tangible products of innovation and adaptation cycles. Much tropical

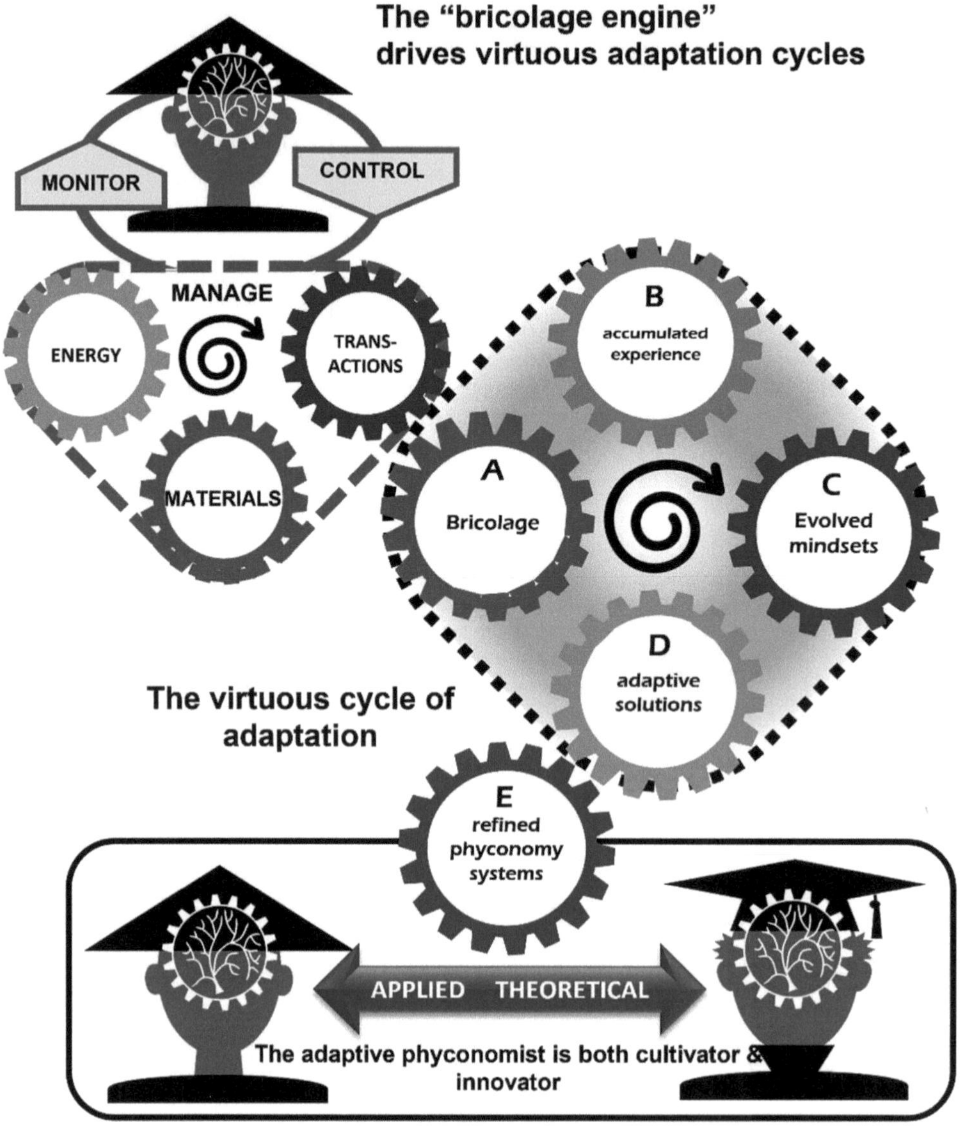

Fig. 19.7 The "bricolage engine" drives virtuous cycles of adaptation. The French notion of "bricolage" is an adaptive approach to phyconomy that is illustrated in this heuristic model. It is a form of iterative trial-and-error, but bricolage has an enhanced connotation of effecting innovation with the tools and materials that are immediately at hand. It is at the core of virtuous cycles of phyconomy innovation

phyconomy was evolved by farmers in the sea, but farmers generally do not preserve their KITS in writing or other forms of documentation. The KITS of farmers are recorded by technicians, scientists, and development professionals. The present document is a contribution toward that undertaking.

Key functions of phyconomists and development professionals are the compilation and analyses of information and data from operating farm systems, evolution and communication of the phyconomy knowledge base among value chain stakeholders, participation in development and production of useful tools, and roles played in formulating practical phyconomy solutions as theory leads to actions.

Seaweed scientists and development professionals have opportunities to travel the world, to see farming and other value chain actions being undertaken, and to learn by direct communication with farmers and other seaweed chain stakeholders. They observe farmers refining methods in the sea; they develop hypotheses to explain what they see; they spread the word in the phyconomy community (Fig. 19.9); and they strive to move forward.

Unfortunately, seaweed-based value chains are still relatively small by global standards, so phyconomy has never attracted the magnitude of research and development funding that has been lavished on agronomy of staple agricultural crops.

19.7 Evolution of Phyconomy Mindsets

Each phyconomist perceives the world through the filters of their personal mindset and undertakes phycotechnology actions based on that mindset. Phyconomists are linked through a global mindscape of people's past, present, and future that includes the wide array of disciplines that revolve

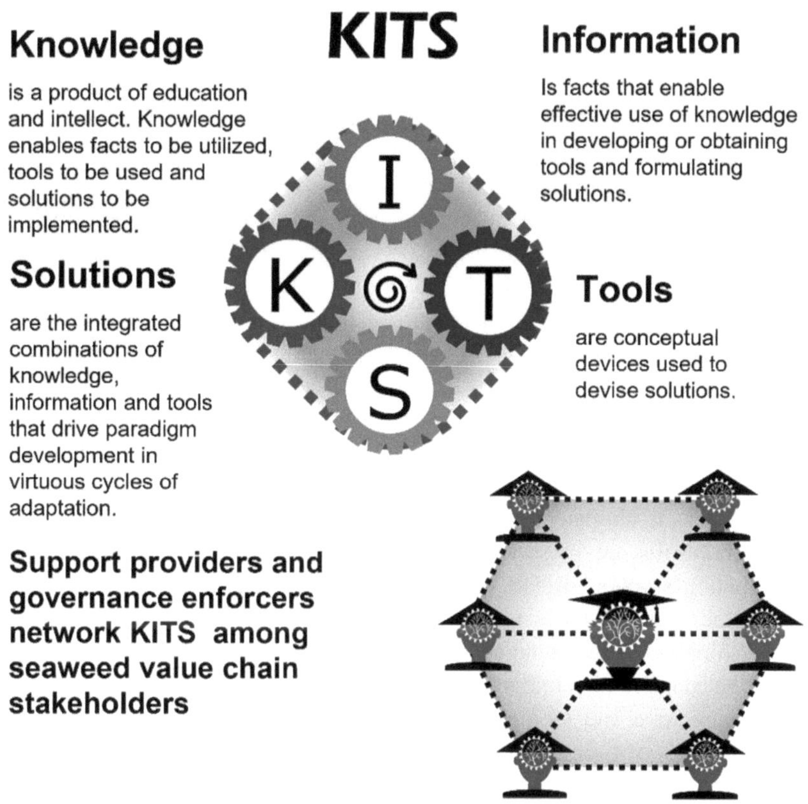

Fig. 19.8 Knowledge + Information + Tools + Solutions (KITS) results in adaptive phyconomy actions. Several coordinated actions result in the performance of functions that influence materials, energy, and transaction flow through adaptive phyconomy systems

around human production and stewardship of living organisms, including phyconomy, agronomy, aquaculture, silviculture, and many others.

For phyconomy to evolve, it is necessary to connect phyconomists on a mindset-to-mindset basis. This is no simple thing because mindsets are very personal. They are influenced by phyconomists' socioeconomic roles in life; phyconomists speak many different languages, and the best phyconomic ideas can be difficult to explain to others. Farmer-phyconomist mindsets are therefore linked through a wide variety of means and media (Fig. 19.9).

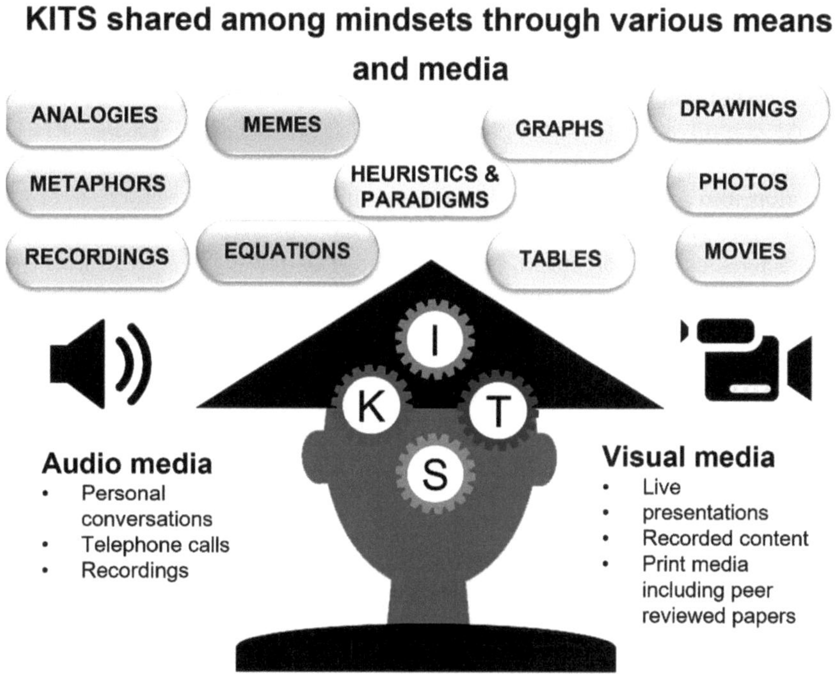

Fig. 19.9 Farmer-phyconomist mindsets link as KITS are conveyed through various means and media. Phyconomy mindsets evolve through the progression of virtuous adaptation cycles and through the assimilation of KITS made available through such means and media

Phyconomy mindsets evolve through the progression of virtuous adaptation cycles and through the assimilation of KITS made available through such means and media. At the time of writing, in a virtual world stimulated by COVID-19 precautions, multimedia approaches were becoming common. For example, the present document was prepared as a book chapter, as a PowerPoint presentation, and as an MP4 video for streaming on Phyconomy.org as presentations in the Tropical Phyconomy Coalition Development Eucheumatoid Seaweeds Webinar on 7–8 July 2021.

Adaptive phyconomy and precision phyconomy differ with respect to the mix of means and media used for communication. The core of precision phyconomy is peer-reviewed, printed scientific literature published in reputable scientific journals or as books from the academic press. For proprietary material, such as patented technology and trade secrets, research results are preserved for the long term in secure information storage formats that are scientifically rigorous. Excerpted material tends to be officially released through professionally crafted audio and video media.

In the adaptive phyconomy sector, information tends to be less quantitative and more qualitative than in the precision sector. "Rules of thumb" and heuristic models are common, but rigorous mathematical models are rare. Much information flows informally among farmers and phyconomists through internet social media, mobile smartphone systems, and workshops conducted by a variety of government and international organizations.

Bridging the gap between the adaptive and precision sectors is the global community of scientists and development professionals discussed below.

19.8 The Essence of Precision Phyconomy

The heart of precision phyconomy lies in the belief that if we could control every parameter associated with a seaweed strain, we would be able to control what the strain would develop into, or in essence, its phenotype. In this context, phenotype is defined as the sum of all the observable characteristics of an organism.

The phenotype is essentially determined by two factors, as seen in the following simplistic equation:

$$P = G \times E$$

where
P is the Phenotype
G is the Genotype
E represents Environmental factors
x represents the interactions between the Genotype and the Environment.

The genotype, or the genetic material with which a crop strain starts its life cycle, plays a huge role in the final output of the plant, whether it is larger fruit, more grains, or bigger leaves, depending on what humans value. However, this alone is not enough, as the environment also plays a huge role. For example, drought conditions or a flood could change the final output entirely. Apart from these physical parameters, there is a host of biological factors in the environment that affect the Phenotype (including pests, microbes, and viruses) that play a significant role in how the final output of a strain selection process would look.

While the complexity of the relationship between the genotype and the environment goes far beyond a simple multiplication operation, the equation does represent the importance of these two factors in developing the phenotype.

The axiom of precision phyconomy lies in the assumption that through rigorous control over both the environment and the genotype, it is possible to obtain the desired phenotype in any crop. The essence of precision phyconomy is that rigorous scientific studies are undertaken to test hypotheses relevant to phyconomy; the results are presented in peer-reviewed scientific literature or as information retained by scientific institutions; and phycotechnology based on the studies is implemented in commercial production systems.

The lack of capacity to undertake extensive phyconomy research and development in tropical ocean seaweed farms has resulted in a situation where phyconomists are, figuratively, drowning in a sea of untested hypotheses. As precision techniques are blended into evolving phyconomy systems, seaweed farming can move up the scales illustrated in Figs. 19.2 and 19.3. Good phyconomy science has been done, and further scientific investigations can inject more science to complement the arts of adaptive phyconomy in tropical ecoscapes.

19.9 Precision Phycotechnologies Transitioning to Adaptive Systems

While precision phyconomy favors rigorous control of the environment to obtain the right output, it remains that the greater the degree of control, the greater the cost involved. Controlling these factors at a laboratory scale or even at a pond scale may be possible, despite being expensive. However, controlling the environment at scale in the open sea is simply impractical.

What is needed then is to borrow from precision phyconomy approaches and apply them to adaptive phyconomic methods that are already in practice. Central to this is accepting that there will always be factors beyond our control. Managing the few factors we can control, and monitoring and predicting the factors beyond our control, gives seaweed farmers much power in managing their farms. For factors beyond farmer control, knowing the reasons behind them and predicting their onset gives farmers time and knowledge to develop appropriate interventions to mitigate their effects.

Examination of the applied phycology literature reveals that the volume of publications relevant to precision phyconomy has grown exponentially over the past few decades. What follows is a review of some highlights from precision phycotechnology R & D initiatives that are increasingly impacting adaptive phyconomy systems.

19.9.1 Precision Phyconomy Methodologies for Phenotype Control

Precision Phyconomy aims to obtain a specific phenotype by increasing the degree of control over every factor associated with the plant. We will now classify these approaches based on which factor is being controlled.

This section describes approaches that seek to control the genetic material of the seaweed crop and its associated microbiome, thereby controlling the traits that will be manifested. Such approaches have been honed and developed in land agriculture for centuries, even before Gregor Mendel began to elucidate the patterns behind the inheritance of traits. By using these methods, we can select for advantageous traits and develop better "strains" that would grow well in the target environment. An advantage of this approach is that once a good strain has been developed, it can be replicated across multiple sites with very little cost, especially in the case of tropical red seaweeds, which are vegetatively propagated.

The current situation is that propagation is exclusively through vegetative cuttings rather than through spores generated through life cycles. The reality is that "nurseries" for seaweeds have simply been nearby farms that happened to have viable planting material, rather than formal nursery operations and infrastructure. Selective breeding is yet to be used for strain improvement in tropical seaweeds, and genetic modification methods have been rejected as unacceptable.

Below, we review some of the approaches that have been taken to manage the genotype of the crop.

19.9.1.1 Strain Identification

Most, if not all, cultivated tropical red seaweed cultivars are vegetative strains selected by farmers in the field by processes that went unrecorded. Cultivars spread regionally and globally, usually also by processes that went unrecorded. Most farmers do not know the exact taxonomic identity of the crops they grow.

There is a wide range of genetic diversity to be found across seaweed cultivars and in wild populations of tropical seaweeds. Each of these unique strains has its own comfort zones, and matching the right cultivar to the right site can make a huge difference in farm productivity. Tropical seaweeds are notorious for their phenotypic plasticity. Very often, the same strain will look very different when moved from one site to another, changing its color, thallus thickness, and even the degree of branching. Even to the experienced eye, it is difficult to discern whether the specimen being collected is of the same strain or even of the same species.

Thanks to developments by several research groups, it is now possible for the taxonomic relationships of seaweed crop species to be defined at a molecular level using both sequencing and DNA barcoding of the cox2–3 and plastidal RuBisCo spacers (e.g., Lim et al. 2013; Zuccarello et al. 2006). However, beyond the species level, protocols that identify cultivar strains uniquely are yet to be standardized. There is some promise in this field using DNA "fingerprinting" techniques that apply technologies such as ISSR-PCR fingerprinting (Chin et al. 2017). Attempts have also been made to identify the ploidy of specimens being identified (Zitta et al. 2012). These methods need to be standardized and adapted so that researchers around the world can build a database of available strains. This will also enable the survey and mapping of the distribution and abundance of specific cultivars throughout seaweed farming regions.

19.9.1.2 Maintenance of Unique Characterized Strains

Discovery and identification of unique strains is not enough. What is also essential is a means for storing useful strains for posterity. This allows future farmers to access a diverse genetic pool to find strains best suited to their environment. Currently, strains are being maintained among farmers themselves. Under such circumstances, it is only a matter of time until a major climatic event can lead to strains simply vanishing. While strain maintenance in tiny culture vessels such as conical flasks or beakers is effective, strains require much maintenance and a long-term commitment to funding support. Despite these hurdles, there are pockets where such strain libraries have been maintained for over a decade (Narvarte et al. 2022). Developing means of cryopreserving such plants would significantly reduce infrastructure requirements and maintenance burdens for holding stocks of germplasm, and efforts are underway to develop cryopreservation facilities.

19.9.1.3 Tissue Culture and Micropropagation

Several initiatives have introduced tissue-cultured and micropropagated strains of known taxonomic identity through nurseries into commercial farm systems. Early on, in 2003, Dr. Reddy identified means for inducing callus growth in *Kappaphycus* by growing it in a solid medium (Reddy et al. 2003), and there seems to be promise for obtaining better-growing strains by using such methods.

Similarly, with micropropagation techniques, it is now possible to grow a large quantity of biomass from a small piece of tissue, again showing promising growth in the sea. However, it has been observed in a few cases that the beneficial traits brought in by these methods do not last more than a few generations of vegetative propagation in the sea, the exact cause of which seems unknown.

19.9.1.4 Quarantine and Cultivar Distribution Protocols

Quarantine protocols and protocols for distributing cultivars among regions have been developed and applied on a limited basis (e.g., Sulu and others). There is always the danger of introducing alien species and pathogens into an existing farm if quarantine systems are not maintained. Pests such as *Neosiphonia* seem to be endemic to certain seaweed-growing regions currently, and care should be taken that such pests do not spread out to new regions through infected propagules. There is much literature on the appropriate ways in which this can be done, drawing upon the rich history of aquaculture.

19.9.1.5 Selective Breeding of Seaweeds

Selective breeding has been applied to some temperate zone seaweeds (Hu et al. 2021). Methods such as genetic breeding, haploid breeding, DNA fingerprinting, mutation breeding, and polyploid breeding have long been applied to produce better quality strains in temperate kelps, especially in China, Japan, and Korea. Understanding the entire life cycle of a species, as well as a good understanding of the protocols and media to enable its breeding in the laboratory, has been crucial in this respect. With the tools to perform breeding being well studied, breeders can focus on identifying desired traits and breeding toward them in their laboratory.

It remains to be demonstrated how such methods may be applied to tropical red seaweeds. The entire life cycles of tropical red seaweeds such as *Kappaphycus* have been demonstrated in nursery conditions, but there are hurdles in the selective breeding of seaweeds. One is that the induction of reproduction in these plants has not yet been identified, and spores are obtained by waiting for the plants to become fertile on their own. Another is that this "spontaneous" induction of fertility has not yet been seen in commercially cultivated cultivars, which prevents phyconomists from using the advantageous traits of commercial cultivars while breeding with other specimens. Work is being done to identify the tools and techniques that enable predictable sexual reproduction in red seaweeds. Positive results will provide a good platform for strain development.

19.9.1.6 Microbiome Management

Seaweeds are not standalone entities because they are heavily influenced by a microbiome comprised of diverse microorganisms inside and outside the thallus. It is increasingly being seen that the phenotype of seaweeds is also heavily dependent on the associated microbiome. Studies show (Grueneberg et al. 2016) that simply by varying an associated microbe, *Ulva mutabilis* exhibited a variety of distinctly different morphologies.

For tropical seaweeds, it has been seen that there are both beneficial and parasitic microorganisms working in associated microbiome communities, and species composition may shift toward beneficial organisms that increase seaweed growth. For example, several nitrogen-fixing microbes have been identified in red seaweeds, which could be useful in oligotrophic waters to supplement low nitrogen levels. Improved responses in seaweed growth have been observed by selectively inoculating seaweeds with nitrogen-fixing microbes.

Seaweeds themselves have been shown to exude their own unique cocktail of metabolites, which selectively favor certain microbiome populations over others. Some beneficial microbes may be epiphytic, residing on the seaweed surface, and others may be endophytic, living within seaweed tissue. Identifying and understanding how these organisms behave may provide an interesting means of producing high-performance cultivars.

What a farmer perceives as a cultivar is truly the seaweed itself combined with its own unique microbiome community. Developing the ability to tune this population based on site, season and weather conditions would go a long way toward building robust seaweed farms. This is no easy task and needs a variety of labs to work on the local biodiversity of microbes available, screen them for their effects on seaweed crops, and after screening, maintain and develop seaweed + microbiome complexes that seaweed farmers can use.

19.10 Controlling and Monitoring the Seaweed Crop Environment

Every seaweed strain expresses its phenotype in relation to the environment to which it is exposed. By providing a favorable environment, we can allow the plant to express its phenotype to the fullest extent. This comes with its own share of challenges, which will be described in the following sections. For instance, there may be some parameters that may be prohibitively expensive to control, and in these cases, monitoring these parameters and having mitigation measures in place would be the best option. In general, the environmental parameters involved can be broadly divided into three categories: physical, chemical, and biological. We will describe how precision phyconomy deals with these parameters.

19.10.1 Physical Ocean Parameters

Unlike soil on land, the oceans are a dynamic system that changes from moment to moment, and these changes have a profound impact on the organisms residing there. Parameters such as waves, temperature, and currents have a direct impact on seaweed growth.

19.10.1.1 Water Motion

It is known that many seaweeds thrive well in regions with good water motion, which is a combination of currents, waves, and local turbulence. The greater the water motion around the seaweeds, the smaller the diffusion layer around the tissues, which allows seaweed fronds to effectively absorb nutrients and exchange metabolites with the external environment. Most of the time, higher water motion results in good growth rates. However, if it goes too high, the plants stand the risk of breaking and sinking into the sea. In such conditions, a farmer would do well to plant smaller pieces of seaweed so that the amount of seaweed breakage is minimized. Of course, this has its downsides because the crop growth rate may be high, but lower planted biomass results in a lower yield of biomass per farm unit.

19.10.1.2 Turbidity

Another parameter that could potentially impact seaweed growth is turbidity. While there seems to be no direct correlation between the degree of turbidity and seaweed growth, there are many postulated indirect effects. For instance, a high plankton load in the water may cause turbidity, which may result in increased filter-feeding fouling organisms that could grow on the seaweed as a substrate (e.g., barnacles). At the same time, the presence of phytoplankton in the water is also indicative of the presence of nutrients in the water, which is beneficial to seaweed, except that phytoplankton compete with the seaweed crop for nutrients. Sometimes the presence of silt or sand suspended in the water can result in some of it settling on the seaweed or on farm structures, which may require many cleaning operations later. The authors have seen seaweed thriving both in crystal clear waters as well as in low visibility, turbid waters, so turbidity alone cannot be taken to impact seaweed health. In combination with other factors, there may be an effect on seaweed growth, which makes it a parameter worth tracking.

19.10.1.3 Temperature

Temperature is a parameter that plays a very important role in seaweed growth and impacts seaweed farmers around the world. Metabolic pathways of all organisms have adapted to a unique temperature range, and the warm summer temperatures seen in the tropics are already on the borderline of mortality for many cultivated strains. A change in just a few degrees can mean life or death for the entire crop, so temperature data can be an important tool in making farm decisions. This becomes even more important in today's global warming scenario, where maximum recorded temperatures increase every year in many farm regions.

19.10.1.4 Salinity

Another parameter affecting seaweeds is the salinity of seawater. Sudden drops in salinity may be seen in regions near rivers, where fresh water entering the sea can decrease the salinity, which could hurt the seaweeds. Additionally, on hot and windy days and in shallow regions, excessive heat can cause large water evaporation and increase the salinity of the water. All sea plants have a distinct salinity comfort zone inside which they thrive. Understanding the salinity of a region can help phyconomists understand the stresses faced by the crop. Some genera, such as *Gracilaria,* are resistant to large variations in salinity, while this is not the case for *Kappaphycus*. The duration of salinity changes often matters as much as the extent of the change in salinity, with most seaweeds being able to survive short bursts of salinity changes.

19.10.2 Chemical Ocean Parameters

19.10.2.1 pH

The pH, a measure of the alkalinity of the sea, has been known to affect the physiology of all marine organisms and is an indirect measure of the carbon dioxide present in the sea. pH levels fluctuate during the day, rising to their highest levels when photosynthesis is at its peak. Measurement of pH is an important function related to understanding the carbon cycles that are germane to climate change and "blue carbon" processes.

19.10.2.2 Dissolved Oxygen

Dissolved oxygen (DO) gives aquaculturists an idea of the oxygen present in seawater. At the peak of photosynthesis, around the middle of the day, DO is at its highest, and it falls to its lowest around night-time when all organisms utilize this oxygen to respire and generate carbon dioxide in the process. The spatial distribution of DO in a farm can provide clues about whether crop growth is CO_2 limited, which means that while photosynthesis can fix more CO_2, there simply isn't enough available in proximity to the plant, since all the neighboring plants have depleted this resource faster than CO_2 can dissolve through the ocean surface.

19.10.2.3 Nutrients

Nutrients comprise a broad set of parameters that include macronutrients such as nitrogen, phosphorus, and potassium (NPK), and micronutrients present in seawater, including zinc, molybdenum, iron, and many other elements. If essential nutrients are scarce or present at toxic levels, this can impair the growth of the plant. Monitoring the range of elements that impact seaweed crops is not an easy task, with a wide variety of assays needed to understand the environment in totality. To make matters worse, since seawater keeps moving around, a site reading at one minute may be very different from the reading a few minutes later. Seaweed genera, however, have lived in this climate of uncertainty for millions of years, so they have developed coping mechanisms, such as absorbing nitrogen and hoarding it when it is available through surge uptake mechanisms (Dy and Yap 2001).

19.10.3 Biological Ocean Parameters

19.10.3.1 Plankton

Plankton comprise the huge diversity of microorganisms inhabiting ocean photic zones and comprise both phototrophic phytoplankton and heterotrophic zooplankton. While there is no direct causality between plankton profiles and seaweed growth, plankton are great indicators of oceanic biomass activity. For instance, a high nutrient load can lead to a sudden plankton bloom. If these nutrients are beneficial to seaweeds, seaweed farmers will see a corresponding increase in crop growth. However, blooms of microalgae can lead to a variety of problems for seaweed crops. A high plankton load in the water can lead to the proliferation of bivalve mollusks and other filter feeders that settle over the seaweed surface in search of attachment points. Once attached, they grow and try to cover as much of the surface as possible, resulting in epizoan-infested seaweeds and hence diminished crop yields.

19.10.3.2 Grazers

Seaweed farms are habitats for fish and other marine animals, and they can act as fish aggregation devices. In certain regions and certain seasons, the largest visitors tend to be herbivorous fish, such as rabbitfish, which come and nibble away at the seaweed tips. Depending on the visitor, the method of eating changes. Rabbitfish tend to nibble, snails and urchins tend to plane the seaweed, while turtles simply chomp off the whole plant in one bite. As seaweed farmers, we need to live in harmony with the citizens of the sea, and much of the seaweed is simply given away as a "fish tax." Interestingly, fish do not find all seaweeds equally palatable. Some are favored more than others. Knowledge of the kind of grazers attacking the farm helps a farmer decide on the right kind of crops to plant.

19.10.3.3 Weeds and Epiphytes

Epiphytes are plants that grow over the crop seaweed surface and very often compete with the cultivar frond for light and nutrients, hence reducing the thallus to a diseased state. Apart from these, there are also weed species that foul farm substrata such as ropes, lines, and farm infrastructure, and such fouling may influence seaweed growth. Most weed and fouling species are highly seasonal in nature and arrive and disappear with certain seasons. It is necessary to monitor their growth to understand the best means of mitigating the damage caused by them.

19.10.3.4 Crop Diseases

The poorly understood "ice-ice syndrome" remains a substantial risk factor for farmers of eucheumatoid seaweeds and has caused crop losses in several areas (Ward et al. 2020). The only practical reaction of farmers to maladies and diseases has been to crop out their farms and replant under conditions presumed to be disease-free. Monitoring and documenting these diseases is the first step toward finding solutions.

19.11 Monitoring and Controlling the Environment

19.11.1 Monitoring the Environment

Many environmental parameters can at least be monitored, even if they cannot be controlled, and informed farming decisions can be taken. For example, a rise in temperature may require outcropping of the current strain and introducing a hardier, temperature-tolerant strain. Changes in salinity patterns may require salinity-tolerant strains and species. Even with physical parameters, an increase in water motion may imply the need to use more durable farms, as well as strains that can withstand the high-energy environment. Monitoring these parameters requires some infrastructure, as described below.

19.11.1.1 Analytical Tools and KITS

Almost all environmental parameters can be sampled using analytical tools and solutions. For example, the Grashof method for nitrate estimation is well known, and turbidity can be measured utilizing the Tyndall effect. Many of these tools and solutions are well described in the scientific literature, but many still require access to expensive equipment (e.g., spectrophotometers) and need trained personnel to operate them, making them out of reach for most farm operations. Developing an easy-to-use KITS for environmental studies allows phyconomists to understand the environment in which they are working. Such approaches work well when the parameters are those that do not change much on a daily basis and can be measured once a week or longer (e.g., turbidity and salinity). Other parameters, such as temperature and pH, which vary during the day, require continuous monitoring, and it is simply not practical for these samples to be analyzed by a person every hour.

19.11.1.2 Ocean Data Stations Equipped with Sensors

For environmental parameters where a large diurnal variation is seen, having a sensor coupled with a data logger in the sea is an excellent choice. Today, installing a data buoy is an expensive affair, with prices skyrocketing as you add more sensors to the system. Much of the cost lies in the emphasis on reliability in ocean data platforms, since these platforms have traditionally been meant to capture data in rough environments with very little human contact for years. In a seaweed farm, however, having people access a data buoy on the farm is quite easy, allowing us to build on cheaper components.

With the introduction of hardware platforms such as the Arduino and the Raspberry Pi, coupled with the growing online community of electronics enthusiasts, much knowledge is being generated and shared to tackle these problems. Rather than building one expensive data collection unit that is bound to fail catastrophically, building multiple low-cost units with redundancies makes for a robust farm data strategy.

While the electronics involved in logging the data and transmitting the data become far more widespread, there is much work to be done in the field of developing reliable, low-cost sensors and their associated software. Buying sensors from a local e-commerce website may be tempting, but care should be taken that such data should not be used unless thoroughly validated. The final analysis is only as good as the quality of data we manage to collect. As they say, "garbage in, garbage out."

19.11.1.3 Crop Health Monitoring

Monitoring crop health is an often-overlooked aspect of monitoring. However, this is an all-important function since crop conditions must be correlated with environmental data collected from the sea. Knowing the times when seaweed grows best, or when sudden die-out events occur, allows phyconomists to go through the environmental data collected and build a hypothesis on why such events may have happened. If we can identify, with a reasonable level of confidence, the reason behind crop issues, we can use these insights to develop phyconomy solutions.

Crop health monitoring can be performed through regular spot checks of seaweed across different parts of the farm and documenting their health. Site-wise farm data on yields and growth rates also provide a clue as to how the plants are faring. Drone studies can also be used to identify spatial variations in crop health.

19.11.1.4 Data Analytics for Seaweed Farming

Beyond the raw data obtained from these sensors and systems, it is essential to provide an interpretation of these results to phyconomists. For instance, for temperature data, a sensor may be set to record the temperature every 10 min in a particular location in a farm, and there may be a dozen or so sensors scattered over the entire site. Giving a data dump to a farmer about the temperature measured at each time point would overwhelm the farmer with information. Instead, providing the maximum and minimum readings over the day, as well as a color-coded geospatial view of these max/min variations over their farm area, provides the farmer with an understanding of how the environment near their farm is behaving. Teasing a story out of volumes of information and presenting it in an easily palatable format to a seaweed farmer is what constitutes data analytics. The farmer can then correlate this with the seaweed growth he has been seeing to see how various environmental factors impact his farm.

19.11.1.5 Moving Toward Predictive Analytics

It is not enough to understand why our farms were affected by a certain phenomenon. Instead, we would like to know how one can predict and plan to mitigate such damages. Combining various streams of ocean data with weather and oceanographic data and bathymetry, we can identify ranges of features related to global weather patterns such as monsoons, El Niño/La Niña cycles, and impending storms and then provide data products that can be understood by farmers to help drive data-based decisions. This involves modeling and incorporating other streams of historical and current data from weather feeds and satellite data to predict climatic shifts that may impact seaweed farming.

19.11.2 Controlling the Environment

Controlling environmental parameters in the sea is a very tough challenge, and many attempts have been made to control some of them. Some phyconomists have adopted the methodology of submerging the entire farm as a means of managing stormy conditions while providing nutrients to the seaweed farm. Since most of the wave action resulting from windy weather resides near the surface, submerging farm units to greater depths protects them from adverse weather conditions, and if the farm can be submerged beyond the photic zone (the depth until which light penetrates in the ocean), it can potentially tap into a large reservoir of nutrients stored there. It was shown that kelp cultivated through depth cycling produced four times more biomass than the control. [Navarrete et al. 2021]. However, these technologies come with their own challenges, especially in terms of capital and operational costs. It remains to be seen whether the increase in productivity will offset the additional engineering involved.

19.12 Precision Methods for Farm Management

Apart from aiming to manage the phenotype of the plant, precision methodologies are also applied to make it easier to manage seaweed farms, and these can be broadly categorized as described below.

19.12.1 Productivity Improvement

Current situation:

Adaptive planning systems for tropical red seaweeds utilize virtually no mechanization or automation. Crop care methods are manual operations conducted on an ad hoc basis, while planting and harvesting involve tedious manual methods, including individual attachment of cultivar cuttings to ropes.

Precision methodologies under development:

One example of developing methodology is mechanical planters and harvesters based on the replacement of single ropes by tubenets (e.g., SeaCombine systems for eucheumatoid seaweeds by Sea6 Energy). Gear and net/line designs for such systems are in the process of refinement and automation. Increasing the speed at which planting and harvesting can be done allows the entire operation to be done offshore, close to where the seaweed farms are. This reduces the amount of time spent carrying crop biomass to shore and back into the water after tying, thereby reducing the number of hours the seed material remains out of the water. This also improves the health of the seed material planted.

19.12.2 Farm Monitoring

Current situation:

Farm monitoring today is done on an *ad hoc* basis, with farmers having a vague idea of which farms are ready to harvest and which need to be planted. While this method works well when the farms are small, it becomes much harder as the farms grow in size.

Precision technologies under development:

a. Drone technology and other GPS-based technologies are increasingly being applied to seaweed crop logging and crop management (e.g., Neish 2019). Precedents for phyconomy development follow precedents established in the field of precision agronomy, where drone-generated survey results can be followed by automated application of treatments by spray drones, and where harvest or planting machinery can be operated autonomously with the use of GPS guidance systems. Seaweed-based crop-care products are among those that can be applied using precision agronomic technology.
b. Efforts are underway to develop farm monitoring protocols so that all farms are labeled and the entire history of planting, harvesting and maintenance is stored in a cloud database. Such systems will give farmers the tools to optimize and plan farm operations to obtain the best results.

19.12.3 Value Chain Development

Current situation:

It has long been recognized that value chains based on adaptive phyconomy systems are rooted in productive ecoscapes where individual farmers operate as members of coastal communities. The evolution of phyconomy systems on commercial farms has resulted from virtual cycles of adaptation that involve collaboration among practicing farmers, scientists, technologists, and development professionals, as discussed in the previous sections of the present document.

Precision methodology under development:

Phyconomy development is inextricably intertwined with socioeconomic development in productive ecoscapes, so the initiatives itemized in this chapter can be applied to value chain and market system development even as they are applied to phyconomy development.

19.12.4 Biosecurity, Ecology, Microbiome, and Integrated Management

Current situation:

Crop-care regimes known as TLC (tender, loving care) are beginning to be undertaken in view of biosecurity principles (e.g., Kambey et al. 2021). The ecology of seaweed ecoscapes and the significance of microbiomes to seaweed productivity are poorly understood, while seaweed farming has developed as the primary productivity base for diverse integrated aquaculture strategies that are known by various acronyms, including IMTA (integrated multitrophic aquaculture): IAAS (integrated agriculture aquaculture systems), ISIAS (integrated silviculture (mangrove) aquaculture systems), IGWAS (integrated green water aquaculture systems), IPUAS (integrated peri-urban aquaculture systems), IFAS (integrated fisheries aquaculture systems), SEAS (sustainable ecological aquaculture systems), ITAS (integrated temporal aquaculture systems); ISAS (integrated sequential aquaculture systems), and many more. Meanwhile, in production ecoscapes, seaweed farms are essential components of integrated coastal area management (ICAM), so an adaptive phyconomy has a key role in ICAM planning and development (e.g., Neish 2020b).

Precision methodology under development:

a. As biosecurity, microbiome, and ecology issues become better understood, adaptive phyconomy protocols can be improved along lines that approximate precision techniques.
b. Mechanization, automation, remote sensing, and cyber-technologies are increasingly providing tools that can be applied to the biosecurity and integrated management of seaweed farms.

Many of the methods described here are already standard practices in land agriculture, and we phyconomists have a long way to go before we approach the level of precision achieved in agriculture. Through observation, bricolage, and sheer tenacity, adaptive phyconomists have been able to develop tropical seaweed farming to the extent that is seen today. We hope that by borrowing from the tools of precision phyconomy, phyconomy can continue to steadily advance from adaptation to precision.

Acknowledgments The authors have had the privilege of working with thousands of seaweed farmers and with numerous colleagues from both the private and public sectors. For the present chapter, special thanks are extended to contributors to the Phyconomy.org Tropical Phyconomy Collaboration Development eucheumatoid seaweeds workshop of July 2021. We hope that this chapter does justice to their work, and we apologize for any errors or omissions. We are especially indebted to Sea Six Energy, which employed the authors at the time of writing and has been a major force in support of Phyconomy.org.

References

Chin GJ, Mohamad MZ, Maili S, Yong WT, Rodrigues KF (2017) ISSR-PCR fingerprinting of *Kappaphycus* and *Eucheuma* (Rhodophyta, Gigartinales) seaweed varieties from Sabah, Malaysia. Trans Sci Technol 4:420–425

Doty MS, Alvarez VB (1981) *Eucheuma* farm productivity. In: Fogg GE, Jones WE (eds) Proceedings of the eight international seaweed symposium. The Marine Science Laboratory, Menai Bridge, Hawaii, pp 688–691

Dy D, Yap H (2001) Surge ammonium uptake of the cultured seaweed, *Kappaphycus alvarezii* (Doty) Doty (Rhodophyta: Gigartinales). J Exp Mar Biol Ecol 265:89–100. https://doi.org/10.1016/S0022-0981(01)00325-2

Grueneberg J, Engelen AH, Costa R, Wichard T (2016) Macroalgal morphogenesis induced by waterborne compounds and bacteria in coastal seawater. PLoS One 11(1):e0146307. https://doi.org/10.1371/journal.pone.0146307

Holling CS (1978) Adaptive environmental assessment and management. John Wiley & Sons, New York

Holling CS (1993) Investing in research for sustainability. Ecol Appl. 3(4). Pp. 552-555 © 1993 by ecological Society of America

Hu ZM, Shan TF, Zhang J, Zhang QS, Critchley AT, Choi HG, Yotsukura N, Liu FL, Duan DL (2021) Kelp aquaculture in China: a retrospective and future prospects. Rev Aquac 13:1324–1351. https://doi.org/10.1111/raq.12524

Hurtado AQ, Neish IC, Critchley AT (2019) Phyconomy: the extensive cultivation of seaweeds, their sustainability and economic value, with particular reference to important lessons to be learned and transferred from the practice of eucheumatoid farming. Phycologia 58(5):472–483

Kambey CLB, Campbell I, Cottier-Cook E, Bin MD, Nor RA, Kassim A, Sade A, Lim P-E (2021) Seaweed aquaculture: a preliminary assessment of biosecurity measures for controlling disease and pest outbreaks of a *Kappaphycus* farm. J Appl Phycol. [in press] 33: 3179

Lee KN (1999) Appraising adaptive management. Conserv Ecol 3(2)3 Northwest Oregon State Forests Management Plan Jan. 2001. http://www.consecol.org/vol3/iss2/art3/

Lim PE, Tan J, Phang SM, Nikmatullah A, Hong DD, Sunarpi H, Hurtado AQ (2013) Genetic diversity of *Kappaphycus* Doty and *Eucheuma* J. Agardh (Solieriaceae, Rhodophyta) in Southeast Asia. J Appl Phycol. https://doi.org/10.1007/s10811-013-0197-y

Morris T (2008) Adaptive management. Slide presentation online by tom Morris, associate professor, soil fertility specialist, University of Connecticut. thomas.morris@uconn.edu

Morris T, Friedman S, Blackmer T (2011) Adaptive management strategies: improving watershed Management at Farm and Watershed Scales. Slide presentation online, Newark, Delaware thomas.morris@uconn.edu

Navarrete IA, Kim DY Wilcox C, Reed DC, Ginsburg DW, Dutton JM, Heidelberg J, Raut Y, Wilcox BH (2021) Effects of depth-cycling on nutrient uptake and biomass production in the giant kelp *Macrocystis pyrifera*, renewable and sustainable energy reviews, volume 141, 110747, 1364–0321, https://doi.org/10.1016/j.rser.2021.110747

Narvarte BCV, Genovia TGT, Hinaloc LAR, Roleda MY (2022) Growth, nitrate uptake kinetics, and biofiltration potential of eucheumatoids with different thallus morphologies. J Phycol 58: 12–21. https://doi.org/10.1111/jpy.13229

Neish IC (2019) A day in the life of a future tropical seaweed farmer. Paper presented at the 23rd International Seaweed Symposium, International Convention Center, Jeju, Korea

Neish IC (2020a) Art, Max, Vic and origins of tropical marine phyconomy: with special reference to India's role in evolving marine phyconomy from art to science. Paper presented at the India International Seaweed Expo and Summit 2020, NIOT, Chennai

Neish IC (2020b) Adaptive phyconomy for sustainable management of coastal ecoscapes in Indonesia: How adaptive phyconomy supports holistic, sustainable marine ecoscape development as a component of integrated coastal area management (ICAM). Video presentation at MARSAVE 2020 Second Marine Resilience and Sustainable Development International Symposium, Hasanuddin University, Makassar, South Sulawesi, Indonesia, October 2020

Reddy CRK, Kumar GRK, Siddhanta AK, Tewari A, Eswaran K (2003) In vitro somatic embryogenesis and regeneration of somatic embryos from pigmented callus of *Kappaphycus alvarezii* (Doty) Doty (Rhodophyta Gigartinales). J Phycol 39(3):610–616. https://doi.org/10.1046/j.1529-8817.2003.02092.x

Salafsky N, Boshoven J, Burivalova Z, Dubois NS, Gomez A, Johnson A, Lee A, Margoluis R, Morrison J, Muir M, Pratt SC, Pullin AS, Salzer D, Stewart A, Sutherland WJ, Wordley CFR (2019) Defining and using evidence in conservation practice. © 2019 The Authors. Conservation Science and Practice published by Wiley Periodicals, Inc. on behalf of the Society for Conservation Biology

Scoones I (1998) Sustainable rural livelihoods: a framework for analysis. IDS working paper no. 72. Brighton, UK, Institute of Development Studies, University of Sussex

Scoones I, Leach M, Smith A, Stagl S, Stirling A, Thompson J (2007) Dynamic systems and the challenge of sustainability. In: STEPS working paper 1. STEPS Centre, Brighton. [electronic resource available at http://www.ids.ac.uk/go/bookshop Sept, 2009]

Taleb NN (2012) Antifragile–things that gain from disorder. Random House Publishing Group

Voss JP, Bornemann B (2011) The politics of reflexive governance: challenges for designing adaptive management and transition management. Copyright © 2011 by the author(s). Published here under licence by the resilience Alliance. Ecol Soc 16(2):9. [online] URL: http://www.ecologyandsociety.org/vol16/iss2/art9/

Walters CJ (1986) Adaptive management of renewable resources. Biological resource management. A series of primers on the conservation and exploitation of natural and cultivated ecosystems. Wayne M. Getz, series editor. University of California, Berkeley

Ward GM, Faisan JP, Cottier-Cook EJ, Gachon C, Hurtado AQ, Lim PE, Matoju I, Msuya FE, Bass D, Brodie J (2020) A review of reported seaweed diseases and pests in aquaculture in Asia. J World Aquacult Soc 51:815–828. https://doi.org/10.1111/jwas.12649

Zitta CS, Oliveira EM, Bouzon ZL, Hayashi L (2012) Ploidy determination of three *Kappaphycus alvarezii* strains (Rhodophyta Gigartinales) by confocal fluorescence microscopy. J Appl Phycol 24(3):495–499. https://doi.org/10.1007/s10811-011-9704-1

Zuccarello GC, Critchley AT, Smith J, Sieber V, Bleicher Lhonneur G, West JA (2006) Systematics and genetic variation in commercial *Kappaphycus* and *Eucheuma* (Solieriaceae, Rhodophyta). J Appl Phycol 18:643–651. https://doi.org/10.1007/s10811-006-9066-2

Seaweed Health Problems: Major Limiting Factors Affecting the Sustainability of the Seaweed Aquaculture Industry in the Philippines 20

Joseph P. Faisan Jr and Anicia Q. Hurtado

Abstract

The Philippines remains one of the top eucheumatoid seaweed producers worldwide. However, the reported seaweed health problems affecting farms have been one of the reasons for the decline in seaweed production in the country. Outbreaks of "ice-ice" disease (IID), epiphytic pests, and grazing problems have significantly reduced biomass yields and compromised carrageenan quality, directly affecting the livelihood of tens of thousands of families dependent on seaweed farming. In this chapter, we listed seaweed health problems based on type-association with the host seaweed plant and highlighted factors contributing to farm problems. Additionally, recommendations are presented that are relevant to the protection of the long-term sustainability of the eucheumatoid seaweed farming industry in the Philippines.

Keywords

Seaweed · Health · Sustainability · Aquaculture · Philippines

20.1 Introduction

Traditionally collected from the wild, seaweeds have emerged as one of the largest contributors to the aquaculture sector worldwide in the last decade. A three-fold increase in global farmed seaweed production was observed between 2010 and 2019, from 10.6 million metric tonnes (Mt) to 34.74 Mt., respectively (FAO 2021). The exponential increase in production could be attributed to the rise in demand, as the uses of seaweeds were no longer limited to the food sector, but expanded to many industries, including pharmaceutical, cosmetics, and nutraceutical applications (Bixler and Porse 2011; Porse and Rudolph 2017), and seaweeds were even regarded as a possible tool for carbon sequestration to reduce global climate change (Duarte et al. 2017; Ortega et al. 2019).

Eucheumatoid red seaweeds, including several species of *Kappaphycus* and *Eucheuma*, are mainly produced in Southeast Asia, where the majority of the production is centered. In 2019, Indonesia, the Philippines, and Malaysia produced 11.48 Mt (FAO 2021). This production constitutes 33.5% of the world's total seaweed production in the same year (Cai et al. 2021). The high economic demand for carrageenan, a hydrocolloid polysaccharide extract derived from eucheumatoid seaweeds, has resulted in farming expansion and adoption in many tropical and subtropical countries (Therkelsen 1993).

The Philippines was the largest producer of eucheumatoid seaweeds but was overtaken by Indonesia in the mid-2000s. Seaweed production in the Philippines peaked in 2011 (1.84 Mt) but has not fully recovered since then (BFAR 2019). Several factors contributed to the decline in production. However, the most pressing issue could be reported outbreaks of diseases and pests affecting eucheumatoid seaweed farms. In this paper, we identified the common health problems observed on seaweed farms in the Philippines and the factors that contributed to these problems. We then recommend mitigation strategies that could result in sustainable industry growth with protected livelihoods for many stakeholders dependent on seaweeds.

20.2 Economic Importance of Seaweeds

Farmed seaweeds is the leading aquaculture subsector in the Philippines. In 2019, seaweed production in the country (Fig. 20.1) comprised 64% of the total aquaculture

J. P. Faisan Jr (✉)
Southeast Asian Fisheries Development Center Aquaculture Department, Tigbauan, Iloilo, Philippines
e-mail: jfaisan@seafdec.org.ph

A. Q. Hurtado
Integrated Services for the Development of Aquaculture and Fisheries (ISDA), Inc, Jaro Iloilo City, Philippines

A. T. Critchley et al. (eds.), *Tropical Phyconomy Coalition Development*, Developments in Applied Phycology 11,
https://doi.org/10.1007/978-3-031-47806-2_20

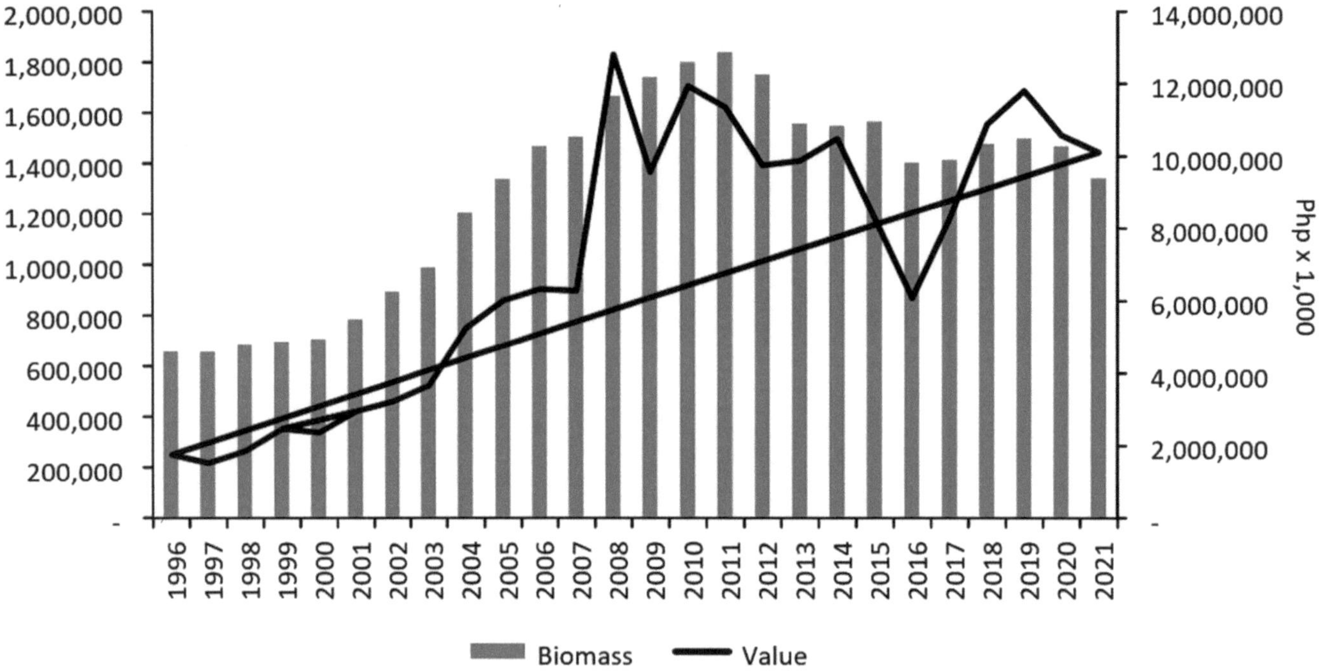

Fig. 20.1 Seaweed biomass production and value in the Philippines between 1996 and 2021 (Source: PSA 2022)

production, making it the top aquatic commodity in volume and second in export value, amounting to $250 million (BFAR 2020). The Philippines remains one of the top seaweed producers globally, producing 1.50 Mt in 2019 and the fourth-largest contributor of farmed seaweeds worldwide (BFAR 2020; FAO 2021). Additionally, the country is the second-largest producer of carrageenan seaweeds after Indonesia and the eighth-largest producer of agar seaweed, in addition to being the top country in producing the green alga *Caulerpa* (1.1 Mt) (FAO 2021). In 2021, the country produced 1.34 Mt of dried seaweed with a gross value of ₱10.1 billion (~$202.8 million; $1 = ₱50).

Eucheumatoid red seaweeds, including *Kappaphycus* (*K. alvarezii* and *K. striatus*), commercially known as "cottonii", and *Eucheuma* (*E. denticulatum*), commercially known as "spinosum", are the leading farmed seaweeds in the country. They are highly valued for their carrageenan content. Carrageenan is a marine hydrocolloid polysaccharide with commercial value. The Philippines processed eucheumatoid seaweed products either as alkali-treated chips (ATC), semirefined carrageenan, or refined carrageenan. The green alga *Caulerpa* spp. was mainly used as direct human food (BFAR 2020).

Out of the 17 geographic regions of the Philippines, 15 are engaged in seaweed cultivation. The bulk of the production comes from the Bangsamoro Autonomous Region for Muslim Mindanao (BARMM), including Sulu and Tawi-Tawi, where the traditional areas for seaweed production are located. BARMM is followed by Region IV-B (Palawan and Mindoro), Region 9 (Zamboanga Peninsula), Region 6 (Iloilo and Antique), and Region 7 (Cebu and Bohol). The farms are located in coastal areas (near-shore) and offshore areas (BFAR 2020).

Widespread cultivation of seaweeds around the country benefited many rural coastal communities and contributed to job generation, where livelihood opportunities mainly involved fishing (Hurtado 2013; Valderrama et al. 2013). Seaweed farms around the country are primarily family-managed income-generating activities, with household members involved in different capacities during preparation, farming, and harvesting (Hurtado 2013). More than 200,000 families are engaged in the seaweed industry in the country, not to mention downstream activities, including post-harvesting, marketing, trading, and processing (export), which generate socio-economic impacts on these communities (Pedrosa 2017). However, after the country recorded the highest production in 2011, a steady decline in seaweed biomass production was observed (BFAR 2011; PSA 2022).

Among the factors contributing to the decline in seaweed production were the reported outbreaks of seaweed health problems on farms (Ward et al. 2020). "Ice-ice" disease (IID), epiphytic pests (macro-epiphytes and epiphytic filamentous algae, EFA), and grazing are the most common yield-limiting seaweed problems observed on farms.

20.3 Seaweed Health Problems

The common health problems affecting farmed eucheumatoid seaweeds and their associations with the host plant are listed in Table 20.1 and illustrated in Fig. 20.2.

20.3.1 "Ice-Ice" Disease (IID)

"Ice-ice" disease (IID) is the most common health problem reportedly affecting eucheumatoid seaweed farms. IID-infected seaweeds manifested loss of pigmentation followed by softening of the affected thallus tissues and, in severe cases, detachment of diseased plants from the cultivation lines (type 1 association). The name ice-ice is derived from the observations of farmers in the Philippines, who described the whitening observation on the thallus of the seaweed plant.

IID is one of the most prevalent seaweed health problems observed on eucheumatoid farms. In the Philippines, previous reports showed its occurrence in many farms around the country (Uyenco et al. 1981; Largo et al. 1995; Hurtado et al. 2006; Solis et al. 2010; Alibon et al. 2019). A recent survey conducted in major seaweed farms in the Philippines between 2018 and 2019 (Faisan Jr et al. 2021) showed IID incidence in all the farms surveyed, albeit with varying degrees of infection. The survey results revealed the widespread occurrence of IID encompassing different geographic regions of the Philippines and highlighted the need to mitigate the impacts of this problem.

Seaweeds affected by IID have been reported to be caused by pathogenic microbial complexes, including gram-negative bacteria (*Vibrio* sp., *Cytophaga-Flavobacterium* complex, *Alteromonas*, *Pseudoalteromonas*, and *Aurantomonas*) and fungi (*Aspergillus ochraceus*, *A. terreus*, *Phoma* sp.) (Largo et al. 1995; Solis et al. 2010; Syafitri et al. 2017). *Vibrio* spp. have also been shown *in vitro* to induce more severe IID in healthy tissue in the presence of environmental stressors, such as water salinity and elevated water temperature (Largo et al. 1995; Azizi et al. 2018). IID-affected seaweed thalli often disintegrate from the main culture line, where the farmed seaweed is attached, resulting in biomass losses and thereby a loss of income to farmers. Additionally, IID-infected seaweed resulted in reduced carrageenan quality, which may result in a lower market price (Mendoza et al. 2002).

20.3.2 Epiphytic Pests

20.3.2.1 Macro-Epiphytes

Macro-epiphytes are seaweed species found entangled or loosely attached to cultivation sites other than farmed seaweeds (type association 2) (Table 20.1). Common macro-epiphytes found on farms include *Sargassum* (brown alga), filamentous *Ulva* (green alga), and *Gracilaria* (red alga) (Faisan Jr et al. 2021). Entangling of these macro-epiphytes may result in compromised growth of farmed seaweeds through competition for nutrient uptake from the water and photosynthesis. However, the observation of macro-epiphytes on seaweed farms could be seasonal in occurrence. In the Philippines, annual vegetative peaks of the brown alga *Sargassum* are observed from October to December (Trono and Largo 2019), which is concordant with the high prevalence of pest fouling observed in farmed seaweeds in major farming areas in the country (Faisan Jr et al. 2021). It could be that the high prevalence observed was due to macro-epiphyte buoyancy (i.e., *Sargassum*) easily carried by the water current and trapped by the farmed seaweeds or the cultivation lines. However, other factors, such as slow water movement and anthropogenic eutrophication, might contribute to the proliferation of these macro-epiphytes. Additionally, the fouling incidence of macro-epiphytes on farmed seaweeds might be less likely to be

Table 20.1 Type associations of important seaweed health problems affecting eucheumatoid red seaweed crops in the Philippines (Adapted from Faisan Jr et al. 2021)

Association with the host seaweed	Pest or disease	Description	References
Type 1	"Ice-ice" disease (IID)	Loss of pigmentation followed by softening of the thallus, and detachment of diseased plants from seaweed cultivation lines	Doty and Alvarez (1975); Largo et al. (1995); Arasamuthu and Edward (2018)
Type 2	Macro-epiphyte algae	Seaweeds that grow on the surface of seaweed plants (loosely attached or entangled), e.g., *Sargassum*, *Ulva*, *Gracilaria*	Leonardi et al. (2006); Ingle et al. (2018)
Type 3	Epiphytic filamentous algae (EFA)	Red seaweeds that grow on the surface of seaweed plants, penetrating from the cortical to the medullary layers, causing mechanical damage to the host plant, e.g., *Melanothamnus* (=previously known as *Neosiphonia*), *Polysiphonia*	Ask (1999); Vairappan (2006); Pang et al. (2011); Tsiresy et al. (2016)
Type 4	Grazing	Mechanical damage or absence of tips on the soft tissues at the apex of the thallus	Ask (1999); Pang et al. (2015)

Fig. 20.2 Seaweed health problems observed on seaweed farms in the Philippines, including "ice-ice" disease (**a**), epiphytic pests: macro-epiphytes, including filamentous green *Ulva* (**b**), brown alga, *Sargassum* (**c**) and epiphytic filamentous algae, EFA (**c**, **d**), and grazing (**f**) (Photos by JP Faisan Jr and AQ Hurtado)

observed in shallow water farms due to the constant tidal fluctuations that make the macro-epiphytes easily detached from the farmed seaweed or cultivation lines (Faisan Jr et al. 2021).

20.3.2.2 Epiphytic Filamentous Algae (EFA)

Epiphytic filamentous algae (EFA) are epiphytic pests affecting farmed seaweeds by attachment to the host seaweed plant (type 3 association). EFA infect the host plant through intercellular penetration, causing mechanical damage to the cortical and medullary cells (Ask 1999; Vairappan 2006; Pang et al. 2011; Tsiresy et al. 2016). The target thallus region of EFA infection has not been fully studied; however, based on the study of Yamamoto et al. (2012), the soft tissues of the apex thallus region are most susceptible to epiphyte colonization. Additionally, Faisan Jr et al. (2021) reported increased susceptibility to epiphyte colonization, suggesting that the initial entry of epiphytes occurs in young tissues. When observed microscopically, EFA penetrated up to the cortical cells and created injury to the host plant. EFA-affected seaweeds are often observed to have a bumpy, mound-like structure where the filaments of the EFA are attached to the host seaweed (Vairappan 2006). The mechanical damage caused by EFA penetration creates an opening for opportunistic pathogens to infect the plant once EFA dies off. Red epiphytic algal pests commonly affecting eucheumatoid seaweeds belong to the genera *Melanothamnus* (=previously known as *Neosiphonia*) (Hurtado et al. 2006; Vairappan 2006; Vairappan et al. 2008) and *Polysiphonia* (Tsiresy et al. 2016). Outbreaks of EFA on farms reportedly result in reduced biomass and compromised carrageenan quality (Correa and McLachlan 1994; Hurtado et al. 2006; Vairappan et al. 2008; Ali et al. 2018, 2020).

20.3.3 Grazing

Signs of grazing on farmed seaweeds are often observed, as manifested by the absence of tips on the soft tissues of the apical region of the seaweed thalli (type association 4). Grazing incidence is commonly caused by herbivore fishes (i.e., siganids, parrot fish, surgeon fish), sea urchins, and sea turtles (Ask 1999). However, grazing on farmed seaweeds is observed to occur on a seasonal basis, particularly by juvenile herbivorous fish (Pang et al. 2015). Grazing incidence may result in a shift in the microbial community structure of the affected seaweed thallus (Tan et al. 2020), compromising the health status of farmed seaweeds. The damaged thallus caused by grazing exposes potential pathogenic microbes, leading to infection. Heavy damage caused by grazing could lead to loss of biomass and failure of production, especially during the critical phase of nursery culture when seaweed seedlings are used to expand farming areas.

20.4 Factors Contributing to Seaweed Health Problems

20.4.1 Fluctuations in Environmental Parameters

The occurrence of diseases and pests affecting farmed seaweeds has been reportedly attributed to fluctuating environmental conditions. Key environmental parameters, including temperature, salinity, irradiance, water movement, pH, inorganic nutrients, and combinations thereof, may have influenced the health status of farmed seaweeds. Thus, any fluctuation in these abiotic parameters influences the physiological fitness of the seaweed crop, making it more susceptible to disease and epiphytic pest problems.

Elevated water temperature, salinity, and pH levels have been attributed to the increased incidence of IID on farms (Arasamuthu and Edward 2018; Msuya and Porter 2014; Azizi et al. 2018). In the Philippines, fluctuations in environmental parameters outside of the optical culture conditions, including elevated water temperature observations, coincided with the reported increase in IID incidence (Alibon et al. 2019; Vince Cruz-Abeledo et al. 2019). Additionally, an outbreak of EFA was observed in farms with high temperature, high irradiance, and slow water movement observations (Hurtado et al. 2006). In Malaysia, the incidence of EFA was observed to be higher during the dry months when the seawater temperature and salinity increased from 27 to 31 °C and from 28 to 34 psu, respectively (Vairappan 2006).

The shift in environmental conditions and its influences on farmed seaweeds could be felt more significantly in shallow water areas (i.e., fixed off-bottom farms), as the fluctuations in this area are more pronounced than in deep water farms. However, the study by Faisan Jr et al. (2021) reported the widespread occurrence of IID even in deep water areas (i.e., employing hanging long-line, multiple raft long-line, or triangular methods of farming), which may suggest that fluctuations in environmental conditions, either short- or long-term, have a negative consequence on the health status of the seaweed crop, notwithstanding the farm location.

20.4.2 Extreme Climatic Events and Natural Calamities

Changes in environmental parameters are further exacerbated by extreme climatic events, such as the occurrence of tropical cyclones or typhoons. The Philippines is located in the typhoon belt area, and approximately 20 tropical cyclones enter through the Philippine Area of Responsibility (PAR) annually, nine of which cross the country (Cinco et al. 2016). Typhoons bring strong winds, high rainfall levels, and storm surges to farming areas, destroying many farms, especially in shallow water areas. Furthermore, the loss of seaweed biomass and farming materials prevents continuous seaweed farming year-round (Valderrama et al. 2013; Hurtado 2013). Aside from typhoons, other natural calamities, such as earthquakes and volcanic eruptions, disrupt farming activities and constitute a significant threat to farming communities that depend on seaweed cultivation as their source of livelihood.

Additionally, prolonged environmental anomalies, including El Niño and La Niña phenomena, further aggravate the problems observed on farms and have negative consequences for seaweed crops. The impacts of climate change on farmed eucheumatoid seaweeds are not only limited to physical damage to the seaweed crop but also have deleterious effects on the seaweed's eco-physiological, reproductive, and metabolic processes (Largo et al. 2017).

20.4.3 Deterioration of Seaweed Seedlings

Vegetative propagation, also known as the cut and plant technique, is the most commonly used and conventional method in seaweed farming. This method uses healthy seaweed thalli as "seedling" material for subsequent cropping. However, the repetitive use of this method could result in the slow growth of seaweeds and make the seaweeds "less vigorous" (Hayashi et al. 2017) and more susceptible to diseases and pests, as stress responses are influenced by changing environments attributed to climate change.

20.4.4 Incoherent Biosecurity Legislation and Policy

Exponentially increasing demand for eucheumatoid seaweeds and seaweed by-products has led to the expansion of farming areas in many tropical and subtropical countries (Sulu et al. 2004; Pickering 2006; Hurtado et al. 2014; Msuya and Porter 2014; Hayashi et al. 2017; Shanmugam et al. 2017; Alemañ et al. 2019). However, the unregulated introduction of non-native seaweed cultivars inevitably resulted in the transfer of seaweed "hitchhikers", including disease and pest agents, leading to outbreaks and crop losses. Compared to other important aquatic commodities, not to mention terrestrial commodities, seaweed biosecurity measures in many seaweed-growing countries have been absent, scanty, or not strictly implemented (Mateo et al. 2020). Many seaweed biosecurity policies are lacking in seaweed-producing countries, especially in developing countries (Cottier-Cook et al. 2016). Additionally, Campbell et al. (2019) reported significant challenges in biosecurity policies in the seaweed industry, including inconsistent terminology for the inclusion of seaweeds in regulatory frameworks, limited guidance for the responsibility of implementating biosecurity measures, insufficient evidence to develop disease and pest-specific policies, and a lack of a coherent approach to seaweed biosecurity risk management in international policies. These problems are mirrored in the national biosecurity-related regulations and policies in seaweed aquaculture in the Philippines. In relation to these challenges, Mateo et al. (2020) emphasized the key gaps in the legislation and policies governing the Philippine seaweed industry, including the lack of seaweed-specific frameworks, lack of binding policies for seaweed aquaculture, limited biosecurity approaches, lack of competent authority, limited involvement of experts in framework development, and insufficient guidance for the use of the precautionary principle.

Biosecurity strategies should be laid out and implemented starting from sourcing seaweed "seedlings" up to harvest. The use of tissue culture seedlings from biosecured facilities should be considered. However, the sourcing of disease- and pest-free seedlings remains a major limitation in the Philippines, especially in far-flung farming areas in the country. The screening process and presence of a quarantine facility in major seaweed farming areas in the country are limited, if not totally non-existent. One of the problems faced by the Philippine seaweed industry is the lack of regulatory mechanisms (Hurtado et al. 2021). There are only very few policies involving seaweeds and only the updated version of the Code of Good Aquaculture Practices for Seaweeds (GAqP-S), where seaweed and disease management are mentioned (PNS/BAFS 208: 2021).

20.5 Recommendations

In the Philippines, the use of vegetative propagation methods through direct cutting from selective seaweed stock remains the most widely used and cost-effective farming practice. However, the repeated use of the same cultivar inadvertently results in reduced seaweed robustness, compromising growth and carrageenan yields. The loss of vigor in a limited number of farmed seaweed cultivars further leads to increased susceptibility to disease and pest problems. The continuous use of vegetative cuttings from the limited number of seaweed varieties resulted in lower genetic diversity (Zuccarello et al. 2006; Halling et al. 2013). The need for new cultivars from farmed and wild seaweeds has never been urgent (Brakel et al. 2021). A survey of seaweed cultivars in Southeast Asia revealed the genetic diversity of many haplotypes of *Kappaphycus* spp. and *Eucheuma denticulatum* (Lim et al. 2014). Recently, Hinaloc and Roleda (2021) reported new progenies from wild seaweeds, which might be possible sources of cultivars in the future.

Mateo et al. (2021) listed the farm management and biosecurity measures on farms to include (1) proper zonation of farm areas, (2) rigorous screening of propagules, (3) acclimation of seedlings from an outside source, (4) removal and collection of macro-epiphytes, (5) strengthening of main lines, culture lines, and anchors, (6) inputs of organic nutrients, disinfection of cultivation ropes, and (7) fallowing. Additionally, Hurtado et al. (2021) highlighted policy recommendations aimed at ensuring the long-term sustainability of the Philippine seaweed industry, including (1) promoting the conservation of wild populations to preserve genetic diversity, (2) strengthening infrastructure for the consistent supply of robust and biosecured seedlings, (3) enhancing biosecurity measures, (4) integrating the seaweed sector into climate change adaptation planning, and (5) steering multistakeholder partnerships.

Seaweed health problems are one of the causative factors in the decline of eucheumatoid red seaweed production in the Philippines. These problems have been exacerbated by changes in meteorological patterns, resulting in extreme weather and prolonged climatic events that directly affect the health status of farmed seaweeds. The widespread prevalence of these problems at the farm level warrants extensive studies to determine the factors that affect seaweed susceptibility to these yield-limiting diseases and pests. The many reports of fluctuations or abrupt changes in key environmental parameters could be a predisposing factor for the occurrence of seaweed health problems. As the changes in weather patterns become more frequent and intense, as observed in recent years, they could significantly impact the sustainability of seaweed aquaculture in the country. Proactive and immediate mitigation strategies should be implemented to prevent the further spread of these problems and compromise the future and sustainability of the seaweed aquaculture industry in the Philippines.

References

Arasamuthu A, Edward JKP (2018) Occurrence of ice-ice disease in seaweed *Kappaphycus alvarezii* at gulf of Mannar and Palk Bay, southeastern India. Indian J Geo Mar Sci 47:489–492

Alemañ AE, Robledo D, Hayashi L (2019) Development of seaweed cultivation in Latin America: current trends and future prospects. Phycologia 58(5):462–471

Ali MKM, Critchley AT, Hurtado AQ (2020) The impacts of AMPEP K + (*Ascophyllum* marine plant extract, enhanced with potassium) on the growth rate, carrageenan quality, and percentage incidence of the damaging epiphyte *Neosiphonia apiculata* on four strains of the commercially important carrageenophyte *Kappaphycus*, as developed by micropropagation techniques. J Appl Phycol 32:1907–1916. https://doi.org/10.1007/s10811-020-02117-0

Ali MKM, Yasir S, Critchley AT, Hurtado AQ (2018) Impacts of *Ascophyllum* marine plant extract powder (AMPEP) on the growth, incidence of the endophyte *Neosiphonia apiculata* and associated carrageenan quality of three commercial cultivars of *Kappaphycus*. J Appl Phycol 30:1185–1195

Alibon RD, Gonzales JMP, Ordoyo AET, Echem RT (2019) Incidence of ice-ice disease associated with *Kappaphycus alvarezii* in the seaweed farms in Zamboanga peninsula, Mindanao, Philippines. SSR Inst Int J Life Sci 5(1):2148–2155

Ask EI (1999) Cottonii and spinosum cultivation handbook. FMC Food Ingredients Division, Philadelphia, pp 1–50

Azizi A, Mohd Hanafi N, Basiran MN, Teo CH (2018) Evaluation of disease resistance and tolerance to elevated temperature stress of the selected tissue-cultured *Kappaphycus alvarezii* Doty 1985 under optimized laboratory conditions. 3 Biotech 8:321. https://doi.org/10.1007/s13205-018-1354-4

Bixler H, Porse H (2011) A decade of change in the seaweed hydrocolloids industry. J Appl Phycol 23:321–335

Brakel J, Sibonga RC, Dumilag RV, Montalescot V, Campbell I, Cottier-Cook EJ, Ward G, Le Masson V, Liu T, Msuya FE, Brodie J, Lim P-E, Gachon CMM (2021) Exploring, harnessing and conserving marine genetic resources towards a sustainable seaweed aquaculture. Plants People Planet 3(4):337–349. https://doi.org/10.1002/ppp3.10190

Bureau of Fisheries and Aquatic Resources BFAR (2011) Philippine fisheries profile 2011. Bureau of Fisheries and Aquatic Resources, Department of Agriculture, Quezon City, Philippines, pp 1–70. https://www.bfar.da.gov.ph/publication.jsp?id=42#post Accessed on 15 March 2022

Bureau of Fisheries and Aquatic Resources BFAR (2019) Philippine fisheries profile 2019. Bureau of Fisheries and Aquatic Resources, department of agriculture, Quezon City, Philippines, pp 1–72. https://www.bfar.da.gov.ph/publication.jsp?id=2375#post Accessed on 04 February 2022

Bureau of Fisheries and Aquatic Resources BFAR (2020) Philippine fisheries profile 2020. Bureau of Fisheries and Aquatic Resources, Department of Agriculture, Quezon City, Philippines, pp 1–72. https://www.bfar.da.gov.ph/publication.jsp?id=2383#post Accessed on 09 March 2022

Cai J, Lovatelli A, Aguilar-Manjarrez J, Cornish L, Dabbadie L, Desrochers A, Diffey S, Garrido, Gamarro E, Geehan J, Hurtado A, Lucente D, Mair G, Miao W, Potin P, Przybyla C, Reantaso M, Roubach R, Tauati M, Yuan X (2021) Seaweeds and microalgae: an overview for unlocking their potential in global aquaculture development. FAO fisheries and aquaculture circular no. 1229. FAO, Rome. https://doi.org/10.4060/cb5670en

Campbell I, Kambey C, Mateo J, Rusekwa SB, Hurtado A, Msuya F, Cottier-Cook EJ (2019) Biosecurity policy and legislation for the global seaweed aquaculture industry. J Appl Phycol 32:2133. https://doi.org/10.1007/s10811-019-02010-5

Cinco TA, de Guzman RG, Ortiz AMD, Delfino RJP, Lasco RD, Hilario FD, Juanillo EL, Barba R (2016) Observed trends and impacts of tropical cyclones in The Philippines. Int J Climatol 36(14): 4638–4650. https://doi.org/10.1002/joc.4659

Correa JA, McLachlan JL (1994) Endophytic algae of *Chondrus crispus* (Rhodophyta). V. Fine structure of the infection by *Acrochaete operculata* (Chlorophyta). Eur J Phycol 29:33–47. https://doi.org/10.1080/09670269400650461

Cottier-Cook EJ, Nagabhatla N, Badis Y, Campbell M, Chopin T, Dai W, Fang J, He P, Hewitt C, Kim GH, Huo Y, Jiang Z, Kema G, Li X, Liu F, Liu H, Lu Q, Luo Q, Mao Y, Msuya FE, Rebours C, Shen H, Stentiford GD, Yarish C, Wu H, Yang Z, Zhang J, Zhou Y, Gachon CMM (2016) Safeguarding the future of the global seaweed aquaculture industry. United National University (INWEH) and Scottish Association for Marine Science Policy Brief, pp 1–12

Doty MS, Alvarez VB (1975) Status, problems, advances and economics of *Eucheuma* farms. Mar Technol Soc 9(4):30–35

Duarte CM, Wu J, Xiao X, Bruhn A, Krause-Jensen D (2017) Can seaweed farming play a role in climate change mitigation and adaptation? Front Mar Sci 4:1–8. https://doi.org/10.3389/fmars.2017.00100

Faisan JP Jr, Luhan MRJ, Sibonga RC, Mateo JP, Ferriols VMEN, Brakel J, Ward GM, Ross S, Bass D, Stentiford GD, Brodie J, Hurtado AQ (2021) Preliminary survey of pests and diseases of eucheumatoid seaweed farms in The Philippines. J Appl Phycol 33(4):2391–2405. https://doi.org/10.1007/s10811-021-02481-5

FAO (2021) FAO global fishery and aquaculture production statistics (FishStatJ). Accessed on March 9, 2022. www.fao.org/fishery/statistics/software/fishstatj/en

Ingle KN, Polikovsky M, Chemodanov A, Goldberg A (2018) Marine integrated pest management (MIPM) approach for sustainable seagriculture. Algal Res 29:223–232

Halling C, Wikström SA, Lilliesköld-Sjöö G, Mörk E, Lundsør E, Zuccarello GC (2013) Introduction of Asian strains and low genetic variation in farmed seaweeds: indications for new management practices. J Appl Phycol 25:89–95

Hinaloc LAR, Roleda MY (2021) Phenotypic diversity, growth and sexual differentiation on the progeny of wild *Kappapphycus alvarezii* (Gigartinales, Florideophyceae). Phycologia. https://doi.org/10.1080/003188884.2021.1946307

Hayashi L, Reis RP, dos Santos AA, Castelar B, Robledo D, de Vega GB, Msuya FE, Eswaran K, Yasir SM, Ali MKM, Hurtado A (2017) The cultivation of *Kappaphycus* and *Eucheuma* in tropical and subtropical waters. In: Hurtado AQ, Critchley AT, Neish IC (eds) Tropical seaweed farming trends, problems and opportunities. Springer Nature, Cham, Switzerland, pp 55–90

Hurtado AQ, Critchley AT, Trespoey A, Bleicher-Lhonneur G (2006) Occurrence of *Polysiphonia* epiphytes in *Kappaphycus* farms at Calaguas is., Camarines Norte, Philippines. J Appl Phycol 18:301–306

Hurtado AQ (2013) Social and economic dimensions of carrageenan seaweed farming in The Philippines. In: Valderrama D, Cai J, Hishamunda N, Ridler N (eds) Social and economic dimensions of carrageenan seaweed farming, Fisheries and aquaculture technical paper no. 580. FAO, Rome, pp 91–113. http://www.fao.org/3/a–i3344e.pdf

Hurtado AQ, Gerung GS, Suhaimi Y, Critchley AT (2014) Cultivation of tropical red seaweeds in the BIMP-EAGA region. J Appl Phycol 26:707–718

Hurtado AQ, Luhan MR, Ferriols VME, Faisan J, Mateo J, Sibonga R, Suyo JG (2021) Towards a robust and resilient seaweed aquaculture in The Philippines. Policy Brief #1

Largo DB, Fukami K, Nishijima T (1995) Occasional pathogenic bacteria promoting ice–ice disease in the carrageenan–producing red algae *Kappaphycus alvarezii* and *Eucheuma denticulatum* (Solieriaceae, Gigartinales, Rhodophyta). J Appl Phycol 7:545–554. https://doi.org/10.1007/BF00003941

Largo DB, Chung IK, Phang SM, Gerung GS, Sondak CFA (2017) Impacts of climate change on *Eucheuma-Kappaphycus* farming. In: Hurtado A, Critchley A, Neish I (eds) Tropical seaweed farming trends, problems and opportunities, Developments in applied phycology, vol 9. Springer, Cham, pp 121–129

Leonardi PI, Miravalles AB, Faugeron S, Flores V, Beltrán J, Correa JA (2006) Diversity, phenomenology and epidemiology of epiphytism in farmed *Gracilaria chilensis* (Rhodophyta) in northern Chile. Eur J Phycol 41(2):247–257

Lim PE, Tan J, Phang SM, Rahiman AA (2014) Genetic diversity of *Kappaphycus* Doty and *Eucheuma* J. Agardh (Solieriaceae, Rhodophyta) in Southeast Asia. J Appl Phycol 26:1253–1272

Mateo JP, Campbell I, Cottier-Cook EJ, Luhan MRJ, Ferriols VMEN, Hurtado AQ (2020) Analysis of biosecurity-related policies governing the seaweed industry of The Philippines. J Appl Phycol 32(3):2009–2022. https://doi.org/10.1007/s10811-020-02083-7

Mateo JP, Faisan Jr. JP, Sibonga RC, Suyo JGB, Luhan MRJ, Ferriols VMEN, Hurtado AQ (2021) Farm management and biosecurity measures of Eucheumatoids: cultivars, pest and disease, risks and risk managements. United Kingdom Research and Innovation-Global Challenge Research Fund (UKRI-GCRF); Institute of Aquaculture, University of the Philippines Visayas-GlobalSeaweedSTAR-Philippines

Mendoza WG, Montano NE, Ganzon-Fortes ET, Villanueva RD (2002) Chemical and gelling profile of ice-ice infected carrageenan from *Kappaphycus striatum* (Schmitz) Doty "sacol" strain (Solieriaceae, Gigartinales, Rhodophyta). J Appl Phycol 14:409–418

Msuya FE, Porter M (2014) Impact of environmental changes on farmed seaweed and farmers: the case of Songo Songo Island, Tanzania. J Appl Phycol 26:2135–2141

Ortega A, Geraldi NR, Alam I, Kamau AA, Acinas SG, Logares R, Gasol JM, Massana R, Krause-Jensen D, Duarte CM (2019) Important contribution of macroalgae to oceanic carbon sequestration. Nat Geosci 12:748–754. https://doi.org/10.1038/s41561-019-0421-8

Pang T, Liu J, Liu Q, Lin W (2011) Changes of photosynthetic behaviors in *Kappaphycus alvarezii* infected by epiphyte. Evid Based Complementary Altern 2011:1. https://doi.org/10.1155/2011/658906

Pang T, Liu J, Liu Q, Li H, Li J (2015) Observations on pests and diseases affecting a eucheumatoid farm in China. J Appl Phycol 27: 1975–1984

Pedrosa AA (2017) A regional scientific meeting attaining sustainable development goals: Philippine fisheries and other aquatic resources 20/20. Current status of Philippine seaweed industry (Powerpoint presentation) https://nast.ph/level.php/downloads/category/108–day–1–march–13–2017?download=346:4–plenary–2–mr–pedrosa–iii. Accessed on 04 February 2022

Philippine National Standards/Bureau of Agriculture and Fisheries Standards (PNS/BAFS) (2021) Philippine National Standard Code of Good Aquaculture Practices (GAqP) for Seaweed No. 208. http://www.bafs.da.gov.ph/

Philippine Statistics Office (PSA) (2022). Aquaculture production of seaweeds. https://openstat.psa.gove.ph Accessed on 26 January 2022

Pickering T (2006) Advances in seaweed aquaculture among Pacific Island countries. J Appl Phycol 18:227–234

Porse H, Rudolph B (2017) The seaweed hydrocolloid industry: 2016 updates, requirements, and outlook. J Appl Phycol 29:2187–2200

Shanmugam M, Sivaram K, Rajeev E, Pahalawattaarachchi V, Chandraratne PN, Asoka JM, Seth A (2017) Successful establishment of commercial farming of carrageenophyte *Kappaphycus alvarezii* Doty (Doty) in Sri Lanka: economics of farming and quality of dry seaweed. J Appl Phycol 29(6):3015–3027

Solis MJL, Draeger S, dela Cruz TEE (2010) Marine-derived fungi from *Kappaphycus alvarezii* and *K. striatum* as potential causative agents of ice-ice disease in farmed seaweeds. Bot Mar 53(6):587–594

Sulu R, Kumar L, Hay C, Pickering T (2004) Kappaphycus seaweed in the Pacific: review of introductions and field testing proposed quarantine protocols. Secretariat of the Pacific Community, Noumea, p 84

Syafitri E, Prayitno SB, Ma'ruf WF, Radjasa OK (2017) Genetic diversity of the causative agent of ice-ice disease of the seaweed *Kappaphycus alvarezii* from Karimunjawa Island, Indonesia. IOP Conference Series: Earth and Environmental Science 55(1):012044

Tan TT, Song SL, Poong SW, Ward GM, Brodie J, Lim PE (2020) The effect of grazing on the microbiome of two commercially important agarophytes, *Gracilaria firma* and *G. salicornia* (Gracilariaceae, Rhodophyta). J Appl Phycol 32:2549–2559. https://doi.org/10.1007/s10811-020-02062-y

Therkelsen GH (1993) In: Whistler RL, Bemiller JN (eds) Carrageenan. Industrial gums, 3rd edn. Academic Press, pp 145–180. https://doi.org/10.1016/B978-0-08-092654-4.50011-5

Trono GC, Largo DB (2019) The seaweed resources of The Philippines. Bot Mar 62(5):483–498. https://doi.org/10.1515/bot-2018-0069

Tsiresy G, Preux J, Latriva T, Dubois P, Lepoint G, Eeckhaut I (2016) Phenology of farmed seaweed *Kappaphycus alvarezii* infestation by the parasitic epiphyte *Polysiphonia* sp. in Madagascar. J Appl Phycol 28:2903–2914

Vairappan CS (2006) Seasonal occurrences of epiphytic algae on the commercially cultivated red alga *Kappaphycus alvarezii* (Solieraceae, Gigartinales, Rhodophyta). J Appl Phycol 18:611–617

Vairappan CS, Chung CS, Hurtado AQ, Msuya FE, Lhonneur GB, Critchley A (2008) Distribution and symptoms of epiphyte infection in major carrageenophyte-producing farms. J Appl Phycol 20:477–483

Valderrama D, Cai J, Hishamunda N, Ridler N (2013) Social and economic dimensions of carrageenan seaweed farming. Fisheries and aquaculture technical paper no. 580. FAO, Rome, p 204

Vince Cruz-Abeledo CC, Alvero APS, Erabo DDR (2019) Seaweed biodiversity and temperature fluctuations of Calatagan Bay, Verde Island passage. Journal of Fisheries Science 1(1):26–30. https://doi.org/10.30564/jfsr.v1i1.890

Uyenco FR, Saniel LS, Jacinto S (1981) The ice-ice problem in seaweed farming. In: Levring T (ed) Proc Int Seaweed Symp, vol 10, pp 625–630

Ward GM, Faisan JP Jr, Cottier-Cook EJ, Gachon C, Hurtado AQ, Lim PE, Matoju I, Msuya FE, Bass D, Brodie J (2020) A review of reported seaweed diseases and pests in aquaculture in Asia. J World Aquacult Soc 51(4):815–828. https://doi.org/10.1111/jwas.12649

Yamamoto K, Yoshikawa S, Ohki K, Kamiya M (2012) Unique distribution of epiphytic *Neosiphonia harveyi* (Rhodomelaceae, Rhodophyta) along Sargassacean hosts. Phycol Res 60:70–75

Zuccarello GC, Critchley AT, SmithJ SV, Lhonneur GB, West JA (2006) Systematics and genetic variation in commercial *Kappaphycus* and *Eucheuma* (Solieriaceae, Rhodophyta). J Appl Phycol 18(3–5):643–651

Antimicrobial and Growth-Promoting Properties of Cultured Seaweeds Confer Resistance and Attraction to *Ice-Ice* Disease-Causing Bacteria: A Proposed Seaweed-Bacteria Pathosystem Model

21

Danilo B. Largo, Kimio Fukami, Masao Adachi, Flower E. Msuya, and Masao Ohno

Abstract

Cultivated seaweeds, especially the eucheumatoids and gracilarioids, suffer from frequent incidence of *ice-ice* disease outbreaks, which affect their health, growth, and product quality. Microbial load (both epiphytic and endophytic) is observed to be one to two orders of magnitude higher in seaweeds with ice-ice condition as compared to healthy fronds. Acetone extracts of healthy *Kappaphycus alvarezii* showed inhibition of antibacterial activity of up to 40% of bacterial strains used in the bioassay, consisting of Gram-positive and Gram-negative strains. Microbial inhibition is higher in extracts from healthy cultivars than from cultivars with ice-ice disease. Meanwhile, the filtrates from the culture medium used in culturing the seaweed also showed growth-promoting ability on the same bacterial strains used in the antimicrobial assay. These episodes of growth inhibition and growth promotion of microorganisms in cultivated seaweeds could determine whether or not the seaweed host can be at risk of microbial infection leading to an ice-ice disease. A model of seaweed-microbe pathosystem based on boom-or-bust episodes is therefore proposed for farmed seaweeds, which means that seaweed cultivars that remain healthy protect themselves from bacterial attack through the production of antimicrobial compounds but enhance bacterial growth when they are weak through the release of nutrient-rich exudates.

Keywords

Antimicrobial · Growth-promoting · Ice-ice disease · *Kappaphycus alvarezii* · Seaweed-bacteria pathosystem

21.1 Introduction

Ice-ice disease continues to wreak havoc in the seaweed farms of the Philippines, Indonesia, Tanzania, and many other countries where eucheumatoid species have been introduced during the last six decades. The association of microorganisms in the development of ice-ice disease in eucheumatoids has been confirmed by several studies, particularly those caused by bacteria (Largo et al. 1995b; Riyaz et al. 2020, 2021) and, to a minor extent, those caused by marine fungi (Solis et al. 2010). The interaction of these microorganisms with the seaweed host, particularly with the farmed species *Kappaphycus alvarezii* and *Eucheuma denticulatum* in southeast Asia, has also been previously studied by these authors and several others, as one that reflects a natural relationship between microbes and host in an aquatic environmental change under certain circumstances of seasonal conditions (Largo et al. 1995a). Field observations (Largo et al. 2020) and experimental verification studies (Largo et al. 1995a) show that stressful environmental factors, such as high seawater temperature occurring during dry summer months and low salinity during the wet season, have been found to promote the development of ice-ice disease in the farm, and in the presence of naturally occurring opportunistic pathogens (Largo et al. 1995b). The farmed *Eucheuma/Kappaphycus* cultured in the Philippines was observed to harbor bacteria in the range of 10^3–10^5 CFU g^{-1} wet wt. for healthy thalli and from 10^6 to 10^7 CFU g^{-1}

D. B. Largo (✉)
Marine Biology Section, Department of Biology, School of Arts and Sciences, University of San Carlos, Cebu City, Philippines
e-mail: dblargo@usc.edu.ph

K. Fukami
Kochi Study Center, The Open University of Japan, Kochi, Japan
e-mail: fukami-k@ouj.ac.jp

M. Adachi
Laboratory of Aquatic Environmental Science, Faculty of Agriculture, Kochi University, Nankoku City, Kochi, Japan
e-mail: madachi@kochi-u.ac.jp

F. E. Msuya
Zanzibar Seaweed Cluster Initiative, United Republic of Tanzania, Zanzibar, Tanzania

M. Ohno
The Seaweed Resources Research Institute Co., Ltd, Kochi, Japan

A. T. Critchley et al. (eds.), *Tropical Phyconomy Coalition Development*, Developments in Applied Phycology 11,
https://doi.org/10.1007/978-3-031-47806-2_21

wet wt. for ice-ice diseased thalli (Largo et al. 1995b). A subsequent field survey of bacteria associated with the farmed eucheumatoids in Philippine seaweed farms also showed a relatively high percentage (59%) of bacterial isolates from ice-ice-diseased thalli compared to 41% from healthy thalli (Butardo et al. 2003). The survey further showed that out of 165 bacterial isolates, 10 potential isolates showed significant agarolytic and carrageenolytic activities, indicating some potentially pathogenic bacteria just lurking among non-pathogenic forms. The elucidation of this natural relationship between microorganisms and seaweed host is a knowledge gap that remains to be resolved amidst the several findings of microorganisms associated with carrageenophytes (e.g., Syafitri et al. 2017; Kopprio et al. 2021; Ward et al. 2021) that have harmed the carrageenan industry for the past several decades amidst a backdrop of global warming and climate change.

We observed that ice-ice disease not only occurs in eucheumatoids but is also reported in the fleshy red alga *Gracilaria* spp. (Lavilla-Pitogo 1992; Correa and Flores 1995; Weinberger et al. 1997; Nasmia et al. 2014; Zainuddin et al. 2019), which they also refer to as "white rot disease" or "rotten thallus syndrome" or simply as "whitened thallus". Here, we propose a natural defense mechanism carried out by the seaweed host and evidence of an antimicrobial property against colonizing microorganisms that could predispose the seaweed to ice-ice disease, based on the available information and our personal works.

Several seaweed species have been generally found to produce antimicrobial compounds, which have practical applications in reducing microbial contamination in food (Cabral et al. 2021), but the extent of this ability to ward off microorganisms from the seaweed host that could otherwise cause the ice-ice condition is not well-known and is the main subject of this chapter.

21.2 Seaweed Culture and Determination of Ice-Ice Index

Data from the culture of *K. alvarezii*, *E. denticulatum*, and *Gracilaria* spp. (*G. changii*, *G. heteroclada*, *G. lemaneiformis* and *Gracilaria* sp.) conducted in Tosa Bay and a semi-outdoor concrete tank at the Usa Marine Biological Institute, Kochi University, Japan, were used to demonstrate the population dynamics of bacteria under different growth conditions of the cultivars. To determine the health of seaweed, cultivars of each species were vertically hung from a floating platform inside the said bay and in a land-based concrete tank. Algal growth was monitored every 1–2 weeks for both field and tank cultivars. Measurements of daily growth rate (DGR) followed the procedure described in Ohno et al. (1994) and Largo et al. (1995a). Ice-ice index is measured as the degree of occurrence of whitened parts using the formula:

$$\text{Ice}-\text{ice Index} = \frac{\text{No.of ice}-\text{ice diseased parts}\times\text{No.of diseased cultivars}}{\text{Total No.of cultivars}} \times 100$$

Results of the seaweed culture showed that the growth of *Kappaphycus alvarezii* and *Eucheuma denticulatum* followed a "boom-bust" cycle during summer-autumn period, with peak growth observed in summer when water temperature in the field reached 28 °C, and slow to negative growth towards autumn when water temperature declined to 14 °C (data not shown). The growth of both species in the tank, which had comparatively lower water exchange, did not perform well, growing only to <3% d^{-1} and 5% d^{-1} or less in summer for *K. alvarezii* and *E. denticulatum*, respectively, compared to cultivars in the field, which grew up to 8% and 5% d^{-1}, respectively (Figs. 21.1a, f and 21.2a, e). In the case of *Gracilaria* species, cultivars of *G. lemaneiformis* registered a peak growth of 10% d^{-1} in late autumn (November), while *G. changii* and *G. heteroclada* (originally from Vietnam) both showed maximum growth of 7% d^{-1} in summer (August–September). *Gracilaria* sp., on the other hand, had no marked seasonal pattern of daily growth rate, which managed to grow up to 5% d^{-1} (Fig. 21.3f).

Occasional outbreaks of ice-ice disease were observed in all the species during lowest or declining growth rates in both field and tank cultures, with the highest ice-ice index of 100% recorded in all species. *G. changii* was observed to develop ice-ice from December to January and in both summer and winter seasons for *Gracilaria* spp. (Fig. 21.3g). Ice-ice resulted in premature fragmentation of *Gracilaria* spp., whereas thallus loss in *G. lemaneiformis* was due to intense grazing by amphipods rather than ice-ice disease, as reflected in the irregular growth rate of this species in the tank cultivation (Fig. 21.4f).

In general, bacterial numbers in non-growing season of the seaweeds were higher than those in the growing season. Temperature plays a major influence on the growth performance, especially of these species of tropical origin, i.e., a high growth rate around summer and a low growth rate observed towards the winter season. As most, if not all, of the algae were incapable of tolerating low temperatures in winter, a total collapse with ice-ice disease appearance on all the cultivated species occurred. A distinction between ice-ice as a result of bacterial infection and that as a result of winter die-back is difficult to make, as both conditions (especially in an advanced stage of the disease) in the field and tank cultivars appeared to be similar. Therefore, ice-ice that occurred during the growing season for *K. alvarezii*,

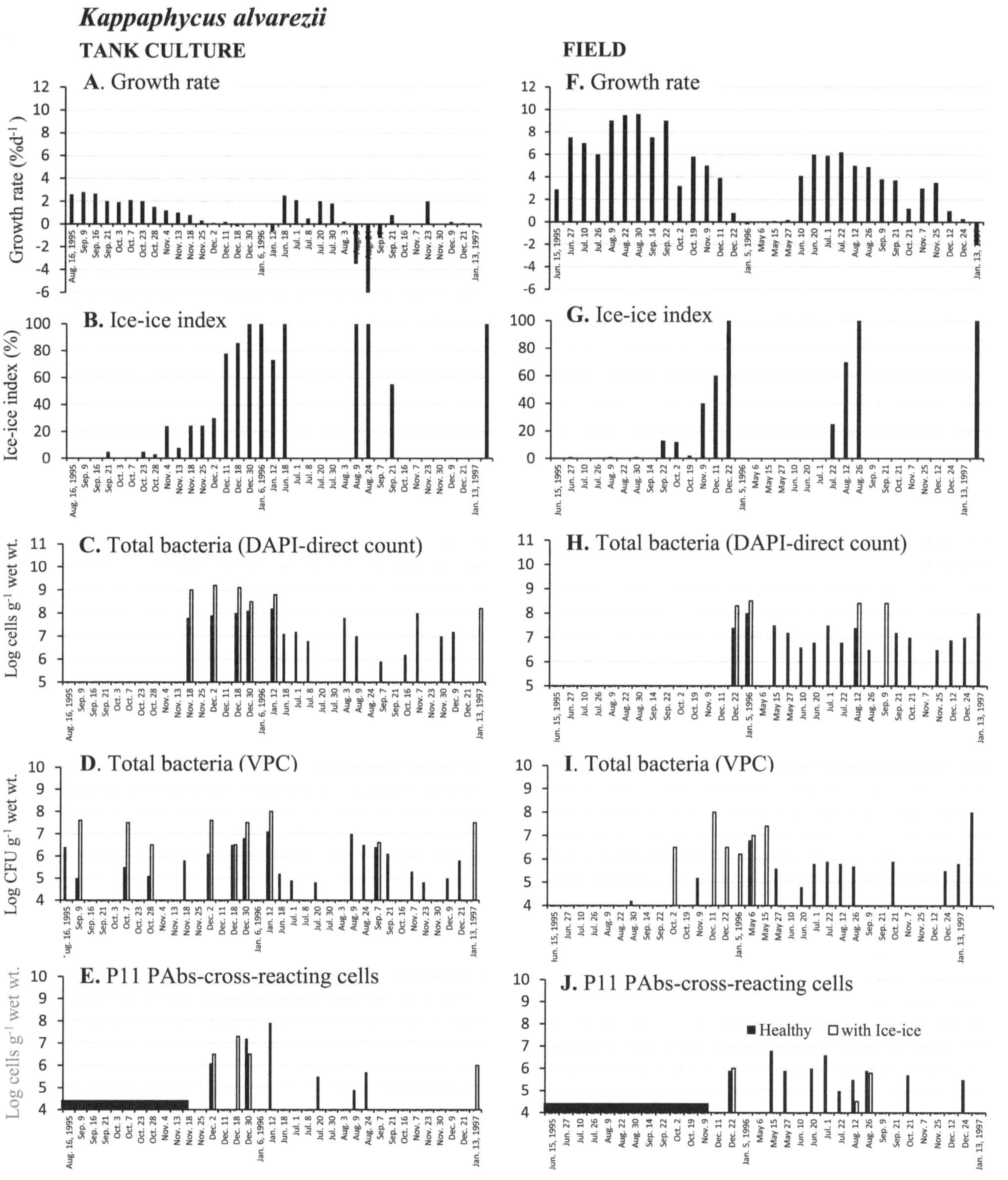

Fig. 21.1 Daily growth rates and ice-ice index of seaweeds and population dynamics of heterotrophic bacteria and native form of bacteria similar to strain P11 during culture of *Kappaphycus alvarezii* in a semi-outdoor concrete tank and in the field (Uranouchi Inlet). Dates with horizontal bar mean polyclonal antibodies (PAbs) against ice-ice-inducing bacterium *Vibrio* sp. P11 has not been applied yet. *CFU* colony-forming unit, *DAPI* 4′,6-diamidino-2-phenylindole fluorescent dye used for direct counting, *VPC* viable plate count. Legend in panel J applies to all graph panels

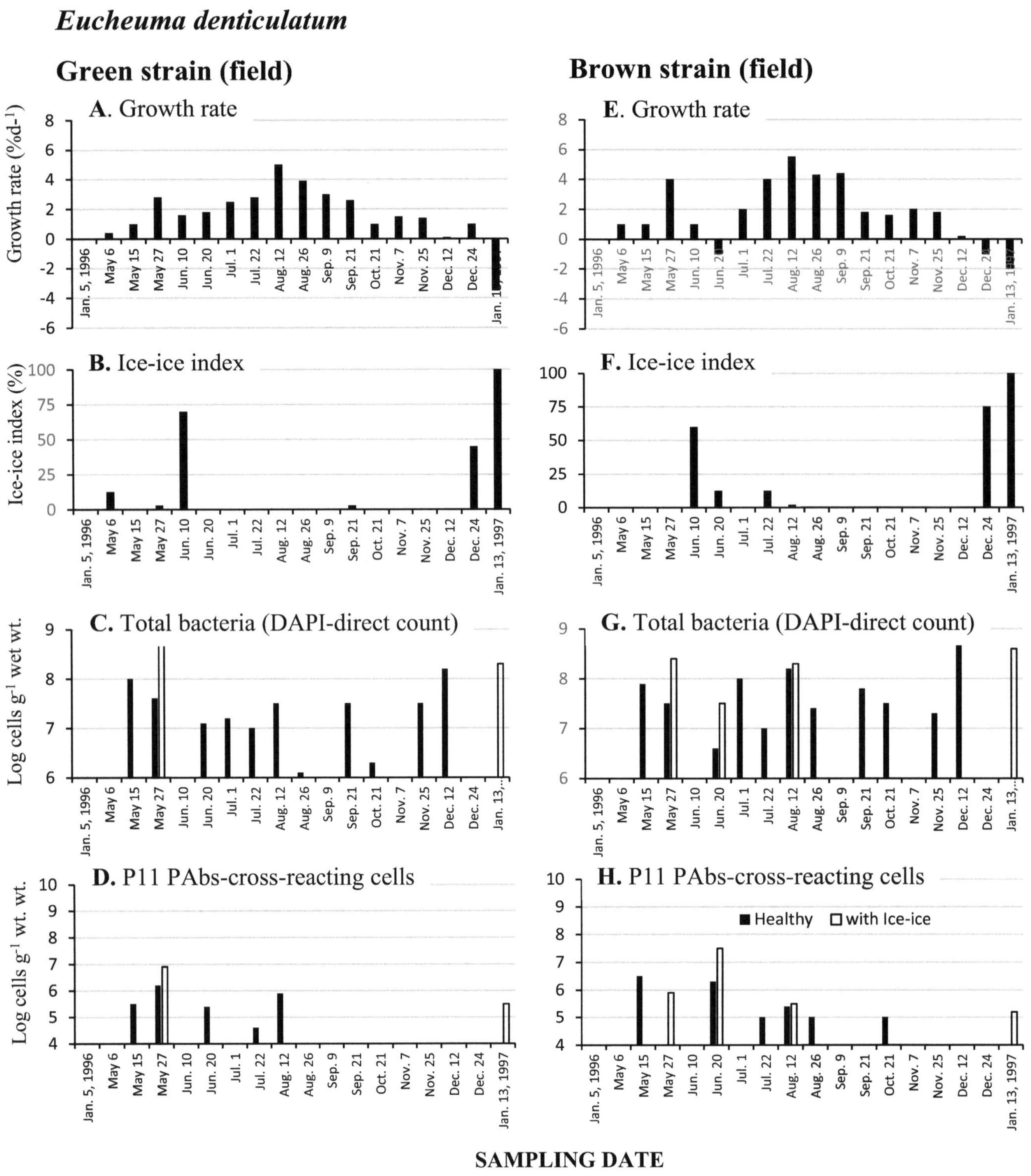

Fig. 21.2 Daily growth rates and ice-ice index of seaweeds and population dynamics of heterotrophic bacteria and native form of bacteria similar to strain P11 during culture of *Eucheuma denticulatum* in Tosa Bay. *DAPI* 4′,6-diamidino-2-phenylindole fluorescent dye used for direct counting, *VPC* viable plate count. Legend in panel H applies to all graph panels

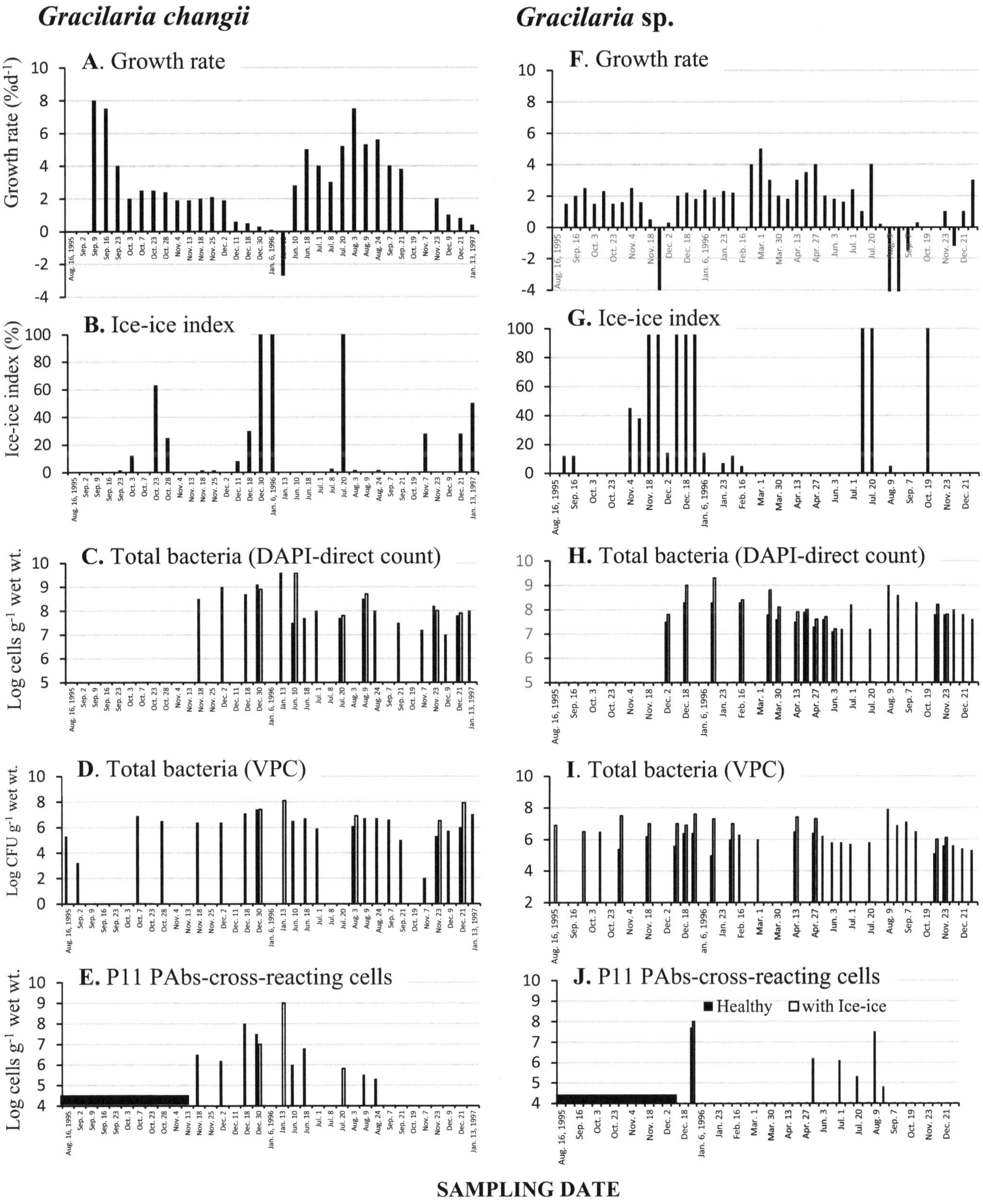

Fig. 21.3 Daily growth rates and ice-ice index of seaweeds and population dynamics of heterotrophic bacteria and native form of bacteria similar to strain P11 during culture of *Gracilaria changii* and *Gracilaria* sp. in a semi-outdoor concrete tank. Dates with horizontal bar mean polyclonal antibodies (PAbs) against ice-ice-inducing bacterium *Vibrio* sp. P11 has not been applied yet. *CFU* colony-forming units, *DAPI* 4′,6-diamidino-2-phenylindole fluorescent dye used for direct counting, *VPC* viable plate count. Legend in panel J applies to all graph panels

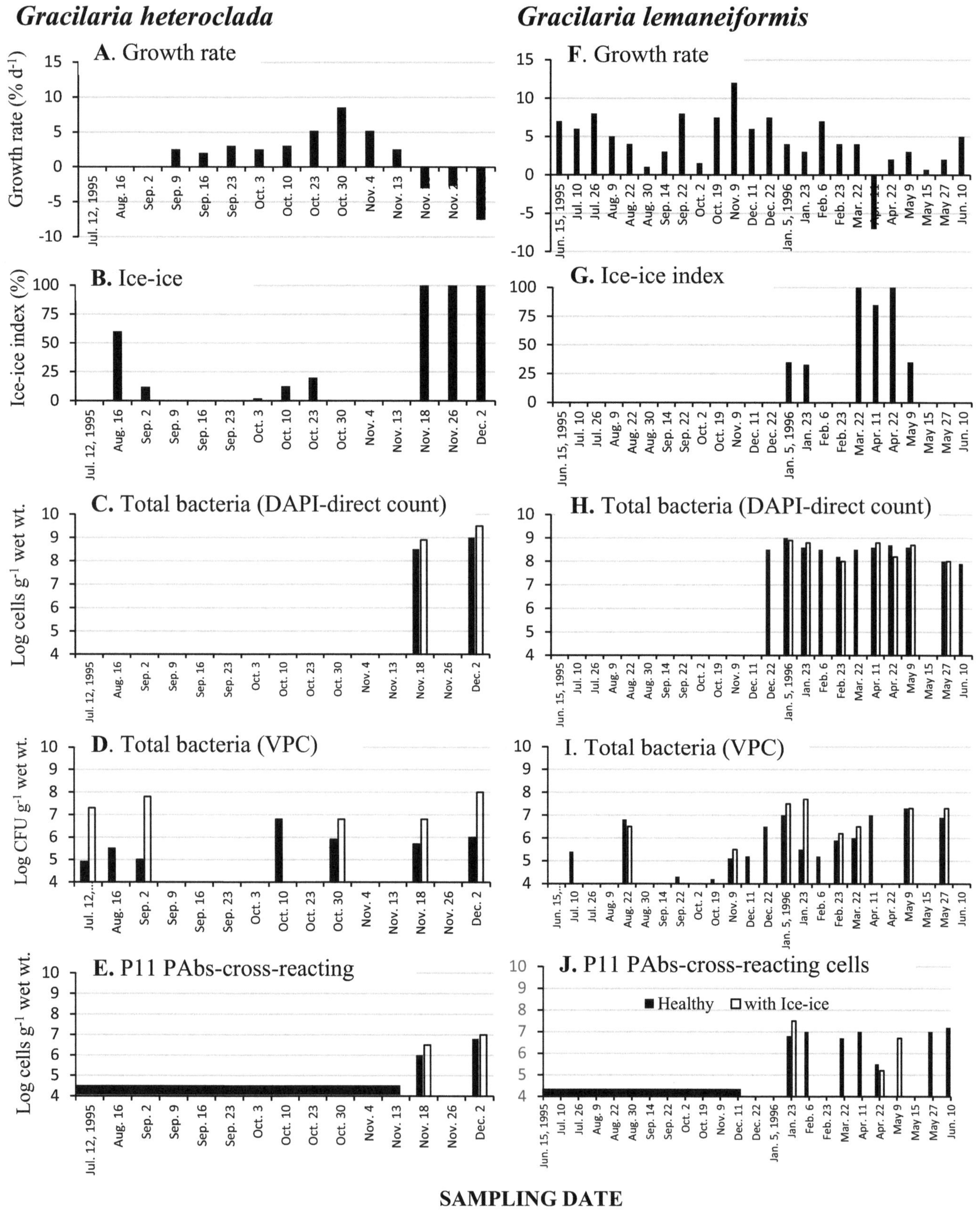

Fig. 21.4 Daily growth rates and ice-ice index of seaweeds and population dynamics of heterotrophic bacteria and native form of bacteria similar to strain P11 during culture of *Gracilaria heteroclada* and *G. lemaneiformis* in a semi-outdoor concrete tank. Dates with horizontal bar mean polyclonal antibodies (PAbs) against ice-ice-inducing bacterium *Vibrio* sp. P11 has not been applied yet. *CFU* colony-forming units, *DAPI* 4′,6-diamidino-2-phenylindole fluorescent dye used for direct counting, *VPC* viable plate count. Legend in J applies to all graph panels

E. denticulatum, *G. changii*, and *G. heteroclada* (i.e., late spring to early autumn) is suggested to be microbially induced, while ice-ice occurring during the non-growing season (late autumn to early spring) could be considered as low-temperature induced. Since little annual variation in temperature is expected in tropical waters, from where most of the algal species were obtained, opportunistic pathogenic bacteria could play a significant role in the development of ice-ice disease in these waters. Although bacterial cells cross-reacting with polyclonal antibodies (PAbs) against P11 (based on the method described in Largo et al. 1998), a pathogenic strain (Largo et al. 1995b), were detected in seaweed thalli cultured in Tosa Bay throughout the year, the physiological activity of this strain cannot be ascertained as far as the field/tank data are concerned. Based on 0.3% ferric oxide-trypticase-yeast agar (FeTY agar) culture (Fukami et al. 1992), P11 has a range of temperature requirements from 15 to 30 °C for growth (data not shown). Therefore, the condition of P11 occurring during winter in the field, where temperatures could reach 14 °C or below, could be in a critical physiological threshold. This bacterium could be inactive during the winter season, and it is likely that psychrophilic or psychrotolerant bacteria are the ones that are more active.

In experiments investigating the behavior of P11 between stressed and non-stressed cultivars during *in vitro* inoculation of this strain (Largo et al. 1999), it was clear that growth of this bacterium is stimulated when *K. alvarezii* was under environmental stress. The PAbs, which positively detected the specific bacteria in both healthy and diseased cultivars in all seaweed species investigated, suggest that native bacteria like P11 are normally associated with these seaweeds. However, the results of the field and tank culture showed that the development of ice-ice in the field during stressful season did not correlate well with the observed abundance of P11, as it was observed even during actively growing season of the seaweeds (e.g., field-cultivated *K. alvarezii*; Fig. 21.1). Although positively reacting bacteria with PAbs were not necessarily P11, the result that there was no apparent relationship between PAb-binding bacteria and ice-ice disease suggests that these native bacteria are facultative or opportunistic pathogens in promoting ice-ice disease.

21.3 Determination of Bacterial Population Dynamics

Seaweed samples for bacterial enumeration, obtained simultaneously with samples for the growth measurements, were taken aseptically from healthy and ice-ice diseased parts of cultivars from both field and tank setups. They were then brought to the laboratory within 30 minutes of collection and processed immediately to maintain the microbial population's integrity. The abundance of culturable bacteria as viable plate count (VPC) was determined by the agar spread plate method as previously described (Largo et al. 1995b), while the number of bacterial colonies was determined by a combined direct counting method with 4′6-diamidino-2-phenylindole (DAPI) for total number of bacteria and immunofluorescence antibody (IFA) method as described in Largo et al. (1997), specifically for *Vibrio* sp. P11, which has been shown experimentally to trigger ice-ice condition in *K. alvarezii*. Duplicate samples, each with two subsamples, were counted separately. A minimum of 150 microscopic fields per sample were counted for the number of *Vibrio* sp. P11. If no such cells were detected, the sample was marked as negative for native P11.

Total bacteria on seaweed tissues using DAPI direct counting method showed density to range from 10^6 to 10^8 cells g^{-1} wet wt. in both *K. alvarezii* and *E. denticulatum* during the growing periods of both species, whereas total bacterial counts in ice-ice diseased parts of both species were one to two orders of magnitude higher, at 10^7–10^9 cells g^{-1}, than those of healthy cultivars (Figs. 21.1c, h and 21.2c, g). Viable plate counts (VPC) using the agar spread plate method for *K. alvarezii* (no data available for *E. denticulatum*) had bacterial counts one to two orders of magnitude lower than the DAPI direct count (VPC = 10^5–10^8 CFUs g^{-1}), indicating that the majority of the cells can still be active at any time while the seaweed is in culture. In the case of *Gracilaria* sp., bacterial densities were observed to be higher in healthy cultivars in all *Gracilaria* species than in the healthy eucheumatoids, ranging from 10^7–10^9 cells g^{-1} wet wt. (VPC = 10^4–10^7 CFUs g^{-1}), whereas ice-iced diseased cultivars of *Gracilaria* spp. had bacterial density slightly higher, but still within a range of 10^7 to 10^9 cells g^{-1} wet wt. (VPC = 10^6–10^8 CFUs g^{-1}) of algal tissue.

Polyclonal antibodies (PAbs) prepared against strain P11 detected native bacteria in both healthy and ice-ice diseased cultivars of all seaweed species grown in tank and field at any season (Figs. 21.1e, j, 21.2d, h, 21.3e, j, and 21.4e, j). (The polyclonal antibodies against P11 became applicable only after November 18, 1996, when the antiserum was developed; therefore, dates prior to this period have no data to show for P11.) P11 PAbs cross-reacting cells occurred more often in healthy parts, especially during non-growing season of the species. However, their numbers from ice-ice diseased parts were mostly higher than those in healthy parts. Approximately 77% of ice-ice diseased parts found to have PAbs-cross-reacting cells had a higher number of the latter than those from healthy parts (approx. 23%; Figs. 21.1e, j, 21.2d, h, 21.3e, j, and 21.4e, j).

21.4 Antibacterial Property of *K. alvarezii* During Growing and Non-growing Period

Dried samples of *K. alvarezii* collected monthly from the above-mentioned field culture during the peak growing period in summer and during the growth-declining period in autumn were used for the extraction of the antibacterial compound using organic solvents. Of the four solvents initially tested, namely, diethyl ether, methanol, ethanol, and acetone, acetone proved to be a more efficient solvent than the others and was therefore used in the subsequent extraction. Ten (10)-g dried samples were cut into fine pieces, placed each into a 100-mL Erlenmeyer flask, and 100 mL of acetone (reagent grade) was poured into it, sealed with paraffin film, and placed in a rotatory shaker at a speed of 100–120 rpm for 48 h. The crude extracts were then separated from the seaweed debris by filtration through a GF/C Whatman glass filter under vacuum. All filtrates were evaporated under low pressure until only a 1-mL volume remained. The crude extracts were temporarily stored at −80 °C until use.

The crude extract was then assayed for antibacterial activity against a total of 24 bacterial pathogens (3 Gram-positives: P1, P67, *Escherichia coli*; and 21 Gram-negatives: P2, P10, P11, P12, P25, P50, P57, P77, P81b, P97, P100, P108, JPNCLSE-1, JPNCLSE-2, *Staphylococcus epidermidis*, *Pseudomonas fluorescens*, *Aeromonas hydrophila*, *Vibrio (Listonella) anguillarum*, *V. parahaemolyticus*, *V. fisheri*, *E. coli* ATCC 11775). Fourteen of these strains included the suspected ice-ice-promoting bacterium (*Vibrio* sp. P11) originally isolated from a seaweed farm in Bohol, Philippines, two strains isolated from seaweeds experimentally cultured in southern Japan (JPNCLSE-1, JPNCLSE-2), and six known fish pathogens (*Staphylococcus epidermidis*, *Ps. fluorescens*, *Aeromonas hydrophila*, *Vibrio anguillarum*, *V. parahaemolyticus*, *V. fischeri*) from a culture collection, were also included for comparison. Each bacterial strain was pre-cultured in 0.3% FeTY agar. After 24–48 h, a loopful of this culture was transferred into liquid FeTY medium (pH 7.5) and grown to exponential phase. One hundred microliter of each culture was spread-plated on 1.5% agar. Three-millimeter diameter discs made of No. 1 Whatman filter were soaked in 10 μL of the seaweed extract, while some discs soaked in solvent only served as a negative control. The discs were allowed to dry in the air and were aseptically placed into the bacterial lawn of the agar plate (in duplicates). For comparison, commercially available antibiotic discs (Penicillin G, 50 μg/disc, Streptomycin, 50 μg, Kanamycin, 50 μg, and Polymyxin B, 100 units/disc, all manufactured by Showa Yakuhin Kako & Co. Ltd, Tokyo, Japan) were added into the same bacterial lawn at 1–2 discs per plate to serve as positive control. Evaluation of the antibacterial activity was made after 24 h, when bacterial colonies were visible enough on the agar plate. Zones of inhibition, shown as a clear hallow around each disc, were measured using a ruler and duplicate values (in mm) for each extract and control were averaged.

Crude extracts of *K. alvarezii* showed antibacterial property against several bacterial pathogens ranging from 5 to 40% of the strains assayed being inhibited, depending on the date the seaweed was harvested during its culture period in Tosa Bay (Fig. 21.5). No zone was observed in the control. The highest percentage (40%) of strains was inhibited by the extract from young branches of seaweed material collected on September 9, 1995. In comparison, the same bacterial strains assayed showed that 40–90% of the strains were inhibited by standard antibiotics, with Kanamycin and Streptomycin inhibiting most of the bacterial strains.

Based on measurements of the zone of inhibition, the antibacterial activity from the *K. alvarezii* extracts ranged from an average of 0.5–4 mm. These were relatively narrow when compared to those obtained from standard antibiotics, which ranged from 0.5 to 23 mm. Crude extracts from field-cultured seaweed showed fewer than 17% (4 out of 24 strains) activity (Fig. 21.6a), while 50% (12 out of 24 strains) were observed to be sensitive to the crude extracts from tank-cultured seaweed (Fig. 21.6b). In contrast, all 24 bacterial strains (100%) reacted to the commercial antibiotics (Fig. 21.6c). Except for a slight inhibition (1 mm) against *Vibrio* sp. P11 by the seaweed extract from the young tips of *K. alvarezii* collected on Septtember 9, 1995, extracts from different collecting dates had generally no, if not very weak,

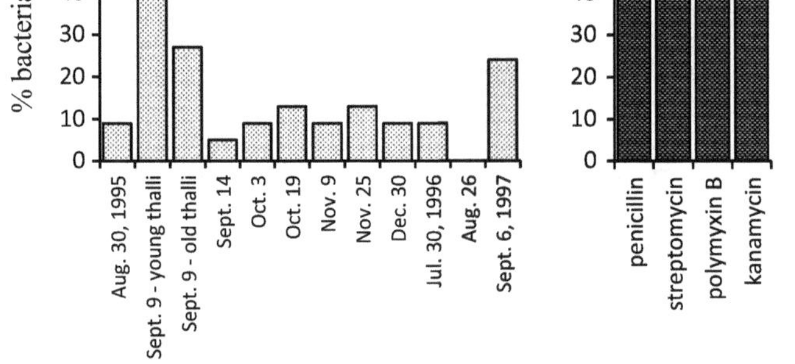

Fig. 21.5 Bioassay of bacteria showing percentage of strains inhibited by acetone crude extract from *K. alvarezii* harvested at different sampling dates from 1995 to 1997 (**a**) in comparison to commercial antibiotics (**b**)

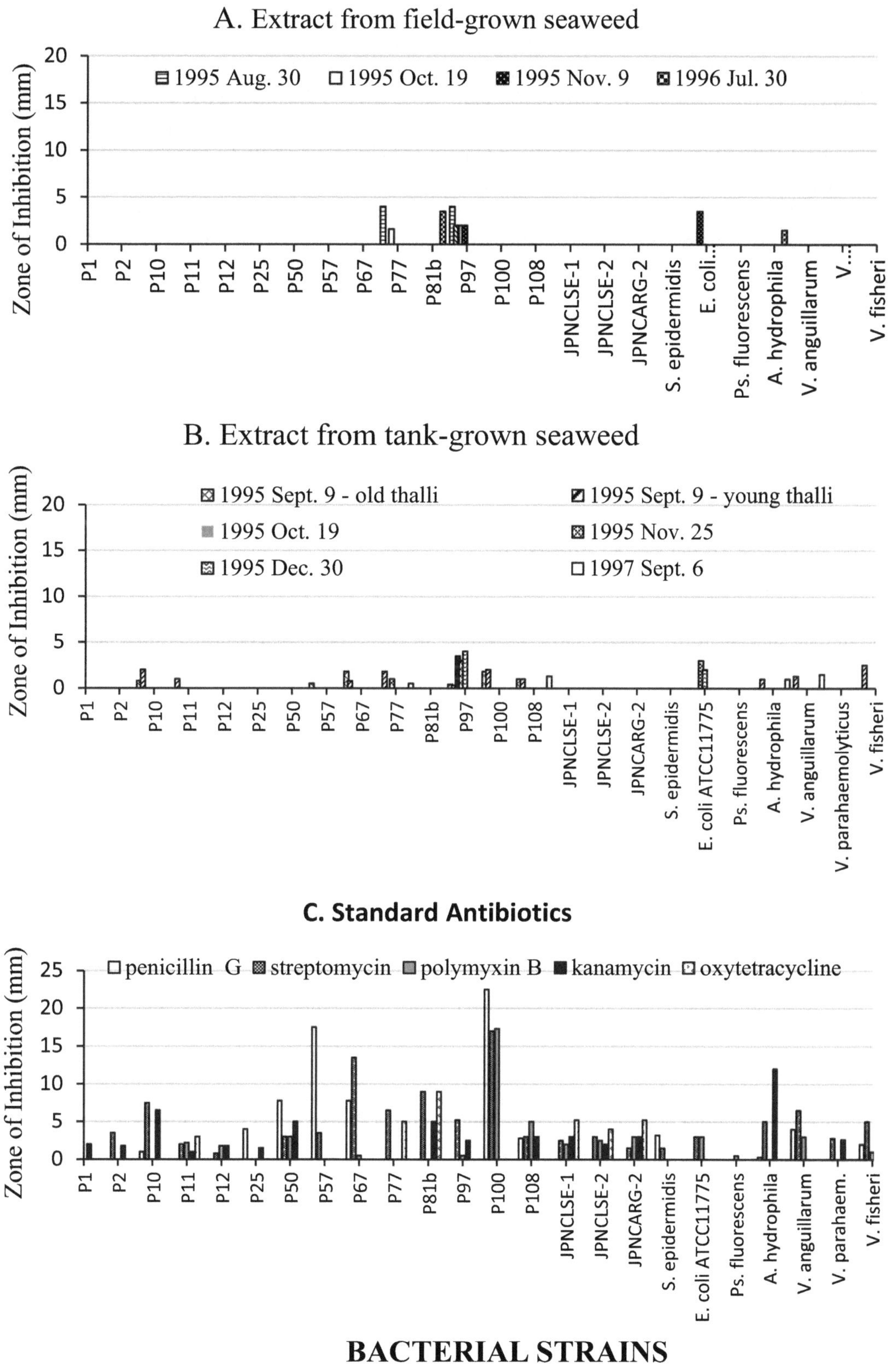

Fig. 21.6 Antibacterial activity based on zone of inhibition around sensitivity discs containing acetone crude extract of *Kappaphycus alvarezii* collected from field (**a**) and tank (**b**) at various dates, compared to commercial antibiotics (**c**) against several bacterial strains originally isolated from *Kappaphycus*/*Eucheuma* ("P-strains") and fish. Discs containing acetone solvent only (control) did not show growth inhibition of the same strains (data not shown)

inhibition against this bacterium. Most of the Gram-negative bacteria, including strains originally isolated from eucheumatoids (P10, P77, P81b, P97, P108) and fish pathogens (*V. fischeri*, *V. anguillarum*, *V. hydrophila*, and *E. coli* ATCC 11775) were inhibited. This could be attributed to Gram-negative bacteria's outer membrane, believed to produce certain exogenous elicitors that trigger macroalgae like *E. denticulatum* (Mtolera et al. 1995, 1996) and *Solieria chordalis* (Ar Gall et al. 2008) to release defensive compounds like hydrogen peroxide. Of the three Gram-positive bacteria isolated from the Philippine-grown seaweed (P1, P67, P81b), only P67 was inhibited, which could be due to an absence of such elicitors found in Gram-positives. Other defensive mechanisms, such as the production of anti-quorum sensing compounds found in some micro- and macroalgae, cannot be also discounted.

21.5 Growth-Promoting Property of Spent Culture Medium Used for *K. alvarezii* in Batch System

Common observations that algae in general have growth-promoting properties, as reviewed by Goecke et al. (2013) and Kouzuma and Watanabe (2015), have not been tested in eucheumatoids for its associated bacteria, which normally thrive on low-molecular-weight organic compounds. To test the ability of *K. alvarezii* to promote growth of bacteria associated with it, filtrates of spent culture media from *K. alvarezii*, which are basically made of autoclaved seawater and collected after certain days of algal culture (1 day, 1 week, 2 weeks and 4 weeks), were inoculated with ten (10) different bacterial strains, namely P1, P11, P14, P19, P25, P50, P57, P77, P81b, and J5. The culture media were first filtered through a 0.2 μm pore size Nuclepore membrane filter under vacuum. Half of the media was un-supplemented with any organic nutrient, while the other half was supplemented with low molecular weight organic compounds consisting of 100 μM D-glucose, 100 μM sodium acetate, and 1 mM each of amino acid mixture containing alanine, glycine, serine, and leucine. This is to find out the bacterial response between the un-supplemented filtrate (plain seawater with only the released algal metabolic and/or exudative products) and the supplemented filtrate. The filtrates with supplements are designated here as SSW (+) ("seaweed-seawater plus"), while filtrates without supplements are designated as SSW (−) ("seaweed-seawater minus"). All filtrates were dispensed into 5-mL screw-capped glass tubes before bacteria were inoculated. Forty-eight-hour cultures of the bacterial strains in their late logarithmic phase were each inoculated into the tubes, and their relative growth, as a function of time of up to 12 days, was measured with an absorbance meter at 460 nm wavelength.

Results showed that the un-supplemented filtrate (SSW-) from spent *K. alvarezii* culture media promoted the growth of all ten selected bacterial strains assayed, namely: P1, P11, P14, P19, P25, P50, P57, P77, P81b, and J5 (Fig. 21.7). However, relative growth varied with the age of the culture medium used. Filtrate from one-day-old culture medium showed delayed growth in 6 out of 10 bacterial strains tested, including strain P11—a suspected ice-ice-promoting bacterium. Other bacterial strains (e.g., P1, P25) that were not immediately promoted (1–3 days after inoculation) in 1-day-old filtrate showed early growth promotion (i.e., immediately one day after inoculation) in 1-week, 2-week, and 4-week-old un-supplemented filtrates. Supplementation of filtrates (SSW+) from all culture media of different ages with low molecular weight organic compounds seemed to promote growth early in all bacterial strains being assayed (Fig. 21.7), and such growth was sustained up to 11 days of incubation in the filtrate for all bacterial strains except for P19, which stopped growing after 10 days in culture.

The bioassay of filtrates and crude extracts of seaweeds against several strains of bacteria supported the generally held view of the growth-promoting and inhibiting properties of seaweeds on associated microorganisms, and this also holds true for the eucheumatoids and gracilarioids on bacteria. All the bacterial strains assayed were promoted by the seaweed extracellular products even without organic supplementation. These extracellular products include secondary metabolites and exopolymers exuded onto the surface of the alga (Goecke et al. 2010). However, since the time at which growth was inhibited varied with the age of the seaweed culture medium from which the filtrates were obtained for the assay, it is suggested that metabolic products released by the seaweed may vary over time in terms of both chemistry and quantity. Obviously, filtrate from a 1-week-old culture promoted all the bacterial strains, including P11, earlier and later than other filtrate ages. Since the growth of all bacterial strains improved when the filtrates were supplemented with low-molecular-weight organic carbon, it is suggested that nutrients in the supplement could also be derived in a more utilizable amount in the filtrates from the 1-week-old seaweed culture than from those of other culture ages. Jones (1982) suggested that the release of specific algal extracellular products, such as photosynthetically-derived organic carbon (PDOC), to which many bacteria are likely to respond, may vary with the phase of growth and environmental conditions. Fukami et al. (1990) also reported that the extracellular organic carbon (EOC) released by phytoplankton may vary with season, depending on the predominant species, and that such changes can have an effect on a "selective media" for the bacterial community. Furthermore, Riquelme et al. (1989) also observed that the EOC released by the phytoplankton, *Asterionella glacialis*, selectively stimulated and inhibited bacterial growth. The findings from these

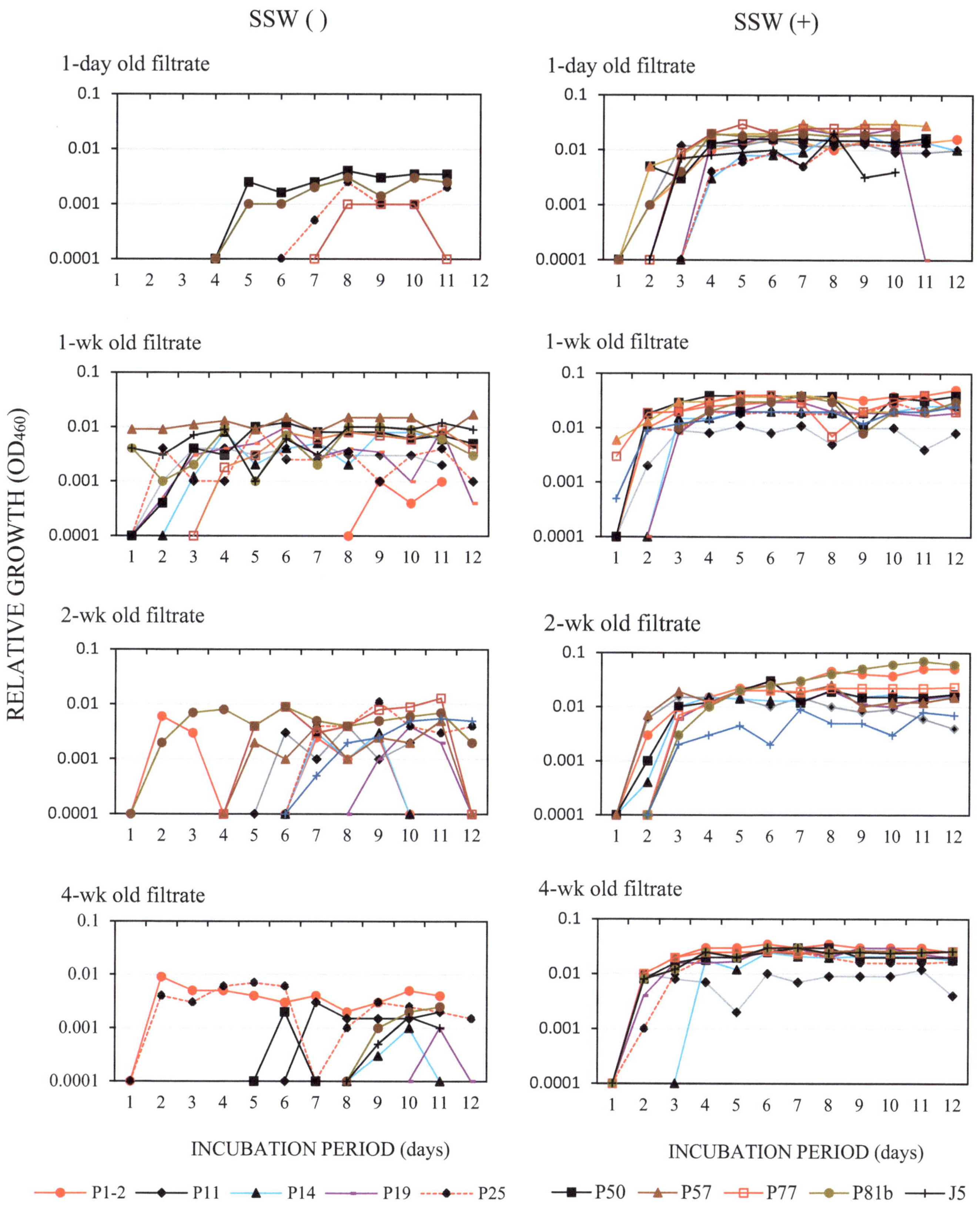

Fig. 21.7 Relative growth (OD_{460}) of selected bacterial strains on filtrates of spent seawater culture medium from *Kappaphycus alvarezii* at various culture ages without (SSW−) and with supplementation (SSW+) of low-molecular-weight organic compounds. Legend at the bottom applies to all graph panels

earlier studies suggested the significant influence of EOC on the successional changes in bacterial communities. On the other hand, with increasing limitation of nutrient in older cultures, algae may release exudates from autolytic algal cells rather than PDOC, which could explain the reduction in the number as well as the relative growth of strains in the filtrate from a 2- to 4-week-old seaweed culture medium. The growth of P11 diminished in filtrates from cultures older than 2 weeks, probably either due to exhaustion of specific algal products it requires or susceptibility against the antibacterial products in algal exudates. This is verified by the decline in the growth of P11 after more than a week in batch culture (Largo et al. 1999).

The antibacterial property in the solvent extract of *K. alvarezii* was demonstrated against several strains of bacteria. Since there was only one occasion that the crude extract of *K. alvarezii* showed inhibition against the known ice-ice promoting bacterium *Vibrio* sp. P11, it shows therefore that the extract has a weak inhibitory activity against this bacterium (Fig. 21.6b). This finding, therefore, substantiates the earlier observation of the ability of *K. alvarezii* to support the growth of P11, which, in turn, enhances its pathogenic ability to cause the ice-ice disease under stressful conditions (Largo et al. 1995a). It can be suggested that strain P11's ability to promote the ice-ice disease hinges not only on its ability to resist against the seaweed's antibacterial activity but also on the possibility that *K. alvarezii* may reduce its own ability to produce an antibacterial factor against most bacteria during extremely stressful periods. The variation in antibacterial activity may also depend on the condition of the seaweeds from which the extracts were obtained. Variability of antimicrobial substances with respect to season and growth condition was observed in the antimicrobial assay of extracts of seaweed harvested at various time intervals during the culture period (Fig. 21.6a, b). Based on the review by Pesando (1990), the production of antimicrobial substances by the same species is variable, which may depend on the collecting sites, the stage of the alga's active growth, as well as their sexual maturity. Collens and Pedersen (1990), in a separate study on the same seaweed species, attributed the ice-ice problem in *Eucheuma* to the production and accumulation of halogenated compounds during stressful conditions resulting when water exchange is low. Although this might hold true in seaweeds grown in stagnant water, wherein a build-up of this compound is possible, their study may not be able to explain the fact that ice-ice prevails even in a well-flushed coral reef–seaweed farm area (D. Largo, personal observation). Furthermore, contrary to the findings of this study, the release of brominated and hydrogen peroxide compounds observed by Collens and Pedersen (1990) and by Mtolera et al. (1995, 1996) during stressful conditions of the seaweed appears to have no inhibitory effects on P11 as well as on other bacterial strains, especially the Gram-positives. The primary involvement of opportunistic biological agents capable of producing carrageenan-lysing enzymes, such as *Vibrio* sp. P11, is therefore a more plausible explanation of the ice-ice disease development.

21.6 Proposed Model for Seaweed-Bacteria Pathosystem

This review provides evidence of the interaction between bacteria and cultured seaweeds, particularly the eucheumatoids and gracilarioids, which are the sources of carrageenan and agar, respectively. At least one species of seaweed (*Kappaphycus alvarezii*) was demonstrated, through a bioassay of its extract, to produce antimicrobial factors that prevent the growth of some bacterial strains. This could be an indication that the seaweed has a natural defense mechanism against potential bacterial colonizers, which is a chemical interaction between a macroalga and bacteria (Goecke et al. 2010). While this may be the case, seaweeds have also been shown here to produce substances that promote the growth of bacteria inoculated to filtrate from spent culture medium of *K. alvarezii* and that this growth-promoting ability depends on the length of time the medium was used to culture the seaweed, such that a one-day-old culture medium has limited growth-promotion ability compared to a 1-week, 2-week, or 4-week-old medium. This would indicate that *K. alvarezii* releases extracellular enzymes that promote the growth of bacteria and that the longer the medium has been used to culture the seaweed, the more the amount of growth-promoting substances the seaweed has accumulated over a period of time, as shown by the sustained growth of most bacteria for up to 11–12 days in a 1-week-old culture medium. However, the growth of some bacterial strains is short-lived in a non-supplemented filtrate with low-molecular weight organic compounds, in contrast to when filtrates were supplemented with low molecular weight organic compounds, which sustained bacterial growth even in a 1-day-old culture medium. The release of extracellular substances is usually an indication that the seaweed is under nutrient limitation—a kind of stress that weakens the seaweed's defenses against bacterial colonization. Based on these observations as a premise, we propose a seaweed-bacteria pathosystem model that could explain the development of ice-ice disease in eucheumatoids (Fig. 21.8). The model explains that healthy seaweeds have a natural protection against bacterial colonization by the production of antimicrobial compounds in the form of secondary metabolites (e.g., brominated compounds, Collens and Pedersen 1990; Mtolera et al. 1996; hydrogen peroxide, Mtolera et al. 1995), but under prolonged stressful conditions for algal growth, they lose this ability to prevent bacterial attack from occurring, while releasing growth-enhancing extracellular organic

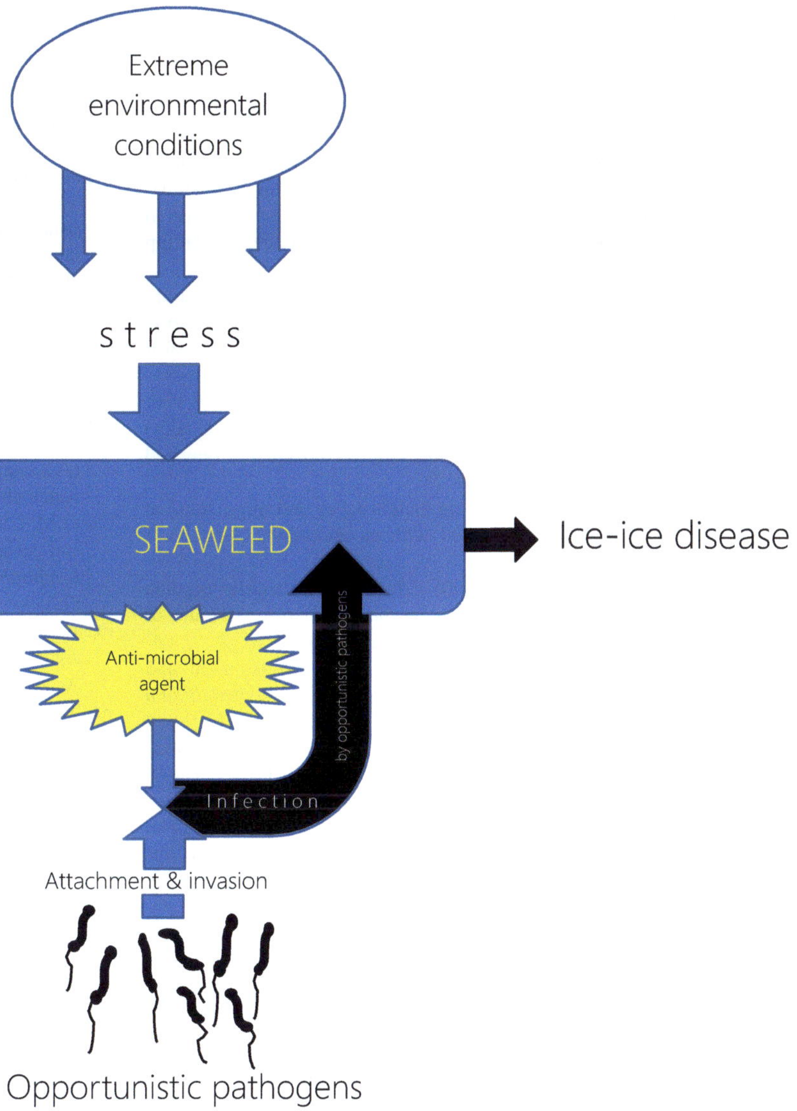

Fig. 21.8 A proposed seaweed-bacteria pathosystem model of the ice-ice disease based on the interaction between seaweed (*Kappaphycus*) and bacteria as influenced by environmental stress factors

compounds that allow growth of bacteria, especially opportunistic ones, to establish themselves on a weakening thallus. Bacterial strains may have varying degrees of influence on the seaweed host, as they (bacteria) themselves may compete, the reason some bacterial strains appear and tend to dominate, while others are inhibited and disappear, as shown in the inoculation of different bacterial strains on filtrates from seaweed culture medium of different culture ages (Fig. 21.7). This observation is shared by several studies cited by Goecke et al. (2010), including that of Steinberg et al. (2002), which states that most metabolites, such as carbohydrates, amino acids, peptides, and proteins, are inducers of microbial colonization. Moreover, certain bacterial strains have host-specific preference, which could explain why some strains have no reaction to extracts from the seaweed. The explanation that ice-ice disease occurs during stressful conditions in seaweed farms, and those seaweeds, under stress, could lose their ability to produce antimicrobial agents, allowing some opportunistic pathogens to colonize and infect the seaweed, thus the disease formation, can be represented by this model. Finally, whether this model also applies to other farmed seaweeds requires validation of their ability to produce antimicrobial and growth-promoting factors. So far, this has been fully demonstrated in *Kappaphycus alvarezii* in this review.

Acknowledgments The authors are grateful to Dr. Alan T. Critchley for inviting us to contribute this chapter to this book, and to the organizers of the Tropical Phyconomy Coalition Development (TPCD)—1, in which most part of this paper was presented during its Workshop on Advances in Eucheumatoid Research (www.phyconomy.org) in June 2021.

References

Ar Gall EA, Cornec L, Deslandes E (2008) Oligocarrageenans and tissue-dependant oxidative burst in *Solieria chordalis* (Rhodophyceae, Gigartinales). Phycol Res 56:7–12

Butardo VM, Ganzon-Fortes ET, Silvestre VA, Bacano-Maningas MBB, Montaño MNE, Lluisma AO (2003) Isolation and classification of culturable bacteria associated bacteria with *ice-ice* disease in *Kappaphycus* and *Eucheuma* (Rhodophyta, Solieriaceae). Philipp Scient 40:223–247

Cabral EM, Oliveira M, Mondala JRM, Curtain J, Tiwari BK, Garcia-Vaquero M (2021) Antimicrobials from seaweeds for food applications. Mar Drugs 19:211. https://doi.org/10.3390/md19040211

Collens J, Pedersen M (1990) Production of haloamine from *Eucheuma* species. In: Mshigeni KES, Bolton J, Critchley AT, Kiangi G (eds) Proceedings of 1st international workshop on sustainable seaweed research and development, pp 69–73

Correa JA, Flores V (1995) Whitening thallus decay and fragmentation in *Gracilaria chilensis* associated with an endophytic amoeba. J Appl Phycol 7:421–425

Fukami K, Ohara S, Ishida Y (1990) Effect of extracellular organic carbon (EOC) from phytoplankton on the community structure of oligotrophic bacteria. Arch Hydrobiol Beih 34:43–47

Fukami K, Yuzawa A, Nishijima T, Hata Y (1992) Isolation and properties of a bacterium inhibiting the growth of *Gymnodinium nagasakiense*. Nipp Suisan Gakk 58(6):1073–1077

Goecke F, Labes A, Wiese J, Imhoff JF (2010) Chemical interactions between marine macroalgae and bacteria. Mar Ecol Prog Ser 409: 267–300

Goecke F, Thiel V, Wiese J, Labes A, Imhoff JF (2013) Algae as an important environment for bacteria – phylogenetic relationships among new bacterial species isolated from algae. Phycologia 52: 14–24

Jones AK (1982) The interaction of algae and bacteria. In: Bull AT, Slater JH (eds) Microbial interactions and communities. Academic, London, pp 189–247

Kopprio GA, Cuong LH, Luyen ND, Duc TM, Ha TH, Huong LM, Gardes A (2021) Carrageenophyte-attached and planktonic bacterial communities in two distinct bays of Vietnam: Eutrophication indicators and insights on ice-ice disease. Ecol Indic 121. https://doi.org/10.1016/j.ecolind.2020.107067

Kouzuma A, Watanabe K (2015) Exploring the potential of algae/bacteria interactions. Curr Opin Biotechnol 33:125–129

Largo DB, Fukami K, Nishijima T, Ohno M (1995a) Laboratory-induced development of the ice-ice disease of the farmed red algae *Kappaphycus alvarezii* and *Eucheuma denticulatum* (Solieriaceae, Gigartinales, Rhodophyta). J Appl Phycol 7:539–543

Largo DB, Fukami K, Nishijima T (1995b) Occasional pathogenic bacteria promoting ice-ice disease in the carrageenan-producing red algae *Kappaphycus alvarezii* and *Eucheuma denticulatum* (Solieriaceae. Gigartinales, Rhodophyta). J Appl Phycol 7:545–554

Largo DB, Fukami K, Adachi M, Nishijima T (1997) Direct enumeration of total bacteria from macroalgae by epifluorescence microscopy as applied to the fleshy red algae *Kappaphycus alvarezii* (Doty) Doty and *Gracilaria* spp. (Rhodophyta). J Phycol 33:554–557

Largo DB, Fukami K, Adachi M, Nishijima T (1998) Immunofluorescent detection of ice-ice disease-promoting bacterial strain *Vibrio* sp. P11 of the farmed macroalga, *Kappaphycus alvarezii* (Gigartinales. Rhodophyta). J Mar Biotechnol 6:178–182

Largo DB, Fukami K, Nishijima T (1999) Time-dependent attachment mechanism of bacterial pathogen during ice-ice infection in *Kappaphycus alvarezii* (Gigartinales, Rhodophyta). J Appl Phycol 11:129–136

Largo DB, Msuya FE, Meñezes A (2020) Understanding diseases and control in seaweed farming in Zanzibar. FAO Fisheries and Aquaculture Technical Paper No. 662. FAO, Rome. doi:https://doi.org/10.4060/ca9004en

Lavilla-Pitogo CR (1992) Agar-digesting bacteria associated with "rotten thallus syndrome" of *Gracilaria* sp. Aquaculture 102:1–7

Mtolera MSP, Collen J, Pedersen M, Semesi A (1995) Destructive hydrogen peroxide production in *Eucheuma denticulatum* (Rhodophyta) during stress caused by elevated pH, high light intensities and competition with other species. Eur J Phycol 30: 289–297

Mtolera MSP, Collen J, Pedersen M, Ekdahl A, Abrahamson K, Semesi A (1996) Stress-induced production of volatile halogenated organic compounds in *Eucheuma denticulatum* (Rhodophyta) caused by elevated pH and high light intensities. Eur J Phycol 31:89–95

Nasmia N, Syamsuddin R, Rantetondok A, Zainuddin EN (2014) Characterization and identification of bacteria isolated from seaweed *Gracilaria verrucosa* (Linn., 1758) infected by ice-ice. Int J Aquacult 4(23):1–6

Ohno M, Largo DB, Ikumoto T (1994) Growth rate, carrageenan yield and gel properties of cultured kappa-carrageenan producing red alga *Kappaphycus alvarezii* (Doty) Doty in the subtropical waters of Shikoku, Japan. J Appl Phycol 6:1–5

Pesando D (1990) Antibacterial and antifungal activities of marine algae. In: Akatsuka I (ed) Introduction of applied phycology. SPB Academic Publishing bv, The Hague, pp 3–26

Riquelme CE, Fukami K, Yuzaburo I (1989) Growth responses of bacteria to extracellular products of bloom algae. Nipp Suisan Gakk 55:349–355

Riyaz SUM, Prabhu NS, Nalini S, Tejaswini G, Christy HJ, Inbakandan D (2020) Microbiome identification from ice-ice disease infected *Kappaphycus alvarezii* using 16S rRNA gene sequencing analysis and in silico docking studies against carrageenan degrading bacteria. Biocatal Agric Biotechnol 27:101707

Riyaz SUM, Inbakandan D, Manikandan D, Bhavadharani P, Elson J, Shrinivasa Prabhu N, Stalin Dhas T, Nalini S, Bavani Latha M, Simal-Gandara J (2021) Microbiome in the ice-ice disease of the farmed red algae *Kappaphycus alvarezii* and degradation of extracted food carrageenan. Food Biosci 42:101138

Solis MJL, Draeger S, dela Cruz TEE (2010) Marine-derived fungi from *Kappaphycus alvarezii* and *K. striatum* as potential causative agents of ice-ice disease in farmed seaweeds. Bot Mar 53:587–594

Steinberg PD, de Nys R, Kjelleberg S (2002) Chemical cues for surface colonization. J Chem Ecol 28:1935–1951

Syafitri E, Prayitno SB, Ma'ruf WF, Radjasa OK (2017) Genetic diversity of the causative agent of ice-ice disease of the seaweed *Kappaphycus alvarezii* from Karimunjawa island, Indonesia. IOP Conference Series: Earth and Environmental Science 55. doi:https://doi.org/10.1088/1755-1315/55/1/012044; Accessed 2 March 2022

Ward GM, Kambey CSB, Faisan JP, Tan P-L, Daumich CC, Matoju I, Stentiford GD, Bass D, Lim P-E, Brodie J, Poong S-W (2021) Ice-ice disease: an environmentally and microbiologically driven syndrome in tropical seaweed aquqaculture. Rev Aquac 001:1–26, DOI: https://doi.org/10.1111/raq.12606; Accessed 8 February 2022

Weinberger F, Hoppe H, Friedlander M (1997) Bacterial induction of a fast necrotic response in *Gracilaria conferta* (Rhodophyta). J Appl Phycol 9:277–285

Zainuddin EN, Anshary H, Huyyirnah H, Hiola R, Baxa DV (2019) Antibacterial activity of *Caulerpa racemosa* against pathogenic bacteria promoting "ice-ice" disease in the red alga *Gracilaria verrucosa*. J Appl Phycol 31:3201–3212

Novel Methods for Protecting *Kappaphycus alvarezii* from Herbivores: An Overview of Development and Economic Prospects

22

Ma'ruf Kasim, La Sara, Nurdiana, Ernaningsih, Andi Tamsil, Wardha Jalil, Tamar Mustari, and Mudian Paena

Abstract

Kappahycus alvarezii is widely cultivated by Indonesian people living in coastal areas. The use of simple cultivation methods and low investment in *K. alvarezii* cultivation has encouraged an increase in the number of seaweed farmers. Farmers mainly use the longline method of seaweed cultivation. One of the most severe problems seaweed farmers face is the attack of fish and turtle pests. Such attacks are associated with the longline method, which is an open method that provides opportunities for grazing of the seaweed crop by fish and turtles. Confinement innovation is urgently needed to protect seaweed from consumption by various fish and turtles. In 2018, cage innovations were developed and tested with various variants, including horizontal net (Horinet), vertical net (Vertinet), long net, and basket net. This chapter explored the technical effectiveness and economic feasibility of using Horinet and Vertinet. The results showed that the growth of *K. alvarezii* on the Horinet, Vertinet, and longline was 5.0, 4.6, and 4.2%/day, respectively. Economic analysis of the Horinet and Vertinet systems indicated that their implementation in seaweed cultivation shows economic feasibility. From the value of the revenue/cost (R/C) ratio and other economic feasibility indices, the Horinet and Vertinet methods seemed to be feasible for commercial development as effective solutions for protecting seaweed crops from fish and turtle pests.

M. Kasim (✉) · L. Sara · Nurdiana
Faculty of Fishery and Marine Science, Halu Oleo University, Kendari, Southeast Sulawesi, Indonesia

Ernaningsih · A. Tamsil
Faculty of Fisheries and Marine Science, Muslim University of Indonesia, Makassar, Southeast Sulawesi, Indonesia

W. Jalil · T. Mustari
Faculty of Fisheries and Marine Science, Dayanu Iksanuddin University, Baubau, Southeast Sulawesi, Indonesia

M. Paena
Research Institute for Coastal Aquaculture and Fisheries Extension, Makassar, Indonesia

Keywords

Horizontal net · Vertical net · Longline · Seaweed · *Siganus* sp.

22.1 Introduction

Indonesia is one of the countries with the largest *Kappaphycus alvarezii* production in the world (Langford et al. 2022). Seaweed production in Indonesia is mostly based on cultivation because the carrying capacity of coastal waters for cultivation is excellent and comprehensive. Indonesia's seaweed cultivation area has been estimated at 6,846,800 km (BPS 2021). This coastal area produces seaweed species for export commodities, including *Gracilaria* sp. and *K. alvarezii*. Several Indonesian regions have cultivated seaweed, including the provinces of Bali, Lampung, West Nusa Tenggara, East Nusa Tenggara, Riau, Sulawesi, and Maluku (Kasim et al. 2022). Seaweed farming in Indonesia commenced with local cultivars of *Eucheuma denticulatum*, but *Kappaphycus* seaweed farmed in Indonesia was initially imported from the Philippines and known as the Tambalang strain. Experimental growth of this strain showed that it could be propagated through cultivation (Hurtado and Agbayani 2000). Eventually, the most widely cultivated species of eucheumatoid seaweeds farmed in Indonesia were *K. alvarezii*, *E. striatus*, and *E. denticulatum* (Sulu et al. 2004). These species have been widely cultivated in several countries, including the Philippines, where they have been commercially farmed since 1971 (Doty and Alvarez 1975). Cultivated eucheumatoid seaweed production increased rapidly following initial cultivation successes in 1971, with an initial biomass of less than 1000 tonnes dry weight increasing to 100,000 tonnes dry weight within a few years (Doty and Alvarez 1975). Two countries in Southeast Asia where *K. alvarezii* is intensely cultivated are the Philippines and Indonesia (Dawes 1998). Coastal communities in several other tropical countries in Southeast Asia also cultivated

A. T. Critchley et al. (eds.), *Tropical Phyconomy Coalition Development*, Developments in Applied Phycology 11,
https://doi.org/10.1007/978-3-031-47806-2_22

K. alvarezii with high production. Each country has several commonly used cultivation techniques. However, the most commonly applied cultivation systems include longlines, raft systems, and off-bottom and on-bottom systems. The longline cultivation method and other methods are simple methods of tying seaweed on ropes to be cultivated for up to 40 days before harvesting (Kasim et al. 2019). However, the seaweed industry continues to face various problems. One of the serious problems in seaweed cultivation using open methods is the attack of herbivorous fish pests (Kasim and Mustafa 2017; Kasim et al. 2017). For example, seaweed cultivation areas in Mediterranean coastal waters have experienced an increase in herbivore fish populations correlated with the advance of certain invasive macroalgal species (Tomas et al. 2011). In India, especially on Krusadai Island, seaweed cultivation, such as *K. alvarezii* and *E. denticulatum*, has experienced decreased production by at least 10% due to predatory pests such as *Siganus javus* (Rabbitfish), *Acanthurus* sp. (surgeon fish), *Cetoscarus* sp. (Parrotfish) and sea urchin *Tripneustes* sp. (Ganesan et al. 2006). The development of seaweed cultivation methods is still very slow; since 1980, people in Southeast Asia and Indonesia have continued to use the longline cultivation method (Adnan and Porse 1987; Luxton 1993; Reis et al. 2015). This is a conventional method and has many drawbacks, including susceptibility to pest attacks and to the emergence of ice-ice syndrome (Kasim and Asnani 2012; Kasim et al. 2016). Herbivorous pest attacks and attachment of epifauna or epiphytes are still commonly observed in cultivated areas in Southeast Asia, which can reduce seaweed production (Ganesan et al. 2006; Johnson and Gopakumar 2011). Even in Indonesia, seaweed production has decreased by up to 60% of the total production per harvest (Kasim et al. 2017). Several countries, such as the Philippines and Brazil, have started to conduct seaweed maintenance efforts to protect seaweed from pests by using traditional bow ropes (Reis et al. 2015; Hurtado-Ponce 1992). Hurtado-Ponce (1992) recommended the cultivation of *K. alvarezii* seaweed in floating cages to prevent the seaweed from being attacked by herbivores. Modification of cultivation with this protection system is dependent on the location's topography and has been implemented in several cultivation areas in the Philippines. The use of cages for cultivating *K. alvarezii* can be carried out in areas with a high intensity of attack by herbivore fish. However, the economic feasibility of tools to support aquaculture needs to be considered. The cultivation of *E. cottoni* with oysters and snapper is an alternative to the caged confinement system (Biswajit et al. 2009). *Gracilaria gracilis* can be cultured in closed containers to avoid attacks by herbivores such as isopods. The floating cage is one of the technological designs that functions as a container for seaweed cultivation, protecting seaweed from grazing while reducing the potential for disease (Kasim 2016). This method has been used by a small number of farmers in Eastern Indonesia, especially in Southeast Sulawesi, to protect seaweed from attacks by herbivorous fish and turtles (Kasim 2016). Various sizes and shapes were developed as tools that can be used for seaweed cultivation. The basic materials of floating cages included PVC pipes and nets equipped with buoys as markers. Weights were used to attach floating rafts to substrates and to make it easier to place the raft at desired locations. The basic material of rafts is designed to form rectangular boxes of varying sizes. The raft surface is left open, the bottom is covered with a multifilament net (mesh size 1 cm), and the sides of the raft are covered with a multifilament net measuring 2 cm (Kasim 2016). During its development, the floating cage underwent several changes and modifications, primarily designed to make it easier for farmers to cultivate seaweed. The sizes and materials used have improved and are cost-effective. This development was based on field tests/topography results at all appropriate topographical levels. Design adjustments were made with suggestions from seaweed farmers. This was done to adjust the habits and effectiveness of using tools for planting and harvesting seaweeds. Several variants of confinement cultivation tools are available for development, including floating net rafts, floating pouch netting, vertical nets (Vertinet), horizontal nets (Horinet), pound nets (basket nets), long nets, wall nets, box nets, and hanging net rafts (Kasim et al. 2019, 2020, 2022). In general, all these variants were cultivation tools devised to protect seaweed from various pest attacks and reduce the incidence of ice-ice syndrome (Budiyanto et al. 2019; Darma et al. 2021). The principle of the tools is to cultivate seaweed in nets with various designs so that the seaweed is protected from pests of herbivorous fish and turtles. Although they have relatively the same primary function, each of these tools has advantages (Kasim et al. 2020). In this paper, we discuss the results of such research in terms of the development and economic prospects of the floating confinement method, which is one of the tools used to protect seaweed from grazing pests, epiphyte attachment, and seaborne contaminants.

22.2 Materials and Methods

22.2.1 Study Sites and Experimental Tools

Research was conducted at one of the locations for seaweed cultivation in coastal Lekeba, Southeast Sulawesi Province, Indonesia (Fig. 22.1). This location is a seaweed cultivation area that uses horizontal and vertical net tools.

The equipment used in this research was a horizontal net (Horinet) and a vertical net (Vertinet). Measurement of the total growth rate and daily growth rate was carried out by

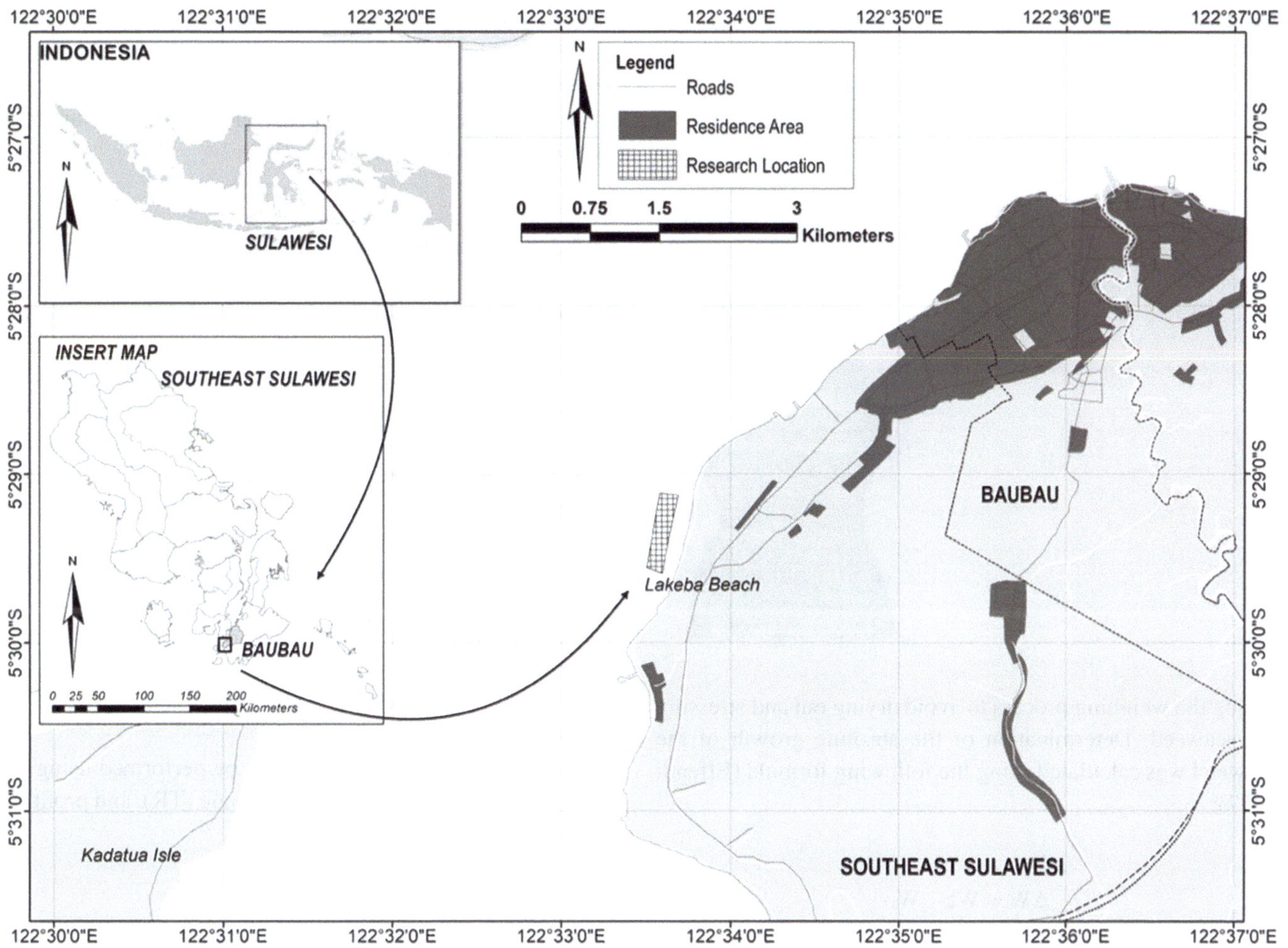

Fig. 22.1 The location of the study site in the seaweed cultivation area of Lakeba Beach, Southeast Sulawesi Province, Indonesia

measuring the weight of the *K. alvarezii* thalli every 7 days. During a 45-day cultivation period, measurements were carried out using a Horinet and a Vertinet. The Horinet was a rectangular box net composed of PVC piping used for framing that supported protective netting. The walls and floor of the box were sheathed with netting (mesh diameter 1–2 cm). The Horinet was 200 × 80 × 60 cm (Fig. 22.2).

The Vertinet is a tool comprising an equilateral rectangular bag that extends towards the bottom of the water from the water's surface. This tool consisted of two main components, namely, upper and lower pipe frames and protective netting. The PVC pipe frame is assembled separately. The primary frame size was 40 × 40 cm, and there was a protective net on the outside with a mesh size of 1 cm in diameter (Fig. 22.3).

22.2.2 Growth Rate Measurement

Growth measurements were carried out by weighing the seaweed thallus spread in horizontal and vertical nets. Weighing was performed every 7 days with five observations over 45 days. This was because *K. alvarezii* has a typical cultivation cycle of 40–45 days. The cultivation process was carried out by growing seaweed within the confinement systems, marking each thallus to facilitate the weighing process. The seaweed was continuously immersed in seawater

Fig. 22.2 A horizontal net (Horinet) was used for seaweed cultivation

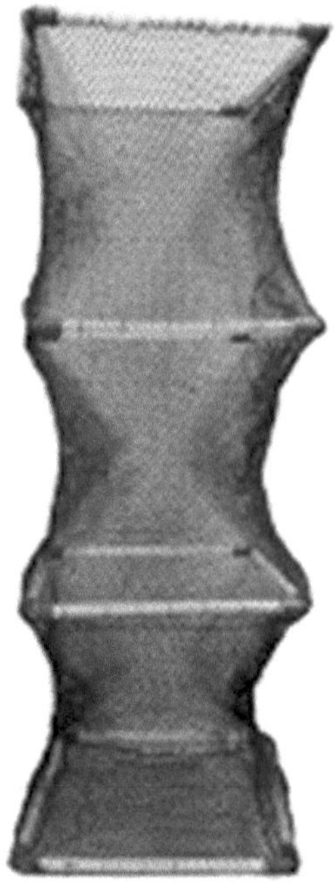

Fig. 22.3 A vertical net (Vertinet) was used for seaweed cultivation

during the weighing process to avoid drying out and stressing the seaweed. Determination of the absolute growth of the seaweed was calculated using the following formula (Effendi 1997):

$$\Delta W = Wt - Wo$$

Description:

ΔW = absolute growth (g)
Wt = average weight of seaweed at the end of the study (g)
Wo = average weight of seaweed at the beginning of the study (g)

Determination of the specific growth rate (SGR) used the following formula (Luhan and Sollesta 2010):

$$SGR = \left(\frac{LnWt/LnWo}{t}\right) \times 100\%$$

Description:

SGR = specific growth rate (% in wet weight per day)
Wt = weight after *t* days (g)
Wo = initial weight (g)
t = time (days)

The carrageenan content was determined to assess the quality of the *K. alvarezii* thallus for each cultivation method. Carrageenan content was also measured for all *K. alvarezii* thalli inside the Horinet, Vertinet, and longline. The age of the analysed thalli was 30 days in cultivation time. Analysis of carrageenan content was carried out using the carrageenan extraction method (Rizal et al. 2016).

22.2.3 Economic Feasibility Analysis

Primary field data collection was carried out by surveying seaweed farmers. Data collection was carried out according to two methods. The first approach was to collect data by interviewing several farmers who use floating cages. Second, in-depth interviews using the snowball sampling approach were conducted with several potential farmers. Data collection was essential to explore the value of technical and economic information in the development and use of the confinement system and to analyse its economic feasibility. The data collected include information on business capital and components in investment and production obtained in one harvest. Analyses undertaken during the assessment of the economic prospects of the confinement systems are described below.

Simple business unit analysis can be performed using the formulas for total cost (TC), total revenue (TR), and profit (π) (Sugiarto 2005):

$$TC = TFC + TVC$$

Description:

TC = *total cost* (Rp)
TFC = *total fixed cost* (Rp)
TVC = *total variable cost* (Rp)

Determination of total revenue (TR) was calculated by the formula:

$$TR = P \times Q$$

Description:

TR = total revenue (Rp)
P = price (Rp)
Q = amount (Quantity) (Rp)

Analysis of the revenue/cost ratio (R/C ratio) was carried out to determine the relative profit of a business in one year against the costs used in activities. The formulation of the revenue cost ratio (R/C ratio) is as follows (Effendi and Oktariza 2006):

$$R/C = \frac{\text{TR}}{\text{TFC} + \text{TVC}}$$

Description:

TR = total revenue (Rp)
TFC = total fixed cost (Rp)
TVC = total variable cost (Rp)

Decision-making:

If the R/C ratio is >1, the business is said to be feasible and profitable.
If the value of the R/C ratio is <1, the business is said to be unfeasible and unprofitable. If the value of the R/C ratio = 1, the business is at break-even (no profit and no loss).

To calculate the break-even point in units, use the formula:

Break-even point (units) = fixed costs/(sales price per unit – variable costs per unit)
Break-even point (sales) = fixed costs contribution margin.

Net present value (NPV) analysis is performed to assess cash flows and determine whether the project is feasible. The formula for NPV is:

$$\text{NPV} = \sum\nolimits_{t=1}^{n} \frac{\text{CF}t}{(1+i)t} - \text{Co}$$

Description:

CFt = cash flow per year in period t
Co = initial investment in year 0 (Rp)
i = discount rate (%)
t = year
n = number of years

Decision-making:

If the NPV is positive, then the investment is accepted. If the NPV is negative, then the investment is rejected.

The internal rate of return (IRR) is the interest rate that will make the total present value of the product equal to the total current value of capital expenditures.

$$\text{IRR} = i1 + \frac{\text{NPV1}}{(\text{NPV1} - \text{NPV2})}(i2 - i1)$$

Description:

$i1$ = interest rate that produces a positive NPV
$i2$ = interest rate that produces negative NPV
NPV1 = positive NPV
NPV2 = negative NPV

Decision-making:

If the IRR is greater than the loan interest, it is accepted/deserved. If the IRR is less than the loan interest, it is rejected/not feasible.

The payback period refers to the amount of time it takes to recover the cost of an investment or how long it takes for an investor to hit break-even.

payback period = initial investment/cash flow per year

22.3 Results

22.3.1 Seaweed Growth in the Horinet and Vertinet

The SGR is the rate of increase in the initial weight of *K. alvarezii* seaweed over 45 days. Measured by an initial seedling weight of 20 g, each treatment used the Horinet and Vertinet with 20 replications for each treatment. The highest and lowest SPGs are shown in Fig. 22.4.

Seaweed growth based on depth showed that depths above 100 cm were better than depths below 100 cm (Fig. 22.5).

The results of the absolute growth analysis (gr) using the Horinet and Vertinet methods were better than those using the longline method (Fig. 22.6).

From the study results, the initial weight of 20 g at the beginning of the day of breeding, after 45 days, changed to 123, 80, and 54 g using a Horinet, Vertinet, and longline, respectively. The carrageenan content of *K. alvarezii* cultivated with Vertinet, Horinet, and longline did not differ significantly (Fig. 22.7).

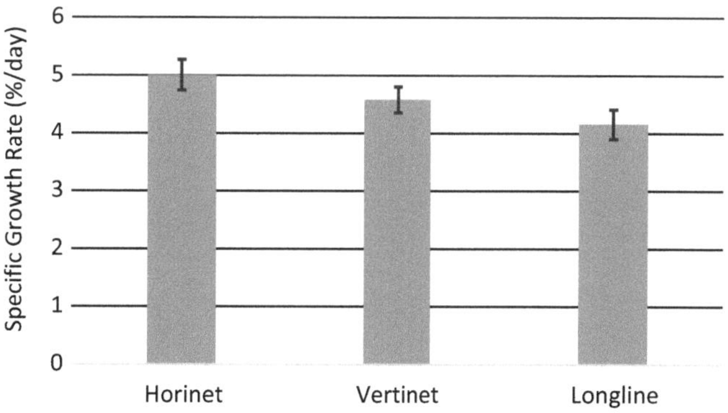

Fig. 22.4 Specific growth rate (%/day) using a Horinet, Vertinet, and longline

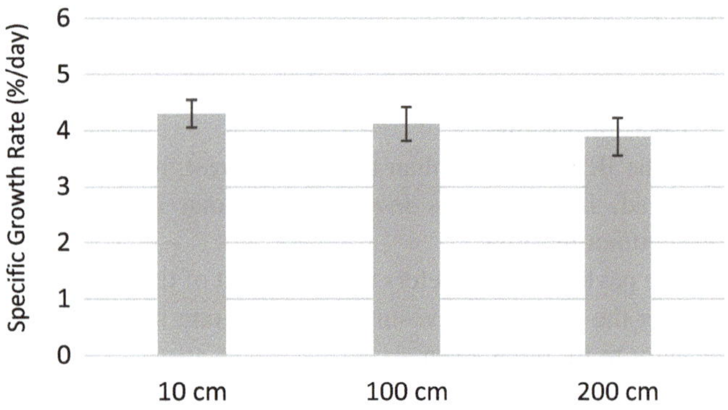

Fig. 22.5 Specific growth rate (%/day) of seaweed in a Vertinet at different depths

22.3.2 Feasibility of the Horinet and Vertinet from a Business Perspective

To determine the business viability of Horinet and Vertinet, we used some simple tests to analyse the R/C ratio. The R/C ratio is a measure of efficiency that compares a company's expenses to its earnings. It considers the cost of revenue and the total revenue. The cost of revenue includes all the expenses of manufacturing, including marketing costs. The total revenue counts the total earnings from sales during a financial period. The IRR is a metric used in financial analysis to estimate the profitability of potential investments. IRR is a discount rate that makes the NPV of all cash flows equal to zero in a discounted cash flow analysis. NPV is one of the methods used to calculate the difference between the present value of an investment and the present value of revenues. If the future net cash receipts are more significant than the present value of the investment, then the business is said to be profitable (Irmayani and Muhammad 2014; Saik 2016). The last is the payback period (PP), which refers to the amount of time it takes to recover the cost of an investment or how long it takes for an investor to reach a break-even point. Account and fund managers use the payback period to determine whether to go through with an investment. Using the previously described formula, we obtained the R/C ratio, IRR, and PP values, as shown in Table 22.1.

22.4 Discussion

A Horinet method is a cultivation tool that utilizes sea level as a living medium for seaweed. *K. alvarezii* cultured with the Horinet were wrapped in polyethylene bags so that predators, such as barongan fish and turtles, could not graze on the seaweed. Julizar et al. (2018) explains that the use of an appropriate maintenance system, such as the floating cage, can result in an increase in seaweed production. Syarqawinagan (2017) also stated that seaweed bags are containers or places that facilitate seaweed cultivation through methods that have been developed.

The Vertinet method utilized the marine water column as a living medium for seaweed with a depth range of 50–300 cm. In a Vertinet, the seaweed is wrapped in a polyethylene bag that is a barrier to predators and prevents mud particles from directly hitting the seaweed. This method can also stimulate seaweed seedlings to increase growth (Julizar et al. 2018). Planting using the floating raft method is more effective than using the longline method because, in general, the floating raft method reduces water movement, and the intensity of light is sufficient for seaweed growth (Wijayanto et al. 2011).

22.4.1 Technical Potential of Horinet and Vertinet Development

During the cultivation process, the SGR of *K. alvarezii* cultivated using the Horinet, Vertinet, and longline methods progressed well, with average values of 5.0, 4.6, and 4.2%/

Fig. 22.6 Absolute growth (gr) during the study

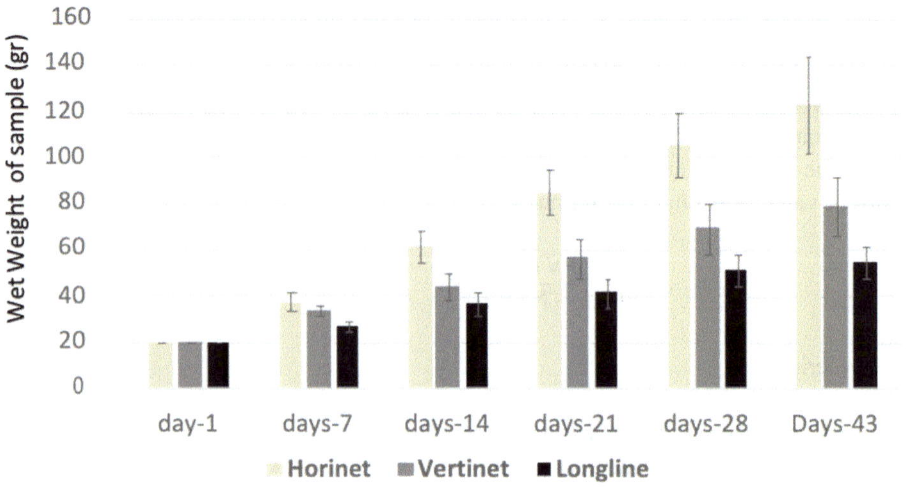

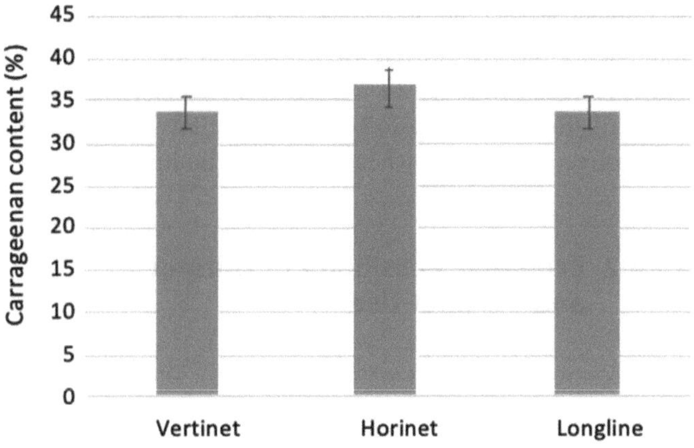

Fig. 22.7 Seaweed carrageenan content in the Vertinet, Horinet, and longline

Table 22.1 Financial feasibility of seaweed cultivation using the Vertinet and Horinet at an estimated 1 Ha/year

No	Results of calculation of the economic viability of seaweed	Vertinet	Horinet
1	R/C Ratio	4.7	1.3
3	IRR	45%	25%
4	PP (Years)	3.5	3.9

day, respectively. Both protection methods had the highest SGR compared to the longline. Bolqiah et al. (2019) reported that the SGR of *K. alvarezii* was 1.34% and 1.57% in the coastal waters of Lakorua, Southeast Sulawesi, Indonesia. Other reports of *K. alvarezii* growth rates include 1.14%/day in the Philippines (Hurtado et al. 2008), 2.0–8.1%/day in Yucatán, Mexico (Munoz et al. 2004), and 2.33–5.22%/day in the Vellar Estuary, India (Thirumaran and Anatharaman 2009). According to Nursyahran and Reskiati (2013), the first week is the phase of the start of vegetative growth. In this condition, the plant undergoes the development of mature tissue cells, so that the weight continues to increase. Mutalib and Rahman (2018) explained that the decrease in SGR was due to the low growth rate, decreased thallus weight, and the increasing age of seaweed maintenance. The cause of the decrease in SGR can be related to competition in obtaining nutrients and absorption of sunlight in the photosynthesis process. Gultom et al. (2019) stated that the denser the seaweed cultivated on one rope, the lower the growth. Amin et al. (2005) suggested that the decrease in SGR is due to the saturation of cell division. Seaweed that has undergone an adaptation process then undergoes a cell growth phase that causes slow growth. According to Gultom et al. (2019), the greater the growth weight of seaweed, the more concentrated the plant density will be. This is because the growth of new shoots and enlarged branches causes the space to become narrower, resulting in competition between plants for nutrients and sunlight.

SGR was influenced by changes in the weight and growth of the seaweed thallus. The present study showed differences in the increase in the wet weight of seaweed in the three different methods during the study. Abdan and Ruslaini (2013) stated that one of the factors affecting growth is the number of thalli. From an economic standpoint, Poncomulyo et al. (2006) noted that an SGR of 3%/day is considered quite profitable; similarly, according to Safia et al. (2019), the SGR of seaweed, considered quite profitable, is 3% weight gain per day or 21% per week. Syahlun et al. (2013) reported that a seaweed cultivation operation is categorized as good if the SGR is approximately 2%.

Absolute growth results from the final weight minus the initial weight of seaweed. The absolute growth of the mean values seen in this study on Horinet, Vertinet, and longline at initial weights of 20 gm changed to 123, 79, and 54 gm, respectively. This difference can be seen from the total weight in one harvest at 45 days. Yusup et al. (2017) explained that the growth of *K. alvarezii* in the coastal waters of Tanjung Tiram grew by 28.4 gm from the initial weight of a 5 g thallus. Faisal et al. (2012) found growth of *K. alvarezii* in the range of 41.48–49.63 gm. The use of the confinement method provided good growth conditions. The biomass in the present study had an initial weight of 20 gm. Small initial seedlings tended to grow fastest because *K. alvarezii* can obtain nutrients evenly. According to Tiwa et al. (2013), the SGR of seaweed is strongly influenced by the weight of the initial seaweed seeds to be cultivated. In the beginning, the seed weight was small, giving faster growth results because there was no competition between the thalli in obtaining nutrients. Indriani and Sumiarsi (2003) stated that nutrient fulfillment greatly affects growth. Sufficient space for seaweed causes an increase in the growth rate because the supply of nutrients takes place optimally. Small cuttings have a faster regeneration process than those with large weights.

The absolute growth graph shows that the thalli of *K. alvarezii* increased every week. This was due to the Horinet and Vertinet support system. The Horinet method had a rigid shape and seemed to protect the seaweed from pests, human activities, and the physical movement of waves. The natural condition of the seaweed attached to the substrate allows the seaweed to be squeezed between the nets and continue to grow well. In some cases, strong currents can break the seaweed thallus. However, with Horinet and Vertinet, the broken thallus will remain in the cage and will not affect the total daily weight change. In contrast, it does not provide protection from pests and only binds one side of the seaweed thallus. Continuous physical friction due to currents and waves can cause the seaweed thallus to break. The broken thallus will be washed away by the current, reducing the overall weight of the seaweed thalli.

Confined seaweeds receive good protection from attacks by herbivorous fish and turtles that choose to graze on the younger seaweed thalli (Kasim et al. 2017; Kasim and Asnani 2012). Cultivation using the floating raft method is more effective than the longline method because, in general, *K. alvarezii* is more able to obtain adequate light intensity and is less influenced by water movement caused by strong currents (Wijayanto et al. 2011). Usually, in the oceans, currents will carry solid objects, including garbage that will stick to the thallus and interfere with the photosynthesis process (Parenrengi et al. 2011). A decrease in photosynthetic activity can reduce daily activities and impact seaweed's ability to obtain nutrients (Sardianti and Widiastuti 2010). The highest absolute growth of *K. alvarezii* seaweed was obtained using the Horinet method, followed by the Vertinet and longline methods. The difference in absolute average growth in each method was because planting using the Horinet and Vertinet methods was more effective than using the longline method. The light that enters the bottom of the water is also a good determining factor for seaweed. Growth at three depths with a Vertinet (10, 100, 200 cm) did not significantly differ. The light can enter at a depth of up to 200 cm, which allows the seaweed to grow well. Space in the sea is also suitable for seaweed growth if there is good light. Faisal et al. (2012) found the growth of seaweed *K. alvarezii* at a higher surface of 41.48–49.63 gm. In the value of the weight change, it was also seen that the seaweed experienced a significant change in the surface waters at a depth of approximately 20 cm. Reported weight change values include 4.45–7.30 gm (Hamid 2009), 6.16–10.68 gm (Masyahoro and Mappiratu 2010), and 6.33–16.67 gm (Ladunta 2015). Good seaweed growth occurs because seaweed optimally utilizes sunlight for photosynthesis and nutrients. The lower light penetration ability can cause the low growth rate of seaweed with increasing sea level depth (Aditya et al. 2001; Wati 2019). This study reported that the slightly lower value at a depth of up to 200 cm on the Vertinet is the result of an uneven lighting process. The growth of *K. alvarezii* was maximal in Horinet. The growing space provided by Horinet for *K alvarezii* is better than that provided by Vertinet. Widyorini (2010) stated that light provides energy for photosynthesis, so the ability of light to penetrate at a certain depth will determine the existence of aquatic organisms. Sardianti and Widiastuti (2010) also stated that increased photosynthesis could increase the ability of marine algae to obtain nutrients. Merisca (2013) added that seaweed achieves good growth when it receives sufficient light and nutrients from its environment. When the amount of nutrients in the environment is low or limited, seaweed will absorb little, causing its growth to be low or stunted. Liang et al. (2013) reported that the depth of *Sargassum* sp. in the water, which is closely related to light transmission, must be adjusted to support a faster growth rate. The causative factors are light intensity, water temperature, and salinity, which significantly affect the rate of photosynthesis in seaweed. Vairappan et al. (2007) also stated that an important factor influencing the growth rate of seaweed is the difference in light received by the seaweed at different depths. This condition will affect the expansion of the new cell wall, which does not change appreciably when the light is limiting.

22.4.2 Economic Feasibility of Horinet and Vertinet Use

The economic feasibility of seaweed cultivation depends on many factors. One significant factor is the price that farmers get to sell their harvest. The right price will provide more profit for seaweed farmers. Farmers expect a good price for *K. alvarezii*. In the last 10 years in Indonesia, the selling price of seaweed has fluctuated greatly. The selling price range is under $1USD/kg dry. This price depends on local conditions and the needs of seaweed farmers (Langford et al. 2022).

By setting the appropriate price, this study tries to determine the feasibility of the business carried out by seaweed farmers in Southeast Sulawesi, Indonesia, especially by using Horinet and Vertinet. Horinet and Vertinet had R/C ratio values of 1.3 and 4.7, respectively. Based on these values, their business applications were judged to be feasible because the R/C value was more than one. Other research at Lakeba Beach, Baubau City, Southeast Sulawesi, reported R/C ratios of 2.9 (Sumarni 2015), 1.28 (Nursima and Nurdinana 2021), and 4.73 (Suryadin 2021) for longline, Horinet, and Vertinet, respectively, demonstrating feasibility. It is not surprising that research conducted in the same location using different methods resulted in different R/C ratio values.

The internal rate of return (IRR) value of Horinet at Lakeba Beach is 25%, with an interest rate of 12.5%. This business can be feasible because it is based on the stipulation that if the IRR is greater than the loan interest, then the business is feasible to develop (Nursima and Nurdinana 2021). Agrand (2019) explained the feasibility of cultivating *E. cottonii* using the long line method in Poso Pesisir waters, Poso Regency, Central Sulawesi Province, Indonesia. The IRR value was 50% and had an interest rate of 10%, illustrating the profitability of the business. The PP value of *K. alvarezii* cultivation using the horizontal net method was 3.9. The business was said to be feasible because the PP was smaller than the investment age of the business, which was 5 years. Agrand (2019) had a PP of 2.9 and a business investment life of 3 years. Irmayani and Muhammad (2014) studied a seaweed operation in Mallasoro Village, Bangkala District, Jeneponto Regency and reported that the IRR value obtained (47.67%) was greater than the bank interest rate used (40%). This indicated that this business could be feasible because it was based on the stipulation that if the IRR was greater than the loan interest, then the business was profitable. Similar research results by Kasim et al. (2021)

concluded that small-scale cultivators mostly used horizontal systems. This method had better efficiency and effectiveness compared to similar businesses. The horizontal net method had good economic opportunities. The efficiency value obtained was an average of approximately 3.207 with an effectiveness value of 0.00032 kg per USD. The income of farmers on Lakeba Beach was 3037 USD per year. The use of horizontal net cultivation methods provided good production values and profits for farmers.

Cultivation of *K. alvarezii* is a small business for coastal communities that is economically profitable. The small capital used will be able to provide benefits for the lives of farmers. In countries that cultivate *K. alvarezii*, such as Indonesia, the Philippines, Tanzania, India, Mexico, and the Solomon Islands, seaweed cultivation is a profitable business because the business is engaged on a small scale involving a labor-intensive production system (Cai et al. 2013; Valderrama et al. 2015). The use of this method is one of the efforts developed by seaweed farmers to increase the harvest value and the selling value of the harvested seaweed. The development of this economic value will significantly affect the lives of seaweed farmers. The longline business has been around for more than 20 years and continues to grow. As an alternative, confinement methods such as Horinet and Vertinet to increase crop production are feasible approaches. In general, the cultivation of *K. alvarezii* is profitable. The FAO conducted research and obtained test results regarding the social and economic dimensions of *K. alvarezii* cultivation (Msuya 2013; Robledo et al. 2013). In the cultivation of *K. alvarezii*, there is a large difference in the economic performance of production (Valderrama et al. 2015). This business is very dependent on the size of the area owned by the farmer. Variations in land size significantly affect crop yields and labor (Hurtado and Agbayani 2002). Research on the economic feasibility of seaweeds in the North Sea found that offshore seaweed production was not economically viable. Sensitivity analysis shows that farmers' income in one harvest must increase by approximately 300% to be profitable (van den Burg et al. 2016). In contrast, the cultivation of *K. alvarezii*, if managed properly, can be profitable. Seaweed farmers are motivated to perform aquaculture business with the hope of good and better profits than other fishery businesses. A farmer will benefit from increasing his skills and ability to manage his cultivated area to obtain funds for his daily activities (Mariño et al. 2019; Steenbergen et al. 2017).

22.5 Conclusion

The development of protective cultivation using special tools such as Horinet and Vertinet can result in increases in absolute growth and specific growth of *K. alvarezii* that are greater than those observed on longlines. The new Horinet and Vertinet innovations were first introduced and developed by seaweed farmers in several areas, especially in Indonesia. These two innovations effectively protected the seaweed from fish and turtle pests. The best indicator of their success is the high rate of daily growth, reaching 6%/day. The economic aspect was one of the analytical tools to provide recommendations for feasibility. The Horinet and Vertinet methods have a business feasibility value that deserves to be developed. The analysis showed that a return on investment can occur in three years or less. The profit value is very good, so this tool is feasible as an innovation for the future cultivation of *K. alvarezii* in Indonesia.

Acknowledgments The authors are grateful to the Ministry of Education, Culture, Research, and Technology for the grant to conduct this study. The authors are also thankful to the research institutions and community service and the Faculty of Fishery and Marine Science, Halu Oleo University, for providing the facilities for this research.

References

Abdan AR, Ruslaini (2013) Effect of plant distance on growth and content of seaweed carrageenan (*Eucheuma spinosum*) using the long line method. Jurnal Mina Laut Indonesia 3:113–123. (In Indonesian)

Aditya TW, Yuwan P, Sudjiharno (2001) Site selection of seaweed cultivation technology (*Kappaphycus alvarezii*). Technical guidance of the Ministry of Marine Affairs and Fisheries. Directorate General of Aquaculture. Lampung Sea Cultivation Center, 16–22 p (In Indonesian)

Adnan H, Porse H (1987) Culture of *Eucheuma cottonii* and *Eucheuma spinosum* in Indonesia. Hydrobiologia 151:355–358

Agrand LD (2019) Analysis of the development of seaweed cultivation with the long line method in Poso Pesisir District, Poso Regency, South Sulawesi Province. Universitas Terbuka. Jakarta (In Indonesian)

Amin M, Rumaya TP, Femmi NF, Kemur D, Suwitra IK (2005) Study of seaweed (*Eucheuma cottonii*) cultivation with different planting systems and seasons in Bangkep Regency, Central Sulawesi. Jurnal Pengkajian dan Pengembangan Teknologi Pertanian 8:282–291. (In Indonesian)

Biswajit D, Suresh KM, Syda RG (2009) Integration of seaweed (*Kappaphycus alvarezii*) and pearl oyster (*Pinctada fucata*) along with Asian seabass (*Lates calcarifer*) in open sea floating cage off Andhra Pradesh coast, National Fisheries Development Board, http://eprints.cmfri.org.in/6070/1/10._Bis.pdf; access on 12 May 2020

Bolqiah S, Kasim M, Afu LOA (2019) The correlation of oceanographic factors on seaweed growth with the floating net raft method in Lakorua Waters, Mawasangka Tengah District, Central Buton Regency. Journal Sapa Laut 3:25–36. (In Indonesian)

BPS (Badan Pusat Statistik) (2021) Statistik Indonesia 2021. Retrieved from https://www.bps.go.id/publication/2021/02/26/938316574c78772f27e9b477/statistik-indonesia-2021.html

Budiyanto, Kasim M, Abadi SY (2019) Growth and carrageenan content of local and tissue culture seed of *Kappaphycus alvarezii* cultivated in floating cage. AACL Bioflux 12(1):167–178

Cai J, Hishamunda N, Ridler N (2013) Social and economic dimensions of carrageenan seaweed farming: a global synthesis. In

Valderrama D, Cai J, Hishamunda N, Ridler N (eds) Social and economic dimensions of carrageenan seaweed farming. Fisheries and Aquaculture Technical Paper No. 580. FAO, Rome, pp 5–59

Darma TA, Kasim M, Ruslaini NI, Rahman A, Muskita W, Hamzah M, Balubi AM, Patadjai RS, Jalil W (2021) Bacteria population on *Eucheuma denticulatum* (Rhodophyta) thallus infected by ice-ice disease cultivated on horizontal net cage. AACL Bioflux 14:399–411

Dawes CJ (1998) Marine botany, 2nd edn. John Wiley and Sons Inc, Canada. http://books.google.co.id/

Doty MS, Alvarez VB (1975) Status, problems, advances, and economics of Eucheuma farms. Mar Technol Soc J 9:30–35

Effendie MI (1997) Fisheries biology. First printing. Nusatama Pustaka Foundation, Yogyakarta, 163p

Effendi I, Oktariza W (2006) Fisheries agribusiness management. Penebar Swadaya Publishing, Jakarta. (In Indonesian)

Faisal LO, Patadjai RS, Yusnaini (2012) Growth of Seaweed (*Kappaphycus alvarezii*) and Boronang Fish (*Siganus guttatus*) cultivated together in cages. Thesis, Aquaculture Study Program, Faculty of Fisheries and Marine Sciences. Halu Oleo University (In Indonesian)

Ganesan MS, Thiruppathi N, Sahu N, Rengarajan V, Veer-agurunathan JB (2006) In situ observations on preferential grazing of seaweeds by some herbivores. Curr Sci 91:1256–1260

Gultom RC, Dirgayusa IGNP, Puspitha NLPR (2019) Comparison of growth rate of seaweed (*Eucheuma cottonii*) using coculture and monoculture cultivation systems in Geger Coastal Waters, Nusa Dua, Bali. J Mar Res Technol 2:8–16. (In Indonesian)

Hamid A (2009) Effect of initial seed weight with floating method on daily growth percentage of seaweed (*Eucheuma cottonii*). Thesis Department of Biology. Faculty of Science and Technology. UIN. Malang (In Indonesian)

Hurtado AQ, Agbayani RF (2000) The farming of the seaweed *Kappaphycus*. Aquaculture extension manual no. 32. Aquaculture Department, Southeast Asian Fisheries Development Center, Tigbauan, Iloilo, Philippines. 32p

Hurtado AQ, Agbayani RF (2002) Deep-sea farming of *Kappaphycus* using the multiple raft, long-line method. Bot Mar 45:438–444

Hurtado AQ, Critchley AT, Trespoey A, Bleicher-Lhonneur G (2008) Growth and carrageenan quality of *Kappaphycus striatum* var. sacol grown at different stocking densities, duration of culture and depth. J Appl Phycol 20:551–555

Hurtado-Ponce AQ (1992) Cage culture of *Kappaphycus alvarezii* var. tambalang (Gigartinales, Rhodophyceae). J Appl Phycol 4:311–313

Indriani H, Sumiarsi E (2003) Cultivation, processing and marketing of seaweed. Penebar Swadaya Publishing, Jakarta, p 78p. (In Indonesian)

Irmayani SY, Muhammad N (2014) Feasibility analysis of seaweed cultivation in Mallasoro Village, Bangkala District, Jeneponto Regency. Jurnal Bisnis Perikanan 1:17–28. (In Indonesian)

Johnson B, Gopakumar G (2011) Farming of the seaweed *Kappaphycus alvarezii* in Tamil Nadu coast - status and constraints. Mar Fish Infor Serv 208:1–5

Julizar S, Kurnia A, Kasim M (2018) Comparison of growth rate *Kappaphycus alvarezii* result of tissue cultures with different depths which are kept in Jaring Kantung Apung. Jurnal Media Akuatika 3: 749–759

Kasim M (2016) MacroAlgae, biology, ecology, utilization, and cultivation studies. Penebar Swadaya Publishing, Jakarta, p 156p. (In Indonesian)

Kasim M, Asnani (2012) Determination of the generative reproduction season and preference for spore placement, *Eucheuma cottoni*. J Ilmu Kelautan 17:209–216. (In Indonesian)

Kasim M, Mustafa A (2017) Comparison growth of *Kappaphycus alvarezii* (Rhodophyta, Solieriaceae) cultivation in floating cage and longline in Indonesia. Aquacult Rep 6:49–55. https://doi.org/10.1016/j.aqrep.2017.03.004

Kasim M, Mustafa A, Munier T (2016) The growth rate of seaweed (*Eucheuma denticulatum*) cultivated in longline and floating cage. AACL Bioflux 9:291–299

Kasim M, Jamil MR, Irawati (2017) Occurrence of macro-epiphyte on *Eucheuma spinosum* cultivated on floating cages. AACL Bioflux 10: 633–639

Kasim M, Marlia A, Balubi AM, Jalil W (2019) Succession of epiphyte on thallus of *Kappaphycus alvarezii* (Rhodophyta) in horizontal net cage culture. AACL Bioflux 12:1710–1717

Kasim M, Balubi AM, Mustafa A, Nurdin R, Patadjai RS, Jalil W (2020) Floating cage: a new innovation of seaweed culture, emerging technologies, environment and research for sustainable aquaculture, Qian Lu and Mohammad Serajuddin, IntechOpen. doi:https://doi.org/10.5772/intechopen.90035

Kasim M, Balubi AM, Astuti O, Rahman A, Patadjai RS, Muskita W, Takwir A, Ruslaini B, Jalil W (2021) Comparison between the growth of *Kappahycus alvarezii* (Rhodophyta) seed from tissue culture and clone selection cultivated using horizontal net. Egypt J Aquat Res 47(2):179–184

Kasim M, Balubi AM, Jalil W, Astuti O, Bahtiar, Muskita W, Nurdiana, La Ola O, Rahman A, Patadjai RS, Budiyanto, Pangerang UK, Ruslaini (2022) Seaweed cultivation technologies in Indonesia: current trends and future prospects. In: Ranga Rao A, Ravishankar GA (eds) Sustainable global resources of seaweeds, vol 1. Springer, Cham. https://doi.org/10.1007/978-3-030-91955-9_5

Ladunta H (2015) Growth of marine algae *Kappaphycus alvarezii* using verticulture methods at different depths in Boalemo Regency. Thesis Department of Aquaculture, Faculty of Fisheries and Marine Sciences. UNG, Gorontalo. (In Indonesian)

Langford A, Zang J, Waldron S, Julianto B, Siradjuddin I, Neish I, Nuryartono N (2022) Price analysis of the Indonesian carrageenan seaweed industry. Aquaculture 550:737828. https://doi.org/10.1016/j.aquaculture.2021.737828

Liang Z, Sun X, Wang F, Wang W, Liu F (2013) Impact of environmental factors on photosynthesis and respiration of young seedling of *Sargassum thunbergii* (Sargassaceae Phaeophyta). Am J Plant Sci 4:27–33

Luhan MRJ, Sollesta H (2010) Growing the reproductive cells (carpospores) of the seaweed, *Kappaphycus striatum*, in the laboratory until outplanting in the field and maturation to tetrasporophyte. J Appl Phycol 22:579–585

Luxton DM (1993) Aspect of farming and processing of *Kappaphycus* and *Eucheuma* in Indonesia. Hydrobiologia 60:365–371

Mariño M, Breckwoldt A, Teichberg M, Kase A, Reuter H (2019) Livelihood aspects of seaweed farming in Rote Island, Indonesia. Mar Policy 107. https://doi.org/10.1016/j.marpol.2019.103600

Masyahoro, Mappiratu (2010) Growth response at various seed depths and harvesting age of seaweed *Eucheuma cottonii* in Palu Bay Waters. Media Litbang Sulteng 3:104–111. (In Indonesian)

Merisca N (2013) The acclimatization of *K. alvarezii* seaweed from tissue culture at different densities in an aquarium in a greenhouse. Thesis Bogor Agricultural University, Faculty of Fisheries and Marine Sciences. IPB University. 73p (In Indonesian)

Msuya FF (2013) Social and economic dimensions of carrageenan seaweed farming in the United Republic of Tanzania. In: Valderrama D, Cai J, Hishamunda N, Ridler N (eds) Social and economic dimensions of carrageenan seaweed farming. Fisheries and Aquaculture Technical Paper No. 580. FAO, Rome, pp 115–146

Munoz J, Freile-Pelegrin Y, Robledo D (2004) Mariculture of *Kappaphycus alvarezii* (Rhodophyta Solieriaceae) color strain in tropical water of Yucatan, Mexico. Aquaculture 239:161–177

Mutalib Y, Rahman AS (2018) Growth and carrageenan content of *Kappaphycus alvarezii* seaweed at dosage of local microorganisms (Moles) of Maja Fruit. J Blue Ocean 2:1–8

Nursima KM, Nurdinana A (2021) Feasibility analysis of sea grass cultivation (*Kappaphycus alvarezii*) using horizontal net (case study in Katobengke Village Betoambari District Baubau City). J Fish Socioecon 6:157–163. (In Indonesian)

Nursyahran, Reskiati (2013) Increased growth rate of Seaweed Thallus (*Kappaphycus alvarezii*) immersed in rice water with different concentrations. Jurnal Balik Diwa 4(2):13–18. (In Indonesian)

Parenrengi A, Rachmansyah, Suryati E (2011) Cultivation of carrageenan-producing seaweed (Karaginophy). Revision Edition. Brackish Water Aquaculture Research Institute, Marine and Fisheries Research and Development Agency, Ministry of Fisheries and Marine Affairs. Jakarta. 54p (In Indonesian)

Poncomulyo T, Maryani H, Kristiani L (2006) Seaweed cultivation and management. PT. Agromedia Pustaka. Surabaya (In Indonesian)

Reis RP, Pereira RR, Geromel de Góes H (2015) The efficiency of tubular netting method of cultivation for *Kappaphycus alvarezii* (Rhodophyta, Gigartinales) on the southeastern Brazilian coast. J Appl Phycol 27:421–426

Rizal M, Mappiratu AR, Razak R (2016) Optimization of semi refined carrageenan (SRC) production from seaweed (*Eucheuma cottoni*). Kovalen J 2:33–38

Robledo D, Gasca-Leyva E, Fraga G (2013) Social and economic dimensions of carrageenan seaweed farming in Mexico. In Valderrama D, Cai J, Hishamunda N, Ridler N (eds) Social and economic dimensions of carrageenan seaweed farming. Fisheries and Aquaculture Technical Paper No. 580. FAO, Rome, pp 185–204

Safia W, Budiyanti, Musyrif (2019) Growth and carrageenan of seaweed (*Eucheuma cottoni*) using the development of the hanging raft method at different depths. Multidiscip Synergy Sci Technol 2: 101–108

Saik F (2016) Analysis of business benefits of seaweed cultivation (*Kappaphycus alvarezii*) using floating cage method in the Lakeba beach of Baubau city, Thesis, Department of Fisheries Agribusiness, Faculty of Fisheries and Marine Science, Halu Oleo University, Kendari (In Indonesian)

Sardianti N, Widiastuti MI (2010) Growth and production of seaweed *Eucheuma cottonii* at different planting depths. Jurnal Media Litbang Sulteng III (1):21–26

Steenbergen DJ, Marlessy C, Holle E (2017) Effects of rapid livelihood transitions: examining local codeveloped change following a seaweed farming boom. Mar Policy 82:216–223. https://doi.org/10.1016/j.marpol.2017.03.026

Sugiarto (2005) Microeconomics. A comprehensive study. PT Gramedia Pustaka Utama Publishing, Jakarta. (In Indonesian)

Sulu R, Kumar L, Hay C, Pickering T (2004) *Kappaphycus* seaweed in the Pacific: review of introductions and field testing proposed quarantine protocols. Secretariat of the Pacific Community, Noumea, 84p

Sumarni (2015) Feasibility analysis of seaweed (*Kappaphycus alvarezii*) cultivation using long line method and floating net raft. Southeast Sulawesi. Undergraduate Thesis. Faculty of Fisheries and Marine Sciences, Halu Oleo University (In Indonesian)

Suryadin (2021) Financial feasibility analysis of seaweed cultivation business (case study on vertinet equipment in Lakeba village, Betoambari district, Baubau City), Thesis, Department of Agribusiness, Faculty of Fisheries and Marine Sciences, Halu Oleo University, Indonesia, 85p (In Indonesian)

Syahlun, Rahman A, Ruslaini (2013) Growth test of seaweed (*Kappaphycus alvarezii*) brown strain using verticulture method. Mina Laut Indonesia 1:122–132. (In Indonesian)

Syarqawinagan (2017) The influence of seaweed bag on the growth of seaweed (*Eucheuma cottoni*) in The Waters of Simeulue Regency. Jurnal Ilmiah Mahasiswa Kelautan dan Perikanan Unsyiah 2:277–285. (In Indonesian)

Thirumaran G, Anatharaman P (2009) Daily growth rate of field farming seaweed *Kappaphycus alvarezii* (Doty) Doty ex P. Silva in Vellar Estuary. World J Fish Mar Sci 1:144–153

Tiwa R, Mondoringin L, Salindeho I (2013) Growth of *Kappaphycus alvarezii* seaweed at differences in initial depth and weight in Talengen Waters, Sangihe Islands Regency. J Budidaya Perairan 1: 63–68. (In Indonesian)

Tomas F, Cebrian E, Ballesteros E (2011) Differential herbivory of invasive algae by native fish in the Mediterranean Sea. Estuar Coast Shelf Sci 92:1–8. https://doi.org/10.1016/j.ecss.2010.12.004

Vairappan CS, Chong CS, Hutardo AQ, Soya FE (2007) Distribution and symptoms of epiphyte infection in major carrageenophyte-producing farms. J Appl Phycol. https://doi.org/10.1007/s10811-007-9299-8

Valderrama D, Cai J, Hishamunda N, Ridler N, Neish IC, Hurtado AQ (2015) The economics of *Kappaphycus* seaweed cultivation in developing countries: a comparative analysis of farming systems. Aquac Econ Manag 19:251–277

van den Burg SWK, van Duijn AP, Bartelings H, van Krimpen MM, Poelman M (2016) The economic feasibility of seaweed production in the North Sea. Aquac Econ Manag 20(3):235–252. https://doi.org/10.1080/13657305.2016.1177859

Wati N (2019) Comparative study of Macroepiphyte attachment rate on thallus of *Eucheuma denticulatum* at different depths using Verti net in the waters of Tanjung Tiram Village, South Konawe Regency. J Manajemen Sumber Daya Perairan 4:274–291. (In Indonesian)

Widyorini N (2010) Growth analysis of *Gracilaria* sp. in shrimp farms in terms of sedimentation level. J Saintek Perikanan 6:30–36. (In Indonesian)

Wijayanto T, Hendri M, Aryawati R (2011) Study on the growth of *Eucheuma cottonii* seaweed with different planting methods in Kalianda Waters, South Lampung. J Maspari 3:55–57

Yusup S, Kasim M, Balubi AM (2017) Effect of initial weight on the growth and carragenan of seaweed infected by epiphyte reared in the Floating Net Cage. J Media Akuatika 2:509–518

A Phyconomic Game-Changer: Extracts of Selected Brown Seaweeds as Phyco(bio)stimulants for Eucheumatoids

23

Anicia Q. Hurtado, Majid Khan Mahajar Ali, and Alan T. Critchley

Abstract

Seaweed extract-derived biostimulants (i.e., phyco[bio] stimulants—PBSs) have been reported previously as effective organic ingredients in the agricultural and horticultural industries for several decades. However, their commercial applications to large-scale seaweed cultivation (phyconomy) are still in their nascent stage, although several experimental studies have been reported since 2009. Among the PBS, the brown alga *Ascophyllum nodosum* has perhaps been studied the most extensively (and may even be overrepresented) compared to other brown seaweed extracts, e.g., *Durvillaea* spp., *Ecklonia* spp., *Macrocystis* spp., *Laminaria/Saccharina* spp., and *Sargassum* spp. In this chapter, applications of the PBS *Ascophyllum* Marine Plant Extract Powder (AMPEP) are discussed at different phyconomic stages, such as micropropagation, land-sea-based nursery, sea cultivation to include mitigation of pests and diseases, and semi-refined carrageenan quality, of several seaweeds, with a focus on members of the eucheumatoids.

Keywords

Seaweed extracts · Phyco(bio)stimulants · Eucheumatoids

A. Q. Hurtado (✉)
University of the Philippines Visayas, Miag-ao, Iloilo, Philippines

Integrated Services for the Development of Aquaculture and Fisheries (ISDA) Inc., Iloilo City, Philippines

M. K. M. Ali
School of Mathematical Sciences, Universiti Sains Malaysia, Gelugor, Penang, Malaysia

Centre for Marine and Coastal Studies (CEMACS), Universiti Sains Malaysia, Gelugor, Penang, Malaysia

A. T. Critchley
Verschuren Centre for Sustainability in Energy and Environment, Cape Breton, Canada

23.1 Introduction

The regulatory term "biostimulant" is used as a descriptive category for specific inputs to agriculture and is labelled ecologically wholesome and favorable since most are considered biodegradable, non-toxic, non-polluting, and non-hazardous for the environment (Yakhin et al. 2017). Biostimulants are substances of biological origin, or microorganisms, which when applied via root drench, foliar spray, or a combination of both are intended to stimulate natural processes in treated plants, thereby producing nutrient-use-efficiency and growth processes and/or increased tolerance to abiotic and biotic stress. These benefits are irrespective of the nutrient content, i.e., NPK fertilizing effects of the substances (Yakhin et al. 2017). Biostimulants have many formulations with varying ingredients, including humic substances (humic and fulvic acids), seaweed extracts, beneficial bacteria and fungi, chitosans (a soluble version of chitin), protein hydrolysates, and inorganic compounds, e.g., silicon. Algal extracts are called phyco(bio)stimulants (PBS; Hurtado et al. 2021a), as they are derived from selected macroalgae or seaweeds. They are biostimulants rather than fertilizers since they stimulate defense and growth responses when used at appropriate stages for the growth of terrestrial and aquatic plants (e.g., cultured seaweeds).

Seaweed extracts, as plant biostimulants, not rich in NPK, are associated with micronutrients and nitrogenous compounds that can comprise hormones, e.g., cytokinins, auxins, abscisic acid, gibberellins, and other classes of hormone-like compounds, such as sterols and polyamines (Craigie 2011; Wally et al. 2013; Carillo et al. 2020; Lau et al. 2022). Seaweed extracts may also contain compounds such as polysaccharides (i.e., laminarin, alginates, agarans, and carrageenans) and some other nitrogenous compounds, including betaines, depending on the type of seaweed and processing used (Khan et al. 2009; Craigie 2011). The papers of Shukla et al. (2019, 2021, 2022) are just a few reports that show the importance of seaweed extracts in improving plant

A. T. Critchley et al. (eds.), *Tropical Phyconomy Coalition Development*, Developments in Applied Phycology 11,
https://doi.org/10.1007/978-3-031-47806-2_23

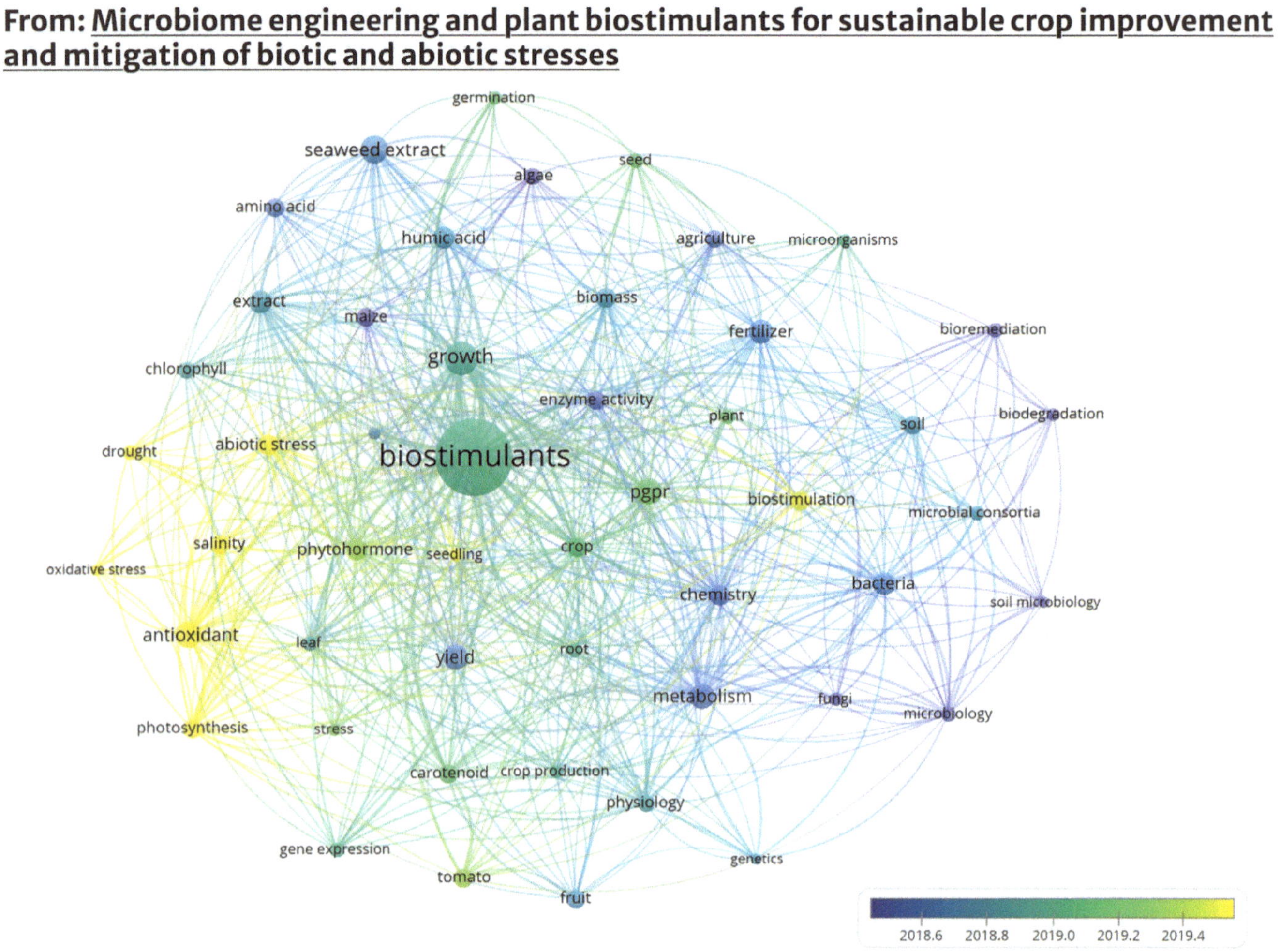

Fig. 23.1 Roles of different phyco(bio)stimulants in agriculture and horticulture (Lau et al. 2022)

growth stress tolerance and both water and nutrient deficiencies.

In 2016, the global, commercial biostimulant share in the market was topped by seaweed extracts at 33%, followed by humic substances (31%), vitamins and amino acids, microbial amendments, and trace minerals at 10%. However, by 2025, it is predicted that amino acids (35%) will overtake seaweed extracts (27%) and humic substances (19%). Interestingly, there was a high growth rate of biostimulant application in Asia and Europe, followed by North and South America, whereas the lowest adoption was in Africa (Markets and Markets 2022).

The first seaweed extracts used as PBS for agricultural (plant) crops were reviewed extensively by Khan et al. (2009), Craigie (2011), Dmytryk and Chojnacka (2018), Górka et al. (2018), Shukla et al. (2019), and Ali et al. (2016). The authors illustrated the benefits of these extracts for sustainable crop production, including (1) stimulators of plant growth and yield and (2) plant immune stimulators that evoke defense responses, which contribute to enhanced resistance of treated plants to several pests, diseases, and abiotic stresses. In some cases, these benefits were linked to the up-regulation of important defense-related genes and pathways, priming defenses against future attacks and (3) stimulators of phytohormonal responses due to their specific components and interactions with plant growth regulation.

Plant biostimulants from different sources, including chitosan and other biopolymers, seaweed extracts, protein hydrolysates, humic and fulvic acids, and microbial constituents (Lau et al. 2022), and their multifaceted impacts on agricultural crops and horticulture are summarized in Fig. 23.1. As illustrated, biostimulants offer a potentially novel approach for the regulation/modification of physiological processes in plants that stimulate growth, mitigate stress-induced limitations, and increase crop yields (Yakhin et al. 2017).

There are many commercial products based on various types of seaweeds and various proprietary extraction processes. To date, there are only three brown seaweed extracts that have specifically been applied and used for their benefits to seaweed cultivation, i.e., used as PBS, namely,

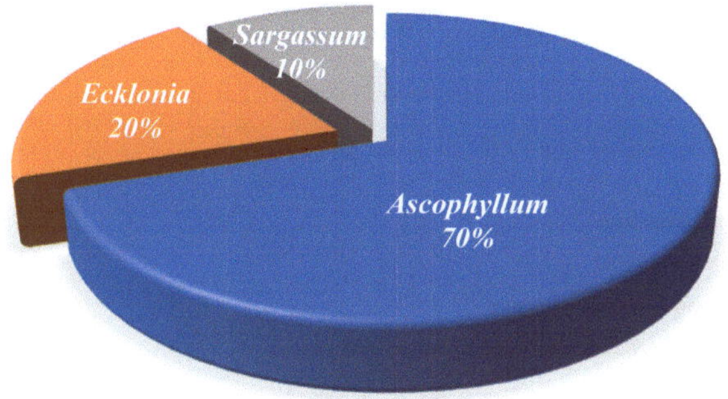

Fig. 23.2 Types of brown seaweed extract phyco(bio)stimulants used in different seaweed studies (Lau et al. 2022)

Ascophyllum nodosum (61%), *Ecklonia maxima* (22%), and *Sargassum* spp. (11%) (Fig. 23.2). The efficacies and benefits of the application of these seaweed extracts have been applied mainly to aid the cultivation of a small number of commercially important seaweeds comprising red (*Eucheumatopsis*, *Gracilaria*, *Kappaphycus*, *Laurencia* and *Neopyropia*), brown (*Saccharina latissima*), and green (*Ulva* sp.) macroalgae (Table 23.1, Fig. 23.3). A total of 28 studies have been reported on the application of PBS in seaweed cultivation, largely with respect to the production of *Kappaphycus*, *Gracilaria*, *Saccharina*, *Ulva*, and *Eucheumatopsis*. These studies were conducted in nine seaweed-producing countries led by Brazil, Malaysia, Korea, the USA , and the Philippines (Fig. 23.4).

23.2 Seaweed Extracts Used in Eucheumatoid Production

23.2.1 Brown Seaweed Extracts

The most common brown seaweeds used as biostimulants for agricultural and horticultural purposes are *Ascophyllum nodosum*, *Durvillaea* spp., *Ecklonia maxima*, *Laminaria digitata*, and *Saccharina* spp. from temperate waters, while *Sargassum* spp. and *Turbinaria* spp. from the tropics have been evaluated. Brown seaweeds are known to have a higher phenolic content than green and red seaweeds (Hardouin et al. 2014), and these compounds have been shown to be involved as secondary metabolites, which are not directly involved in primary processes such as photosynthesis, cell division, or reproduction of algae but are involved in various mechanisms that are protective, such as against the oxidative or cytotoxic effects of UV exposure.

Phenolic compounds from brown seaweeds are commonly characterized as stress-induced compounds involved in chemical protection mechanisms against biotic factors such as contamination due to bacteria and abiotic factors such as UV radiation or metal contamination (Llewellyn and Airs 2010; Stengel et al. 2011). Phenolic compounds are produced in response to oxidative stress and exhibit antioxidative properties, which ensure an efficient cytoprotective system. The antioxidant and free-radical scavenging activities of phenols have essentially been identified in brown seaweeds (Ahn et al. 2004; Athukorala et al. 2007; Heo et al. 2005).

Table 23.1 List of seaweed extracts used in the cultivation of different seaweed species

Type of seaweed extract	*Eucheumatopsis*	*Gracilaria*	*Kappaphycus*	*Laurencia*	*Neopyropia*	*Saccharina*	*Ulva*
Ascophyllum Marine Plant Extract Powder (AMPEP)	Umanzor et al. (2020a)	Souza et al. (2018) Chirapart et al. (2019) De Vega et al. (2020) Dawagne and Jaiswar (2020) Jaiswar et al. (2021)	Hurtado et al. (2009) Borlongan et al. (2011) Yunque et al. (2011) Hurtado et al. (2012) Loureiro et al. (2012) Yong et al. (2014) Loureiro et al. (2014a, b) Suryati et al. (2015) Zuldin and Shapawi (2015) Marroig et al. (2016) Tibubos et al. (2017) Pedra et al. (2017) Ali et al. (2018a, b, 2020a, b) Chirapart et al. (2019) Loureiro et al. (2010)	Souza et al. (2018)	Umanzor et al. (2022a, b)	Umanzor et al. (2019)	Kumar et al. (2013)
Ecklonia (Kelpak)	Umanzor et al. (2020a)					Umnazor et al. (2020b)	Robertson-Anderson et al. (2006)
Sargassum		Cokrowati et al. (2021)					

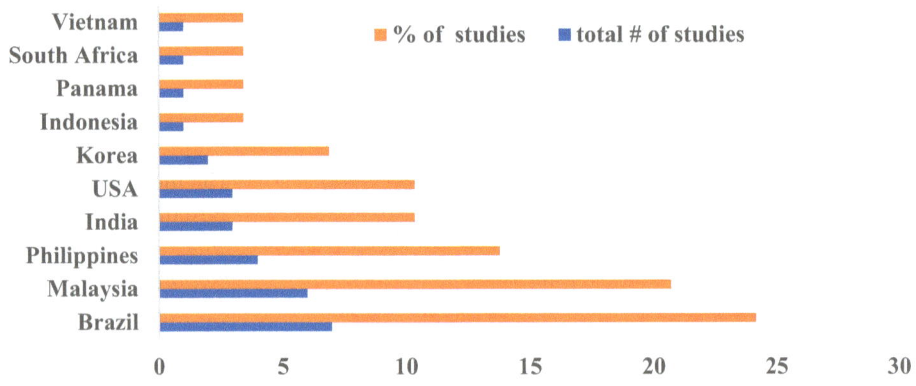

Fig. 23.3 Countries engaged in seaweed extract phyco (bio)stimulant applications to seaweeds

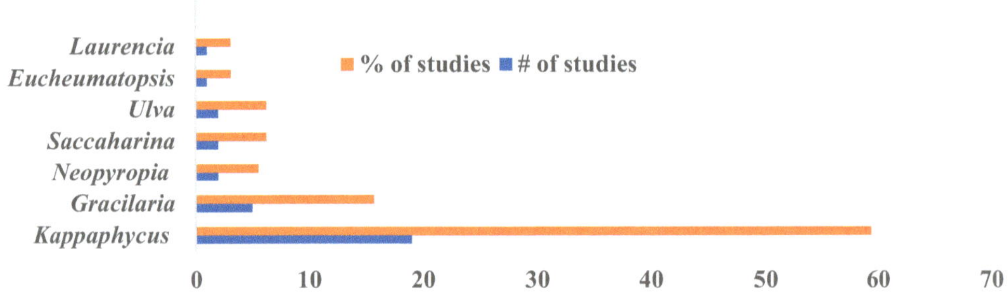

Fig. 23.4 Number of studies performed on different genera of seaweeds using different types of seaweed extract phyco (bio)stimulants

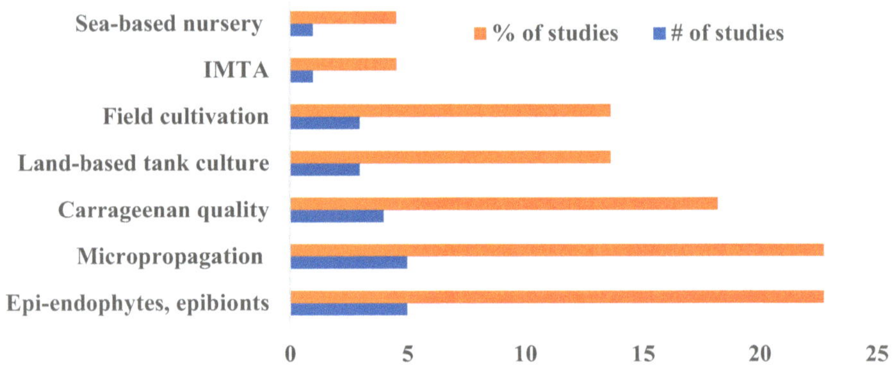

Fig. 23.5 Different areas of *Kappaphycus* studies reported using the brown seaweed extract from *Ascophyllum nodosum*

23.2.2 *Ascophyllum nodosum*

There are several commercial extracts of seaweeds available on the market, particularly of the brown alga *A. nodosum*, from which *Ascophyllum* Marine Plant Extract Powder (AMPEP) is derived, produced using various extraction procedures. For example, AMPEP is a water-soluble extract produced mainly using KOH (open atmosphere) and then dried to a powder, which can then be rehydrated in water or growing solution. There are several manufacturers of AMPEP worldwide. Studies on the effects of extracts of *A. nodosum* on seaweed cultivation report improvements in micropropagation and carrageenan qualities of eucheumatoids (Fig. 23.5). The first report made on the biostimulant effects of this seaweed extract was by Hurtado et al. (2009) using AMPEP for the micropropagation of three varieties of *Kappaphycus alvarezii* (e.g., Kapilaran, Tambalang purple, and Adik-adik) and one variety of *K. striatus* (sacol). The findings revealed that each variety responded differently when supplemented with AMPEP plus the standard plant growth regulators (PGRs: indole-3-acetic acid (IAA) and kinetin) in terms of the number of days required before they formed direct/lateral axes. Thus, adik-adik and tambalang purple formed direct axis shoots after 17–22 days with 1–5 mg L^{-1} AMPEP plus PGRs (IAA and kinetin, 1 mg L^{-1} each), whereas sacol and kapilaran took 25 and 46 days, respectively, to form shoots at 1–3 mg L^{-1} AMPEP plus PGRs (IAA and kinetin at 1 mg L^{-1} each). The reports thus far suggest that low concentrations of AMPEP induce a fast growth rate and mitigate pests and pathogens in seaweed cultures, which mirrors similar benefits when these types of extracts are applied to terrestrial agricultural crops (Juárez-Maldonado et al. 2019).

The report of Hurtado et al. (2009) formed the basis for a wider scope of applications of PBS in seaweed cultivation, particularly that of eucheumatoids. To date, studies span the mitigation of epi-endophytes and epibionts and micropropagation formed 26% of the total, followed by carrageenan quality (18%), land-based tank culture and field cultivation (14%), and sea-based and integrated multi-trophic aquaculture (5%) (Fig. 23.4).

The earliest reports on the application of AMPEP to eucheumatoids showed overall improvement of the cultivated varieties. Benefits included fast growth with robust branches and improved yield, viscosity and gel strength of the extracted carrageenan; in addition, there were reduced percentages of macro-epiphytes and the harmful red, filamentous epi-endophyte *Melanothamnus* sp. (also known as *Neosiphonia/Polysiphonia*), and in the incidence of epibionts (Hurtado et al. 2012; Borlongan et al. 2011; Loureiro et al. 2012; Marroig et al. 2016; Ali et al. 2018b, 2020a).

Initial results on the relative concentration (i.e., g/L of soluble extract solids as a powder) and frequency that the eucheumatoid plants were dipped in AMPEP showed that two or three immersions of seedlings during the entire 45-day growth cycle were not significantly different from a single dipping (Ali et al. unpublished data). However, this was dependent on the month of cultivation. The months of July to September in Semporna, Sabah are described as the wet season and are known to be better growing months; hence, higher growth rates of *Kappaphycus* were obtained during these months.

One of the common responses of seaweeds under stress conditions is the generation of reactive oxygen species (ROS), such as singlet oxygen ($^1O_2$), superoxide anions (O_2^-), hydrogen peroxide (H_2O_2), and hydroxyl radicals (•OH) (Sachdev et al. 2021). ROS refers to short-lived, highly reactive reduced state oxygen atoms or molecules, which when produced in excess oxidize proteins, lipids, DNA and RNA, leading to cell death (Dumanović et al. 2021).

A report on the use of AMPEP as a mitigator of endophytes in Brazil was published by Loureiro et al. (2012). They claimed that the administration of an AMPEP extract enhanced the effects of superficial oxidative bursts (i.e., production of hydrogen peroxide on the surface of the thalli), which can be extremely aggressive not only for the individual plant but also mostly for its epiphytes. The bleaching of the non-corticated portions of the filamentous red epiphyte *Polysiphonia subtilissima* (now described as *Melanothamnus)* thalli in their study was evident, whereas AMPEP treatments protected *K. alvarezii* from hydrogen peroxide evolution. *Ascophyllum* seaweed extract is known to activate systemic defense mechanisms in plants (Santaniello et al. 2017; de Saeger et al. 2020), and similar research is urgently needed for applications to cultivated seaweeds to begin to understand modes of action and then to precisely focus specific applications (at different rates and timings) on different stages of the seaweed cultivation process for their greatest benefits.

There are several reports of various manufactured *A. nodosum* extracts improving the stress tolerance of land plants, especially agricultural crops, as reported by Shukla et al. (2019). Some agricultural crops, such as maize (Basavaraja et al. 2018; Trivedi et al. 2021) and tomatoes (Ali et al. 2016, 2019; Campobenedetto et al. 2021; Hussain et al. 2021), and in particular the model plant *Arabidopsis thaliana* (Rayorath et al. 2008; Khan et al. 2011; Wally et al. 2013; De Saeger et al. 2020), have been studied intensively at the molecular level with respect to the action of biostimulants. Knowledge on the impacts of various seaweed-derived extracts as biostimulants and their mechanisms of action are increasingly required for regulatory and end-user information/education purposes. A particular extract from the red seaweed *Kappaphycus* Seaweed Extract (i.e., KSWE) was found to upregulate the genes involved in IAA, gibberellic acid, and cytokinin signalling while down-regulating the genes involved in abscisic acid biosynthesis (Trivedi et al. 2021). Other key genes that were up- and down-regulated were those involved in photosynthesis and nitrogen metabolism, polysaccharide synthesis, and signal transduction. These gene regulations collectively portray the drought alleviation mechanism of *Kappaphycus* seaweed extract in maize, as the same authors claimed. It is tempting to suggest parallel mechanisms are at play when selected extracts of specific seaweeds are applied to enhance the growth and productivity of cultivated seaweeds. The challenge is to perform the necessary seaweed physiological work required to tease out modes of actions and responses.

There are now reports of the effects of *A. nodosum* extracts on the red seaweeds *Gracilaria* and *Laurencia*. Souza et al. (2019) reported the stimulatory effect of AMPEP on the growth, branching development, and pigment content of *G. caudata* and *L. catarinensis*, which can possibly be applied in large-scale cultivation. Work on another *Gracilaria* species, *G. corticata var. cylindrica* showed that it has the capacity to produce large numbers of plantlets within 4–5 weeks, making it highly attractive for implementation in sustainable seaweed farming (Dawange and Jaisnar 2020). The reports of these authors proved that the use of an *A. nodosum* extract plus plant growth regulators in *G. blodgettii* for micropropagation purposes was effective in producing microplantlets for use as seedlings in field cultivation in Panama (de Vega et al. 2020). The apical, sub-apical, and basal explants of *G. salicornia* showed capacity as sources of seedlings; however, the apical explants clearly showed superior characteristics (Jaiswar et al. 2021). The efficacy of different concentrations of seaweed extract singly and in combination with plant growth regulators on the survival and regeneration of *G. dura* was recently tested in India (Mantri V. Pers. Comm) to determine its modulatory effects on key biological processes at both the biochemical and molecular levels.

Hopefully, this work will be completed within the next 5–10 years, and extracts of seaweeds will play a fuller role in the phyconomic production of seaweeds required for food and carbon-capture purposes. However, to date, no other reports have been made on the role of extracts of seaweeds as PBS for applications in eucheumatoid crops and/or other seaweeds as beneficial phyco-elicitors. This area of research for *Kappaphycus* species and strains (one of the world's largest marine crops) in particular is important, and therefore, new phyconomists are encouraged to do the necessary work required such that all seaweed crops receive their deserved attention with respect to breeding, husbandry production, and control of pests and diseases (Cottier-Cook et al. 2016; Hurtado et al. 2021b).

Further details about brown seaweed extracts used in eucheumatoid production can be found in Hurtado et al. (2021a). This chapter largely described an efficacious, phyconomic, circular blue economy: with special reference to *A. nodosum* and *Kappaphycus* spp., brown and red seaweeds, respectively, which are now the world's largest marine crops with significant socio-economic benefits to those countries within a circum-tropical distribution of the globe. Many of these producing countries have emerging economies, and seaweed cultivation is a socio-economic engine for coastal development and sustainability. The authors in no way suggest that these beneficial properties are exclusive to *Ascophyllum* and *Kappaphycus,* and indeed, the motivation of our recent publications has been to encourage the wider application of many different types of extracts of seaweeds for benefits in the wider field of phyconomic endeavor.

23.2.3 *Ecklonia maxima*

The large brown seaweed (kelp, Laminariales) *Ecklonia maxima* is abundantly found on rocky shores of South Africa and is harvested for the production of liquid, natural growth stimulants for agricultural and horticultural applications. It is mainly applied to roots by direct immersion (e.g., dip or drench) or by soil irrigation systems in different contractions depending on the species and type of production. One such product is Afrikelp (https://afrikelp.com).

Several studies have reported that liquid extracts from *E. maxima* are effective in promoting growth and increasing the crop productivity of some agricultural crops, such as maize (Szczepanek 2018), zucchini squash (Rouphael et al. 2017), and lettuce (Miceli et al. 2021), and in improving tolerance under nutrient stress conditions (Di Stasio et al. 2017).

Only a few studies have reported on the use of *E. maxima* extract in seaweed cultivation. The report of Robertson-Anderson et al. (2006) mainly dealt first with laboratory studies on *Gracilaria gracilis* using Provasoli's enriched seawater (PES) medium in combination with different concentrations of Kelpak® (https://kelpak.com/). *Gracilaria* tips treated with Kelpak® showed significantly better growth than the control. The authors further studied the efficacy of Kelpak® in a pilot commercial-scale *Ulva* and turbot culture. Once a week, the addition of Kelpak® at a low concentration plus the nitrogen-rich effluents coming from the turbot proved to be a good source of nutrients for enhanced growth of *Ulva* compared to a higher concentration of Kelpak® (1: 500) alone. In short, the nutrients coming from the effluents of turbot plus low concentrations of Kelpak® provide a viable approach for the growth of *Ulva* in mariculture.

A recent report of Umanzor et al. (2020a) on *Eucheumatopsis isiformis* grown in land-based nursery systems showed that constant exposure treatment with either Kelpak® or AMPEP produced a greater number of shoots per propagule with no other significant effects. The same authors, in 2020b, reported that juvenile sporophytes of *Saccharina latissima* and *S. angustissima,* when exposed to sub-lethal temperatures at 18 °C, exhibited enhanced thermotolerance when treated with Kelpak®. Seemingly, a low concentration of Kelpak® (0.001 mL $L^{-1)}$ promotes early germination and formation of conchocelis tufts of *Neopyropia yezoensis* at 15 °C; however, a higher temperature at the same concentration of Kelpak® appeared to delay positive its effect on blade formation (Umanzor et al. (2022b). Furthermore, the authors claim that their preliminary results do not show significant evidence that the application of Kelpak® as tested could affect the transition between life stages or provide enhancements on thermotolerance. The efficacy of the liquid extract Kelpak® from *E. maxima* parallels the benefits reported in some terrestrial agricultural crops (Dzugielt and Wadas 2020).

23.2.4 *Sargassum* spp.

Another brown seaweed that has been tested as an effective source of PBS is *Sargassum.* This seaweed is represented by numerous species distributed throughout the temperate and tropical oceans of the world, where they generally inhabit

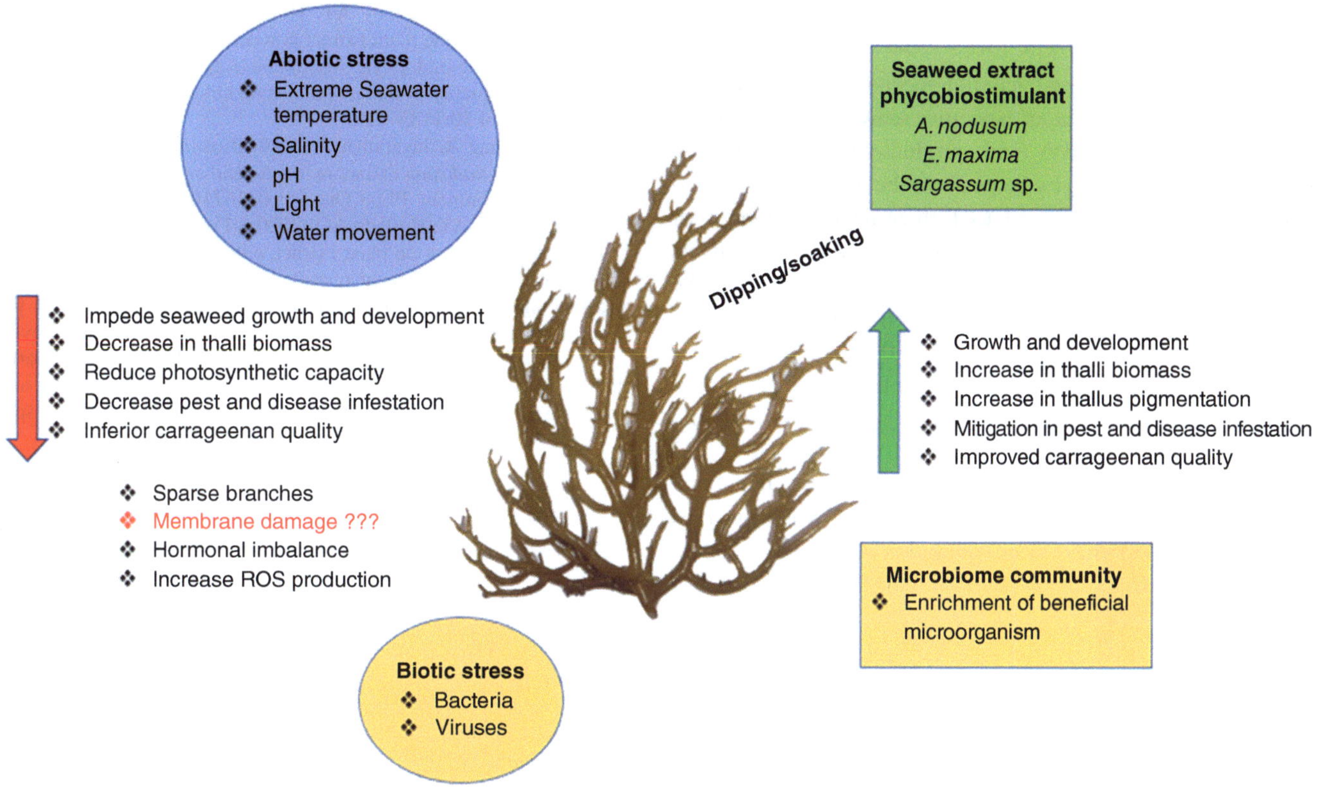

Fig. 23.6 An illustrated diagram of the roles of seaweed extract phyco(bio)stimulants in eucheumatoids (Patterned after Lau et al. 2022)

shallow water and coral reefs and are widely known for their planktonic (free-floating) species. The genus has become infamous in the past decade due to golden brown tides drifting in the Gulf of Mexico and Caribbean, causing unprecedented washups on beaches (Saldarriaga-Hernandez et al. 2021; Chávez et al. 2020; Cabanillas-Terán et al. 2019). Indeed, there are some opportunistic examples of commercial extracts made from this stochastic source of raw materials. In natural beds, *Sargassum* forms a thick bed from the edge of the intertidal to the subtidal region, which serves as an excellent nursery ground for juvenile fish, crustaceans, and molluscs. This seaweed can also be a good source of alginate, which has a wide array of applications, notably in the pharmaceutical and nutraceutical industries (Abdelwahab 2017; Pereira 2018).

The use of an extract from *S. aquafolium* as a PBS was recently described for the cultivation of eucheumatoids (Wouthuyzen et al. 2016). Its use is in keeping with the higher phenolic content of brown seaweeds compared to green and red seaweeds (Hardouin et al. 2014; Mabeau and Fleurence 1993) and the protection they afford against oxidative or cytotoxic effects of UV damage.

Sargassum spp. can be found in the waters of North Sulawesi (Wouthuyzen et al. 2016) and in the waters of Ekas Bay, East Lombok, and West Nusa Tenggara. This seaweed resource is abundant and seemingly untapped in these areas, awaiting exploration. *S. aquifolium* is known to contain carbohydrates (59.51%), fat (8.41%), Ca (3.34%), Fe (0.12%), P (0.18%), N (7.22%), water (12.79%), and ash (12.79%) (Mageswaran and Sivasubramaniam 1984; Cokrowati and Diniarti 2019; Liu et al. 2020). Components present in *S. aquifolium* have the potential to act as phytohormones that can be used as growth stimulants in *K. alvarezii.*

The presence of endogenous auxin (401.89 ppm) and gibberellin (0.89 ppm) in *S. aquifolium* triggers cell division in the thallus and the formation of thallus branching in *K. alvarezii* (Cokrowati et al. 2021). A growth rate of 9.2% for the first 18 days was reported. However, lower growth rates (1–2%) occurred thereafter until day 45, which may be due to prevailing external factors such as slow water movement, which is not optimal during the time of cultivation (Xiao et al. 2019; Nursidi et al. 2017). For further understanding on the role of auxins in a multicellular organism such as seaweeds, the review paper of Bogaert et al. (2022) is worth reading.

The illustration below (Fig. 23.6) summarizes the different roles of seaweed extract PBSs in eucheumatoid growth and development, and other physiological activities as influenced by the health status of the cultured seaweed and the prevailing environmental characteristics impacting the farm site.

23.3 Conclusions

It is tempting to suggest parallel mechanisms are at play when selected extracts of specific seaweeds are applied to enhance the growth and productivity of cultivated seaweeds. The challenge is to perform the necessary seaweed physiological work required to further decipher modes of actions and responses. Inclusion of molecular mechanisms of action/response, as seen in land plants, would also be helpful, but not presently possible. Hopefully, this work will be completed within the next 5–10 years, and extracts of seaweeds will play a fuller role in the phyconomic production of seaweeds required for food and carbon-capture purposes.

The authors in no way suggest that these beneficial properties are exclusive to *Ascophyllum* and *Kappaphycus*; indeed, our motivation in recent publications has been to encourage the wider application of many different types of extracts of seaweeds to benefit the wider field of phyconomic endeavor.

Declarations **Funding**: Writing of the manuscript was partially funded by the Integrated Services for the Development of Aquaculture and Fisheries (ISDA) Inc., Iloilo City Philippines and School of Mathematical Sciences, Universiti Sains Malaysia, Penang, Malaysia; Centre for Marine and Coastal Studies (CEMACS), Universiti Sains Malaysia, Penang, Malaysia.

Contributions: AQ Hurtado prepared the outline, which was reviewed and approved by the coauthors. Writing of the first draft and revisions were performed by all authors.

Competing Interests: The authors declare that they have no competing financial interests or personal relationships that influenced the work reported in this paper.

References

Abdelwahab R (2017) Therapeutical and pharmaceutical applications of seaweeds. In: Nabto E (ed) Biotechnological applications of seaweeds. Nova Science Publishers, Hauppauge, pp 85–116

Ahn CB, Jeon YJ, Kang DS, Shin TS, Jung BM (2004) Free radical scavenging activity of the enzymatic extracts from brown seaweed *Scytosiphon lomentaria* by spin resonance spectrometry. Food Res Int 33:253–258

Ali N, Farerell A, Ramsubhag A, Jayaraman J (2016) The effect of *Ascophyllum nodosum* extract on the growth, yield and fruit quality of tomato grown under tropical conditions. J Appl Phycol 27:2007–2015

Ali N, Farrell A, Ramsubhag A, Jayaraman J (2016) The effect of *Ascophyllum nodosum* extract on the growth, yield and fruit quality of tomato grown under tropical conditions. J Appl Phycol 28:1353–1362

Ali MM, Sani MZB, Hi KK, Yasir SM, Critchley AT, Hurtado AQ (2018a) The comparative efficiency of a brown algal-derived biostimulant extract (AMPEP), with and without supplemented PGRs: the induction of direct, axis shoots as applied to the propagation of vegetative seedlings for the successful mass cultivation of three commercial strains of *Kappaphycus* in Sabah, Malaysia. J Appl Phycol 30:1913–1919

Ali MM, Yashir S, Critchley AT, Hurtado AQ (2018b) Impacts of *Ascophyllum* marine plant extract powder (AMPEP) on the growth, incidence of the endophyte *Neosiphonia apiculata* and associated carrageenan quality of three, commercial cultivars of *Kappaphycus*. J Appl Phycol 30:1185–1195

Ali O, Ramsubhag A, Jayaraman J (2019) Biostimulatory activities of *Ascophyllum nodosum* extract in tomato and sweet pepper crops in a tropical environment. PLoS One 14:e0216710

Ali MK, Critchley AT, Hurtado AQ (2020a) The impacts of AMPEP K + (*Ascophyllum* marine plant extract, enhanced with potassium) on the growth rate, carrageenan quality, and percentage incidence of the damaging epiphyte *Neosiphonia apiculata* on four strains of the commercially important carrageenophyte *Kappaphycus*, as developed by micropropagation techniques. J Appl Phycol 32:1907–1916

Ali MK, Critchley AT, Hurtado AQ (2020b) Micropropagation and sea-based nursery growth of selected commercial *Kappaphycus* species in Penang, Malaysia. J Appl Phycol 32:1301–1309

Athukorala Y, Lee KW, Kim SK, Jeon YJ (2007) Anticoagulant activity of marine green and brown algae collected from the Jeju Island in Korea. Bioresour Technol 98:1711–1716

Basavaraja PK, Yogendra ND, Zodape ST, Prakash R, Ghosh A (2018) Effect of seaweed sap as foliar spray on growth and yield of hybrid maize. J Plant Nutr 41:1851–1861

Bogaert KA, Blomme J, Beeckman T, De Clerck O (2022) Auxin's origin: do PILS hold the key? Trends Plant Sci 27:227–236

Borlongan IA, Tibubos KR, Yunque DAT, Hurtado AQ, Critchley AT (2011) Impact of AMPEP on the growth and occurrence of epiphytic *Neosiphonia* infestations on two varieties of commercially cultivated *Kappaphycus alvarezii*, grown at different depths in the Philippines. J Appl Phycol 23:615–621

Cabanillas-Terán N, Hernández-Arana HA, Ruiz-Zárate MÁ, Vega-Zepeda A, Sanchez-Gonzalez A (2019) *Sargassum* blooms in the Caribbean alter the trophic structure of the sea urchin *Diadema antillarum*. PeerJ 7:e7589

Campobenedetto C, Agliassa C, Mannino G, Vigliante I, Contartese V, Secchi F, Bertea CM (2021) A biostimulant based on seaweed (*Ascophyllum nodosum* and *Laminaria digitata*) and yeast extracts mitigates water stress effects on tomato (*Solanum lycopersicum* L.). Agriculture 11:557

Carillo P, Ciarmiello LF, Woodrow P, Corrado G, Chiaiese P, Rouphael Y (2020) Enhancing sustainability by improving plant salt tolerance through macro- and microalgal biostimulants. Biology (Basel) 9:253

Chávez V, Abigail Uribe-Martínez A, Eduardo Cuevas E, Rodríguez-Martínez RE, van Tussenbroek BI, Francisco V, Estévez M, Celis LB, Monroy-Velázquez LV, Leal-Bautista R, Álvarez-Filip L, García-Sánchez M, Masia L, Silva R (2020) Massive influx of pelagic *Sargassum* spp. on the coasts of the Mexican Caribbean 2014–2020: challenges and opportunities. Water 12:2908

Chirapart A, Khreauthong S, Praiboon J, Rattanasaensri S (2019) Growth and epiphytic responses of *Gracilaria fisheri* to the seaweed extract under controlled-culture conditions. Paper presented at the 23rd International Seaweed Symposium, 28 April–3 May 2019, Jeju, Korea (Abstract only)

Cokrowati N, Diniarti N (2019) Komponen *Sargassum aquifolium* sebagai hormon pemicu tumbuh untuk *Eucheuma cottonii*. Jurnal Biologi Tropis 19:316–321. [Indonesian]

Cokrowati N, Risjani Y, Firdaus M, Andayani S (2021) Accelerated growth of *Kappaphycus alvarezii* using *Sargassum aquifolium* extract and its anatomical characteristics. Biodiversitas 22:5195–5202

Cottier-Cook EJ, Nagabhatla N, Badis Y, Campbell ML, Chopin T, Dai W, Fang J, He P, Hewitt CL, Kim GH, Huo Y, Jiang Z, Kema G, Li X, Liu F, Liu H, Liu Y, Lu Q, Luo Q, Mao Y, Msuya FE, Rebours C, Shen H, Stentiford GD, Yarish C, Wu H, Yang X, Zhang J, Zhou Y, Gachon CMM (2016) Safeguarding the future of the global seaweed aquaculture industry. United National University

(INWEH) and Scottish Association for Marine Science Policy Brief, pp 1–12
Craigie JS (2011) Seaweed extract stimuli in plant science and agriculture. J Appl Phycol 23:371–393
Dawagne P, Jaiswar S (2020) Effects of *Ascophyllum* marine plant extract powder (AMPEP) on tissue growth, proximate, phenolic contents, and free radical scavenging activities in endemic red seaweed *Gracilaria corticata* var. *cylindrica* from India. J Appl Phycol 32:4127–4135
De Saeger J, Van Praet S, Vereecke D, Park J, Jacques S, Han T, Depuydt S (2020) Toward the molecular understanding of the action mechanism of *Ascophyllum nodosum* extracts on plants. J Appl Phycol 32:573–597
de Vega GB, Ulloque CE, Hurtado AQ, Cornish L, Critchley AT (2020) On the efficacy of an *Ascophyllum*-based, soluble extract in association with standard plant growth regulators on the micropropagation of the agarophyte, *Gracilaria blodgettii*, from seaweed farms located at the northern entrance of the Panama Canal. J Appl Phycol 32: 3211–3217
di Stasio E, Rouphael Y, Colla G, Raimondi G, Giordano M, Pannico A, El-Nakhel C, De Pascale SD (2017) The influence of *Ecklonia maxima* seaweed extract on growth, photosynthetic activity and mineral composition of *Brassica rapa* L. subsp. *sylvestris* under nutrient stress conditions. Eur J Hortic Sci 82:286–293
Dmytryk D, Chojnacka K (2018) Algae as fertilizers, biostimulants, and regulators of plant growth. Chapter 10. In: Chojnacka K et al (eds) Algae biomass: characteristics and applications towards algae-based products, developments in applied phycology, vol 8. Springer International Publishing AG, pp 115–122
Dumanović J, Nepovimova E, Natić M, Kuča K, Jácević V (2021) The significance of reactive oxygen species and antioxidant defense system in plants: a concise overview. Front Plant Sci 11:552969
Dzugielt T, Wadas W (2020) Possibility of increasing early crop potato yield with foliar application of seaweed extracts and humic acids. J Cent Eur Agric 21:300–310
Górka B, Korzeniowska K, Lipok J, Wieczorek PP (2018) The biomass of algae and extracts in agricultural production. In: Chojnacka K et al (eds) Algae biomass: characteristics and applications towards algae-based products, developments in applied phycology, vol 8. Springer International Publishing AG, pp 103–114
Hardouin K, Burlot AS, Umami A, Tanniou A, Stiger-Pouvreau V, Widowati I, Bedoux G, Bourgougnon N (2014) Biochemical and antiviral activities of enzymatic hydrolysates from different invasive French seaweeds. J Appl Phycol 26:1029–1042
Heo SJ, Park PJ, Park EJ, Kim SK, Jeon YJ (2005) Antioxidant activity of enzymatic extracts from a brown seaweed *Ecklonia cava* by spin resonance spectrometry and comet assay. Eur Food Res Technol 221:41–47
Hurtado AQ, Yunque DA, Tibubos K, Critchley AT (2009) Use of Acadian marine plant extract powder from *Ascophyllum nodosum* in tissue culture of *Kappaphycus* varieties. J Appl Phycol 21:633–639
Hurtado AQ, Joe M, Sanares RC, Fan PB, Critchley AT (2012) Investigation of the application of Acadian marine plant extract powder (AMPEP) to enhance the growth, phenolic content, free radical scavenging, and iron chelating activities of *Kappaphycus* Doty (Solieriaceae, Gigartinales, Rhodophyta). J Appl Phycol 24:601–611
Hurtado AQ, Neish IC, Ali MKM, Jeffrey Norrie J, Leonel Pereira L, Michalak I, Shukla PS, Critchley AT (2021a) Extracts of seaweeds used as biostimulants on land and sea crops- an efficacious, phyconomic, circular blue economy: with special reference to *Ascophyllum* (brown) and *Kappaphycus* (red) seaweeds. In: Gupta S, van Staden J (eds) Biostimulants for crops from seed germination to plant development. A practical approach. Elsevier Academic Press, London, pp 266–288
Hurtado AQ, Luhan MRJL, Ferriols VMEN, Faisan JP Jr., Mateo JP, Sibonga RC, Suyo, JGB (2021b) Towards a robust and resilient seaweed aquaculture in the Philippines. Policy Brief No. 1. Global SeaweedsSTAR Programme
Hussain HI, Kasinadhuni N, Arioli T (2021) The effect of seaweed extract on tomato plant growth, productivity and soil. J Appl Phycol 33:1305–1314
Jaiswar S, Dawange PS, Thanth C, Mantri VA (2021) Apical, sub-apical, and basal explants of industrially exploited marine red alga *Gracilaria salicornia* exhibited differential response to commercial seaweed-derived plant bio-stimulants. J Appl Phycol 33: 3975–3985
Juárez-Maldonado A, Ortega-Ortíz H, Morales-Díaz AB, González-Morales S, Morelos-Moreno Á, Cabrera-De la Fuente M et al (2019) Nanoparticles and nanomaterials as plant biostimulants. Int J Mol Sci 20:162
Khan W, Rayirath UP, Subramanian S, Jithesh MN, Rayorath P, Hodges DM, Critchley AT, Craigie JS, Norrie J, Prithiviraj B (2009) Seaweed extracts as biostimulants of plant growth and development. J Plant Growth Regul 28:386–399
Khan W, Hiltz D, Critchley AT, Prithiviraj B (2011) Bioassay to detect *Ascophyllum nodosum* extract-induced cytokinin-like activity in *Arabidopsis thaliana*. J Appl Phycol 23:409–414
Kumar M, Reddy CRK, Jha B (2013) The ameliorating effect of Acadian marine plant extract against ionic liquids-induced oxidative stress and DNA damage in marine macroalga *Ulva lactuca*. J Appl Phycol 25:369–378
Lau SE, Teo WFA, Teoh EY, Tan BC (2022) Microbiome engineering and plant biostimulants for sustainable crop improvement and mitigation of biotic and abiotic stresses. Discover Food 2:9
Liu J, Sibusiso L, Qifang W, Mingjiang W, Jong-il C, Haibin T (2020) Pharmaceutical and nutraceutical potential applications of *Sargassum fulvellum*. BioMed Res Intl. https://doi.org/10.1155/2020/2417410
Llewellyn CA, Airs RL (2010) Distribution and abundance of MAAs in 33 species of microalgae across 13 classes. Mar Drugs 8: 1273–1291
Loureiro RR, Reis RP, Critchley AT (2010) *In vitro* cultivation of three *Kappaphycus alvarezii* (Rhodophyta, Areschougiaceae) variants (green, red and brown) exposed to a commercial extract of the brown alga *Ascophyllum nodosum* (Fucaceae, Ochrophyta). J Appl Phycol 22:101–104
Loureiro RR, Reis RP, Berrogain FD, Critchley AT (2012) Extract powder from the brown alga *Ascophyllum nodosum* (Linnaeus) Le Jolis (AMPEP): a 'vaccine-like; effect on *Kappaphycus alvarezii* (Doty) Doty ex P.C. Silva. J Appl Phycol 24:427–432
Loureiro RR, Reis RP, Marroig RG (2014a) Effect of the commercial extract of the brown alga *Ascophyllum nodosum* Mont. on *Kappaphycus alvarezii* (Doty) Doty ex P.C. Silva in situ submitted to lethal temperatures. J Appl Phycol 26:629–634
Loureiro RR, Reis RP, Berrogain FD, Critchley AT (2014b) Effects of a commercial extract of the brown alga *Ascophyllum nodosum* on the biomass production of *Kappaphycus alvarezii* (Doty) Doty ex P.C. Silva and its carrageenan yield and gel quality cultivated in Brazil. J Appl Phycol 26:763–768
Mabeau S, Fleurence J (1993) Seaweed in food products: biochemical and nutritional aspects. Trends Food Sci Technol 4:103–107
Mageswaran R, Sivasubramaniam S (1984) Mineral and protein contents of some marine algae from the coastal areas of Northern Sri Lanka. J Nat Sci Found Sri Lanka 12:179–189
Markets and Markets (2022) Biostimulants Market by Active Ingredient (Humic Substances, Amino Acids, Seaweed Extracts, and Microbial Amendments), Mode of Application (Foliar, Soil Treatment, and Seed Treatment), Form (Liquid and Dry), Crop Type and Region - Global Forecast to 2027. Accessed Nov 2023: https://www.

marketsandmarkets.com/Market-Reports/biostimulant-market-1081.html

Marroig RG, Loureiro RR, Reis RP (2016) The effect of *Ascophyllum nodosum* (Ochrophyta) extract powder on the epibiosis of *Kappaphycus alvarezii* (Rhodophyta) commercially cultivated on floating rafts. J Appl Phycol 28:2471–2477

Miceli A, Vetrano F, Moncada A (2021) Influence of *Ecklonia maxima* extracts on growth, yield, and postharvest quality of hydroponic leaf lettuce. Horticulturae 7:440

Nursidi ASA, Anshary H, Tahya AM (2017) Environmental parameters and specific growth of *Kappaphycus alvarezii* in Saugi Island, South Sulawesi Province, Indonesia. AACL Bioflux 10:698–702

Pedra AGLM, Ramlov F, Maraschin M, Hayashi L (2017) Cultivation of the red seaweed *Kappaphycus alvarezii* with effluents from shrimp cultivation and brown seaweed extract: Effects on growth and secondary metabolism. Aquaculture 479:297–303

Pereira L (2018) Therapeutic and nutritional uses of algae. CRC Press Taylor & Francis Group, Boca Raton, FL

Rayorath P, Mundaya JM, Farid A, Khan W, Palanisamy R, Hankins S, Critchley AT, Prithiviraj B (2008) Rapid bioassays to evaluate the plant growth promoting activity of *Ascophyllum nodosum* (L.) Le Jol. using a model plant, *Arabidopsis thaliana* (L.) Heynh. J Appl Phycol 20:423–429

Robertson-Andersson DV, Leitao D, Bolton JJ, Anderson RJ, Njobeni A, Ruck K (2006) Can kelp extract (KELPAK®) be useful in seaweed mariculture? J Appl Phycol 18:315–321

Rouphael Y, De Micco V, Arena C, Raimondi G, Colla G, De Pascale S (2017) Effect of *Ecklonia maxima* seaweed extract on yield, mineral composition, gas exchange, and leaf anatomy of zucchini squash grown under saline conditions. J Appl Phycol 9:459–470

Sachdev S, Ansari SA, Ansari MI, Fujita M, Hasanuzzaman M (2021) Abiotic stress and reactive oxygen species: generation, signaling, and defense mechanisms. Antioxidants 10:277

Saldarriaga-Hernandez S, Melchor-Martínez EM, Danay Carrillo-Nieves D, Parra-Saldívar R, Iqbal HMN (2021) Seasonal characterization and quantification of biomolecules from S*argassum* collected from Mexican Caribbean coast – a preliminary study as a step forward to blue economy. J Environ Manag 298:113507

Santaniello A, Scartazza A, Gresta F, Loreti E, Biasone A, Di Tommaso D, Perata PA (2017) *Ascophyllum nodosum* Seaweed extract alleviates drought stress in *Arabidopsis* by affecting photosynthetic performance and related gene expression. Front Plant Sci 8:1362

Shukla PS, Mantin EG, Mohd Adil M, Bajpai S, Critchley AT, Balakrishnan PB (2019) *Ascophyllum nodosum*-based biostimulants: sustainable applications in agriculture for the stimulation of plant growth, stress tolerance, and disease management. Front Plant Sci 10:655

Shukla PS, Borza T, Critchley AT, Prithiviraj B (2021) Seaweed-based compounds and products for sustainable protection against plant pathogens. Mar Drugs 19:59

Shukla PS, Yadav NS, Critchley AT, Prithiviraj B (2022) Editorial: biostimulants as an avenue of abiotic stress tolerance improvement in crops. Front Sustain Food Syst 6:908555

Souza JMC, Castro JZ, Critchley AT, Yokoya NS (2018) Physiological responses of the red algae *Gracilaria caudata* (Gracilariales) and *Laurencia catarinensis* (Ceramiales) following treatment with a commercial extract of the brown alga *Ascophyllum nodosum* (AMPEP). J Appl Phycol 31:1883–1888

Souza JMC, Castro JZ, Critchley AT, Yokoya NS (2019) Physiological responses of the red algae *Gracilaria caudata* (Gracilariales) and *Laurencia catarinensis* (Ceramiales) following treatment with a commercial extract of the brown alga *Ascophyllum nodosum* (AMPEP). J Appl Phycol 31:1883–1888

Stengel DB, Connan S, Popper ZA (2011) Algal chemodiversity and biodiversity: source of natural variability and implications for commercial application. Biotechnol Adv 29:483–501

Suryati E, Rosmiati PA, Tenriulo A (2015) *In vitro* growth rate of *Kappaphycus alvarezii* micropropagule and embryo by enrichment medium with seaweed extract. Indo Aquacult J 10:13–17

Szczepanek M (2018) Technology of maize with growth stimulants application. Eng Rural Dev 17:483–490

Tibubos K, Hurtado AQ, Critchley AT (2017) Direct formation of axes in new plantlets of *Kappaphycus alvarezii* (Doty) Doty, as influenced by the use of AMPEP K+, spindle inhibitors and plant growth hormones. J Appl Phycol 29:2345–2349

Trivedi K, Gopalakrishnan VAK, Kumar R, Ghosh A (2021) Transcriptional analysis of maize leaf tissue treated with seaweed extract under drought stress. Front Sustain Food Syst 5:774978

Umanzor S, Shin S, Marty-Rivera M, Augyte S, Yarish C, Kim JK (2019) Preliminary assessment on the effects of the commercail seaweed extract, AMPEP, on growth and thermal tolerance of the kelp *Saccharina* spp. from the Northwest Atlantic. J Appl Phycol 31: 3823–3829

Umanzor S, Jang S, Antosca R, Critchley AT, Yarish C, Kim JK (2020a) Optimizing the application of selected biostimulants to enhance the growth *of Eucheumatopsis isiformis*, a carrageenophyte with commercial value, as grown in land-based nursery systems. J Appl Phycol 32:1917–1922

Umanzor S, Shin SK, Yarish C, Augyte S, Kim JK (2020b) Exploratory evaluation of the effects of Kelpak seaweed extract on cultivated kelp *Saccharina* spp. exposed to sublethal and lethal temperatures. J World Aquat Soc 51:960–969

Umanzor S, Han S, Song HI, Park JS, Critchley AT, Yarish C, Kim JK (2022a) Ascertaining the interactions of brown seaweed-derived biostimulants and seawater temperature on spore release, germination, conchocelis, and newly formed blades of the commercially important red alga *Neopyropia yezoensis*? Algal Res 64:102692

Umanzor S, Han S, Song HI, Park JS, Critchley AT, Yarish C, Kim JK (2022b) Enhancements provided by the use of an *Ascophyllum nodosum* extract can be transferred through archeospores in the red alga *Neopyropia yezoensis* (Ueda) L.-E. Yang & J. Brodie. Aquatic Bot 177:103481

Wally OSD, Critchley AT, Hiltz D, Craigie JS, Han X, Zaharia LI et al (2013) Regulation of phytohormone biosynthesis and accumulation in *Arabidopsis* following treatment with commercial extract from the marine macroalga *Ascophyllum nodosum.* J Plant Growth Regul 32: 324–339

Wouthuyzen S, Herandarudewi SMC, Komatsu T (2016) Stock assessment of brown seaweeds (Phaeophyceae) along the Bitung-Bentena Coast, North Sulawesi Province, Indonesia for alginate product using satellite remote sensing. Procedia Environ Sci 33:553–561

Xiao X, Agusti S, Lin F, Xu C, Yu Y, Pan Y, Li K, Wu J, Duarte CM (2019) Resource (light and nitrogen) and density-dependence of seaweed growth. Front Mar Sci 6:618

Yakhin OI, Lubyanov AA, Yakhin IA, Brown PH (2017) Biostimulants in plant science: a global perspective. Front Plant Sci 54:23–27

Yong WTL, Ting SH, Yong YS, Thien VY, Wong SH, Chin WL, Rodrigues KF, Anton A (2014) Optimization of culture conditions for the direct regeneration of *Kappaphycus alvarezii* (Rhodophyta, Solieriaceae). J Appl Phycol 26:1597–1606

Yunque DAT, Tibubos KR, Hurtado AQ, Critchley AT (2011) Optimization of culture conditions for tissue culture production of young plantlets of carrageenophyte *Kappaphycus*. J Appl Phycol 23:433–438

Zuldin WH, Shapawi R (2015) Performance of red seaweed (*Kappaphycus* sp.) cultivated using tank culture. J Fish Aquat Sci 10:1–12

Stakeholders' Perspectives and Gender Relations as Indicators of Knowledge Systems: Empirical Evidence from the Philippine Seaweed Industry

24

Jee Grace B. Suyo-Diala and Anicia Q. Hurtado

Abstract

Social networks and interactions shape knowledge systems, stakeholder perceptions, and behaviors. The advancement of knowledge systems depends on the nature of social exchanges and the capacity of seaweed industry actors (e.g., farmers) to comprehend and translate their learning into actions. Patron-client system, geographical location, and gender were the key factors that affected access to resources and services, particularly among seaweed farmers. Access was found to be integral for the advancement of knowledge systems because it affects market expansion and mobility and, consequently, access to information, learning opportunities, and financial facilities. Since knowledge acquisition is dependent on social interactions, especially in an industry governed by informal networks, the position of a particular stakeholder group determines the extent to which it can benefit from and influence economic exchanges within and outside of its network. Seaweed farmers remain at the tail end of the value chain because of a lack of access to resources and services, especially those that would help expand their network relations. A strategic expansion of social networks would contribute greatly to the advancement of knowledge systems and adoption of sustainable farming practices as long as stakeholders are provided with access to capacity-building activities that match their needs and aspirations. Stakeholder perceptions and insights provide essential information about the current knowledge base in the seaweed industry and should be utilized, together with scientific knowledge, in the formulation of guidelines, policies, and procedures aimed at fostering sustainable growth of the seaweed industry.

Keywords

Seaweed · Farm · Stakeholder · Gender · Value chain · Access Philippines

24.1 Introduction

Different types of knowledge systems are integral for reshaping public policy and governance toward achieving sustainable development (Blaikie et al. 1997; Cornell et al. 2013; Negev and Teschner 2013; Xavier et al. 2018; Alexander et al. 2019; Wheeler and Root-Bernstein 2020; Williams et al. 2020). Knowledge systems are sets of propositions used by agents and institutions (i.e., communities) to understand and interpret the world. They are transferrable as networks of actors that are able to share information, beliefs, and practices within the networks (Berkes 2009; Cornell et al. 2013), including processes and actions that promote sustainable development (FAO 2009; Van Kerkhoff and Szlezák 2016). Based on these definitions, the process of "learning, knowing, and doing" is contextual because it is influenced and shaped by social factors (i.e., norms, belief systems, values), institutions, and environmental conditions where a particular knowledge system is recognized and used. Although knowledge systems are generally interrelated, they can also be distinguished from each other based on processes of acquisition, types of end-users, and the manner by which information is assessed and verified (Berkes 2009; Negev and Teschner 2013). In fisheries and aquaculture, the use of local knowledge is evident in the manner by which households manage their livelihoods, address various forms of risks, and formulate innovative strategies for improving their economic wellbeing. Wellbeing is generated and developed through peoples' interactions with their environment and their familiarity with resources

J. G. B. Suyo-Diala (✉)
Department of Management, College of Management, University of the Philippines Visayas, Iloilo City, Philippines

A. Q. Hurtado
Integrated Services for the Development of Aquaculture and Fisheries (ISDA) Inc., MacArthur Highway, Tabuc Suba, Jaro, Iloilo City, Philippines

A. T. Critchley et al. (eds.), *Tropical Phyconomy Coalition Development*, Developments in Applied Phycology 11,
https://doi.org/10.1007/978-3-031-47806-2_24

that are culturally and economically important for them. Local knowledge systems tend to be dynamic because communities continually need to adapt to changes and uncertainties (Blaikie et al. 1997; Lam et al. 2020; Williams et al. 2020). These processes allow communities or social groups to update their knowledge bases and to apply and transmit their ideas and practices to others within their social network (Blaikie et al. 1997; Berkes 2009; Cornell et al. 2013; Beaman and Dillon 2018).

The seaweed industry is composed of multiple actors linked through networks that are grounded and strengthened through social and economic transactions. These actors play different roles in the seaweed value chain and therefore possess sets of unique information that may be useful to other stakeholder groups and policy makers (Cai et al. 2013; Hurtado 2013; Valderrama et al. 2015; Nor et al. 2019; Suyo et al. 2020). One of the characteristics of local knowledge is that it is shared by individuals belonging to a particular group (Warburton and Martin 1999), and in the seaweed industry, a wealth of information can be found in the production sector, which is the source of raw materials for the manufacturing of seaweed products (i.e., carrageenan). Farmers and traders possess information about farming and marketing activities as well as about different challenges that affect seaweed biomass and product value. A number of studies have investigated these socioeconomic phenomena in the production sector, particularly those that affect seaweed farmers (i.e., Cai et al. 2013; Radulovich et al. 2015; Valderrama et al. 2015; Hurtado et al. 2017, 2019). There is a wealth of empirical evidence pertaining to how local knowledge is being used by seaweed farmers to improve their production and address different problems that they encounter. These beliefs, perceptions, and practices that have helped seaweed farmers sustain their livelihood can further improve policy discussions, especially about the values, interests, and priorities shared among seaweed farming communities.

The commercial production of red seaweed varieties (i.e., *Kappaphycus* and *Eucheuma*) has been undertaken for more than four decades in the Philippines, making it one of the largest producers of carrageenophytes in the world. Although it has been considered by some investigators to be a "livelihood of last resort" (Fröcklin et al. 2012), studies have shown that it can help address poverty by providing livelihood options to coastal communities because it can provide high returns for low investments (Hurtado 2013; Valderrama et al. 2015). Women have benefited from engaging in seaweed farming/business through increased income and involvement in sociocivic activities, as narrated in the study conducted by Msuya and Hurtado (2017). Seaweed farming, therefore, is not only economically beneficial but also an agent for the empowerment of women, provided that women seaweed farmers are given an opportunity to be fully involved in decision-making and are able to gain leadership experience.

There have been a number of studies that investigated various dimensions of the seaweed industry, but there remains a need to address the uptake of stakeholder insights into scientific and policy discussions. This chapter does not intend to evaluate the importance of different types of knowledge systems, but is intended to promote interest and balance in the utilization of multiple knowledge systems in research and development and in policy making for the promotion of a sustainable seaweed industry. The analysis is centered on relationships shared by various actors in carrageenan seaweed value chains and on how such relationships influenced the manner by which information is shared and developed within the network, specifically for economic transactions and in dealing with risks and uncertainties. The analysis is based on the premise that social networks (i.e., relationships between value chain actors) serve as indicators of how knowledge is developed, utilized, and shared between and among seaweed farmers, traders, and processors. The results suggest that the strength of the existing network determines the extent to which various stakeholder groups are able to prevent or mitigate risks affecting the industry. Phycosecurity concepts were developed from the body of lessons learned during the Phyconomy.org Tropical Phyconomy Collaboration Development (TPCD) eucheumatoid seaweed biology webinar of July 7–8, 2021 (see the audio-visual references section of the present document).

24.2 Materials and Methods

The data and information used in this chapter were generated from studies conducted for the GlobalSeaweedSTAR (GSSTAR), an interdisciplinary research programme on seaweeds that was funded by the UK Research and Innovation—Global Challenges Research Fund (UKRI-GCRF). The programme was carried out for over 4 years (January 2018 to December 2021) by research teams from nine partner institutions in the United Kingdom, the Philippines, Malaysia, Tanzania, and Canada. It focused on four key areas of research, namely, (a) disease and pest infestation; (b) biosecurity practices and policy; (c) algal genetic resources; and (d) socioeconomic resilience (GSSTAR website). This chapter dealt with socioeconomic resilience, as it focused primarily on the importance of stakeholders' perspectives in enriching policy discourses and bridging the gap between technical (i.e., scientific knowledge) and local knowledge. Stakeholder perspectives were investigated by mapping social networks in seaweed-to-carrageenan value chains and were matched with their perceptions of risks that affect the seaweed industry as described by Suyo et al. (2020, 2021a). Information and data embedded in the narratives of seaweed farmers, traders, and processors were used to examine power relations in the seaweed industry and to assess how prevailing networks among supply chain actors influenced and molded their awareness of risks and their corresponding risk management strategies.

Focus group discussion (FGD) was used during the collection of data from seaweed farmers, while traders and processors were interviewed individually using a Key Informant Interview guide. For instance, questions to farmers that pertained to their social networks were asked with intent to understand power relations and gender dynamics in production segments of value chains. Information pertinent to market structure and issues affecting trade, export, and governance mechanisms was obtained from traders and processors. This chapter is generally based on the results of the studies on social networks and risk perceptions and management published by Suyo et al. (2020, 2021a) and highlights the importance of stakeholders' perspectives (i.e., local knowledge) on the policy making and governance of the seaweed industry.

24.3 Results and Discussion

24.3.1 Social and Economic Exchanges as Indicators of Knowledge Systems

The commercial cultivation of eucheumatoids (i.e., *Kappaphycus* and *Eucheuma*) was first introduced and became successful in the Philippines, making it one of the leading sources of raw dried seaweeds and carrageenan products in the last four decades (Doty 1973; Hurtado and Agbayani 2000; Hurtado et al. 2014, 2019). Despite fluctuations in the volume of production due to environmental and social issues (Hayashi et al. 2010, 2017; Hurtado 2013; Hurtado et al. 2019), eucheumatoids have been recorded as one of the main aquaculture products of the Philippines in recent years (Philippine Statistics Authority 2016, 2018). Recent advancements in technology and practices were also projected to contribute positively to the comparative advantage of the Philippine seaweed industry (Valderrama et al. 2015; Buschmann et al. 2017; Hayashi et al. 2017; Hurtado and Critchley 2018; Hurtado et al. 2019). Given these factors and the increasing interest in seaweeds and seaweed products worldwide, a number of value chain analyses have been conducted to understand and promote more efficient production processes. Recent value chain analyses on seaweeds included studies funded through the Philippine Rural Development Project (PRDP) of the Department of Agriculture (i.e., Philippine Rural Development Project 2014a, b, 2015) as part of the series of value chain analyses conducted for key agricultural commodities per region. There were also value chain case studies that focused on certain issues such as disaster impacts (Edo and Zack 2017), labor and employment (Narvaez 2015), and gender inequality (Ramirez et al. 2020).

A conventional seaweed supply chain in the Philippines is composed of three key categories of actors (i.e., farmers, traders, and processors/exporters) that have specific roles in the industry. Seaweeds were sold fresh or as raw dried seaweeds (RDS), mostly to village traders or rarely to town traders/consolidators and/or directly to processors. Even though seaweed farmers earn less by selling fresh seaweeds, it was preferred by those who wanted to reduce waiting time to gain a return on their invested capital, especially among women and those who do not have drying platforms. However, this is not possible in certain regions, such as Tawi-Tawi, where village traders only accept RDS. The moisture content (MC) level was considered an integral determinant of buying price for farmers. The standard MC ranges from 40% up to 47% for *Kappaphycus* and 38% for *Eucheuma* varieties, respectively. In most cases, traders would accept RDS with higher moisture content but with corresponding penalties such as purchase price reductions of 2% up to 10% off the total price per kilo, which equates to approximately USD 0.02 (PhP 1.00) up to USD 0.12 (PhP 6.00). Similar sanctions are imposed if seaweeds are not properly cleaned (e.g., soft plastic ties) or harvested prematurely. Even though there are national standards on RDS, farmers, traders, and processors adjust based on each other's capacities and tend to not adhere to strict rules, especially in terms of RDS quality. This showed that the seaweed industry was market demand-based, so farmers' preferences and decisions were highly influenced by industry standards, more than by national standards.

Because the seaweed industry social network is defined by economic and social transactions between and among seaweed industry actors, network sizes varied. Suyo et al. (2020) illustrated and thoroughly discussed power relations in the seaweed industry based on the social network structures of seaweed farmers and discussed how they are related to access to resources and services. Overall, seaweed farmers were found to have a relatively narrow social network compared to other actors in the industry; hence, they tended to be adversely affected by the uneven distribution of income and opportunities. In a series of value chain analyses funded through the Philippine Rural Development Project (PRDP 2014a, b, 2015), it was discovered that seaweed farmers across the country benefited less from value addition than other actors because they were selling raw products (i.e., fresh or dried seaweeds), and they lacked the capacity to expand or upscale their production; hence, they remained at the lowest tier of the market. Across various segments of seaweed value chains, much attention has been given to seaweed production because it is integral to the sustainability of the seaweed industry. There were three interrelated factors that needed to be considered in pursuit of sustainable development, namely: (a) economic factors such as income and employment; (b) social factors, including degree of empowerment and access to basic needs; and (c) environmental factors such as resource conservation and management strategies that aimed to promote intergenerational equity (United Nations 2018). Research and development initiatives in the seaweed industry matched at least one of these themes, but seaweed farmers remained the key target of development efforts because they were considered to be an important subsector in the industry, and there were numerous risks affecting seaweed production that needed to be addressed (Hayashi et al. 2010; Hurtado et al. 2017; Msuya et al. 2022; Suyo et al. 2020, 2021a). This included studies on farming techniques (Hurtado and Agbayani 2002; Hurtado et al. 2014; Radulovich et al. 2015; Valderrama et al. 2015), varieties of eucheumatoid seaweeds cultivated in tropical and subtropical regions (Brakel et al. 2021; Dumilag et al. 2022), the economic contribution of seaweed farming to seaweed farming communities (Cai et al. 2013; Hurtado 2013; Hurtado et al. 2014; Radulovich et al. 2015; Valderrama et al. 2015; Hayashi et al. 2017), and various socioeconomic issues and constraints faced by the seaweed industry (Hayashi et al. 2010; Hurtado et al. 2017, 2019; Mateo et al. 2020; Suyo et al. 2021a, b). The gender dimension of the seaweed industry has also been explored in pursuit of understanding the degree of involvement of women and men in various segments of the seaweed industry (Besta 2013; Narvaez 2015; Msuya and Hurtado 2017; Kunjuraman et al. 2019; Ramirez et al. 2020; Larson et al. 2021) and in pursuit of knowledge as to how social norms and values influence gender relations and women's participation in value chain activities (Hedberg et al. 2018; Suyo et al. 2020, 2021a).

Seaweed pests (i.e., endophytic and epiphytic microorganisms) and diseases (e.g., ice-ice syndrome) have been long-standing problems faced by seaweed farmers (Hayashi et al. 2010, 2017; Cai et al. 2013; Hurtado 2013; Buschmann et al. 2017; Mulyati and Geldermann 2017; Hurtado et al. 2019; Ramirez et al. 2020; Suyo et al. 2021a). Pests and diseases resulted in production losses ranging from 50% to as much as 100% of the farm crops when farmers do not adopt strategies that prevent or mitigate the spread of infections or infestations. Mateo et al. (2021) and Suyo et al. (2020, 2021a) discussed various methods adopted by farmers to address these recurring issues. These included (a) selecting areas with good water movement; (b) ensuring that water depths were suitable for farming techniques used; (c) selecting varieties of seaweeds that were maximally resilient to pests and diseases; (d) trimming of "infested" portions of seaweeds; or (e) premature harvesting of seaweeds when the spread of infection or infestation was no longer manageable. After harvest, culture lines were inspected, cleaned, and disinfected through sun-drying methods to ensure that they were free from any debris and could be used for the next cropping cycle. Since knowledge and skills pertaining to seaweed farming were gained informally (e.g., learned from family members and developed through experience), risk mitigation strategies remained traditional, and farmers lacked knowledge of the variety of factors that instigated the seasonal occurrence of pests and diseases, such as the introduction of invasive species and lack of disinfection and monitoring procedures (Mateo et al. 2021). Hazards that caused extensive damage or uncertainties (e.g., typhoons and price instability) were found to be challenging and overwhelming for farmers; hence, they resorted to unsustainable practices such as premature harvesting of seaweeds, reduction of the cropping season, or obtaining credit from village traders (Suyo et al. 2021a). The social networks of farmers were relatively narrow compared to other key actors in the seaweed industry (e.g., traders, processors, and exporters), which made risk management challenging. While farmers were able to address the impacts of different types of risks affecting their seaweed farms, these practices were not always effective in sustaining or promoting livelihood advancement. Having strong local ties was found to be adequate in providing short-term solutions to adverse impacts of different risks, such as when farmers are able to seek financial support from village traders to reconstruct their seaweed farms or through sharing of information to mitigate pest and disease infestations. However, it was not enough to address risks with large-scale impacts such as those brought about by weather-related hazards (e.g., typhoon, drought) and economic inequality (e.g., price fluctuations, changes in market trends). According to Suyo et al. (2021a), most risk management initiatives focused on recurring issues such as pest and disease infestations because farmers were relatively familiar with these hazards. Strategies to manage typhoons and droughts tended to be reactive. For example, premature harvesting of seaweeds or engaging in alternative livelihood sources during lean seasons to compensate for seaweed-generated income loss. Overall, farmers were able to respond to risks that directly affected their livelihood but were less capable of responding to issues affecting seaweed prices, input costs, and poor market access. These problems were further compounded by inconsistencies in the adoption or implementation of seaweed-related policies intended to provide guidelines for promoting sustainable seaweed production practices (e.g., adoption of climate-resistant varieties) or support microseaweed business enterprises that could provide increased access to financial resources among key stakeholders in the seaweed industry. For instance, there was a lack of biosecurity-related policies aimed at addressing recurrent and emergent pests and diseases; hence, seaweed farmers adopted strategies formulated through the process of trial and error or learned mostly through informal methods (e.g., information shared by fellow seaweed farmers). Although some farming practices and procedures could help mitigate the impacts of production risks, particularly seaweed pests and diseases, additional mechanisms were needed to further improve farmers' capacity to prepare for, respond to, and recover from a variety of risks. Seaweed farmers' reliance on local networks helped sustain economic relations among the key actors in the seaweed industry; however, this was not sufficient or sustainable because traders and processors also encountered risks and were affected by changes in local market trends and prices related to unstable supply of raw materials.

Stakeholders' narratives revealed an uneven distribution of information in the industry, and the position of each actor in the seaweed supply chain determined the extent of their knowledge and their capacity to adopt sustainable production processes. Seaweed farmers remained at the bottom tier of the industry, and their access to resources was curtailed by their social networks (Suyo et al. 2020); hence, their actions and decisions were largely influenced by perceptions that were rarely validated through scientific processes. While they were knowledgeable of production processes and were able to manage some risks (i.e., pests and diseases), they had poor access to new information and capacity-building activities such as training on seaweed production processes, including postharvest activities such as handling and drying seaweeds. Farmers mostly relied on knowledge obtained from their long experience in seaweed farming to address recurring and eminent risks that affected their seaweed farms (Suyo et al. 2020; Mateo et al. 2021) with higher uncertainty of outcomes. For example, when farmers trimmed infected portions of seaweeds, they did not have the utmost assurance that the infection had been totally eliminated. Similarly, when they disinfected seaweed lines but not other components of seaweed farm structures, they continued to contribute to an increased probability of pest and disease occurrence, especially when other environmental factors (i.e., salinity, temperature) became suitable for opportunistic bacteria to proliferate (Syafitri et al. 2016; Ward et al. 2021; Faisan, J. personal comm.). As seaweed farmers were not provided with enough opportunities to improve their skills or farming techniques, they continued to be reactive to hazards that continually affected their farms even though there were available technologies and mechanisms that could have increased the resilience of cultivated seaweeds to environmental risks such as pests and diseases. This resulted in the adoption of risk-sharing contracts through patron-client systems. Traders and processors were heavily reliant on farmers as sources of raw materials for carrageenan production; hence, they acknowledged the importance of addressing production-related risks by providing financial support, credit/loan, and even saving schemes for farmers who did not have access to formal financial services (Suyo et al. 2021a). Such evidence showed that the segmented nature of the seaweed industry had shaped the perceptions, knowledge, and practices of key value chain actors, especially seaweed farmers. While the current system has been effective in sustaining the seaweed industry as well as in fostering collective response measures for managing risks, it has constrained the opportunities of seaweed farmers to work toward improving their condition because they remain dependent on their informal network for their livelihoods.

24.3.2 Social Networks as Shapers of Knowledge Systems in the Seaweed Industry

The social values of relationships, particularly informal connections among industry actors, are difficult to quantify, and their impacts are usually not discussed in detail in supply chain studies. Additional tools should be used with value chain analyses during the assessment of relationships with respect to their economic values (M4P 2008). This was achieved by combining principles of social network analyses with value chain analyses during assessments of how intricacies in social relations affect and influence stakeholders' perceptions and decisions in various segments of the seaweed industry. An in-depth analysis of stakeholders' narratives uncovered a variety of sources of risks and uncertainties that were relevant to key actors in the seaweed industry. All these provided interesting insights into how knowledge systems, particularly in the production sector, were formed and developed. Moreover, the analytical process underscored the important factors involved in promoting equitable distribution of resources and sustainable development of the seaweed industry. From a socioeconomic perspective, social networks and relationships appeared to be among the key factors that shaped the structure and growth of the seaweed value chain, but further analysis showed that their impacts were complex and needed to be discussed by layers. Figure 24.1 shows that the patron-client system, location of seaweed production sites, and gender were the three key factors that affected access to and distribution of resources and services in seaweed supply chains. These factors then affected market access, mobility, resources and services related to key factors since the economic activities of seaweed farmers were highly intertwined with their social networks.

Information such as farm gate price, industry standards, and seaweed varieties was found to be a valuable resource because these parameters affect production costs and income. Patron-client systems and market access impacted information flow in the seaweed supply chain. The information network in the supply chain appeared to be concentric and hierarchical, with farmers being at the tail end, while traders served as bridges that relayed information and instructions concerning prices and industry standards. Therefore, access

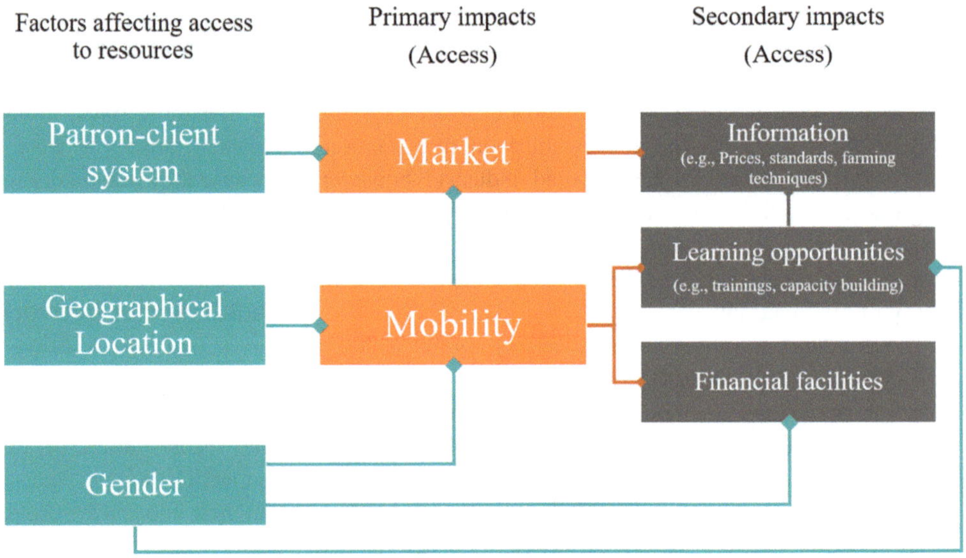

Fig. 24.1 Factors affecting access to resources and their impacts on the seaweed industry

to updates about farmgate prices corresponded to the level of market where farmers sold their harvests. Generally, village traders were the main clients of seaweed farmers and the primary source of information about farmgate prices and industry standards, but farmers in other areas had access to larger markets (i.e., town trader or processors). Other than the patron-client system, which regulated market options among farmers, geographical location was another consideration. For instance, farmers in Bohol were able to sell their seaweeds to town traders, especially if they were able to produce large volumes of RDS, which could offset logistical costs; however, expenses for transport and hauling were relatively higher for farmers in Zamboanga City and Tawi-Tawi, especially for those who resided in far-flung island villages. As most farmers mostly dealt with village traders, they continued to be reliant on information that has been passed on to several tiers and had limited information about factors that influenced price fluctuations.

The geographical location of farmers directly affected mobility (i.e., ease of movement) and consequently constrained market access. Seaweed farmers residing closer to towns had more options than farmers who cultivated seaweeds in far-flung island villages (Fig. 24.2). Barriers to mobility (i.e., geographical location, high logistical cost) consequently affected access to learning opportunities and financial services as well as information (i.e., as an offshoot of limited market access). Capacity-building activities (e.g., training) benefited only certain groups of farmers, particularly men, because of norms and practices that shaped gender power relations within households and in communities. Learning opportunities were closely linked to information access because participation in training and other capacity-building activities enabled seaweed farmers to gain new or additional knowledge about seaweed diseases and pests and improved farming techniques that they could use in their farms. Access to these opportunities enabled industry actors to develop or adopt innovative strategies that would allow them to improve their seaweed biomass.

Gender differences were noted in terms of mobility, access to learning opportunities, and financial facilities. Constraints on women's mobility were twofold: first, in terms of access to seaweed farms, and second, in terms of the market. Norms and values influence women's behavior and decisions, including their involvement in seaweed farming activities. Women were mostly involved in pre- and postharvest activities and less involved in activities commonly associated with men (e.g., farm construction, seaweed planting) or those that entail physical labor such as hauling seaweed lines or sacks full dried seaweeds. Although women reported that they accompanied their husbands when checking their farms, the frequency of farm visits was significantly lower than for men. Labor was often shared among male farmers in the village with or without pay (i.e., as a form of cooperative undertaking). Although all farmers appeared to have limited market access, women were more affected by social and economic imbalances because of their lack of participation in formal learning opportunities (e.g., training) and limitations in terms of travel, such as when transporting seaweeds to markets other than village traders. While capacity-building activities mostly involved men seaweed farmers because they were considered the primary owners of seaweed farms and were deemed to be more involved in all phases of seaweed production, women had more access to financial services (e.g., loans) because they were considered to be more reliable payers than men.

Social relationships influenced knowledge systems and stakeholder behaviors/practices in the seaweed industry. Although formal learning opportunities were provided, farmers had varying chances of participating in these activities because these were provided by various agencies

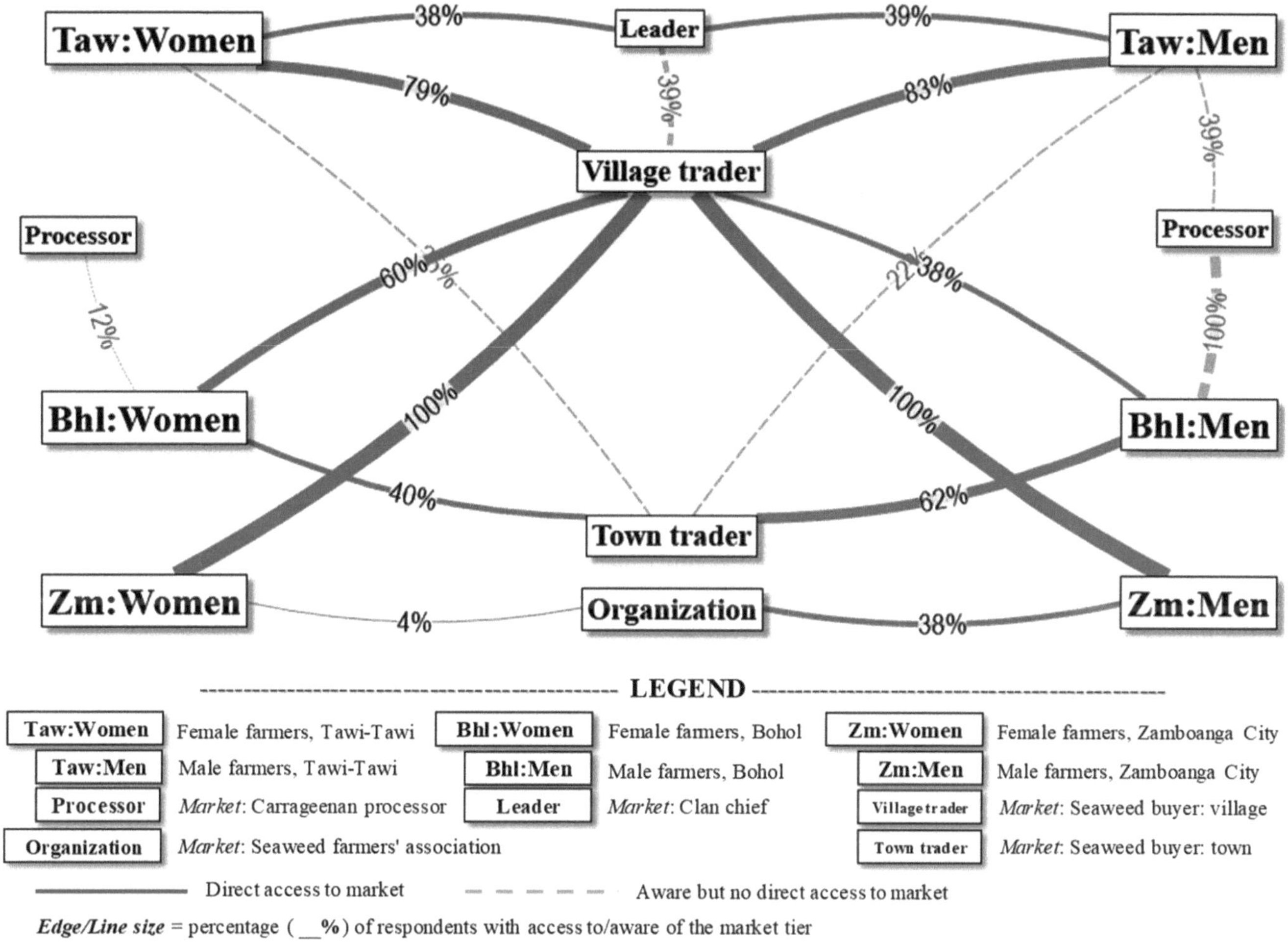

Fig. 24.2 Access to local seaweed trading networks between female and male seaweed farmers in Bohol, Zamboanga City, and Tawi-Tawi (Suyo et al. 2020)

and organizations that tended to target different areas or groups across the regions (Suyo et al. 2020; Mateo et al. 2021). There appeared to be different factors that influenced farmers' access to these opportunities, such as gender (i.e., men tend to have more opportunities than women) and location. For instance, Suyo et al. (2020) reported that male farmers in Zamboanga City and Tawi-Tawi were able to participate in formal training on seaweed farming and farm management, while women mainly relied on information provided by family members. The decision concerning the selection of men as primary recipients of training tend to be inadvertent and could be considered an indication of the systemic inequality between women and men (Kaplinsky and Morris 2003; de la Torre-Castro et al. 2017). Roles tend to be highly gendered within the seaweed supply chain, with physically demanding tasks mainly assigned to men, while women were given less laborious roles, mostly during the preparation of seedstocks and postharvest activities (e.g., cleaning). Farming techniques also created a gender divide since they required different degrees of labor intensity. Mateo et al. (2021) found that deep-sea farming methods were mostly adopted by men because they were more labor intensive than fixed-off bottom methods. Although seaweed farming was considered to be a family enterprise, farm ownership tend to be linked with men, especially with deep-sea farming techniques that required more skills and labor for the installation and maintenance of structures.

Knowledge acquisition is considered a key by-product of social interactions that coevolve with social structure (Fig. 24.3). While individuals interact and communicate with one another, they start to share information that is integral to the completion of different roles or tasks in the community. This means that the type of information shared relates largely to the roles or tasks assumed by an individual because for an item of information to be comprehensible, it needs to match the current level of understanding of the recipient. Similarly, the position of an individual in the social structure determines the type of information or knowledge

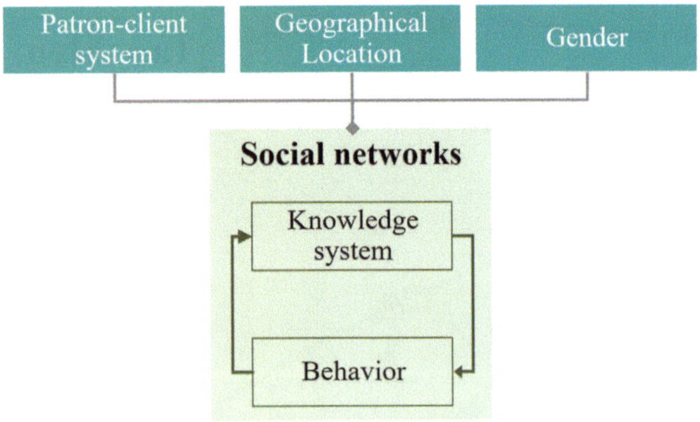

Fig. 24.3 Social networks and their determinants (e.g., patron-client system, geographical location, and gender) shape stakeholders' knowledge system, practices and behavior

that can be acquired (Carley 1986). In the seaweed industry, Suyo et al. (2020) showed that the social structure is highly segmented, with farmers having narrow access and influence in the industry. The expansion of seaweed farmers' social networks is constrained largely by the lack of institutional support and the need to sustain patron-client relations. This does not mean, however, that addressing these two barriers will automatically result in increased access to resources and information. According to Suyo et al. (2021a, b), risks (e.g., diseases and pests, inclement weather, etc.) and uncertainties (e.g., price fluctuations, inconsistent industry standards, etc.) are recurring issues that seaweed farming communities need to address. In fact, informal relationships, such as the patron-client relationship, bridge the economic gap that emanates from weak governance or lack of strong policies and procedures in the seaweed industry. For instance, in a study conducted by Mateo et al. (2020) on biosecurity-related policies in the Philippine seaweed industry, the implementation of regulations, policies, and procedures concerning the management of diseases and pests appeared to be lacking; hence, seaweed disease and pest infestations continue to be a major risk faced by seaweed farmers. Both Mateo et al. (2021) and Suyo et al. (2021a, b) discussed how various actors in the Philippine seaweed industry are mostly left to manage risks and uncertainties using the strategies that they have developed over time, such as how to prevent "ice-ice" syndrome from damaging the whole farm. Such evidence demonstrates how knowledge acquisition, particularly among seaweed farmers, is limited in terms of scope and rate of progression because farmers primarily rely on information shared among members of their social network (i.e., fellow seaweed farmers and traders). Moreover, based on the knowledge acquisition theory discussed by Carley (1986), opportunities to gain new and relevant information are also limited because of the lack of farmers' access to formal learning opportunities, which can help increase their understanding and appreciation of science-based farming techniques and behavior/practices, which will also help further improve the current knowledge base. Once a change has been introduced, changes in the type of information shared, behavior/practices, and even the structure of the industry can therefore be expected.

24.4 Conclusion and Policy Recommendations

The perceptions and behaviors of key actors in the seaweed industry, particularly seaweed farmers, revealed interesting insights about the existing knowledge base at different segments of the industry, including how information is shared and updated to match the current needs and interests of various stakeholder groups. A clear understanding of the current knowledge base through analysis of stakeholders' perceptions and social networks is essential for policymaking because it can help determine what farmers, traders, and processors already know and the mechanisms they have adopted to address the different risks that affect the industry.

To promote more sustainable governance of the public commons (i.e., coastal and marine resources utilized for seaweed farming), there is a need to bridge the gap between scientific and local knowledge systems to ensure that governance mechanisms are based on scientific practices and are attuned to the needs and interests of different stakeholder groups, particularly of women who have less access to resources compared to men. To attain this, it is essential to (1) provide more opportunities for farmers (especially women) to participate in capacity-building activities, particularly training on sustainable seaweed production processes; (2) ensure that different stakeholder groups are well represented in decision-making processes and that they are given a full chance to convey their ideas or opinions; and (3) research and development programs for the development of sustainable farming practices and technology should be paired with assessments of the socioeconomic conditions of the project site, thus enabling the identification of site-specific factors that have shaped local knowledge systems and practices.

References

Alexander SM, Provencher JF, Henri DA et al (2019) Bridging Indigenous and science-based knowledge in coastal and marine research, monitoring, and management in Canada. Environ Evid 8:1–24. https://doi.org/10.1186/s13750-019-0181-3

Beaman L, Dillon A (2018) Diffusion of agricultural information within social networks: Evidence on gender inequalities from Mali. J Dev Econ 133:147–161. https://doi.org/10.1016/j.jdeveco.2018.01.009

Berkes F (2009) Evolution of comanagement: role of knowledge generation, bridging organizations and social learning. J Environ Manag 90:1692–1702. https://doi.org/10.1016/j.jenvman.2008.12.001

Besta NA (2013) Seaweed farming and intra-household gender relations on Songo Songo Island, Tanzania. PhD Thesis, School of International Development University of East Anglia
Blaikie P, Brown K, Stocking M et al (1997) Knowledge in action: local knowledge as a development resource and barriers to its incorporation in natural resource research and development. Agric Syst 55:217–237. https://doi.org/10.1016/S0308-521X(97)00008-5
Brakel J, Sibonga RC, Dumilag RV et al (2021) Exploring, harnessing and conserving marine genetic resources towards a sustainable seaweed aquaculture. Plants People Planet 3:337–349. https://doi.org/10.1002/ppp3.10190
Buschmann AH, Camus C, Infante J et al (2017) Seaweed production: overview of the global state of exploitation, farming and emerging research activity. Eur J Phycol 52:391–406. https://doi.org/10.1080/09670262.2017.1365175
Cai J, Hishamunda N, Ridler N (2013) Social and economic dimensions of carrageenan seaweed farming. A global synthesis. In: Valderrama D, Cai J, Hishamunda N, Ridler N (eds) Social and economic dimensions of carrageenan seaweed farming, fisheries. Food and Agriculture Organization of the United Nations, FAO, Rome, pp 5–59
Carley K (1986) Knowledge acquisition as a social phenomenon. Instr Sci 14:381–438
Cornell S, Berkhout F, Tuinstra W et al (2013) Opening up knowledge systems for better responses to global environmental change. Environ Sci Pol 28:60–70. https://doi.org/10.1016/j.envsci.2012.11.008
de la Torre-Castro M, Fröcklin S, Börjesson S et al (2017) Gender analysis for better coastal management – Increasing our understanding of social-ecological seascapes. Mar Policy 83:62–74. https://doi.org/10.1016/j.marpol.2017.05.015
Doty MS (1973) Farming the red seaweed, *Eucheuma*, for carrageenans. Micronesica 9:59–73
Dumilag RV, Crisostomo BA, Aguinaldo ZA et al (2022) The diversity of eucheumatoid seaweed cultivars in the Philippines. Rev Fish Sci Aquac. https://doi.org/10.1080/23308249.2022.2060038
Edo A, Zack L (2017) Viable insertion in agribusiness value chains? Seaweed farming after Typhoon Yolanda (Haiyan) in Iloilo Province, the Philippines. Singap J Trop Geogr 38:25–40. https://doi.org/10.1111/sjtg.12178
Food and Agriculture Organization (2009) FAO and traditional knowledge: the linkages with sustainability, food security and climate change impacts, Rome, pp 1–16
Fröcklin S, de la Torre-Castro M, Lindström L et al (2012) Seaweed mariculture as a development project in Zanzibar, East Africa: a price too high to pay? Aquaculture 356–357:30–39. https://doi.org/10.1016/j.aquaculture.2012.05.039
Hayashi L, Hurtado AQ, Msuya FE et al (2010) A review of *Kappaphycus* farming: prospects and constraints. In: Israel A, Einav R, Seckbach J (eds) Seaweeds and their role in globally changing environments. Springer, London, New York, pp 255–283
Hayashi L, Reis RP, Santos A et al (2017) The cultivation of *Kappaphycus* and *Eucheuma* in tropical and subtropical waters. In: Hurtado AQ, Critchley AT, Neish IC (eds) Tropical seaweed farming trends, problems and opportunities, developments in applied phychology. Springer International Publishing, Cham, pp 55–90
Hedberg N, von Schreeb K, Charisiadou S et al (2018) Habitat preference for seaweed farming – a case study from Zanzibar, Tanzania. Ocean Coast Manag 154:186–195. https://doi.org/10.1016/j.ocecoaman.2018.01.016
Hurtado AQ (2013) Social and economic dimensions of carrageenan seaweed farming in the Philippines. In: Valderrama D, Cai J, Hishamunda N, Ridler N (eds) Social and economic dimensions of carrageenan seaweed farming. Fisheries and Aquaculture Technical Paper No. 580. FAO, Rome, pp 91–113
Hurtado AQ, Agbayani R (2000) The farming of seaweed *Kappaphycus*. Aquaculture Department, Southeast Asian Fisheries Development Center, Tigbauan, Iloilo
Hurtado AQ, Agbayani RF (2002) Deep-sea farming of *Kappaphycus* using the multiple raft, long-line method. Bot Mar 45:438–444. https://doi.org/10.1515/BOT.2002.044
Hurtado AQ, Critchley AT (2018) A review of multiple biostimulant and bioeffector benefits of AMPEP, an extract of the brown alga *Ascophyllum nodosum*, as applied to the enhanced cultivation and micropropagation of the commercially important red algal carrageenophyte *Kappaphycus alvarezi*. J Appl Phycol 30:2859–2873. https://doi.org/10.1007/s10811-018-1407-4
Hurtado AQ, Gerung GS, Yasir S, Critchley AT (2014) Cultivation of tropical red seaweeds in the BIMP-EAGA region. J Appl Phycol 26: 707–718. https://doi.org/10.1007/s10811-013-0116-2
Hurtado AQ, Critchley AT, Neish IC (2017) Tropical seaweed farming trends, problems and opportunities: focus on *Kappaphycus* and *Eucheuma* of commerce.
Hurtado AQ, Neish IC, Critchley AT (2019) Phyconomy: the extensive cultivation of seaweeds, their sustainability and economic value, with particular reference to important lessons to be learned and transferred from the practice of eucheumatoid farming. Phycologia 58:472–483. https://doi.org/10.1080/00318884.2019.1625632
Kaplinsky R, Morris M (2003) A handbook for value chain research. International Development Research Center, Ottawa
Kunjuraman V, Hossin A, Hussin R (2019) Women in Malaysian seaweed industry: motivations and impacts. Kaji Malaysia 37:49–74. https://doi.org/10.21315/km2019.37.2.3
Lam DPM, Hinz E, Lang DJ et al (2020) Indigenous and local knowledge in sustainability transformations research: a literature review. Ecol Soc 25. https://doi.org/10.5751/ES-11305-250103
Larson S, Stoeckl N, Fachry ME et al (2021) Women's well-being and household benefits from seaweed farming in Indonesia. Aquaculture 530. https://doi.org/10.1016/j.aquaculture.2020.735711
M4P (2008) Making markets work better for the poor: a toolbox for practitioners of value chain analysis. Agricultural Development International, Phnom Penh, pp 1–165
Mateo JP, Campbell I, Cottier-Cook EJ et al (2020) Analysis of biosecurity-related policies governing the seaweed industry of the Philippines. J Appl Phycol 32:2009–2022. https://doi.org/10.1007/s10811-020-02083-7
Mateo JP, Campbell I, Cottier-Cook EJ et al (2021) Understanding biosecurity: knowledge, attitudes and practices of seaweed farmers in the Philippines. J Appl Phycol 33:997–1010. https://doi.org/10.1007/s10811-020-02352-5
Msuya FE, Hurtado AQ (2017) The role of women in seaweed aquaculture in the Western Indian Ocean and South–East Asia. Eur J Phycol 52:482–494. https://doi.org/10.1080/09670262.2017.1357084
Msuya FE, Bolton J, Pascal F et al (2022) Seaweed farming in Africa: current status and future potential. J Appl Phycol 34:985–1005. https://doi.org/10.1007/s10811-021-02676-w
Mulyati H, Geldermann J (2017) Managing risks in the Indonesian seaweed supply chain. Clean Techn Environ Policy 19:175–189. https://doi.org/10.1007/s10098-016-1219-7
Narvaez TA (2015) Seaweeds jobs value-chain analysis in Zamboanga Peninsula, Philippines. Int J Oceanogr Aquac 2:12. https://doi.org/10.23880/ijoac-16000136
Negev M, Teschner N (2013) Rethinking the relationship between technical and local knowledge: toward a multitype approach. Environ Sci Pol 30:50–59. https://doi.org/10.1016/j.envsci.2012.10.011
Nor AM, Gray TS, Caldwell GS, Stead SM (2019) A value chain analysis of Malaysia's seaweed industry. J Appl Phycol. https://doi.org/10.1007/s10811-019-02004-3
Philippine Rural Development Project (2014a) Value chain analysis and competitiveness strategy: Carrageenan seaweed Mindanao. Quezon City, pp 1–91

Philippine Rural Development Project (2014b) Value chain analysis for seaweeds in Bohol, Cebu, and Guimaras. Quezon City, pp 1–54

Philippine Rural Development Project (2015) Seaweed value chain analysis: Palawan. Quezon City, pp 1–64

Philippine Statistics Authority (2016) Fisheries statistics of the Philippines 2013–2015. Quezon City, pp 1–578

Philippine Statistics Authority (2018) Fisheries statistics of the Philippines, 2015 to 2017. Quezon City, pp 1–316

Radulovich R, Neori A, Valderrama D et al (2015) Farming of seaweeds. In: Tiwari BK, Troy DJ (eds) Seaweed sustainability: food and nonfood applications. Academic, San Diego, pp 27–59

Ramirez PJB, Narvaez TA, Santos-Ramirez EJ (2020) Gender-inclusive value chains: the case of seaweed farming in Zamboanga Peninsula, Philippines. Gend Technol Dev 24:110–130. https://doi.org/10.1080/09718524.2020.1728810

Suyo JGB, Le Masson V, Shaxson L et al (2020) A social network analysis of the Philippine seaweed farming industry: Unravelling the web. Mar Policy 118. https://doi.org/10.1016/j.marpol.2020.104007

Suyo JGB, Le Masson V, Shaxson L et al (2021a) Navigating risks and uncertainties: risk perceptions and risk management strategies in the Philippine seaweed industry. Mar Policy 126:104408. https://doi.org/10.1016/j.marpol.2021.104408

Suyo JGB, Le Masson V, Shaxson L, et al (2021b) Bringing to light women's participation in the seaweed industry in the Philippines. In: GAFS Section of the Asian Fisheries Society. https://www.genderaquafish.org/2021/02/27/bringing-to-light-womens-participation-in-the-seaweed-industry/

Syafitri E, Prayitno SB, Ma'ruf WF, Radjasa OK (2016) Genetic diversity of the causative agent of ice-ice disease of the seaweed *Kappaphycus alvarezii* from Karimunjawa island, Indonesia. IOP Conf Ser Earth Environ Sci 55:012044. https://doi.org/10.1088/1755-1315/55/1/012044

United Nations (2018) Transforming our world: the 2030 agenda for sustainable development A/RES/70/1, pp 1–41

Valderrama D, Cai J, Hishamunda N et al (2015) The economics of *Kappaphycus* seaweed cultivation in developing countries: a comparative analysis of farming systems. Aquac Econ Manag 19:251–277. https://doi.org/10.1080/13657305.2015.1024348

Van Kerkhoff L, Szlezák NA (2016) The role of innovative global institutions in linking knowledge and action. Proc Natl Acad Sci U S A 113:4603–4608. https://doi.org/10.1073/pnas.0900541107

Warburton H, Martin A (1999) Local people's knowledge in natural resources research. Natural Resources Institute, Chatnam

Ward GM, Kambey CSB, Faisan JP et al (2021) Ice-Ice disease: an environmentally and microbiologically driven syndrome in tropical seaweed aquaculture. Rev Aquac 14:1–26. https://doi.org/10.1111/raq.12606

Wheeler HC, Root-Bernstein M (2020) Informing decision-making with Indigenous and local knowledge and science. J Appl Ecol 57:1634–1643. https://doi.org/10.1111/1365-2664.13734

Williams PA, Sikutshwa L, Shackleton S (2020) Acknowledging indigenous and local knowledge to facilitate collaboration in landscape approaches- lessons from a systematic review. Land 9:1–17. https://doi.org/10.3390/LAND9090331

Xavier LY, Jacobi PR, Turra A (2018) On the advantages of working together: social learning and knowledge integration in the management of marine areas. Mar Policy 88:139–150. https://doi.org/10.1016/j.marpol.2017.11.026

Understanding the Organism: Insights from *Chondrus crispus* (Rhodophyta) for the Tropical Carrageenan Seaweed Industry

25

Juliet Brodie

Abstract

An exploration of the red alga *Chondrus crispus* was undertaken to demonstrate the importance of understanding the organism that is being studied and how that knowledge might be used in relation to tropical eucheumatoids grown for the carrageenan industry. *C. crispus* provides an excellent model for this because it is a carrageenophyte, and many aspects of its biology have been extremely well studied over many years. The paper covers differences in the isomorphic life history phases and characters that distinguish the gametophyte from the tetrasporophyte phase, including hypothesizing that the presence of structural colour in the gametophyte phase is a protective mechanism of the reproductive structures against damaging UV. It also considers the value of studying the composition of the holobiont (host and all associated prokaryotes and eukaryotes) in relation to morphology and phenology. Eukaryotic green algal endophytes are discussed, and how differences in the host vegetative morphology of *C. crispus* and eucheumatoids influence penetration by these endophytic algae is discussed. Resolving eucheumatoid life histories is proposed as a means of working towards new knowledge and strains to improve cultivation. Whether new UV protective materials can be engineered based on the multilayered lamellae that produce structural colour in *C. crispus* is contemplated. It is concluded that traditional approaches are valuable in understanding the organism and, in conjunction with state-of-the-art techniques, provide a potentially powerful tool with which to address the long-term future of the eucheumatoid seaweed industry.

J. Brodie (✉)
Research, Natural History Museum, London, UK
e-mail: j.brodie@nhm.ac.uk

Keywords

Endophytic eukaryotes · Holobiont · Life history · Phenology · Reproduction · Structural colour

25.1 Introduction

The importance of knowing the organism that is being studied can never be underestimated. Traditional approaches to understanding seaweeds included the study of morphology and anatomy as well as determining life histories either from field observations or once seaweed culture techniques became commonplace in the laboratory. However, it could be argued that such traditional approaches have been somewhat neglected in recent years with the development of new techniques. The molecular revolution has been scientifically transformative, and the application of molecular techniques to seaweeds has totally changed the way we approach the study of seaweeds, opening up many new avenues of research, notably taxonomy, phylogenetics, population genetics and omics (Brodie et al. 2017). Whilst the use of molecular tools is a remarkable shift in the understanding of identity and the relationships between the seaweeds, and indeed organisms as a whole across the tree of life over a relatively short period of time, the need remains to integrate this knowledge with the physical organism. In all aspects of the organisms, including morphology, reproduction, life history, chemistry, mineralogy, ecology and biogeography, there are many answers to be found in relation to interpreting the molecular data, and we need this information now. Furthermore, more than ever, we need a drive to conserve wild seaweeds where their biological characteristics are maintained and not lost through hybridization with or competition from non-indigenous cultivated escapees (Cottier-Cook et al. 2016; Buschmann et al. 2017).

The aim of this paper is to demonstrate why understanding the organism is important and how that knowledge can be applied in related contexts in relation to the seaweeds used in

A. T. Critchley et al. (eds.), *Tropical Phyconomy Coalition Development*, Developments in Applied Phycology 11,
https://doi.org/10.1007/978-3-031-47806-2_25

the aquaculture industry. This paper focuses on the red alga *Chondrus crispus*, a common carrageenophyte in the North Atlantic. The reason for focusing on this species is because many aspects of its biology have been extremely well studied over many years, and while this paper focuses on direct observations, *C. crispus* is also one of the first seaweeds to have its whole genome sequenced (Collén et al. 2013 and references therein). It therefore provides an excellent model with which to explore the topic via an observational re-evaluation of the organism. It also has the potential to provide information with possible application for tropical eucheumatoids, notably *Eucheuma* and *Kappaphycus* species, which are the backbone of the modern carrageenan industry, notably in southeast Asia and Africa (FAO 2020; Msuya et al. 2022). Global eucheumatoid production, while worth USD 1.9 billion in 2018 (FAO 2020), is nevertheless beset by a range of factors, including rising seawater temperatures, reliance on vegetative propagation from a limited number of commercial strains, and an increase in pests and diseases, leading to a decline in the crop in some parts of the world (e.g., Lim et al. 2017, Faisan Jr. et al. 2021, Ward et al. 2019).

25.2 Life History

Chondrus crispus has a triphasic life history, with dioecious haploid female and male gametophytes, a diploid carposporophyte that develops directly on the female gametophyte, and a tetrasporophyte that is morphologically similar in appearance to the gametophyte phase. The life history has been well documented (Chen and McLachlan 1972), but it is worth also considering that there are distinct characteristics relating to structure, cell wall content, development, structural colour and its annual cycle. Whilst both phases have carrageenan in their cell walls, the gametophyte walls typically contain kappa- and iota-carrageenan and the tetrasporophyte lambda-carrageenan.

25.3 Reproductive Structures

The gametophyte and tetrasporophyte phases are described as morphologically isomorphic: a discoid holdfast from which few to many sub-cylindrical stipes arise that expand into compressed or flattened more or less dichotomously branched erect blades. The blade is composed of a cortex of 6–7 cell layers and a medulla of a network of primary and secondary filaments linked by secondary pit connections (Fredericq et al. 1992).

However, there are notable differences in a range of characteristics of the different life history phases, including reproductive anatomy, chemistry, phenology and in relation to the holobiont. While the vegetative gross morphology between the two life history phases can make them virtually indistinguishable, the mechanisms that maintain the stability of such red algal life histories remain a subject of debate with the possibility of each phase exploiting different niches with their habitat (Veira et al. 2021 and references therein).

Anatomically, development of the diploid phase differs from that of the haploid phase, and a very detailed account can be found in Fredericq et al. (1992). In the female gametophyte, numerous carpogonial branches may be present early in development, but only a small proportion develop to maturity, whereas tetrasporangial sori tend to be numerous in the tetrasporophyte blade.

25.4 Structural Colour

It is common on the seashore to see the tips of *C. crispus* fronds displaying a striking blue iridescence. This structural colour is only found in the gametophyte phase and is confined to the tips of the fronds. The colour is produced by the multilayered cuticular structure (lamellae), and the reason that it only occurs at the tips is because there is a reduction in the number and uniformity of cuticle layers in the middle and base of the frond (Chandler et al. 2015, fig. 2, p. 3). This phenomenon also requires *C. crispus* to be wet to provide the required contrast in refractive index between the lamellae. In comparison, Chandler et al.'s (2015) study provides conclusive evidence that the lack of the blue colour in the tetrasporophyte phase can be accounted for by the relatively few lamellae that connect and fuse to form an irregular structure, an observation made over 30 years ago by Craigie et al. (1992). More intriguingly is the question as to why *C. crispus* produces structural colour and why it is confined to the gametophyte phase. Our hypothesis that by removing damaging wavelengths at the blue end of the spectrum, structural colour is a mechanism to reduce the absorption of UV (Chandler et al. 2015) has now been tested (García Fleitas et al., unpublished data), and results indicate that the photonic structure ensures effective functioning of photosynthetic machinery whilst minimizing the impact of harmful wavelengths of light on the reproductive structures. Thus, structural colour plays an important role in the sexual life history phase of this species, knowledge which may be of value in the design of new materials. For red algae, structural colour has only been reported in Rhodymeniophycidae (pers. obs.). Yang et al. (2016) dated the origin of this subclass to between 580–442 MYA, a time when there are a range of strategies to develop more offspring and when the fertilized nucleus in the carpogonium is moved to an auxiliary cell followed by the development of the carposporophyte. In *C. crispus*, the carpogonium consists of a supporting cell (which becomes the auxiliary cell after fertilization and cell fusions) and a three-celled carpogonial branch (Fredericq et al. 1992, fig. 23, p. 549). While carpogonia are initiated at the surface of the thallus, after fertilization and cell fusions, diploid auxiliary cell development occurs deeper down in the thallus and thus is more protected from the effects of harmful UV radiation. The early stages of the development of the male and female reproductive structures occur in the upper parts of the fronds, which is where the structural colour is seen. This suggests that by removing blue light, at least one function of the structural colour is as a mechanism to protect the developing nuclei in the reproductive structures at a vulnerable time in their development. Between c. 520–543 MYA, during the estimated origin of Rhodymeniophycidae, there was great Cambrian animal diversification, which suggests a potential increase in grazing pressure on the algae. It can be hypothesized that a major geological event forced algae onto the shore, which would have been a driver of mechanisms to survive the littoral environment, including high light and UV levels.

25.5 The Holobiont

It is well established that a diverse array of prokaryotic and eukaryotic organisms, the holobiont, is associated with the macroalgae and that the species composition is distinctly different in composition to the aquatic medium in which they live (Egan et al. 2013). *C. crispus* has a wide range of prokaryotes and eukaryotes associated with it, which makes it a potential model in which to study the carrageenophyte holobiont.

It would be valuable to establish a seasonal baseline of the prokaryote (microbiome) and eukaryote composition and the role of the organisms in order to compare changes in relation to environmental change and pests and diseases associated with the macroalgae (e.g., Hollants et al. 2013, Brodie et al. 2016).

25.6 Phenology of the Host

Phenological observations of a population of *Chondrus crispus* in Somerset, UK, over 2.5 years revealed a marked annual cycle of the seaweed (Plumb 1999). Starting in the winter, old, degraded fronds are shed, and new fronds are developing from the basal disc. In spring, the reproductive structures are developing. Evidence that fertilization appears to be a rapid event in the spring comes from the morphological study of *C. crispus* (Fredericq et al. 1992), where it was not possible to find unfertilized carpogonia after the spring. In the summer and autumn, carpospores are released, leaving entry points on the surface of the seaweed for prokaryotes and green and brown microeukaryotes to enter. By autumn, these processes, coupled with grazing, result in many old, damaged fronds that are released from the basal disc, which appears to be long lived and from which the new fronds grow in the winter.

A study of this sequence of events with changes in the microbiome could provide much insight into this relationship and useful information perhaps as early warning systems if applied to the seaweed industry.

25.7 Eukaryotic Endophytes

C. crispus is host to a wide variety of green epi-endophytic algae. In a study of these green algae isolated from a range of macroalgal hosts, Rinkel et al. (2012) discovered 16 species (9 *Ulvella* [as *Acrochaete*], 2 *Bolbocoleon*, 5 unknown Kornmanniaceae) in *C. crispus*. Such organisms are common in many seaweeds and can be generalists or specialists. There is a high level of diversity of these green algae that we are only truly beginning to understand (e.g., Nielsen et al. 2013, 2014). We also know very little about the nature of the relationship between the host and its eukaryotic endophytes. Endophytes have been described as parasitic and pathogenic in cultivated *C. crispus* (Craigie and Correa 1996), but green endophytes are highly integrated into the life history and seasonality of *C. crispus*. It is typical for us to refer to old degraded or rotting fronds as diseased, and this may be the case, particularly in cultivated carrageenophytes (e.g., green rot, Craigie and Shacklock 1989), but this can also be viewed as part of the annual cycle of events in the life history of wild populations.

Green endophytes typically enter the fronds of *C. crispus* in the upper part of the host fronds. Some of the green algae are epi-endophytes that are partially on the surface and partially growing in the cortex. In *Ulvella,* spores germinate on the surface and are able to grow down into the cortex as very slender filaments and then grow out into the medulla between the filaments of *C. crispus*. Brown endophytic algae tend to appear in the fronds later in this sequence after carpospores or tetraspores have been being released for a while. Endophytes can ramify through many areas of the thallus, which tends to deteriorate and then break away from the basal disc. The advantages of an endophytic lifestyle are reduced competition for space and protection from grazing, wave action and desiccation (Rinkel et al. 2012). The advantage to the host is that the presence of endophytes in relation to the deterioration of fronds triggers their release, and in doing so, they are moved by currents; if they are still releasing spores, it is a means of widening their spread.

25.8 Comparison of *Chondrus crispus* and *Eucheuma* Morphology

Both *C. crispus* and *Eucheuma* are carrageenophytes but in different families (Gigartinaceae and Solieriaceae, respectively) and are structurally distinct. *Eucheuma* has a very compact cortex and medulla. Whilst it can have a very well-developed biofilm on the surface, unless the cuticle is disrupted, it appears to be much harder for endophytic algae to penetrate into the cortex. We know that for cultivated *Eucheuma*, pests, such as the red algal epiphyte *Melanothamnus* (Brakel et al. 2021), are increasing, and these damage the surface of the cortex. This appears to make the eucheumatoids more susceptible to endophytes. Preliminary evidence suggests that species include members of the genus *Blastophysa* (pers. obs.). There is much work to be done on the taxonomy of the green endophytes in the eucheumatoids, and we need to make more comparisons both with cultivated and wild species of these carrageenophytes with *C. crispus*.

25.9 Concluding Remarks

A thorough understanding of *C. crispus* reiterates the complexity and evolutionary ingenuity displayed by this little red alga. It also demonstrates the value of returning to the older literature, where many scientific gems and much insight can be found. It is also important to think carefully about how we interpret aspects of the biology of such red species. It is useful to distinguish between those life histories that are isomorphic (as opposed to those that are heteromorphic, i.e., morphologically different), but the distinction in how the different phases function can differ considerably as if distinct entities are subject to different ecological drivers.

Given the knowledge of *C. crispus* life histories, it would be valuable to re-evaluate vegetative and reproductive morphology and to take eucheumatoid species through their life histories in detail, using individuals identified using robust molecular techniques. Ideally, these individuals would be indigenous samples from regions where there is eucheumatoid seaweed aquaculture. This could provide knowledge relevant to the improvement of strains used for cultivation. This will be timely and costly in the short term but should be valuable in the long term in the carrageenan industry. Alongside such studies, conservation measures need to be put in place to ensure the sustainability of wild stocks and consequently the long-term survival of the eucheumatoid seaweed industry.

The discovery of the mechanisms responsible for structural colour in *C. crispus* has the potential to lead to the development of new materials that could be used to protect against UV radiation. Perhaps it could be possible to engineer canopies based on multilayered lamellae that would provide protection from UV and temperature, yet let enough light penetrate and in doing so destress the seaweed crop, making it less prone to pests and diseases.

Understanding the holobiont and its prokaryotic and eukaryotic components potentially offers much promise in seaweed aquaculture. A study of the microbiome of *Chondrus crispus* will be invaluable in obtaining similar data in the eucheumatoids. However, much work is needed to study the causations of correlations between these microorganisms, host morphology, events in life histories and phenology, e.g., seasonality. Documenting the microbiome in eucheumatoid species has the potential to understand the timeline of different processes, and developing cheap and easy-to-use systems to detect, e.g., the presence of particular bacteria in farms, could be valuable in the development of early warning systems of pests and diseases. Documenting microeukaryote endophyte diversity, host-endophyte relationships and impact on morphology and life histories is also greatly needed in eucheumatoids both in the wild and in culture.

To conclude, now is the time for a renaissance of 'traditional' approaches to 'knowing the organism'. Coupled with the state-of-the-art techniques available, they have the potential to provide a powerful tool in addressing many of today's challenges, including a sustainable future for the eucheumatoid seaweed industry.

Acknowledgements I am grateful to Alan Critchley for inviting me to contribute to the Tropical Phyconomy Workshop in 2022.

References

Brakel J, Sibonga R, Dumilag D, Montalescot M, Campbell I, Cottier-Cook E, Ward W, Le Masson V, Liu T, Msuya F, Brodie J, Lim PE, Gachon CMM (2021) Exploring, harnessing and conserving genetic resources towards sustainable seaweed aquaculture. Plants, People, Planet. https://doi.org/10.1002/ppp3.10190

Brodie J, Williamson C, Barker GL, Walker RH, Briscoe A, Yallop M (2016) Characterising the microbiome of *Corallina officinalis*, a dominant calcified intertidal red alga. FEMS Microbiol Ecol 92(8): fiw110. https://doi.org/10.1093/femsec/fiw110

Brodie J, Ball SG, Bouget F-Y, Chan CX, De Clerck O, Cock M, Gachon C, Grossman AR, Mock T, Raven J, Saha M, Smith A, Vardi A, Yoon HS, Bhattacharya D (2017) The omics revolution in algal biology: recent achievements and future perspectives. Trends Plant Sci. pii: S1360-1385(17)30105-X

Buschmann AH, Camus C, Infante J, Neori A, Israel Á, Hernández-González MC, Pereda SV, Gomez-Pinchetti JL, Golberg A, Tadmor-Shalev N, Critchley AT (2017) Seaweed production: overview of the global state of exploitation, farming and emerging research activity. Eur J Phycol 52(4):391–406

Chandler CJ, Wilts BD, Vignolini S, Brodie J, Steiner U, Rudall PJ, Glover BJ, Gregory T, Walker RH (2015) Structural colour in Irish moss, Chondrus crispus. Sci Reports 5:11645. https://doi.org/10.1038/srep11645

Chen LC-M, McLachlan J (1972) Life history of *Chondrus crispus* in culture. Can J Bot 50:1055–2000

Collén J et al (2013) Genome structure and metabolic features in the red seaweed *Chondrus crispus* shed light on evolution of the Archaeplastida. PNAS 110:5247–5252

Cottier-Cook EJ, Nagabhatla N, Badis Y, Campbell M, Chopin T, Dai W, Fang J, He P, Hewitt C, Kim GH (2016) Safeguarding the future of the global seaweed aquaculture industry. United Nations University and Scottish Association for Marine Science. Policy Brief. 12p. https://www.sams.ac.uk/t4-media/sams/pdf/globalseaweed-policy-brief.pdf

Craigie JS, Correa JA (1996) Etiology of infectious diseases in cultivated *Chondrus crispus* (Gigartinales, Rhodophyta). Hydrobiologia 326(327):97–104

Craigie JS, Shacklock PF (1989) Culture of Irish moss. In: Boghen AD (ed) Cold water aquaculture in Atlantic Canada. Canadian Institute Research Regional Development, University of Moncton; Moncton, Moncton, pp 243–270

Craigie JS, Correa JA, Gordon ME (1992) Cuticles from *Chondrus crispus* (Rhodophyta). J Phycol 28:777–786

Egan S, Harder T, Burke C, Steinberg P, Kelleberg S, Thomas T (2013) The seaweed holobiont: understanding seaweed–bacteria interactions. FEMS Microbiol Rev 37(3):462–476

Faisan JP Jr, Luhan MRJ, Sibonga RC, Mateo JP, Ferriols VMEN, Brakel J, Ward GM, Ross S, Bass D, Stentiford GD, Brodie J, Hurtado JQ (2021) Preliminary survey of pests and diseases of eucheumatoid farms in the Philippines. J Appl Phycol. https://doi.org/10.1007/s10811-021-02481-5

FAO (2020) The state of world fisheries and aquaculture 2020. Sustainability in action. Rome. https://doi.org/10.4060/ca9229en

Fredericq S, Brodie J, Hommersand MH (1992) Developmental morphology of *Chondrus crispus* (Gigartinaceae, Rhodophyta). Phycologia 31:542–563

Hollants J, Leliaert F, De Clerck O, Willems A (2013) What we can learn from sushi: a review on seaweed–bacterial associations. FEMS Microbiol Ecol 83:1–16

Lim P-E, Yang L-E, Maggs CA, Brodie J (2017) Advancing the taxonomy of economically important red seaweeds (Rhodophyta). Eur J Phycol 52:438–451

Msuya FE, Bolton J, Pascal F, Narrain K, Nyonje B, Cottier-Cook EJ (2022) Seaweed farming in Africa: current status and future potential. J Appl Phycol 34:985–1005

Nielsen R, Petersen G, Seberg O, Daugbjerg N, O'Kelly CJ, Wysor B (2013) Revision of the genus *Ulvella* (Ulvellaceae, Ulvophyceae) based on morphology and *tuf*A gene sequences of species in culture, with *Acrochaete* and *Pringsheimiella* placed in synonymy. Phycologia 52:37–56

Nielsen R, Gunnarsson K, Daugbjerg N, Petersen G (2014) Description of *Ulvella elegans* sp. nov. and *U. islandica* sp. nov. (Ulvellaceae, Ulvophyceae) from Iceland – a study based on morphology of species in culture and *tuf*A gene sequences. Eur J Phycol 49:60–67

Plumb J (1999) Population dynamics and endophytic flora of *Chondrus crispus* (Rhodophyta): a temporal study. PhD thesis, University of the West of England

Rinkel BE, Hayes PH, Gueidan C, Brodie J (2012) A molecular phylogeny of *Acrochaete* and other endophytic green algae (Ulvales, Chlorophyta). J Phycol 48:1020–1027

Veira VMNCS, Engelen AH, Huanel OR, Guillemin OR (2021) Differential frond growth in the isomorphic haploid–diploid red seaweed *Agarophyton chilense* by long-term in situ monitoring. J Phycol 57:592–605

Ward GM, Faisan JP Jr, Cottier-Cook E, Gachon C, Hurtado AQ, Lim PE, Matoju I, Msuya F, Brodie J (2019) A review of reported seaweed diseases and pests in aquaculture in Asia. J World Aquacult Soc 2019:1–14

Yang EC, Boo SM, Bhattacharya D, Saunders GW, Knoll AH, Fredericq S, Graf L, Yoon HS (2016) Divergence time estimates and the evolution of major lineages in the florideophyte red algae. Sci Rep 6:21361. https://doi.org/10.1038/srep21361

The Center of Excellence for Development and Utilization of Seaweeds, Hasanuddin University (CEDUS-UNHAS): Collaborating on Research and Outreach for the SDGs

26

Kasmiati, Asmi Citra Malina A. R. Tassaka, Amanda Priscella Putri, Nadiarti Nurdin Kadir, and Abigail Mary Moore

Abstract

The Centre of Excellence for Development and Utilization of Seaweed at Universitas Hasanuddin (CEDUS-UNHAS) is the only Centre of Excellence (CoE) under the Indonesian CoE programme that is focused on seaweeds. UNHAS has pioneered seaweed cultivation in South Sulawesi for over two decades, and in 2012, UNHAS formed the Seaweed Research and Development Centre Consortium (KP3RL) with stakeholders from government agencies, other universities, research institutes and the private sector. Focused on seaweed cultivation, including new/improved seaweed varieties and high-quality seeds, the KP3RL strengthened the position of South Sulawesi as the largest producer of seaweeds in Indonesia, contributing approximately one-third of national and 11% of world seaweed production in 2017. In 2018, the KP3RL became CEDUS-UNHAS, partnering with Indonesian and overseas institutions. The vision of the CEDUS-UNHAS is to become a leading centre for seaweed research and development in Indonesia by 2025, aiming to advance and disseminate seaweed science and technology for sustainable development. CEDUS-UNHAS fosters approximately 200 seaweed community groups producing innovative products, with patented inventions related to biosugar, bioethanol, capsules, antiaging cosmetics, high protein seaweed and production technology. Collaboration is a priority for achieving CEDUS-UNHAS objectives, for example, cooperation on seaweed cultivation and utilization of high protein seaweeds as a food for the future, and many other aspects of seaweed-related development, especially through downstream programs.

Keywords

Innovation · Collaboration · Seaweed production · Seaweed products · High-protein seaweed · Sustainable development

26.1 Introduction

In this twenty-first century, humanity faces many challenges in meeting stakeholder needs and moving towards a sustainable future, particularly in terms of food security. The Food and Agriculture Organization of the United Nations (FAO) has predicted a global human population of approximately 10 billion by 2050, increasing the need for food by approximately 60-70% (Godfray et al. 2010). Currently, most food is produced on land, and production is threatened by the decline in land and water availability, including the negative impacts

Authors' Contributions:Kasmiati conceived and drafted the paper and reviewed and edited the manuscript.Abigail Mary Moore translated the paper (Indonesian to English), made contributions and critical revisions to the content, and edited the manuscript.Asmi Citra Malina A.R. Tassaka, Amanda Pricella Putri and Nadiarti Nurdin Kadir made contributions and critical revisions to the content of the manuscript.All authors approved the manuscript and agreed to be accountable for all aspects of the work in ensuring that questions related to the accuracy or integrity of any part of the work are appropriately investigated and resolved.

Kasmiati (✉) · A. C. M. A. R. Tassaka
Center of Excellence for Development and Utilization of Seaweeds, Hasanuddin University, Makassar, South Sulawesi, Indonesia

A. P. Putri
Center of Excellence for Interdisciplinary and Sustainability Science, Hasanuddin University, Makassar, South Sulawesi, Indonesia

N. N. Kadir
Fisheries Resources Management Study Program, Faculty of Marine Science and Fisheries, Hasanuddin University, Makassar, South Sulawesi, Indonesia

Center of Excellence for Marine Resilience and Sustainable Development, Universitas Hasanuddin, Makassar, Indonesia

A. M. Moore
Center of Excellence for Marine Resilience and Sustainable Development, Universitas Hasanuddin, Makassar, Indonesia

Graduate School, Hasanuddin University, Makassar, South Sulawesi, Indonesia

A. T. Critchley et al. (eds.), *Tropical Phyconomy Coalition Development*, Developments in Applied Phycology 11,
https://doi.org/10.1007/978-3-031-47806-2_26

of industrialization on the environment (van Zanten et al. 2016). Meeting the future food needs of humanity calls for innovation, including alternative sources of food. In this context, seaweeds have the potential to make a significant contribution, as many algae can provide sustainable, highly nutritious food, including protein-rich seaweed products (Mohamed et al. 2012).

By the dawn of the twenty-first century, several seaweed taxa were being used as sources of food, especially in Asian countries, with approximately 90% of global production coming from this region (McHugh 2002). Seaweeds have several advantages as a food source, as they are typically rich in minerals, protein, vitamins and fibre as well as functional compounds that can promote consumer health (Cardoso et al. 2015). The use of seaweeds as a source of protein also offers opportunities for reducing the environmental impact of food production and avoids competition for increasingly scarce land for conventional land-based agriculture (Van Audenhaege et al. 2009). The annual protein yield from seaweed cultivation can reach approximately 2.5–7.5 tons per hectare, approximately twice the yield of secondary protein sources such as wheat and pulses. Furthermore, most seaweeds contain essential amino acids that can meet human nutritional needs, and in many regions, aquaculture can enable seaweeds to be farmed and harvested year-round (Pangestuti and Kim 2017).

Seaweeds have been hailed as a promising food of the future (Tamayo Tenorio et al. 2018), with many taxa having protein contents similar to those of soybeans and eggs (Fleurence 1999). Indeed, the red alga *Porphyra yezoensis* can have a protein content of approximately 47% dry weight (Fleurence et al. 2018), higher than that of soy beans (Fujiwara-Arasaki et al. 1984), and is easily digested (Belghit et al. 2017). Protein levels tend to vary between seaweed types, generally highest (approximately 30–40%) in red algae (Rhodophyta), lower (20%) in green algae (Chlorophyta) and lowest (5–11%) in brown algae (Phaeophyta) (Dharmananda 2002; Burtin 2003; Rachmaniar 2005). One red seaweed with a high protein content (approximately 35% dry weight) is *Palmaria palmata* (Morgan et al. 1980).

In Indonesia as a whole, and South Sulawesi in particular, there is a growing research focus on seaweeds, including the identification of the species present as well as exploring their characteristics and potential uses, including their suitability for contributing to sustainable food production. Furthermore, to increase the export value of Indonesian seaweeds, there is a need to develop downstream processes as well as production volume. While Indonesia is second only to China in terms of seaweed export volume, in terms of export value, Indonesia is ranked seventh or eighth. The majority (approximately 80%) of Indonesian production is exported as dried seaweed, which fetches a comparatively low price compared to intermediate or semi-refined products such as carrageenan, agar, and alginate, and added value could be even higher for more highly processed or finished products. Therefore, the Indonesian government has policies and programs to decrease the export of raw material (dried seaweed) and increase the export of a range of processed products. Such a shift requires both basic and applied research, with downstream mechanisms to translate research into processes and products.

The Indonesian Centre of Excellence (CoE) program was established by the Government of the Republic of Indonesia to accelerate science-based development (BRIN 2019). Hosted at leading universities and other institutions across Indonesia, major goals include bridging the gap between researchers/academia and other stakeholders to accelerate and expand the transfer of scientific knowledge to real-world products/applications and problem solving and support the Sustainable Development Goals (SDGs). Established in 2018, the Center of Excellence for Development and Utilization of Seaweeds (CEDUS) is based at Hasanuddin University (UNHAS) in Makassar, South Sulawesi, Indonesia. With the Indonesian name *Pusat Unggulan Iptek Penembangan dan Pemanfaatan Rumput Laut* (PUI-P2RL), the CEDUS-UNHAS is the only centre under the CoE Program with a specific focus on seaweeds or algae.

Several factors supported the choice of UNHAS as the host institution for the CEDUS. One reason is proximity to existing seaweed-related development. Indonesia is the top seaweed-producing country (FAO 2020), and South Sulawesi is the leading Indonesian province in the field of seaweed farming (mariculture and brackish-water pond culture), both in terms of local production and as a regional hub for seaweed farming, trade and processing industries (Nuryartono et al. 2021; Langford et al. 2022), as shown in Fig. 26.1. Furthermore, UNHAS researchers have a track record in seaweed-related research as well as in bringing seaweed knowledge and advances to the wider community through a variety of outreach mechanisms.

26.2 The Road to CEDUS-UNHAS

The CEDUS-UNHAS emerged from a process lasting over 20 years (Fig. 26.2). In the early 1990s, UNHAS seaweed researchers pioneered seaweed farming in the coastal waters of South Sulawesi, especially in the Takalar, Jeneponto and Bantaeng regencies. These activities contributed to the fast growth of seaweed farming at local and national levels and were often implemented in partnership with other organizations. In 2005, UNHAS initiated the formation of the *Konsorsium Rumput Laut* (Seaweed Consortium), with members from academia (Politeknik Negeri Ujung Pandang (PNUP) and Politeknik Negeri Pangkep (PNP)) as well as the South Sulawesi Provincial Marine and Fisheries Service (DKP Sulsel), government research stations (Takalar Centre

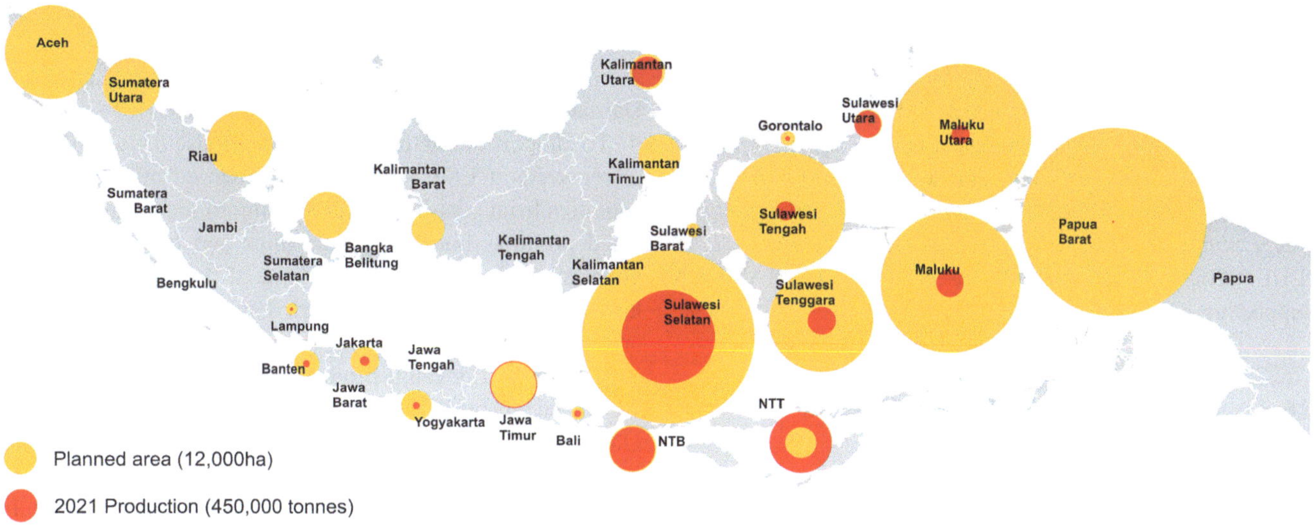

Fig. 26.1 Map of Indonesia showing seaweed production levels and planned seaweed farming extent in 2019; South Sulawesi (Sulawesi Selatan) had the largest production volume (Langford 2024)

for Brackish-water Aquaculture Research (BRBAP Takalar) and Maros Centre for Brackish-water Aquaculture Research and Extension (BRPPBAP Maros)), and the Indonesian Seaweed Association (ASPERLI) representing the private sector. As this Consortium grew and developed, it was formally recognized by the Indonesian Ministry of Education and Culture in 2012 under the name *Konsorsium Penelitian, Pengembangan dan Pemanfaatan Rumput Laut* (KP3RL), which translates as the Consortium for Seaweed Research, Development and Utilization. Under the leadership of Universitas Hasanuddin, the seven KP3RL partners focused their efforts on the development of new seaweed varieties, the production of high-quality seaweed "seeds" for aquaculture, and the development of seaweed farming around the coasts of South Sulawesi.

The Consortium activities helped to strengthen the position of South Sulawesi as a major seaweed farming centre, producing approximately 3.67 tons/year, contributing

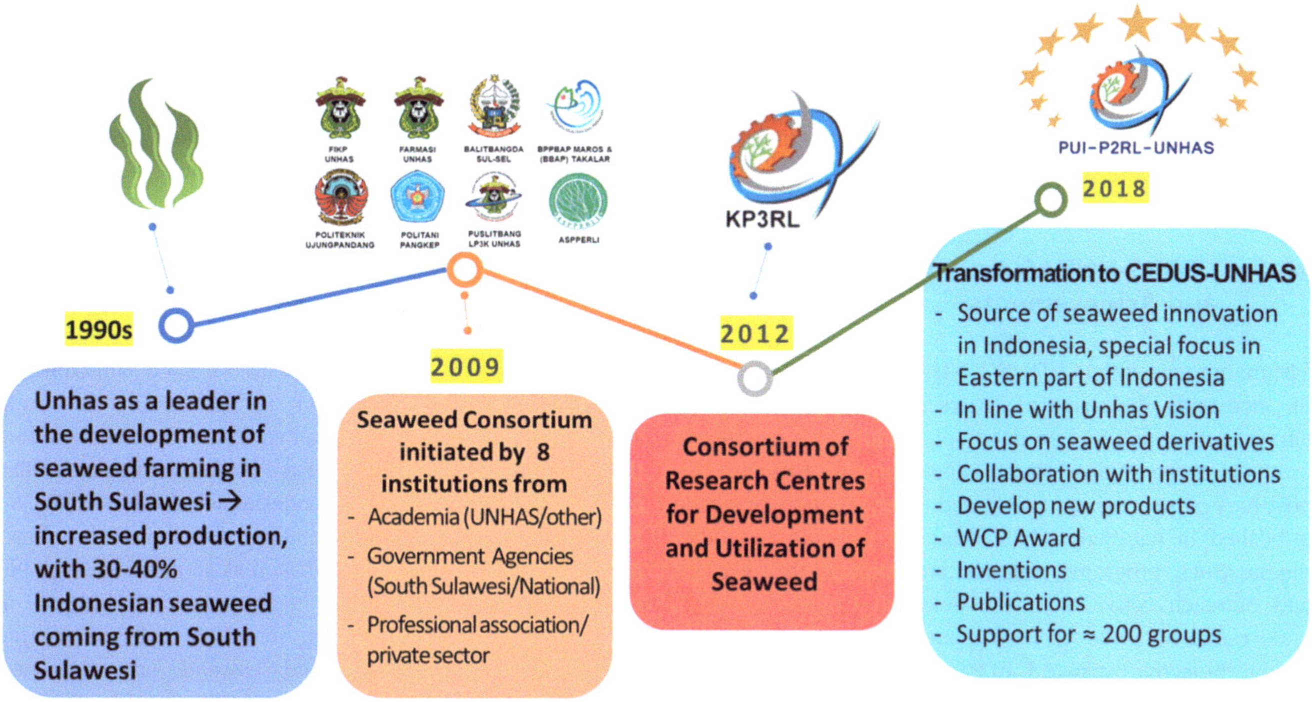

Fig. 26.2 The road to CEDUS-UNHAS

30-35% of Indonesian and approximately 11% of global seaweed production volume, including over 20% of the global supply of carrageenan seaweeds (Nuryartono et al. 2021). Under the aforementioned CoE Program, in 2018, KP3RL was transformed into the current Center of Excellence for the Development and Utilization of Seaweed (CEDUS) in Indonesian *Pusat Unggulan Iptek Pengembangan dan Pemanfaatan Rumput Laut* (PUI-P2RL). The CEDUS is based on the UNHAS with supporting partner institutions.

26.3 CEDUS-UNHAS Vision, Mission, and Structure

The CEDUS-UNHAS is expected to serve as a source of innovation and information to support the seaweed sector in Indonesia, especially in eastern Indonesia. The CEDUS-UNHAS vision is to become the leading seaweed science and technology research and development institution in Indonesia by 2025. To achieve this vision, the mission statements are as follows: (1) Develop mastery of seaweed science and technology to become a prime mover and reference for accelerating national progress and sustainable development in the field of seaweed; (2) Generate added value and maintain biodiversity through mastering technology; (3) Contribute to national capacity building and education through application of science and technology in the field of seaweed biotechnology; and (4) Enhance the performance and governance of research institutions.

As an organization that is largely oriented towards the application of science, with a strong focus on product development, the CEDUS-UNHAS currently has seven innovation divisions (Fig. 26.3). Innovation Division 7 (Socioeconomics and Marketing) also plays a cross-cutting role, supporting the other six divisions.

26.4 CEDUS-UNHAS Activities and Achievements

The main focus of the CEDUS-UNHAS CoE to date has been on increasing the value of seaweeds by improving post-harvest processing and on developing added-value seaweed products through downstream development. Research on post-harvest processing, edible and other products has been published in national and international journals and other media. Such programs continue and expand the research and outreach activities of the Seaweed Consortium in a wide range of fields, including aquaculture (Wenno et al. 2015), bioactive seaweed extracts (Nasmia et al. 2014; Ikbal 2015; Syamsuddin et al. 2016; Zainuddin et al. 2019; Darfiah et al. 2021), functional foods (Zaenab et al. 2020a, b; Utami et al. 2021; Widyaswari et al. 2021; Tassakka et al. 2021), cosmetics (Kasmiati and Amir 2021) and research on seaweed-based antibacterial agents (Zainuddin et al. 2019; Darfiah et al. 2021).

To support the national drive to increase the added value of seaweeds, CEDUS-UNHAS has focused on downstream activities from post-harvest processing to seaweed-based finished products building partnerships and cooperation with government institutions and private sector actors. One collaboration in particular is supported by the National Research and Innovation Agency (BRIN) under a program to facilitate innovation and develop products that could be promoted to and taken up by the public. Under this program, CEDUS-UNHAS is providing technical guidance to community groups producing food products by processing seaweed to increase production capacity and improve product quality. In partnership with the Centre of Technology (CoT), a start-up owned by UNHAS, the CEDUS-UNNHAS has disseminated seaweed processing technology, including providing in-kind assistance in the form of equipment for implementing appropriate technology.

Seaweed-based food products produced through the downstream innovation program include variations on traditional foods where a proportion of grain-based flour (carbohydrate) is replaced by seaweed meal. These include noodles, with 40% of the wheat flour replaced by seaweed; fried and steamed pastries called *pastel* (30% seaweed) and *nastar* (40% seaweed), brownies (30% seaweed), and cendol (75% seaweed). These and other innovative products have been and are being produced at the community group level using appropriate technology, with the aim of increasing group income through sales of the added-value seaweed products. Seaweed sugar made through an enzymatic process has a yield of 60%. However, there is a need for further development of the seaweed sugar production process so that the methods, tools, and materials can be readily used by local community members without special assistance or supervision. Non-food products using seaweeds as an active or key ingredient have also been produced, such as a cleaning agent for dentures, antiaging creams, and bath soap. Active antibacterial compounds from various seaweeds are also used as ingredients in the manufacture of hand sanitizers.

While seaweed production in Indonesia has been dominated by a very small number of species and varieties, there is growing recognition of the potential inherent in the algal biodiversity present in Indonesian waters. The CEDUS-UNHAS contributes to expanding the knowledge base regarding other seaweed taxa as well as supporting diversification in the use of currently cultivated seaweeds. In terms of red seaweeds or Rhodophyta, in addition to the major commodities (i.e., eucheumatoid seaweeds and the genus *Gracilaria*), there are current programs or plans for the genera *Laurencia*, *Asparagopsis*, *Gelidium*, *Halymenia*,

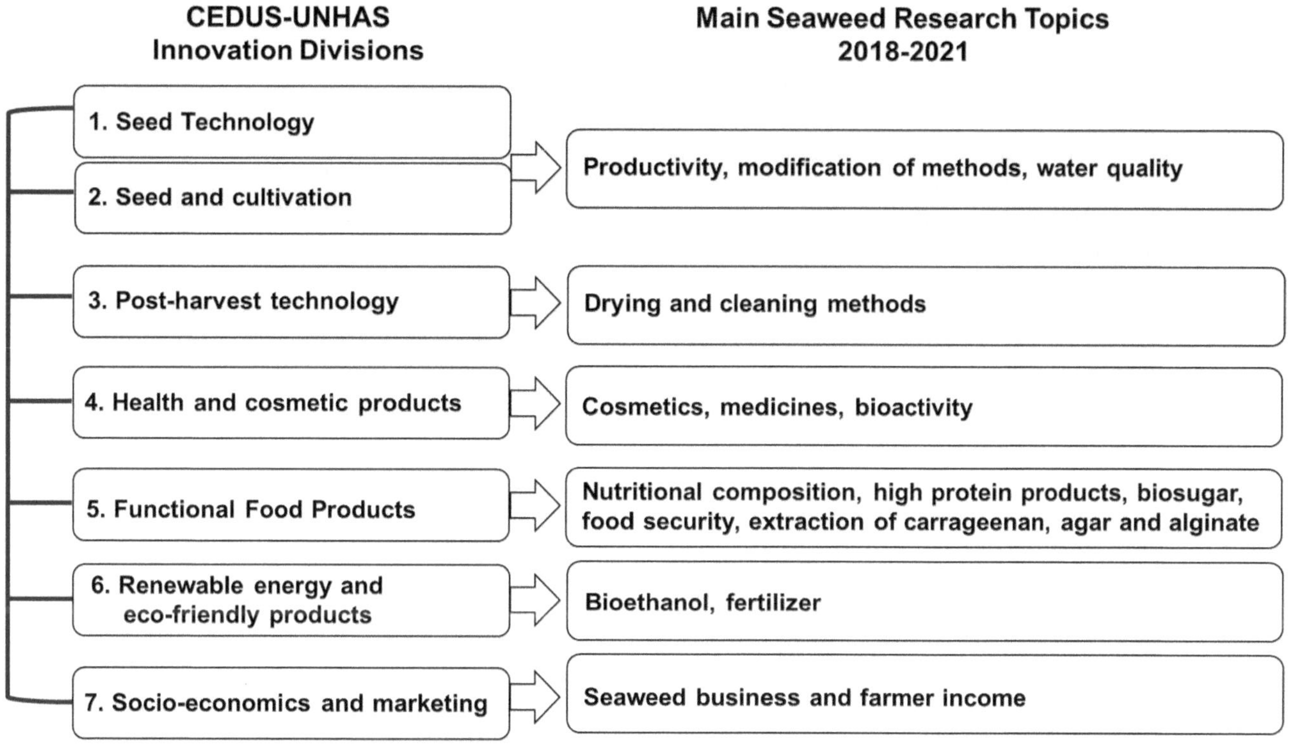

Fig. 26.3 The seven CEDUS-UNHAS Innovation Divisions and Main Research Topics

Galaxaura and *Porphyra*. Other seaweeds include four green seaweed genera (*Codium*, *Ulva*, *Caulerpa* and *Halimeda*) and three brown seaweed genera (*Sargassum*, *Padina* and *Turbinaria*).

As an example, the CEDUS-UNHAS is working with the pharmaceutical company PT Kalbe Farma Indonesia to develop high protein products based on *Porphyra*. *Porphyra* seed material was collected from sites in Ambon and Bali and acclimated to laboratory conditions before transfer to 500 L tanks. This program is still in the early research stages, with the aim of producing high protein biomass for use as an ingredient in a variety of high protein seaweed-based products. The development of *Porphyra* nori has been directly supervised by home industry practitioners from Japan, and the products have already been tested. Similar nori products have also been produced from *Gracilaria changii* and two green seaweeds (*Ulva* sp. and *Caulerpa* sp.).

26.5 CEDUS-UNHAS Collaboration and Outreach: Overview and Invitation

Collaboration is increasingly recognized as a vital key to success in many fields of scientific endeavour, from basic research to science communication and applications. Since its inception, the CEDUS-UNHAS is in itself a microcosm of local-level collaboration and has actively sought opportunities for both in-country and international collaboration. One example is the Word Class Professor (WCP) program. As an awardee of this competitive program, CEDUS-UNHAS was able to invite leading scientists from around the world to share their knowledge and initiate collaboration with CEDUS-UNHAS members. Some of the outputs are briefly described in the previous section, including innovations and joint publications. Together with the UNHAS Graduate School and the Center of Excellence for Marine Resilience and Sustainable Development (MARSAVE CoE), the CEDUS-UNHAS is honoured to serve as co-host to the Tropical Phyconomy Coalition Development (TPCD) interactive workshops. This volume is one of the outputs from the first workshop (TPCD1), focused on eucheumatoid seaweeds, and we look forward to playing an active role in further TPCD workshops on other key groups of algae.

In addition to the above-mentioned collaboration with PT. Kalbe Farma, partners in other in-country collaborations include companies (e.g., PT. Pertamina, PT. Mari Ocean) and other academic institutions (e.g., Politeknik Negeri Nunukan in North Kalimantan). Overseas or multinational partners include Tsukuba University, Japan and the Australian National University, Seatech Energy, and the SNV Netherlands Development Organization from the Netherlands. At the local level, the CEDUS-UNHAS outreach program works with approximately 200 registered community groups in seaweed-related fields. Researchers and community groups work side by side to develop a wide range of new seaweed-based products. The CEDUS-UNHAS has supported the vision and mission of Hasanuddin University through the establishment of a new study program on fisheries processing technology, with a strong focus on seaweed and innovation in the types of seaweeds used as well as the products and processes involved.

As we move forward into an increasingly challenging environment, we are keen to expand our national and international seaweed-related collaborations to help address current and future needs. The CEDUS is especially keen to collaborate on seaweed value-added processing, including the production of intermediate and end (consumer-ready) products.

Acknowledgements The authors thank the Tropical Phyconomy Coalition Development (TPCD) sponsors GenialG, represented by Dr. Philippe Potin, and the TPCD-1 initiators, especially Dr. Iain Neish, Dr. Alan Critchley and Nelson Vadassery, the TPCD-1 team and the volume editors.

Declarations **Funding**: Not applicable

Competing interests: The authors have no competing interests to declare.

Availability of data and material: Data sharing is not applicable to this article, as no datasets were generated or analysed during the study.

Code availability: Not applicable as no code was generated or analysed.

References

Belghit I, Rasinger JD, Heesch S et al (2017) In-depth metabolic profiling of marine macroalgae confirms strong biochemical differences between brown, red and green algae. Algal Res 26: 240–249. https://doi.org/10.1016/j.algal.2017.08.001

BRIN (2019) Siaran Pers Kemenristekdikti Nomor: 202/SP/HM/BKKP/X/2019. https://www.brin.go.id/menristekdikti-indonesia-telah-miliki-81-pusat-unggulan-iptek-utama/. Accessed 28 Jul 2022

Burtin P (2003) Nutritional value of seaweeds. Electron J Environ Agric Food Chem 2:498–503

Cardoso S, Pereira O, Seca A et al (2015) Seaweeds as preventive agents for cardiovascular diseases: from nutrients to functional foods. Mar Drugs 13:6838–6865. https://doi.org/10.3390/md13116838

Darfiah D, Kasmiati K, Latama G (2021) Antibacterial activity and identification of active compounds of seaweed extract *Sargassum* sp., *Halimeda opuntia* and *Halymenia* sp. from Lae-Lae Island of South Sulawesi. Int J Environ Agric Biotechnol 6:187–195. https://doi.org/10.22161/ijeab.66.23

Dharmananda S (2002) The nutritional and medical value of seaweeds used in Chinese medicine. In: Institute for Traditional Medicine, Portland, OR. http://www.itmonline.org/arts/seaweed.htm. Accessed 31 Jul 2022

FAO (2020) The State of World Fisheries and Aquaculture 2020: sustainability in action. Food and Agriculture Organization of the United Nations, Rome

Fleurence J (1999) Seaweed proteins. Trends Food Sci Technol 10:25–28. https://doi.org/10.1016/S0924-2244(99)00015-1

Fleurence J, Morançais M, Dumay J (2018) Seaweed proteins. In: Proteins in food processing, 2nd edn. Elsevier, pp 245–262

Fujiwara-Arasaki T, Mino N, Kuroda M (1984) The protein value in human nutrition of edible marine algae in Japan. Hydrobiologia 116–117:513–516. https://doi.org/10.1007/BF00027735

Godfray HCJ, Beddington JR, Crute IR et al (2010) Food security: the challenge of feeding 9 billion people. Science(80-) 327:812–818. https://doi.org/10.1126/science.1185383

Ikbal M (2015) Uji Toksisitas Ekstrak Metanol Rumput Laut Hijau (*Caulerpa racemosa*) pada Larva Udang Windu (*Penaeus monodon*). Octopus 4:417–421. https://doi.org/10.26618/octopus.v4i2.602

Kasmiati S, Amir N (2021) Pembuatan lotion rumput laut di Desa Aeng Batu Batu, Kabupaten Takalar, Sulawesi Selatan (seaweed lotion processing in Aeng Batu Batu Village, Takalar Regency, South Sulawesi). J Panrita Abdi 5:17–25. https://doi.org/10.20956/jpa.v5i1.9188

Langford Z (2024) Introduction. In: Langford Z (ed) Globalisation and livelihood transformations in the Indonesian seaweed industry. Routledge: New York. https://www.routledge.com/Globalisation-and-Livelihood-Transformations-in-the-Indonesian-Seaweed/Langford/p/book/9781032025469

Langford A, Zhang J, Waldron S et al (2022) Price analysis of the Indonesian carrageenan seaweed industry. Aquaculture 550: 737828. https://doi.org/10.1016/j.aquaculture.2021.737828

McHugh DJ (2002) Prospects for seaweed production in developing countries. FAO Fisheries Circular No. 968

Mohamed S, Hashim SN, Rahman HA (2012) Seaweeds: a sustainable functional food for complementary and alternative therapy. Trends Food Sci Technol 23:83–96. https://doi.org/10.1016/j.tifs.2011.09.001

Morgan KC, Wright JLC, Simpson FJ (1980) Review of chemical constituents of the red alga *Palmaria palmata* (dulse). Econ Bot 34:27–50. https://doi.org/10.1007/BF02859553

Nasmia N, Syamsuddin R, Rantetondok A, Zainuddin EN (2014) Characterization and identification of bacteria isolated from seaweed *Gracilaria verrucosa* (Linn., 1758) infected by ice-ice. Int J Aquac 4:2014. https://doi.org/10.5376/ija.2014.04.0023

Nuryartono N, Waldron S, Tarman K et al (2021) An analysis of the South Sulawesi Seaweed Industry. The Australia-Indonesia Centre

Pangestuti R, Kim S-K (2017) Bioactive peptide of marine origin for the prevention and treatment of non-communicable diseases. Mar Drugs 15:67. https://doi.org/10.3390/md15030067

Rachmaniar (2005) Penelitian Kandungan Kimia Makroalgae Untuk Neutroceuticals dan Agrochemicals. 22

Syamsuddin R, Nasmia RA, Zainuddin EN (2016) *Caulerpa racemosa* (Chlorophyta, Caulerpaceae) extract increases growth and biomass production of *Gracilaria verrucosa* (Rhodophyta, Gracilariaceae). AACL Bioflux 9:1044–1052

Tamayo Tenorio A, Kyriakopoulou KE, Suarez-Garcia E et al (2018) Understanding differences in protein fractionation from conventional crops, and herbaceous and aquatic biomass - consequences for industrial use. Trends Food Sci Technol 71:235–245. https://doi.org/10.1016/j.tifs.2017.11.010

Tassakka ACMAR, Sulfahri, Iqbal Djawad M et al (2021) Biosugar production from marine algae *Kappaphycus alvarezii* by acid hydrolysis. IOP Conf Ser Earth Environ Sci 860:012078. https://doi.org/10.1088/1755-1315/860/1/012078

Utami D, Kasmiati K, Metusalach M (2021) Effect of concentration of *Sargassum polycystum* and fermentation time on thickness and yield of Nata de *Sargassum*. Int J Environ Agric Biotechnol 6:163–167. https://doi.org/10.22161/ijeab.65.21

Van Audenhaege M, Héraud F, Menard C et al (2009) Impact of food consumption habits on the pesticide dietary intake: comparison between a French vegetarian and the general population. Food Addit Contam Part A 26:1372–1388. https://doi.org/10.1080/02652030903031171

van Zanten HHE, Mollenhorst H, Klootwijk CW et al (2016) Global food supply: land use efficiency of livestock systems. Int J Life Cycle Assess 21:747–758. https://doi.org/10.1007/s11367-015-0944-1

Wenno PA, Syamsuddin R, Zainuddin EN (2015) Cultivation of red seaweed *Kappaphycus alvarezii* (Doty) at different depths in South Sulawesi, Indonesia. AACL Bioflux 8:468–473

Widyaswari SG, Metusalach K, Amir N (2021) A review: bioactive compounds of macroalgae and their application as functional beverages. IOP Conf Ser Earth Environ Sci 679:012002. https://doi.org/10.1088/1755-1315/679/1/012002

Zaenab S, Kasmiati K, Sulfahri S, Tassakka ACMAR (2020a) Biosugar production from *Kappaphycus alvarezii* by hydrolysis method using fungus *Trichoderma harzianum*. Int J Environ Agric Biotechnol 4: 865–869. https://doi.org/10.22161/ijeab.54.3

Zaenab S, Tassakka ACMAR, Sulfahri K (2020b) Utilization of double fungal treatment by *Trichoderma harzianum* and *Saccharomycopsis fibuligera* to produce biosugar from red seaweed *Kappaphycus alvarezii*. IOP Conf Ser Earth Environ Sci 575. https://doi.org/10.1088/1755-1315/575/1/012015

Zainuddin EN, Anshary H, Huyyirnah H et al (2019) Antibacterial activity of *Caulerpa racemosa* against pathogenic bacteria promoting "ice-ice" disease in the red alga *Gracilaria verrucosa*. J Appl Phycol 31:3201–3212. https://doi.org/10.1007/s10811-019-01805-w

MIX
Papier aus verantwortungsvollen Quellen
Paper from responsible sources
FSC® C105338

If you have any concerns about our products,
you can contact us on
ProductSafety@springernature.com

In case Publisher is established outside the EU,
the EU authorized representative is:
Springer Nature Customer Service Center GmbH
Europaplatz 3, 69115 Heidelberg, Germany

Printed by Libri Plureos GmbH
in Hamburg, Germany